新工科建设之路

Kotlin移动应用开发技术

白 喆 著

電子工業出版社
Publishing House of Electronics Industry
北京 • BEIJING

内 容 简 介

本书基于 Kotlin 语言对 Android 移动应用程序开发的知识点进行总结，对常用的 SDK 进行分析，并通过实例工程进行演示。本书共 15 章，内容包括 Android 的开发环境、工程创建、工程结构、Kotlin 基础语法、控件与布局、活动与碎片、后台服务与广播、数据存储与共享、多媒体与传感器、HTTP 网络通信、快速开发套件、“粉色辣椒”开发流程等。针对使用技巧、注意事项和相关概念，本书还提供了相应的“提示”。

本书可作为高等院校和高职院校计算机科学与技术、软件工程、网络工程、数字媒体技术等专业 Android 开发课程的参考书，也可供具有 Java 基础的编程爱好者参考。

图书在版编目（CIP）数据

Kotlin 移动应用开发技术 / 白喆著. —北京：电子工业出版社，2021.7

ISBN 978-7-121-41446-6

I. ①K… II. ①白… III. ①JAVA 语言－程序设计－高等学校－教材 IV. ①TP312.8

中国版本图书馆 CIP 数据核字（2021）第 124752 号

责任编辑：张　鑫
印　　刷：北京盛通商印快线网络科技有限公司
装　　订：北京盛通商印快线网络科技有限公司
出版发行：电子工业出版社
　　　　　北京市海淀区万寿路 173 信箱　　邮编：100036
开　　本：787×1092　1/16　印张：33.25　字数：1053 千字
版　　次：2021 年 7 月第 1 版
印　　次：2022 年 7 月第 2 次印刷
定　　价：108.00 元

凡所购买电子工业出版社图书有缺损问题，请向购买书店调换。若书店售缺，请与本社发行部联系，联系及邮购电话：(010)88254888，88258888。

质量投诉请发邮件至 zlts@phei.com.cn，盗版侵权举报请发邮件至 dbqq@phei.com.cn。

本书咨询联系方式：zhangxinbook@126.com。

前 言

PREFACE

Google 的 Android 系统是目前主流的移动设备操作系统之一，具有源代码开源、硬件和开发平台价格低等特点。与 iOS 系统相比，Android 系统对设备兼容性、硬件资源利用率、API 传承性和 App 监管程度的要求较低，而且 Android 系统赋予了开发者和用户更多的权限与选择性。使用 iOS 系统的用户需要通过 App Store 安装 App；而 Android 系统没有指定发布平台，甚至开发者可以在自己的网站上发布 App 的安装包。在 App Store 中发布 App 的审核周期较长，且个人开发者每年需支付一定费用，支付后才可以使用物理设备运行测试，否则只能使用模拟器。在这两个系统上，我都开发过 App，开发难度基本相同，各有优势。

2014 年，我独立开发完成了一个基于“1 公里”半径生活圈的 App——微距，其中包含了社交和购物的功能。我大约用了 10 个月的时间完成了 Android 系统的微距 App 和后台网站开发，又用了 2 个月的时间完成了 iOS 系统的微距 App 开发，感觉在 iOS 系统下开发最“幸福”的地方是无须考虑设备和 API 版本的兼容性，这也是 Android 系统最大的痛点。在微距 App 策划阶段，考虑到对于个人开发者而言，宣传推广和高并发的后台服务器可能是最大的困难，而且积累不了大量的用户就无法发展起来，将微距 App 的社交功能分离出来，又在此基础上增加了用户筛选和评分的功能，形成了一个免费的交友 App——未见。这两个 App 都使用 Java 语言进行开发，开发过程中我积累了不少开发经验和技巧。

随着 Kotlin 语言被确定为 Android 官方首选开发语言，以及 JetPack 库和新版本的 Android 发布，Android 应用的开发环境发生了新的变化。因此，我开始使用 Kotlin 语言和 JetPack 库进行开发。本书将 Kotlin 的主要知识点按照功能进行分类，并将相关类的常用方法和常量进行归纳。由于智能手机普及程度高，所以本书大量使用了高度概括和抽象的方式进行描述，让读者根据功能就能联想到实际使用中的场景，而“提示”中介绍的知识点和使用方法可以弥补读者知识储备的不足。书中的实例工程多以典型的应用场景进行演示，读者掌握方法后可以举一反三，根据需要灵活应用。而 Activity、Fragment、Service、IntentService、BroadcastReceiver 等类的实例工程则以流程的方式进行演示。实例均采用原生代码和 JetPack 库，并没有采用第三方类库，以避免第三方类库的不可预知风险。

本书可作为高等院校和高职院校计算机科学与技术、软件工程、网络工程、数字媒体技术等专业 Android 开发课程的参考书，也可供具有 Java 基础的编程爱好者参考。

学习完本书内容后，理论上读者应该能够开发出具有基本功能的社交、新闻、购物分享、技能分享、拍照、录制视频、音乐播放 App。当然，App 只有这些基本功能是远远不够的，还需要合适的 UI 界面和后台服务器的支持，服务器端推荐使用 Nginx+PHP+MySQL。如果是小团队或个人开发 App，定位、支付、推送、分享、地图、手机验证码、二维码识别、视频、通信等功能的实现建议直接使用第三方提供的服务，这样可以节省大量的开发时间，降低技术难度，减少运营成本。

本书配备素材、工程源代码和基础工程源代码，可通过华信教育资源网(http://www.hxedu.com.cn)下载。源代码使用 Android Studio 4.1.2 和 Gradle 6.6 进行编写。Android Studio 和 Gradle 版本更新比较频繁，读者下载时的版本可能更高，打开工程后根据提示进行更新即可。另外，我还为本书专门建了 QQ 群(群号：653171771，密码：teachol)，加入 QQ 群的读者可以交流学习过程中遇到的问题。

书中的不妥和疏漏之处，欢迎读者批评指正，我的邮箱是 baizhe_22@qq.com，读者可以随时与我联系。

白 喆

2021 年 3 月

CONTENTS

目 录

第 1 章　Android 的基础知识

Android（安卓）的英文原意是机器人，其图标是一个机器人（如图 1-1 所示），来源于法国作家利尔·亚当在 1886 年发表的科幻小说《未来夏娃》，书中将外表像人的机器命名为 Android。学习 Android 先从了解 Android 的历史开始，Android 的诞生与 Andy Rubin 有关，Andy Rubin 被称为 Android 之父。

图 1-1　Android 图标

1.1　Android 与 Andy Rubin

1989 年，Andy Rubin 被一名苹果公司工程师引荐到当时处在第一个全盛时期的苹果公司，参与名为 Magic Cap 的智能手机操作系统开发工作。

1999 年，Andy Rubin 创立了 Danger 公司，开发了一个名为 Hiptop 的类似智能手机雏形的设备，提出了"智能手机"的概念——"支持互联网"和"其上运行着能够实现不同功能的各种应用"。

2002 年，Andy Rubin 在斯坦福大学做了一次讲座，听众中包括 Google 公司的两位创始人 Larry Page 和 Sergey Brin。互联网智能手机的理念深深打动了 Larry Page，尤其是他注意到 Danger 产品上默认的搜索引擎为 Google。

2003 年，Andy Rubin 等人创立了 Android 公司，并组建了 Android 开发团队，注册了 android.com 域名，立志设计一个基于开源思想的移动平台。当时的手机操作系统都是由手机厂商单独开发的，操作系统也是各手机厂商的核心技术，具有很强的封闭性。

2005 年，Andy Rubin 靠自己的积蓄和朋友的支持，艰难地完成了 Android 系统。但在寻找投资方时并不顺利，Andy Rubin 突然想到了 Google 公司的 Larry Page，于是给他发了一封电子邮件。仅仅几周后，Google 公司收购了成立仅 22 个月的 Android 公司。Andy Rubin 成为 Google 公司工程部副总裁，继续负责 Android 项目。

2014 年，Andy Rubin 离开 Google 公司。2015 年，Andy Rubin 创立 Essential 公司，开发 Android 智能手机。两年后发布首款手机 Essential PH-1，但是销量惨淡 2020 年 2 月 12 日，Essential 公司正式宣布停止运营。

1.2　Android 的开发环境

Android 系统刚发布时，Google 公司只提供了 SDK，并没有提供官方的开发环境。开发者只能通过第三方工具进行开发，如 Eclipse 和 IntelliJ IDEA。直到 2013 年，Google 公司才发布了基于 IntelliJ IDEA 的 Android 集成开发工具——Android Studio，支持 Windows（32 位和 64 位）、Mac 和 Linux 操作系统。

1.2.1　Android Studio 的下载

截至 2020 年 7 月 1 日，Android Studio 的最新版本是 4.0.0，下载网站地址为："https://developer.android.google.cn/studio/"。打开网站，单击"DOWNLOAD OPTIONS"链接，跳转至 Android Studio 下载列表（如图 1-2 所示）。

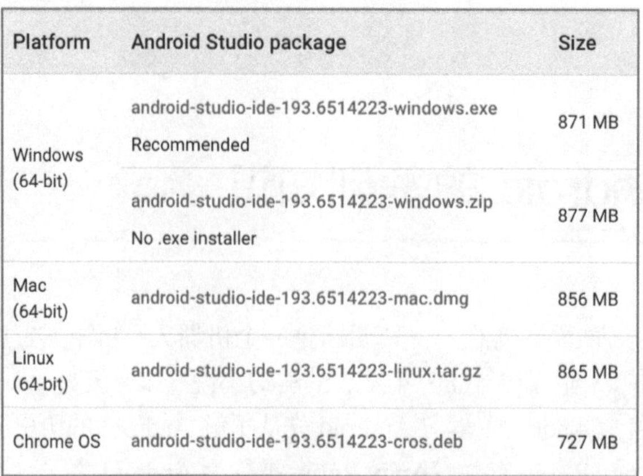

Platform	Android Studio package	Size
Windows (64-bit)	android-studio-ide-193.6514223-windows.exe Recommended	871 MB
	android-studio-ide-193.6514223-windows.zip No .exe installer	877 MB
Mac (64-bit)	android-studio-ide-193.6514223-mac.dmg	856 MB
Linux (64-bit)	android-studio-ide-193.6514223-linux.tar.gz	865 MB
Chrome OS	android-studio-ide-193.6514223-cros.deb	727 MB

图 1-2　Android Studio 下载列表

1.2.2　Android Studio 的安装

1. Windows 版的安装

在 Windows 操作系统中，双击下载的 EXE 格式安装文件，根据安装向导的指示安装 Android Studio 和所有所需的 SDK 工具（如图 1-3 所示）。安装进度完成后，单击“Finish”按钮（如图 1-4 所示）。

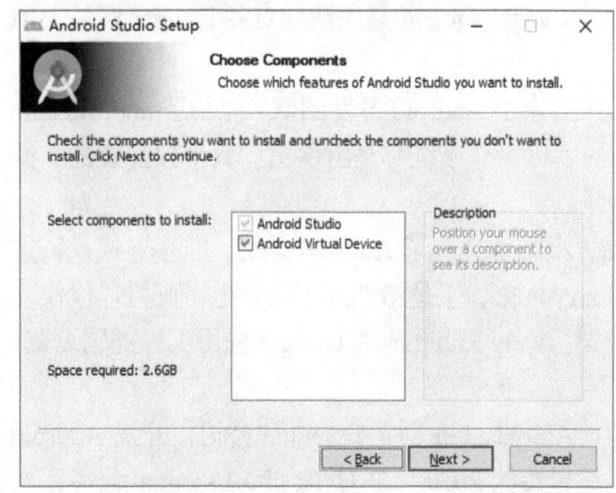

图 1-3　Android Studio 安装向导

Android Studio Setup

Completing Android Studio Setup

Android Studio has been installed on your computer.

Click Finish to close Setup.

Start Android Studio

Android Studio

< Back　Finish　Cancel

图 1-4　Android Studio 完成安装

2. Mac 版的安装

在 Mac 操作系统中，双击下载的 DMG 格式安装文件，在安装界面中将 Android Studio.app 拖曳到 Applications 文件夹中（如图 1-5 所示），即可安装完成。

图 1-5　拖曳安装

1.2.3　Android SDK 的下载安装

由于目前国内无法直接访问 Android 官方网站的部分地址，在 Mac 操作系统中首次运行时会打开“Android Studio First Run”对话框（如图 1-6 所示）。

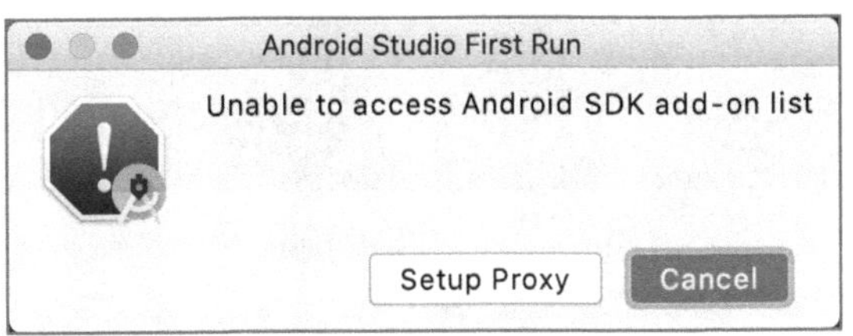

图 1-6　“Android Studio First Run”对话框

单击“Setup Proxy”按钮，打开“Http Proxy”对话框，选择“No proxy”单选按钮(如图 1-7 所示)，再单击“OK”按钮，完成设置。

提示：Http Proxy

无法直接访问 Android 官方网址进行更新时，需要设置代理地址，否则无法进行自动更新和下载。

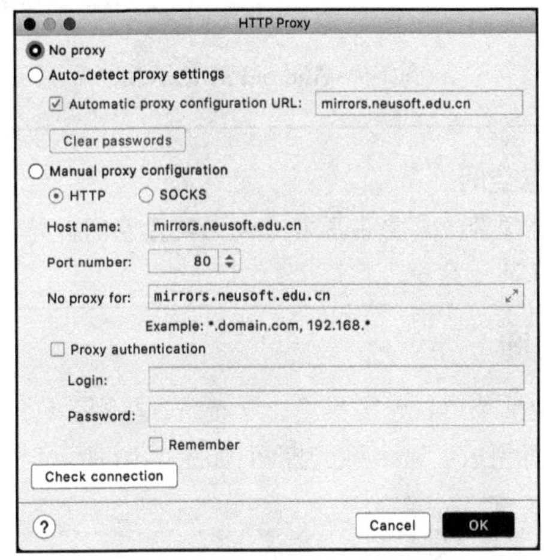

图 1-7　“Http Proxy”对话框

在“Android Studio Setup Wizard”对话框中会提示没有安装 SDK(如图 1-8 所示)，单击“Next”按钮，选择下载的 SDK 程序和保存路径(如图 1-9 所示)，单击“Next”按钮进行安装，完成安装后单击“Finish”按钮。

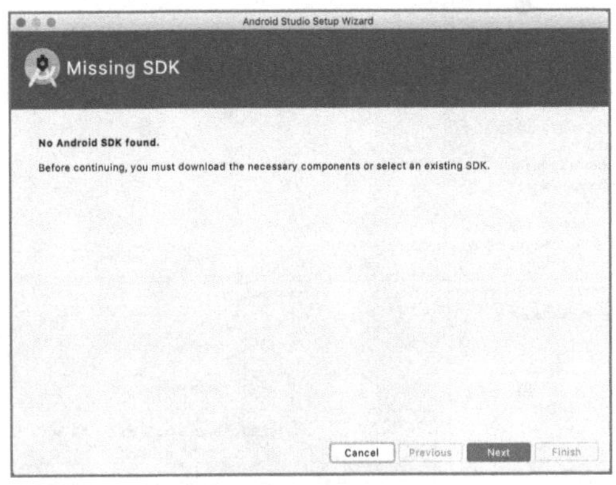

图 1-8　“Android Studio Setup Wizard”对话框

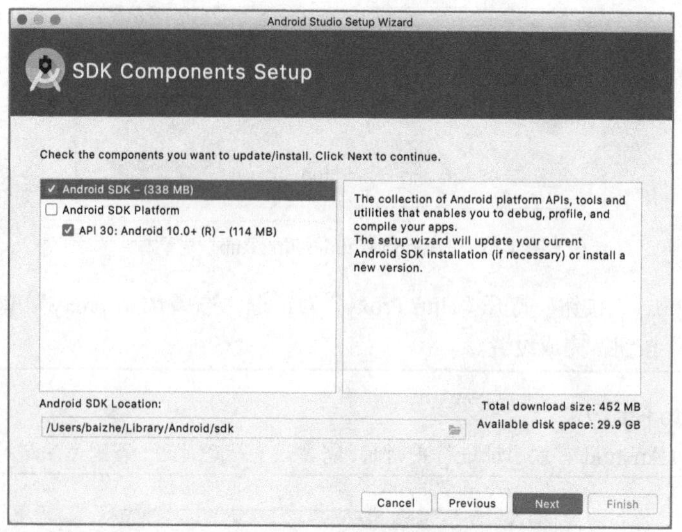

图 1-9　Android SDK 下载

提示：SDK 占用硬盘空间

Android SDK 的更新较为频繁，需要占用非常大的硬盘空间，建议预留 10GB 硬盘空间。如果系统盘空间较小，可以安装在其他盘中。

1.2.4　Android Studio 界面

Android Studio 界面主要包括菜单栏、工具栏、导航条、左侧工具条、工具窗口、编辑器、右侧工具条、运行工具窗口、状态栏等组成。Mac 版的界面(如图 1-10 所示)与 Windows 版的界面稍有不同。

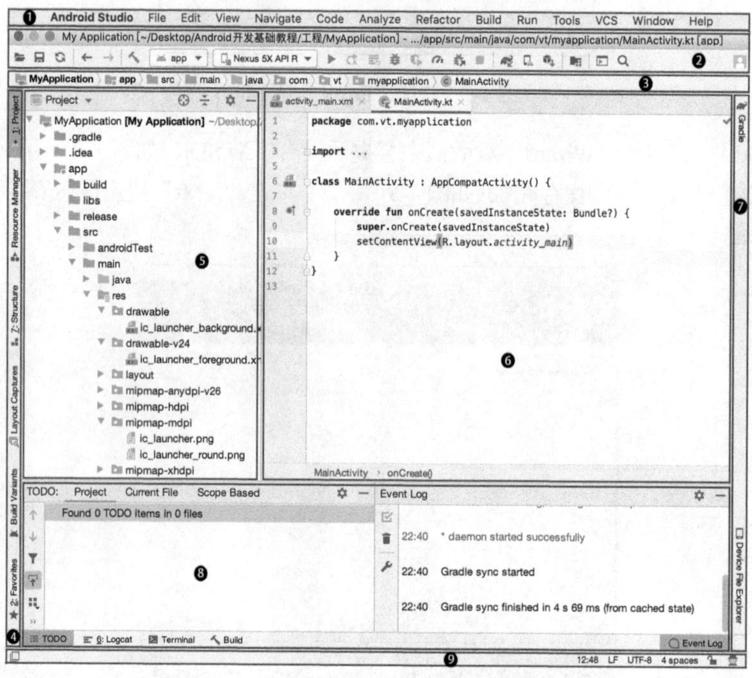

图 1-10　Mac 版的 Android Studio 界面

1. 菜单栏

菜单栏包含文件(File)、编辑(Edit)、视图(View)、导航(Navigate)、代码(Code)、分析(Analyze)、重构(Refactor)、构建(Build)、运行(Run)、工具(Tools)、版本控制系统(VCS)、窗口(Window)、帮助(Help)等功能菜单(如图 1-11 所示)。

Android Studio　File　Edit　View　Navigate　Code　Analyze　Refactor　Build　Run　Tools　VCS　Window　Help

图 1-11　菜单栏

2. 工具栏

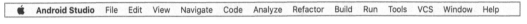

工具栏包含从菜单栏中提取出来的一些常用的功能(如图 1-12 所示)，能够进行快速操作，提高效率。

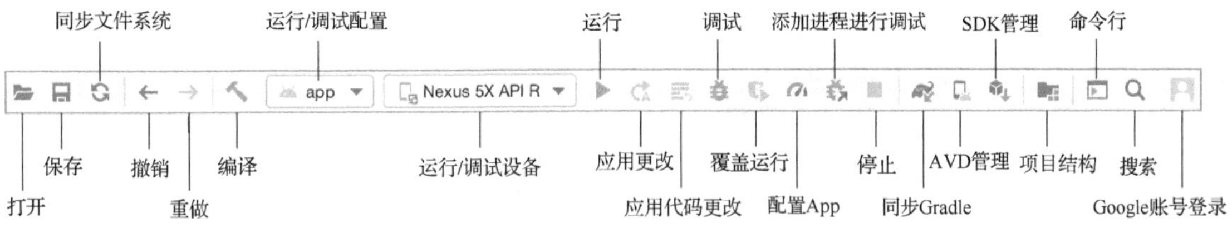

图 1-12　工具栏

3. 导航条

导航条用来辅助查看打开的项目和文件(如图 1-13 所示)，单击文件夹可以快速选择子文件夹或文件。

MyApplication 〉app 〉src 〉main 〉java 〉com 〉vt 〉myapplication 〉MainActivity

图 1-13　导航条

4. 左侧工具条

左侧工具条用来放置窗口的切换按钮(如图 1-14 所示)，包含“Project”、“Resource Manager”、“Structure”、“Layout Captures”、“Build Variants”和“Favorites”视图。

5. 工具窗口

工具窗口最多可以同时显示两个视图(如图 1-15 所示)，单击左侧工具条中的按钮可以进行切换。上面显示“Project”或“Resource Manager”视图，下面显示其余的任意一个视图。

6. 编辑器

编辑器由文件标签栏、左边栏、编辑区和代码定位栏组成(如图 1-16 所示)，是编辑配置信息、代码编写和调试断点设置的区域。

7. 右侧工具条

右侧工具条包含“Gradle”窗口和设备文件浏览器(外接 USB 设备或虚拟设备)，单击后在编辑器右侧显示(如图 1-17 所示)。

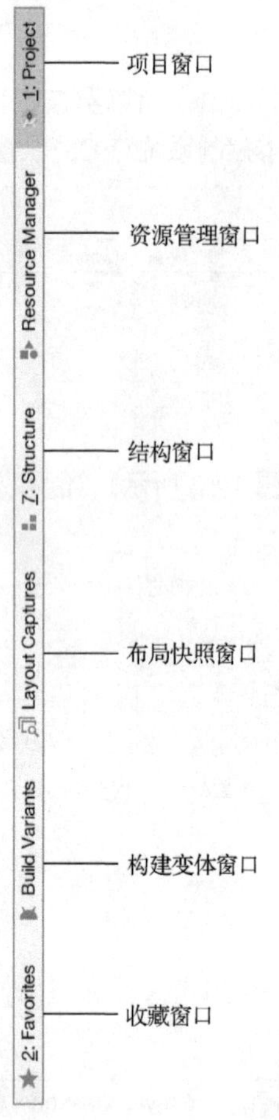

图 1-14　左侧工具条

图 1-15　工具窗口

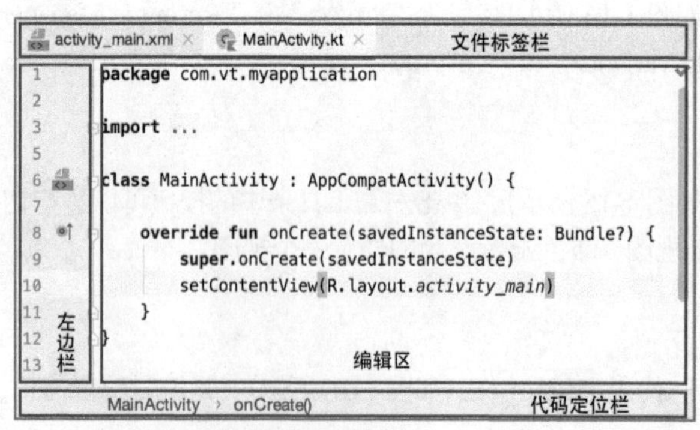

图 1-16　编辑器

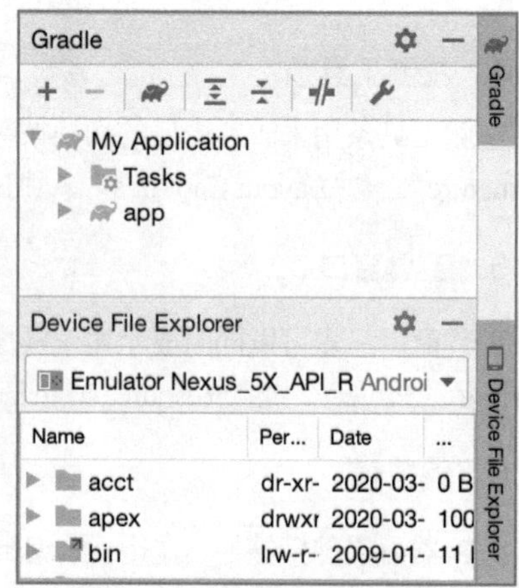

图 1-17　右侧工具条

8. 运行工具窗口

运行工具窗口主要显示 Android Studio 运行过程中输出的信息(如图 1-18 所示)，左侧显示

"Run"、"TODO"、"Logcat"、"Profiler"、"Terminal"或"Build"窗口，右侧显示"Event Log"窗口。

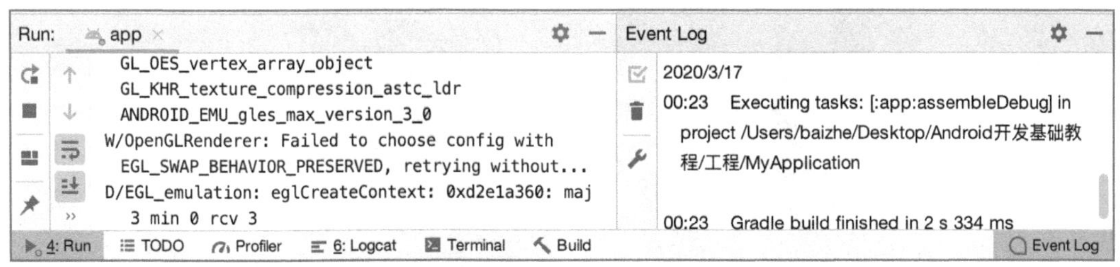

图 1-18　运行工具窗口

9. 状态栏

状态栏通常在界面的底部，主要显示 Android Studio 当前的状态和执行的任务（如图 1-19 所示）。

图 1-19　状态栏

1.2.5　Gradle 更新

Gradle 是 Android Studio 默认的 App 构建工具，它根据构建规则和配置文件自动构建 App，自动构建包括编译、打包等流程。安装 Android Studio 后，还要注意 Gradle 的更新。Gradle 自动更新时并不通过 Android 官网下载更新文件，而是通过 Gradle 网址。但是在国内下载的速度极慢甚至无法下载，可以打开 Gradle 网址（http://services.gradle.org/distributions/）通过手动下载的方式进行更新。

1. Windows 版的手动更新

把下载的新版本 Gradle 解压到 gradle 更新文件夹中（如图 1-20 所示）。然后在 Android Studio 中，选择【File】→【Settings】命令（如图 1-21 所示）。在打开的对话框中，选择左侧的"Gradle"选项。然后选择右侧的"Use local gradle distribution"单选按钮，单击"Gradle home"后面的路径选择按钮，选择刚才解压的 gradle 文件夹，单击"OK"按钮（如图 1-22 所示）。

脑 › 软件 (E:) › Soft › Android › android-studio › gradle ›

名称	修改日期	类型	大小
gradle-2.14.1	2017/2/27 20:27	文件夹	
gradle-3.4.1	2017/3/3 19:53	文件夹	
m2repository	2017/2/27 20:27	文件夹	

图 1-20　更新文件所在的文件夹

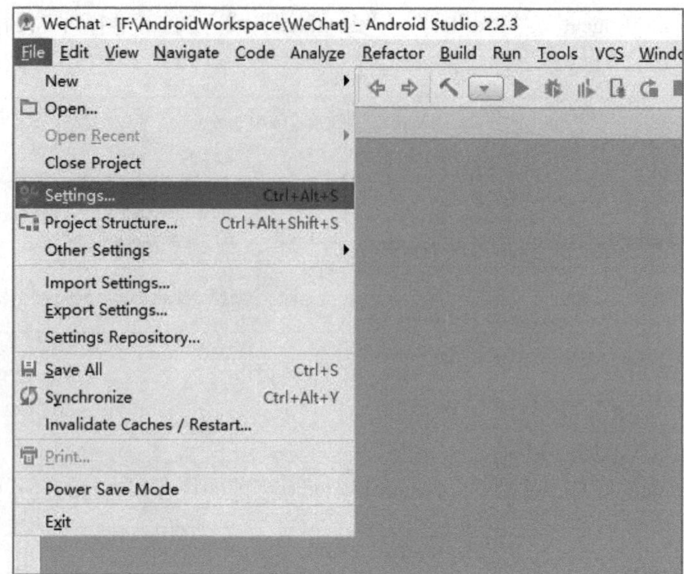

图 1-21 选择【File】→【Settings】命令

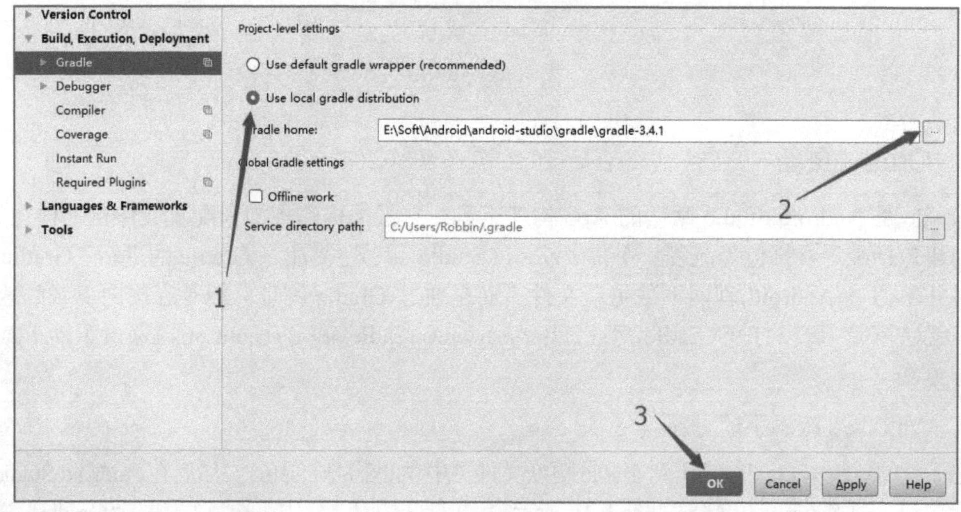

图 1-22 设置本地 gradle 路径

2. Mac 版的手动更新

在启动台中打开终端，输入“open .gradle”命令后回车，打开更新文件所在的文件夹。将手动下载的“gradle-x.x-all.zip”文件放置在相应的文件夹中(如图 1-23 所示)。重启 Android Studio 后会自动进行更新(如图 1-24 所示)。

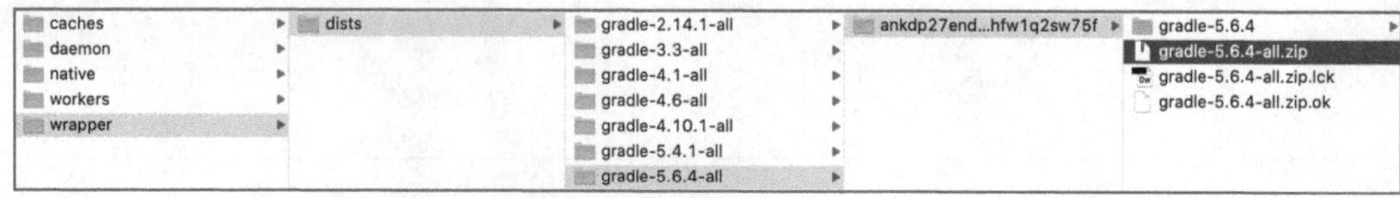

图 1-23 更新文件所在的文件夹

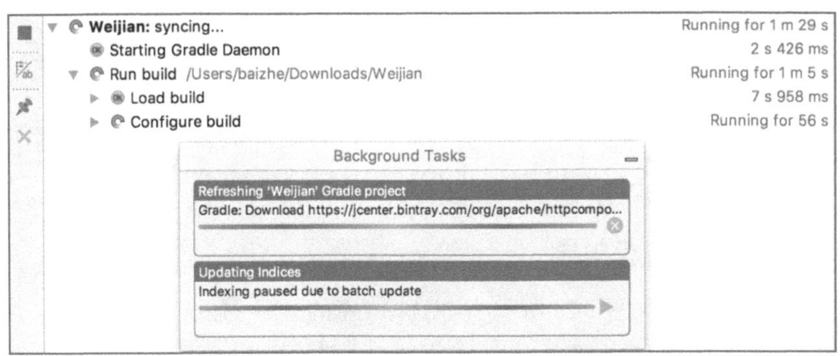

图 1-24　重启 Android Studio 后自动更新 Gradle

1.2.6　重构工程

Android Studio 更新 Gradle 后，打开原有工程后会自动弹出“Plugin Update Recommended”提示框(如图 1-25 所示)。单击后打开“Android Gradle Plugin Update Recommended”对话框，单击“Update”按钮(如图 1-26 所示)，自动重构 Android 工程，详细进度显示在“Build”窗口中。

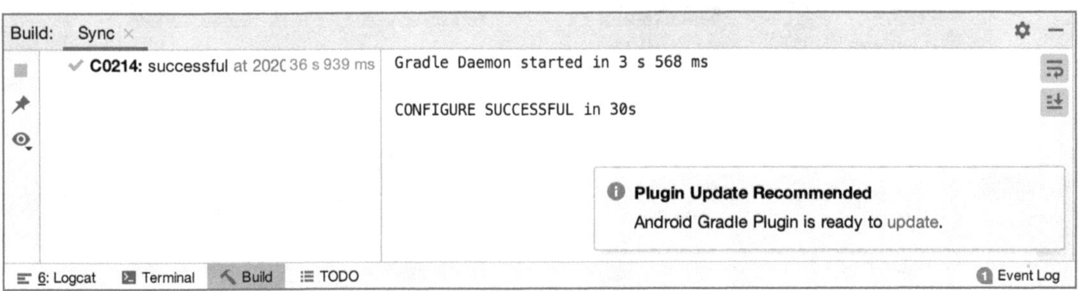

图 1-25　更新 Gradle 的提示

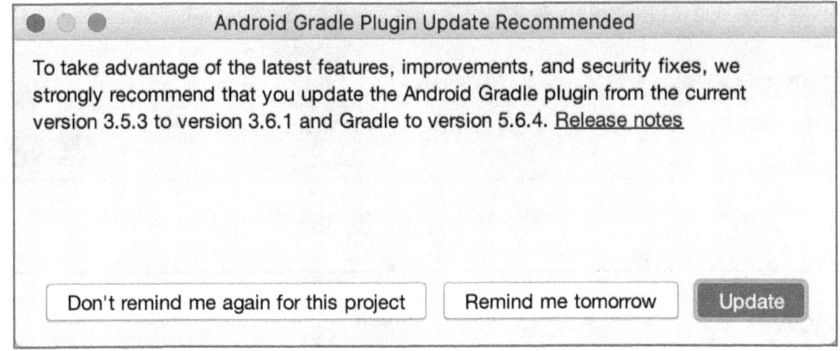

图 1-26　更新 Gradle 的对话框

1.3　创建 Android 工程

1.3.1　Android 工程的新建命令

选择【File】→【New】→【New Project】命令(如图 1-27 所示)，打开“Create New Project”对话框，进行工程创建的向导。

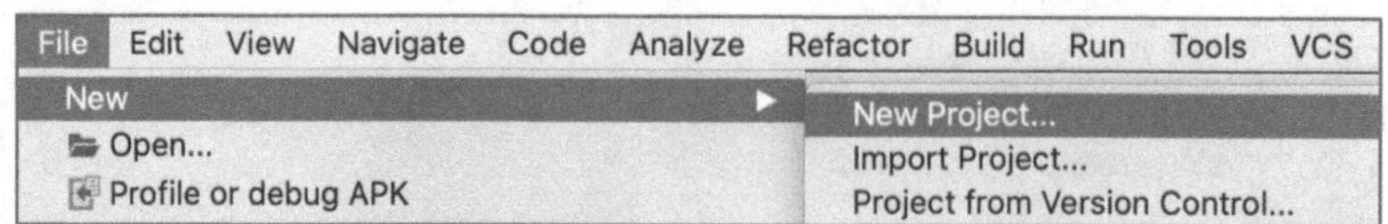

图 1-27 【New Project】命令

1.3.2 Android 工程的创建向导

工程创建向导的第 1 页(如图 1-28 所示)用于选择工程的类型，建议选择“Empty Activity”进行手机项目的 App 开发。若进行单项练习，可以针对练习项目选择相应的类型。

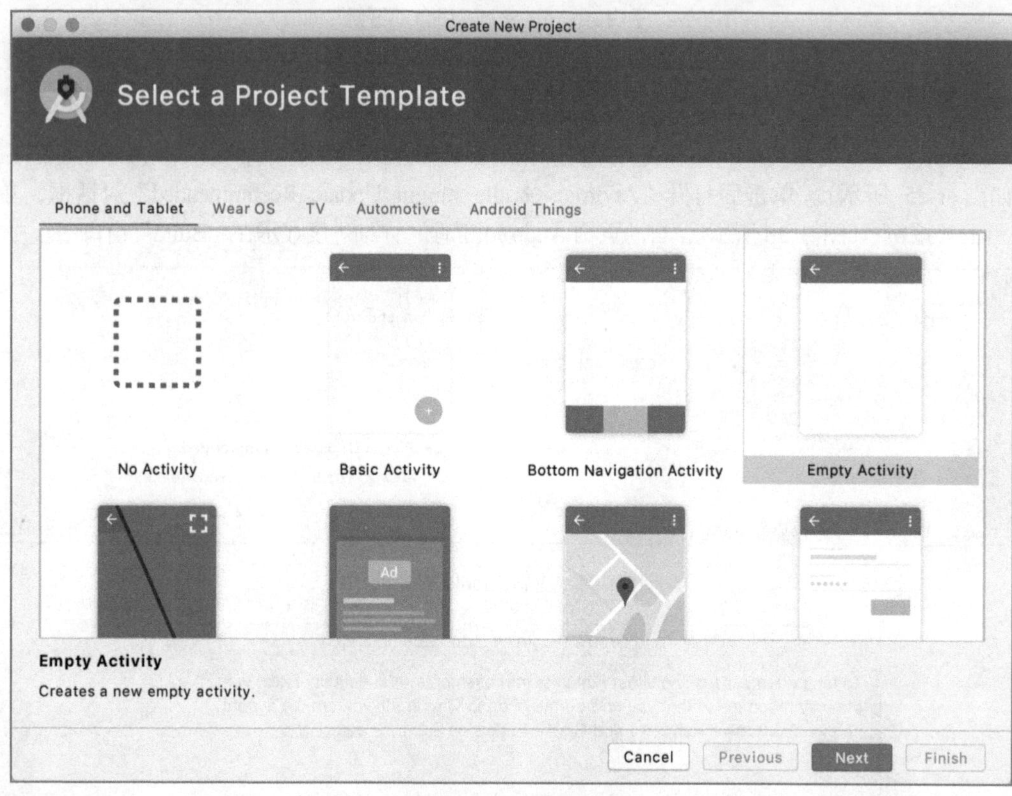

图 1-28 选择工程类型

提示：Activity

Activity 是一个应用组件，用户可与其进行交互，用于绘制用户界面的窗口。窗口通常会充满屏幕，也可不充满屏幕并浮动在其他窗口之上。后续会详细介绍 Activity。

单击“Next”按钮，进入工程创建向导的第 2 页(如图 1-29 所示)，该页面选项如下。

- Name(应用名称)：安装到 Android 设备上后显示的名称。
- Package name(包名)：一般将企业域名倒置顺序后作为前缀，然后加上 App 的英文名。
- Save location(存储位置)：保存工程文件的路径。
- Language(语言)：可以选择 Java 或 Kotlin。
- Minimum SDK(最小 SDK 版本)：可以安装该 App 的 Android 设备最低版本。

● Use legacy android.support libraries（使用传统 Android 支持库）：不勾选时使用 AndroidX 库，API Level 29 及以后的版本无法使用传统 Android 支持库。

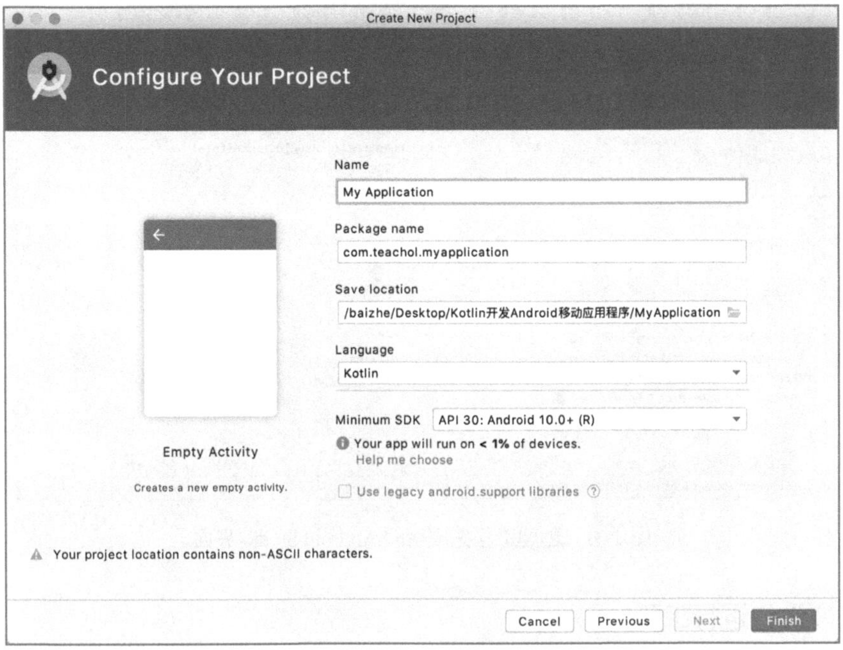

图 1-29　配置工程

单击“Help me choose”按钮，可查看官方统计的各版本使用比例（如图 1-30 所示）。

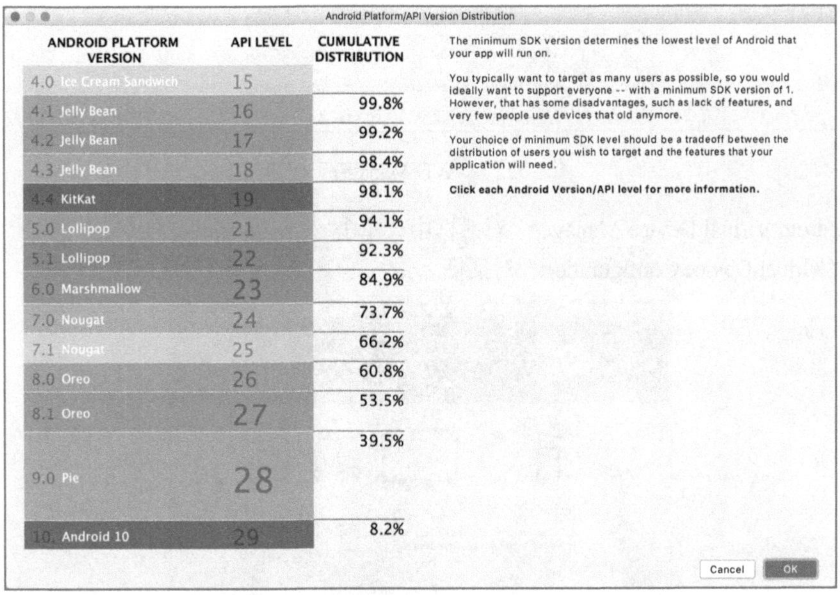

图 1-30　各版本使用比例

单击“OK”按钮，再单击“Finish”按钮完成工程的创建向导，然后显示 Android Studio 界面（如图 1-31 所示）。

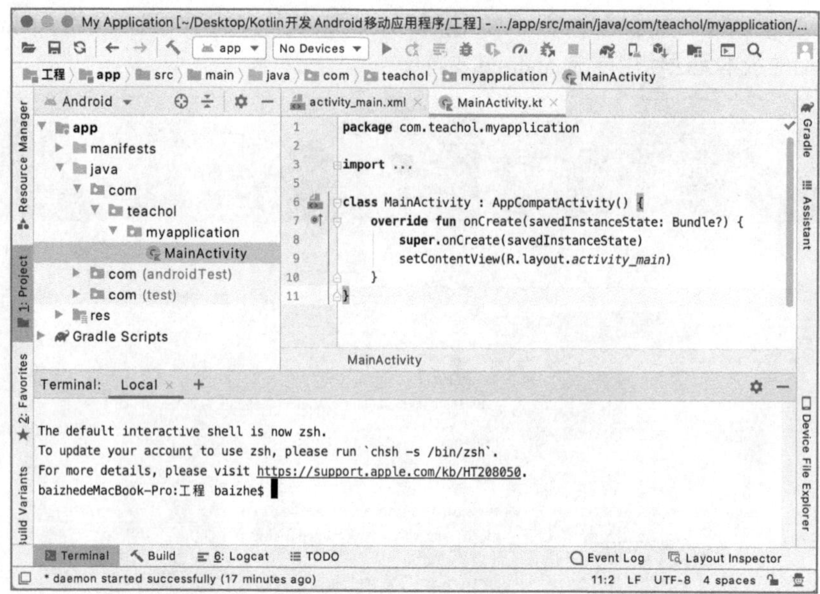

图 1-31　创建工程完成后的 Android Studio 界面

1.3.3　虚拟设备运行工程

1. 创建虚拟设备

在工具栏中，单击“AVD Manager”按钮(如图 1-32 所示)，打开“Android Virtual Device Manager”对话框。

图 1-32　“AVD Manager”按钮

在“Android Virtual Device Manager”对话框中，单击“Create Virtual Device”按钮(如图 1-33 所示)，打开“Virtual Device Configuration”对话框。

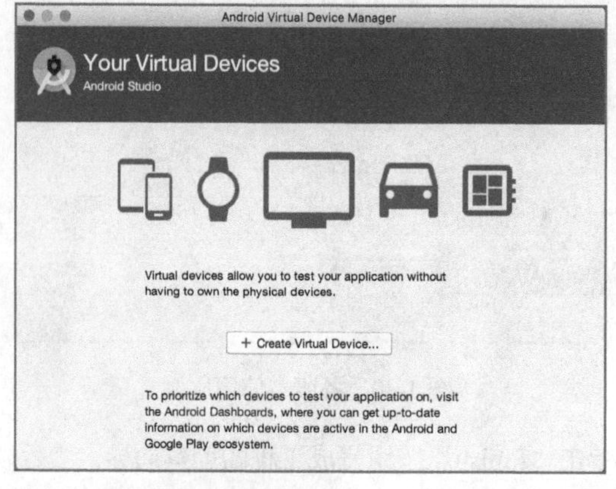

图 1-33　“Android Virtual Device Manager”对话框

在“Virtual Device Configuration”对话框中，显示了虚拟设备的参数列表(如图 1-34 所示)。根据需求选择相应的虚拟设备，然后单击“Next”按钮。

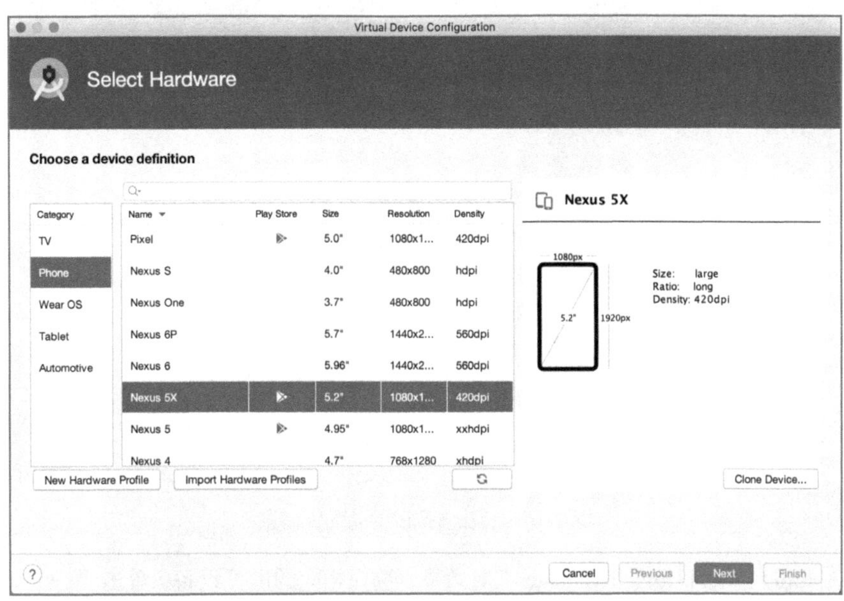

图 1-34　选择虚拟设备

接下来要选择 Android 系统的版本，如果没有下载相应的系统镜像，需要选择“DownLoad”选项下载系统镜像(如图 1-35 所示)。

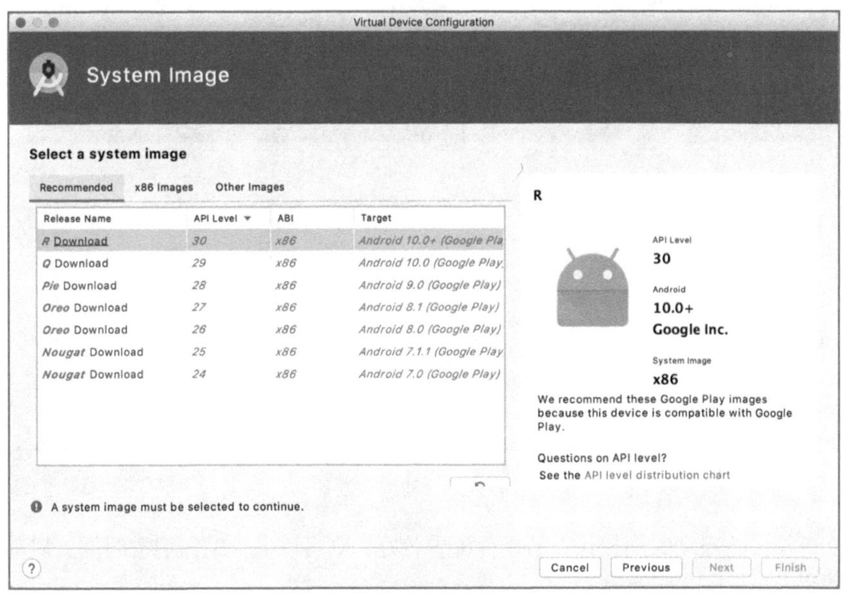

图 1-35　选择系统版本

选择“DownLoad”选项后，会下载 SDK 文件并自动安装。安装完成后，单击“Finish”按钮(如图 1-36 所示)，确认虚拟设备的配置信息，然后单击“Next”按钮完成虚拟设备的创建。

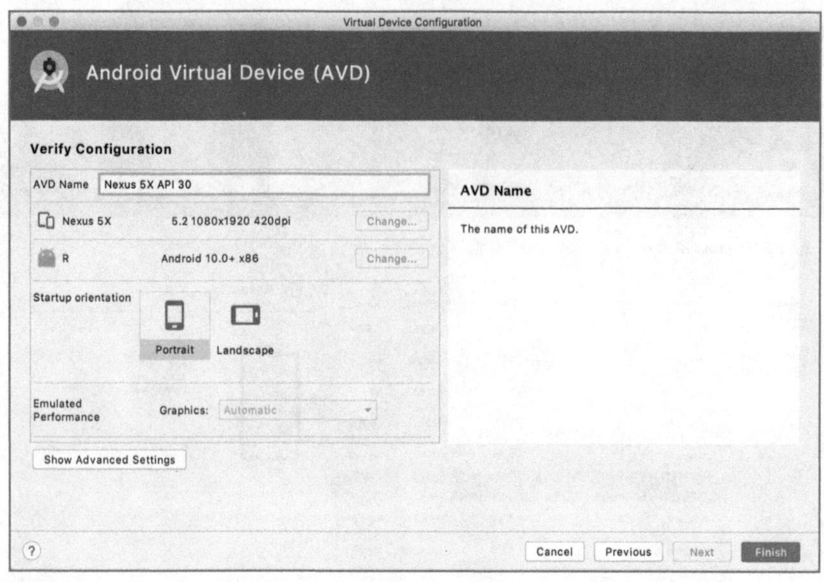

图 1-36　确认虚拟设备的配置信息

返回“Android Virtual Device Manager”对话框，选择相应的虚拟设备，单击“Launch this AVD in the emulator”图标(如图 1-37 所示)，启动虚拟设备。

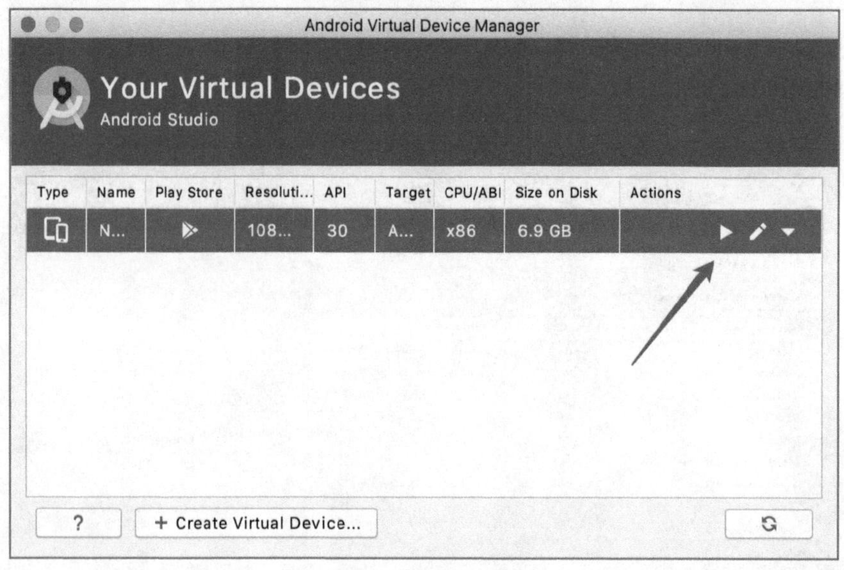

图 1-37　启动虚拟设备

Android 虚拟设备启动时间较长，若硬件配置较好，则可以大大缩短启动时间。启动完成后，显示虚拟设备界面(如图 1-38 所示)。

2. 运行工程

在工具栏中，选择创建的虚拟设备，然后单击“Run 'app' (^R)”按钮(如图 1-39 所示)。工程编译完成后，自动安装在虚拟设备中，然后自动启动运行(如图 1-40 所示)。

图 1-38　虚拟设备界面

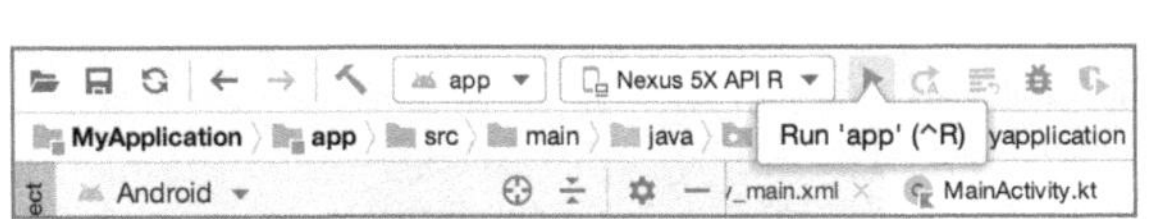

图 1-39　单击“Run 'app' (^R) ”按钮

图 1-40　使用虚拟设备运行工程

1.3.4　物理设备运行工程

使用 USB 线将手机连接计算机，在工具栏的运行/调试设备下拉菜单中会显示连接的手机作为首选项(如图 1-41 所示)。单击“Run 'app' (^R) ”按钮，会编译工程并安装在连接的手机上，然后自动启动运行(如图 1-42 所示)。

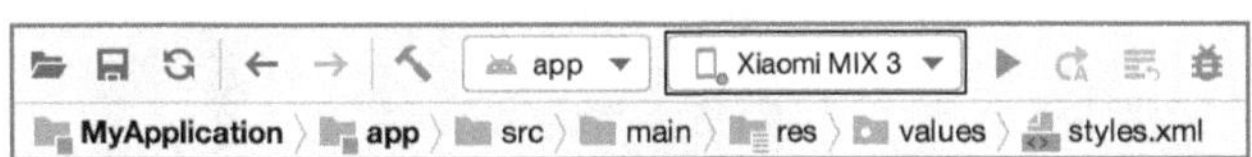

图 1-41　显示连接的手机

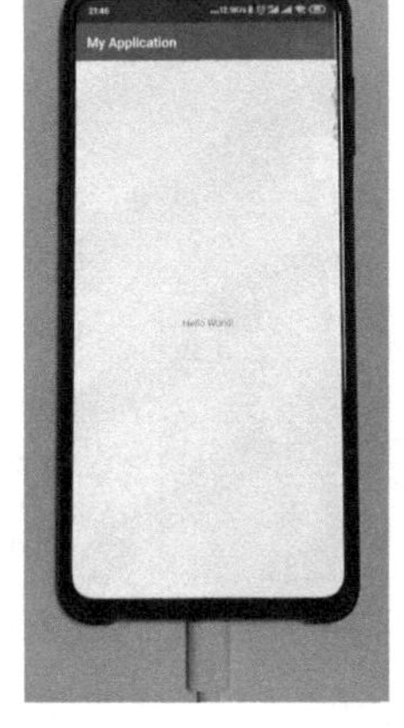

图 1-42　使用手机运行工程

1.3.5　生成签名的 APK 文件

APK 文件是 App 的安装文件，设置 APK 文件签名的目的是让 App 不被恶意者生成的 APK 文件所覆盖安装。在升级 App 时，只有同一签名的 APK 文件才能对 App 进行升级，从而避免恶意覆盖。

选择【Build】→【Generate Signed Bundle/APK】命令。打开“Generate Signed Bundle or APK”对话框，选择“APK”单选按钮，单击“Next”按钮(如图 1-43 所示)，进入下一个页面。

单击“Create new”按钮，打开“New Key Store”对话框，设置签名文件密码、签名密码、有效年限和证书内容(如图 1-44 所示)，然后单击“OK”按钮完成签名文件的创建。

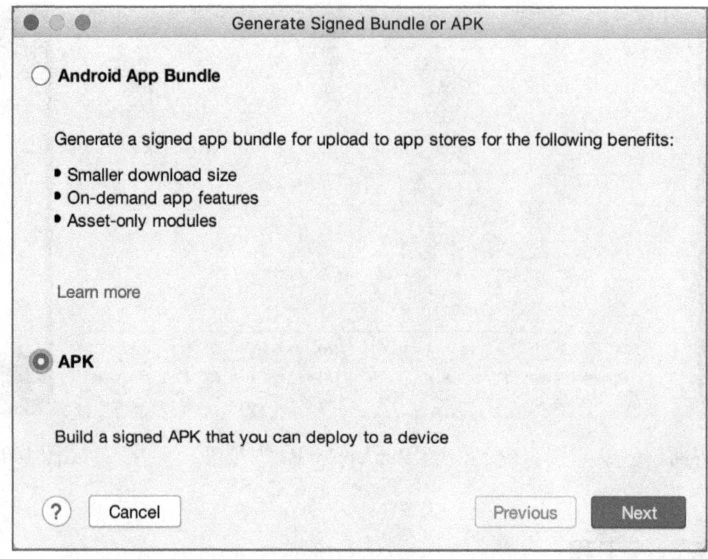

图 1-43　选择“APK”单选按钮

图 1-44　“New Key Store”对话框

创建签名文件后，在“Generate Signed Bundle or APK”对话框中会自动输入签名路径、签名路径密码、签名别名和签名密码(如图 1-45 所示)，单击“Next”按钮，进入下一个页面。在“Build Variants”框中选择“release”选项,勾选“Signature Versions”中的“V2(Full APK Signature)”复选框(如图 1-46 所示)，单击“Finish”按钮生成签名的 APK 文件。

图 1-45　自动输入签名文件信息

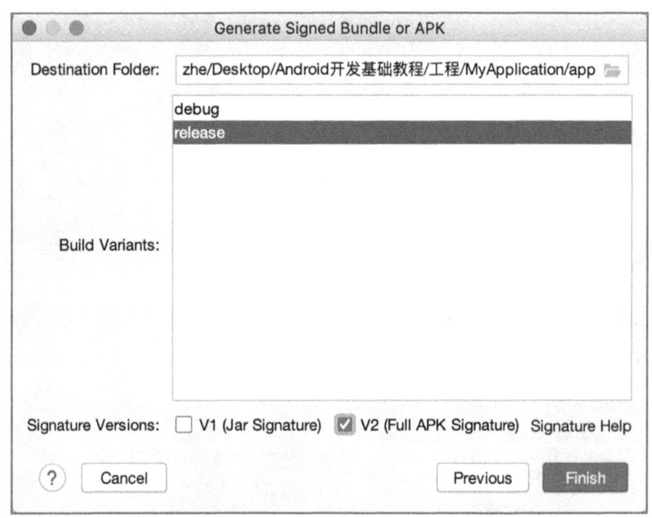

图 1-46　选择完整版和完整签名

提示：防反编译

对 APK 文件进行反编译可以还原源代码，为了防止源代码被恶意使用，商业上使用的 App 还会进行防反编译处理。常用的防反编译方法包括混淆策略、整体 Dex 加固、拆分 Dex 加固、虚拟机加固等，推荐使用阿里聚安全、腾讯云应用乐固、爱加密、娜迦、梆梆等防反编译软件。

1.4　Android 的工程结构

1.4.1　Project 视图

Project 视图是以工程管理的形式显示文件的视图（如图 1-47 所示），主要包括 AndroidManifest.xml 文件（App 的配置信息）、build.gradle 文件（Gradle 配置文件）、res 文件夹（资源文件）和 java 文件夹（源代码和测试代码）。

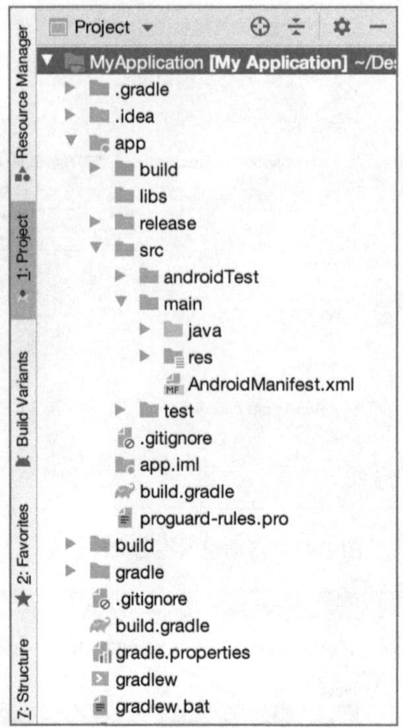

图 1-47　Project 视图

1.4.2　AndroidManifest.xml 文件

AndroidManifest.xml（如图 1-48 所示）是 Android App 的配置文件，包含 App 的基本信息，声明程序中的 Activity、ContentProvider、Service 和 Receiver，指定 permission 和 instrumentation 等。

AndroidManifest.xml

```xml
<?xml version="1.0" encoding="utf-8"?>
<manifest xmlns:android="http://schemas.android.com/apk/res/android"
    package="com.vt.myapplication">

    <application
        android:allowBackup="true"
        android:icon="@mipmap/ic_launcher"
        android:label="My Application"
        android:roundIcon="@mipmap/ic_launcher_round"
        android:supportsRtl="true"
        android:theme="@style/AppTheme">
        <activity android:name=".MainActivity">
            <intent-filter>
                <action android:name="android.intent.action.MAIN" />

                <category android:name="android.intent.category.LAUNCHER" />
            </intent-filter>
        </activity>
    </application>

</manifest>
```

图 1-48　AndroidManifest.xml 文件

1．<manifest>标签

<manifest>标签是 AndroidManifest.xml 的根节点，用于设置 xmlns:android 和 package 属性，还包括一个<application>标签。

- xmlns:android 属性：设置 Android API 的命名空间，用于识别标签的属性。
- package 属性：设置 App 的包名。

2. <application>标签

<application>标签用于声明 App，包含每个 App 组件所声明的子元素，以及能够影响所有组件的属性。

- android:allowBackup 属性：设置 App 数据的备份和恢复功能，默认值为 true，可以通过 adb backup 和 adb restore 对 App 数据进行备份和恢复。
- android:icon 属性：设置一个 Drawable 资源作为 App 的普通图标(如图 1-49 所示)。
- android:label 属性：设置 App 的名称。
- android:roundIcon 属性：设置一个 Drawable 资源作为 App 的圆形图标(如图 1-50 所示)。

图 1-49　默认普通图标

图 1-50　默认圆形图标

- android:supportsRtl 属性：设置是否支持从右到左的布局，默认值为 true。当 build.gradle 文件中的 targetSdkVersion 值设置为 17 或更高时，可以使用 RTL(right-to-left)布局。
- android:theme 属性：设置界面皮肤主题。

3. <activity>标签

<activity>标签用于设置 Activity 的属性，Activity 必须被声明在<manifest>标签中，没有被声明的 Activity 不会被调用。android:name 属性设置声明的 Activity 的类名。

4. <intent-filter>标签

<intent-filter>标签用于指定 Activity、Service 或 Broadcast Receiver 能够响应的 Intent 对象类型，通过<action>、<category>和<data>子标签进行描述。

5. <action>标签

<action>标签用于给 Intent 过滤器添加子标签，<intent-filter>标签必须包含一个或多个<action>标签。如果不包含<action>标签，就不存在 Intent 过滤器。当 android:name 属性设置为 android.intent.action.MAIN 时，该 Activity 作为应用默认启动的 Activity。

6. <category>标签

<category>标签用于向 Intent 过滤器添加类别名称，android:name 属性的默认值为 CATEGORY_DEFAULT。

1.4.3　build.gradle 文件

工程中包含 2 个 build.gradle 文件，分别是根目录下的顶级(top level)build.gradle 文件和 app 目录下的模块级(module level)build.gradle 文件。

顶级 build.gradle 文件位于工程根目录下，用于配置适用于工程中所有模块的构建设置。默认情况下，顶级 build.gradle 文件使用 buildscript {}代码块来定义工程中所有模块公用的 Gradle 存储区和依赖项。

模块级 build.gradle 文件位于 project/module 目录下，用于配置适用其所在模块的构建设置，通过配置这些构建设置来提供自定义打包选项。

1.4.4 res 文件夹

Android 的资源文件存储在工程的 res 文件夹内（如图 1-51 所示），Android Studio 3 将资源分类存储在 4 类文件夹（前缀名为 drawable、layout、mipmap 和 value）中。drawable-v24 文件夹的后缀 v24 代表 API 的版本，mipmap-mdpi 文件夹的后缀 mdpi 表示 Android 设备的屏幕分辨率。

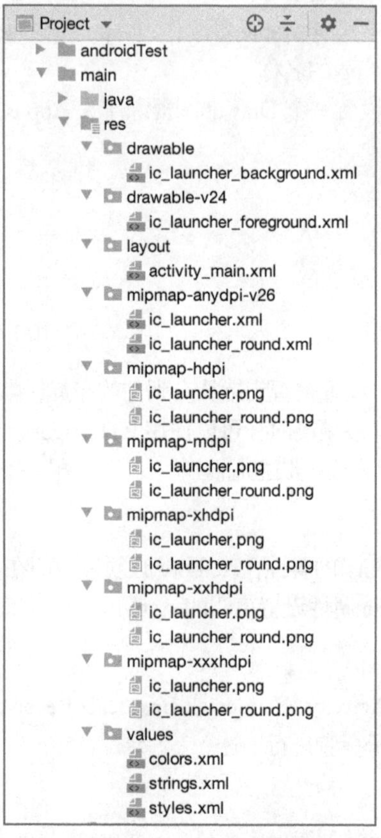

图 1-51　res 文件夹结构

- drawable：用于存储图像和 XMLDrawable 文件，XMLDrawable 文件分为 AnimationDrawable、BitmapDrawable、ClipDrawable、ColorDrawable、GradientDrawable、NinePatchDrawable、RotateDrawable、StateListDrawable、ShapeDrawable 和 TransitionDrawable 等类型。
- layout：用于存储界面布局文件。
- mipmap：用于存储图标文件，默认包含自动生成的 Launcher Icons（桌面图标），还可以存储 Action Bar and Tab Icons（桌面图标）和 Notification Icons（通知图标）。
- values：用于存储 colors.xml、strings.xml 和 styles.xml 等文件。

第 2 章 Kotlin 基础

Kotlin 是一种现代多平台应用的静态类型编程语言，与 Java 一样，都需要编译成.class 文件后通过虚拟机执行代码，因此 Kotlin 可以与 Java 共存。

2.1 简　　介

2011 年，JetBrains 公司发布了新的开源语言——Kotlin，但是直到 2016 年才发布 1.0 正式版。JetBrains 公司是捷克的一家高科技公司，在俄罗斯圣彼得堡和美国波士顿分别设立了分公司。Kotlin 一词源于圣彼得堡以西 30 千米的科特林岛（英文名为 Kotlin Island）。2017 年，Google I/O 全球开发者大会上，Google 公司宣布 Kotlin 成为 Android 官方开发语言。Android Studio 是基于 IntelliJ IDEA 的开发工具，IntelliJ IDEA 也是 JetBrains 公司的产品。

Kotlin 的语法较 Java 而言更加简洁，通常代码量更少，但是生成的 APK 文件可能会更大。Kotlin 不仅可用于 Android 开发，还可用于服务器端开发、JavaScript 开发、原生开发和数据科学。

> **提示：编程语言的选择**
>
> 选择哪种编程语言，通常需要在综合考虑扩展性、第三方类库、兼容性、开发成本等因素后才决定。编程中最重要的从来都不是语言本身，而是通过语言实现的算法和数据结构。只要掌握了一种面向对象编程语言，再学习另一种语言的语法通常就只需要几天时间。而在项目中运用算法和数据结构知识的能力，则需要长期的积累和实践。

2.2 变　　量

变量是计算机语言中存储数据的抽象概念，通过变量名进行赋值和读取数据。

2.2.1 变量的命名规则

变量名即变量的名称，命名时需要遵守一定的规则。在 Kotlin 中，变量的命名规则如下。

- 变量名必须由字母、数字和下画线组成，如 abc、a1、_width。
- 变量名的第一个字符不能使用数字，如 3c、28d。
- 变量名中的英文字母区分大小写，如 name 和 Name 是两个不同的变量名。
- 不能使用关键字作为变量名，如 for、else、if。

> **提示：关键字**
>
> 关键字又称保留字，是编程语言中语法规范所使用的单词，不能作为变量名、类名、函数名、包名和命名空间等使用。

2.2.2 变量的命名方法

除要遵守变量的命名规则外，变量的命名还应方便代码阅读。**Kotlin 的官方类库使用了驼峰式命**

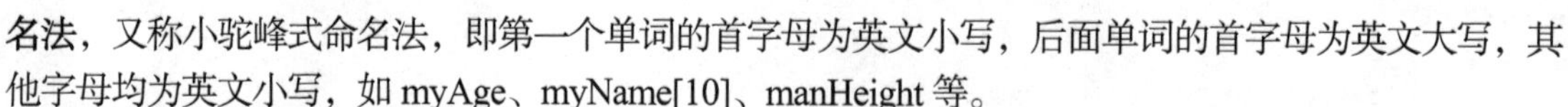

名法，又称小驼峰式命名法，即第一个单词的首字母为英文小写，后面单词的首字母为英文大写，其他字母均为英文小写，如 myAge、myName[10]、manHeight 等。

2.2.3 变量的声明

变量在使用前必须要进行声明，同时设置变量类型，还可以对变量进行初始化赋值。变量可以声明为可变变量和不可变变量（即常量），**var（variable 的简写）用于声明可变变量，val（value 的简写）用于声明不可变变量**。

语法格式：

```
var|val identifier [: type][ = value]
```

- [: type]是指数据类型，为可选项。**当有初始化赋值时可省略，系统自动判断数据类型。**
- [= value]是对变量进行初始化赋值，为可选项。当指定数据类型时可暂时不进行初始化赋值。

提示：语法格式符号

在本书的语法格式中，“|”表示多选一的并列关系，“[]”表示可选项。这两个语法格式符号用于帮助理解语法，实际程序代码中并不会出现。

2.3 数据类型

广义上讲，面向对象编程语言中的所有类都是一种数据类型。狭义上讲，数据类型是指 Byte、Short、Int、Long、Float、Double、Boolean、Char 和 String 等类型。数据类型按照存储方式的不同分为基本数据类型和引用数据类型。

2.3.1 基本数据类型

在 Kotlin 中，**基本数据类型共有 8 种：Byte、Short、Int、Long、Float、Double、Boolean 和 Char。**每种类型对应一个类，可以直接赋值并进行初始化。其默认值、取值范围和位数均有所不同，如表 2-1 所示。除 Boolean 和 Char 类型外，其余类型都支持二进制、十进制和十六进制。Char 类型不能直接使用 ASCII 码值进行赋值，因此也不能和数字直接进行运算。

表 2-1　基本数据类型

数据类型	默认值	取值范围	位数	备注
Byte	0	–128~127	8	
Short	0	–32768~32767	16	
Int	0	–2^31~2^31–1	32	
Long	0L	–2^63~2^63–1	64	L 是 long 数值的后缀
Float	0.0f	3.4e–45~1.4e38	32	f 是 float 数值的后缀
Double	0.0d	4.9e–324~1.8e308	64	d 是 double 数值的后缀
Boolean	false	true/false	1	
Char		\u0000~\uffff	16	可以使用转义字符，不能使用 ASCII 码值进行赋值

提示：进制

进制是指带进位的计数方法，默认使用的是十进制计数方法。在编程中常用的进制还包括二进

制和十六进制，二进制数以 0b 开头，十六进制数以 0x 开头。例如，十进制数 11 用二进制数表示为 0b1011，用十六进制数表示为 0xb。

2.3.2　基本数据类型的自动转换

在混合运算中，不同数据类型的数据先转换为同一种数据类型，转换是从低级到高级的（Byte>Short>Char>Int>Long>Float>Double），然后进行运算。**转换前数据类型的位数要低于转换后数据类型的位数**，例如，16 位的 Short 类型可以自动转换为 32 位的 Int 类型，32 位的 Float 类型可以自动转换为 64 位的 Double 类型。

在 C0201 工程中，演示了运算中基本数据类型的自动转换，输出结果如图 2-1 所示。

```
/java/com/teachol/c0201/MainActivity.kt
10  val b: Byte = 1
11  val s: Short = 10
12  val c = 'a'
13  val i = 100
14  val l = 1000L
15  val f = 1.0f
16  val res = b + s + c.toFloat() + i + l + f
17  println("b+s+c+i+l+f=$res")
```

第 10、11 行 Byte 和 Short 类型变量必须要声明数据类型，否则会被自动判断为 Int 类型，其他类型的变量会根据赋值的数据类型来决定变量的数据类型。第 16 行 c 变量需要使用 toFloat()方法将字符 'a' 转换为 Float 类型的 ASCII 码值，然后才能参与加法运算。第 17 行$res 表示将 res 变量存储的数值转换为字符串，然后将其合成到新字符串中，合成后的新字符串通过 println()语句输出到 Logcat 窗口中。

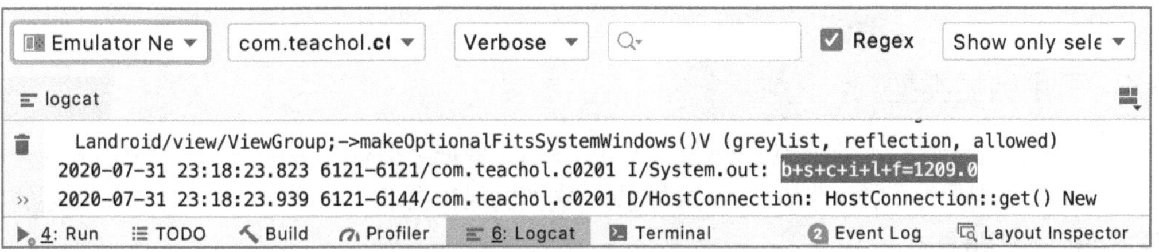

图 2-1　Logcat 窗口输出结果

2.3.3　基本数据类型的强制转换

基本数据类型提供了转换数据类型的方法，可以明确转换后的数据类型，避免了自动转换过程中的不可控性。

在 C0202 工程中，演示了常用的强制数据类型转换方式，输出结果如图 2-2 所示。

```
C0202 工程: MainActivity.kt
val testString1: String = "407"
val testInt1: Int = testString1.toInt()
println("字符串转换成整数：  " + testInt1) //字符串转换成整数
val testInt2: Int = 407
val testString2: String = testInt2.toString()
println("整数转换成字符串：  " + testString2) //整数转换成字符串
val testBoolean1: Boolean = true
```

```
20 val testString3: String = testBoolean1.toString()
21 println("布尔值转换成字符串： " + testString3) //布尔值转换成字符串
23 val testString4: String = "true"
24 val testBoolean2: Boolean = testString4.toBoolean()
25 println("字符串转换成布尔值： " + testBoolean2) //字符串转换成布尔值
27 val testFloat1: Float = 4.07f
28 val testInt3: Int = testFloat1.toInt()
29 println("浮点数转换成整数： " + testInt3) //浮点数转换成整数
31 val testInt4: Int = 407
32 val testFloat2: Float = testInt4.toFloat()
33 println("整数转换成浮点数： " + testFloat2) //整数转换成浮点数
```

第 20 行将 Boolean 类型转换为 String 类型时，真值转换为"true"，假值转换为"false"，而不是 1 或 0。只有字符串"true"(不区分英文大小写)可以使用显式转换的方式转换为 Boolean 类型的真值，其余字符串都返回假值。

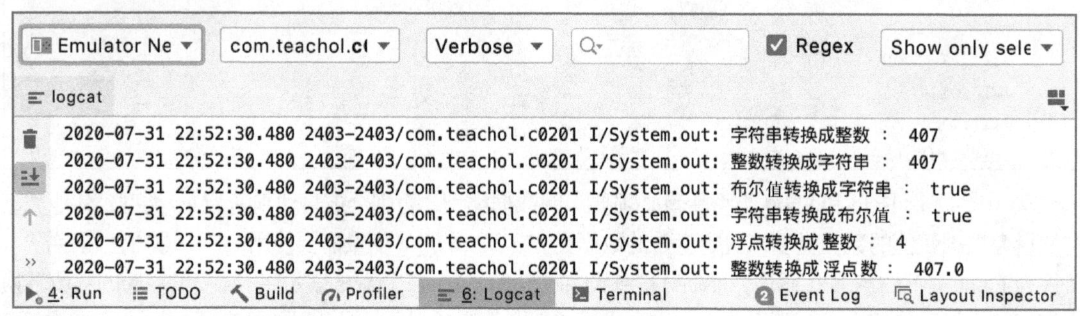

图 2-2　Logcat 窗口输出结果

2.3.4　引用数据类型

引用数据类型是除基本数据类型外的所有类型。引用数据类型有两块存储空间，引用数据类型的变量在栈(Stack)中保存了类实例在内存中的地址，类实例保存在堆(Heap)中。而基本数据类型的变量存储空间被分配到栈中。

提示：栈和堆

Heap 是 Stack 的一个子集。Stack 存取速度比 Heap 高，可共享存储数据，但是其中数据的大小和生存期必须在运行前确定。Heap 是运行时可动态分配的数据区，数据不共享，大小和生存期都可以在运行时再确定。

2.4　运算符和位运算

运算符是执行特定的数学或逻辑操作的符号，常用的运算符包括算术运算符、关系运算符、逻辑运算符和赋值运算符。

2.4.1　算术运算符

算术运算符是完成基本算术运算所使用的运算符。除可以使用算术运算符外，还可以使用基本数据类型的方法实现相同的功能，表 2-2 所示为算术运算符和等价方法。

表 2-2　算术运算符和等价方法

运算符	等价方法	描述
a+b	a.plus(b)	加法
a-b	a.minus(b)	减法
a*b	a.times(b)	乘法
a/b	a.div(b)	除法
a%b	a.rem(b)	取余
a++/++a	a.inc()	自增
a--/--a	a.dec()	自减

在 C0203 工程中，演示了如何使用等价方法实现算术运算符的功能，输出结果如图 2-3 所示。

C0203 工程：MainActivity.kt

```
10 var a:Int = 10
11 var b:Int = 20
12 var c:Int = 25
13 var d:Int = 25
14 var s:Int
15 println("算术运算符:")
16 s = a.plus(b)
17 println("a + b = " + s)
18 s = a.minus(b)
19 println("a - b = " + s)
20 s = a.times(b)
21 println("a * b = " + s)
22 s = b.div(a)
23 println("b / a = " + s)
24 s = b.rem(a)
25 println("b % a = " + s)
26 s = c.rem(a)
27 println("c % a = " + s)
28 s = a.inc()
29 println("a++   = " + s)
30 s = a.dec()
31 println("a--   = " + s)
32 s = d++//先赋值，后加1
33 println("d     = " + d)
34 println("d++   = " + s)
35 s = ++d//先加1，后赋值
36 println("d     = " + d)
37 println("++d   = " + s)
```

第 32 行将 d 变量先赋给 s 变量，然后 d 变量加 1。第 35 行将 d 变量先加 1，再将其赋给 s 变量。

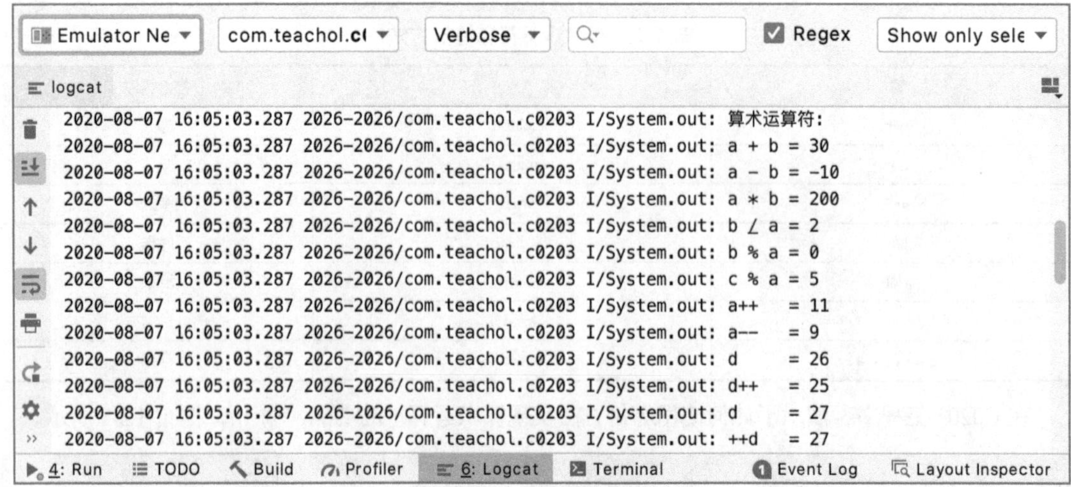

图 2-3　Logcat 窗口输出结果

2.4.2　关系运算符

关系运算符是用于判断大小或是否相等的运算符，运算结果只有 true 和 false 两种。除使用关系运算符外，还可以使用基本数据类型的方法实现相同的功能，如表 2-3 所示。

表 2-3　关系运算符和等价方法

运 算 符	等 价 方 法	描　述
a==b	a.equals(b)	如果两个操作数相等，则返回 true；否则，返回 false
a!=b	! a.equals(b)	如果两个操作数不相等，则返回 true；否则，返回 false
a>b	a.compareTo(b)>0	如果左操作数大于右操作数，则返回 true；否则，返回 false
a<b	a.compareTo(b)<0	如果左操作数小于右操作数，则返回 true；否则，返回 false
a>=b	a.compareTo(b)>=0	如果左操作数大于或等于右操作数，则返回 true；否则，返回 false
a<=b	a.compareTo(b)<=0	如果左操作数小于或等于右操作数，则返回 true；否则，返回 false

在 C0204 工程中，演示了如何使用等价方法实现关系运算符的功能，输出结果如图 2-4 所示。

C0204 工程：MainActivity.kt

```
10  var a:Int = 10
11  var b:Int = 20
12  println("关系运算符:")
13  println("a = = b 的运算结果为 " + (a.equals(b)))
14  println("a != b 的运算结果为 " + (!a.equals(b)))
15  println("a > b  的运算结果为 " + (a.compareTo(b)>0))
16  println("a < b  的运算结果为 " + (a.compareTo(b)<0))
17  println("b >= a 的运算结果为 " + (b.compareTo(a)>=0))
18  println("b <= a 的运算结果为 " + (b.compareTo(a)<=0))
```

第 13、14 行调用 equals() 方法来比较两个元素是否对象相等。第 15 ~ 18 行调用 compareTo() 方法来比较数值大小，a 大于 b 返回 1，a 小于 b 返回-1，a 等于 b 返回 0，然后使用关系运算符进行判断。

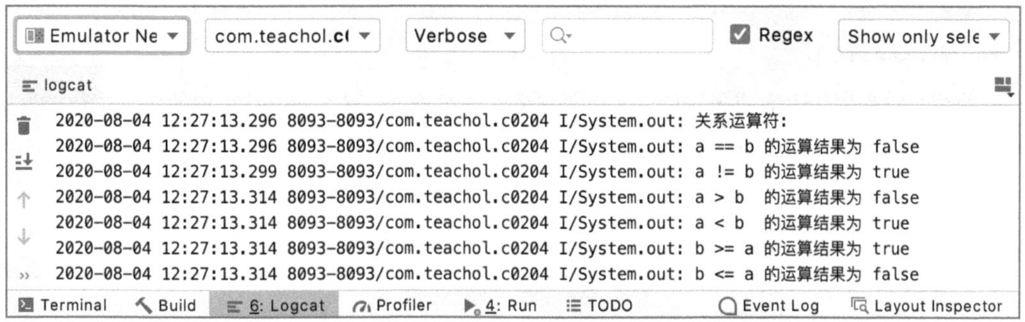

图 2-4　Logcat 窗口输出结果

2.4.3　逻辑运算符

逻辑运算符是用于判断逻辑关系的运算符，该运算符的操作数的值必须是 true 或 false，运算结果只有 true 和 false 两种，如表 2-4 所示。

表 2-4　逻辑运算符

运　算　符	描　　述
!a	逻辑非：反转操作数的逻辑状态。如果条件为 true，则逻辑非运算将得到 false
a&&b	逻辑与：当且仅当两个操作数都为 true，结果为 true
a||b	逻辑或：如果任何两个操作数中任何一个为 true，则结果为 true

在 C0205 工程中，演示了逻辑运算符的使用方法，输出结果如图 2-5 所示。

```
C0205 工程：MainActivity.kt
10  var e:Boolean = true
11  var f:Boolean = false
12  var r:Boolean
13  println("逻辑运算符:")
14  r = e && f
15  println("e && f = " + r)
16  r = e || f
17  println("e || f = " + r)
18  r = !(e && f)
19  println("!(e && f) = " + r)
```

第 18 行先对 e 变量和 f 变量进行逻辑与运算，然后对计算结果进行逻辑非运算，最后将计算结果赋给 r 变量。

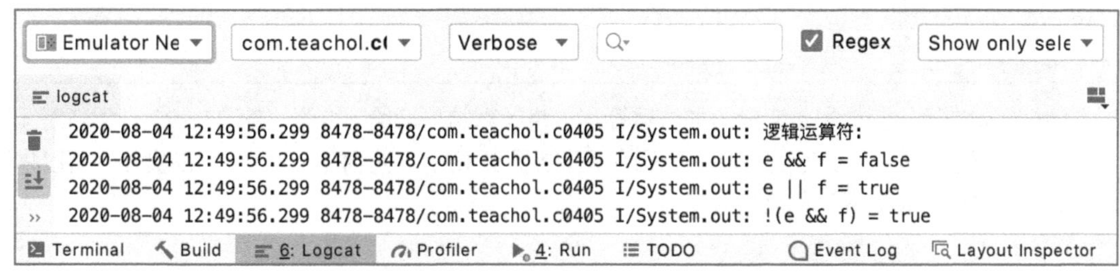

图 2-5　Logcat 窗口输出结果

2.4.4 赋值运算符

赋值运算符是用于赋值或先进行算术运算再赋值的运算符，除赋值外均可以使用基本数据类型的方法实现相同的功能，如表 2-5 所示。

表 2-5 赋值运算符

运 算 符	等 价 方 法	描 述
a=b		赋值：将右操作数的值赋给左操作数
a+=b	a.plusAssign(b)	加和赋值：将左操作数和右操作数相加后赋给左操作数
a-=b	a.minusAssign(b)	减和赋值：将左操作数和右操作数相减后赋给左操作数
a*=b	a.timesAssign(b)	乘和赋值：将左操作数和右操作数相乘后赋给左操作数
a/=b	a.divAssign(b)	除和赋值：将左操作数和右操作数相除后赋给左操作数
a%=b	a.remAssign(b)	取模和赋值：将左操作数和右操作数取模后赋值给左操作数

在 C0206 工程中，演示了赋值运算符的使用方法，输出结果如图 2-6 所示。

C0206 工程：MainActivity.kt

```
10  var a:Int = 10
11  var b:Int = 20
12  var c:Int
13  println("赋值运算符:")
14  c = a + b
15  println("c = a + b，运算结果c的值为 " + c)
16  c += a
17  println("c += a    ，运算结果c的值为 " + c)
18  c -= a
19  println("c -= a    ，运算结果c的值为 " + c)
20  c *= a
21  println("c *= a    ，运算结果c的值为 " + c)
22  a = 10
23  c = 15
24  c /= a
25  println("c /= a    ，运算结果c的值为 " + c)
26  a = 10
27  c = 15
28  c %= a
29  println("c %= a    ，运算结果c的值为 " + c)
```

第 16 行先进行 c 变量和 a 变量的加法运算，再将运算结果赋给 c 变量。第 22、23 行重新对 a 变量和 c 变量赋值。第 24 行 c 变量除以 a 变量的运算结果是 1.5，但是由于 c 变量是 Int 类型，所以将运算结果取整转换为 Int 类型后再赋给 c 变量。

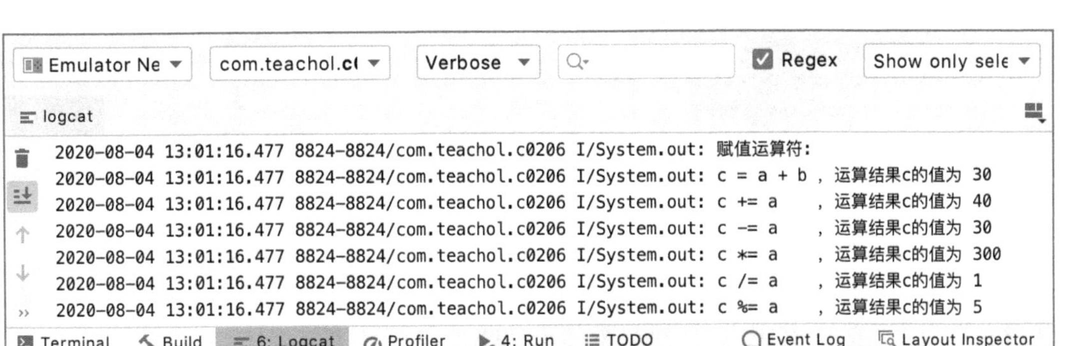

图 2-6　Logcat 窗口输出结果

2.4.5　运算符优先级

在一个表达式中出现多个运算符时，按照运算符优先级的高低顺序依次执行。运算符的优先级与关联性有关，如表 2-6 所示。关联性是指运算符与其左右两侧操作数的组合顺序，对同一个运算符先根据优先级高级进行关联性判断。

表 2-6　运算符优先级

优 先 级	运 算 符	关 联 性
1	++, --	右到左
2	+, –, !, ++, --	左到右
3	*, /, %	左到右
4	+, –	左到右
5	>, >= , <, <=	左到右
6	== , !=	左到右
7	&&	左到右
8	||	左到右
9	=, +=, –=, *=, /=, %=	右到左

2.5　字　符　串

字符串是指由数字、字母和符号组成的一串字符序列。

2.5.1　字符串字面量

字符串字面量可以分为普通字符串和原始字符串两种形式。普通字符串使用一对双引号将字符串引起来，原始字符串使用一对三个双引号将字符串引起来。普通字符串可以使用转义字符，不可以直接换行；而原始字符串不能使用转义字符，可以直接换行。

提示：字面量

字面量(Literal)是表达源代码中一个固定值的表示法。字面量分为字符串字面量(String Literal)、数组字面量(Array Literal)和对象字面量(Object Literal)。

在 C0207 工程中，演示了普通字符串和原始字符串的赋值方法，输出结果如图 2-7 所示。

C0207 工程：MainActivity.kt

```
11 var str1 = "Hi!\nMy Sweet" //普通字符串
13 var str2 = """Hi! //原始字符串
14 My Sweet"""
15 println(str1)
16 println(str2)
```

第 11 行普通字符串使用一对双引号将字符串引起来，字符串中的\n 是换行的转义字符。第 13、14 行原始字符串使用一对三引号将字符串引起来，不支持转义字符，包含转义字符的字符串也不会进行转义。第 14 行不能使用缩进排版代码，否则缩进也会成为字符串的一部分。

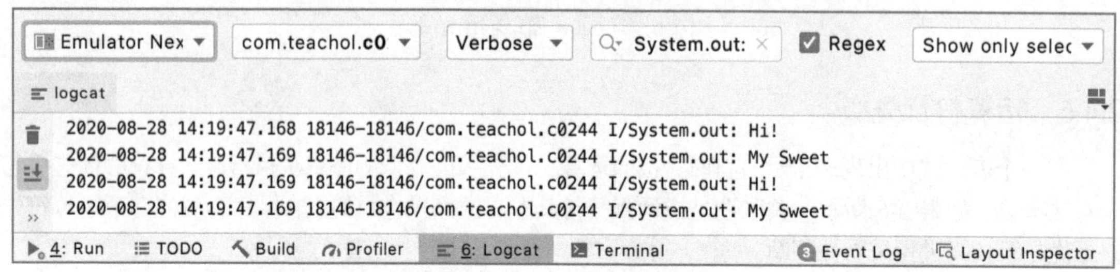

图 2-7　Logcat 窗口输出结果

2.5.2　String 类

String 类用于存储不可变字符串，将在堆内存中占据一个固定的内存空间，String 类的实例无法改变，其与 val 关键字声明的不可变变量是两个概念。String 类的实例重新赋值时，会重新分配内存空间，原有内存空间变成垃圾内存，由 Java 虚拟机负责自动回收。

提示：""和 null 的区别

""是一个长度为 0 且占内存空间的空字符串，在内存中分配一个空间，可以调用方法。

null 既不是对象也不是一种类型，仅是一种特殊的值。null 可以将其赋予任何引用类型，也可以被转换为任何引用类型。null 不能赋给基本数据类型变量，可以赋值给引用类型变量。调用赋值 null 的引用类型变量的非静态方法时，会抛出 null 异常。

String 类只提供了一个属性——length，用于获取字符串的长度，返回值的数据类型为 Int 类型。String 类及其继承的父类提供了大量方法对字符串进行相应的操作，如表 2-7 所示。

表 2-7　String 类的常用方法

类型和修饰符	方法及其描述
inline String	String (bytes: ByteArray) 使用 bytes 参数生成字符串
inline String	String (bytes: ByteArray, offset: Int, length: Int) 使用 bytes 参数生成字符串，返回该字符串以 offset 参数为起始位置、以 length 参数为长度的子字符串
actual inline String	String (chars: CharArray) 使用 chars 参数生成字符串
actual inline String	String (chars: CharArray, offset: Int, length: Int) 使用 chars 参数生成字符串，返回该字符串以 offset 参数为起始位置、以 length 参数为长度的子字符串

续表

类型和修饰符	方法及其描述
actual Int	compareTo (other: String, ignoreCase: Boolean = false) 按字典顺序与 other 参数比较大小，ignoreCase 参数指定是否不区分英文大小写
actual String	concatToString () 将 str 参数连接到此字符串的结尾
actual Boolean	endsWith (suffix: String, ignoreCase: Boolean = false) 是否以 suffix 参数为后缀结束，ignoreCase 参数指定是否不区分英文大小写
actual Boolean	equals (other: String?, ignoreCase: Boolean = false) 与 other 参数比较是否相等，ignoreCase 参数指定是否不区分英文大小写
inline String	format (format: String,vararg args: Any?) 根据 args 参数的格式字符串返回格式化后字符串。format 参数中使用转换符进行替换。例如%1d，1 表示第 1 个 vararg 参数，d 表示该 vararg 参数是十进制整数；%2s，2 表示第 2 个 vararg 参数，s 表示该 vararg 参数是字符串
open Int	hashCode () 返回哈希码
Int	indexOf (string: String, startIndex: Int = 0, ignoreCase: Boolean = false) 返回 string 参数从 startIndex 参数指定的索引开始在此字符串中第一次出现处的索引，ignoreCase 参数指定是否不区分英文大小写；如果不存在，则返回–1
inline String	intern () 返回规范化的表示形式。当调用 intern () 方法时，如果字符串池已经包含一个等于此 String 对象的字符串，则返回字符串池中的字符串；否则，将此 String 对象添加到字符串池中，并返回此 String 对象的引用
Int	lastIndexOf (string: String, startIndex: Int = lastIndex, ignoreCase: Boolean = false) 返回从 startIndex 参数指定的索引开始反向搜索 string 参数在此字符串中最后一次出现处的索引，ignoreCase 参数指定是否不区分英文大小写
actual String	replace (oldValue: String, newValue: String, ignoreCase: Boolean = false) 返回一个新字符串，新字符串是通过用 newValue 参数替换此字符串中出现的所有 oldValue 参数得到的，ignoreCase 参数指定是否不区分英文大小写
List<String>	split (vararg delimiters: String, ignoreCase: Boolean = false, limit: Int = 0) 以 delimiters 参数作为分隔符将此字符串拆分成 limit 参数长度的字符串数组，ignoreCase 参数指定是否不区分英文大小写
actual Boolean	startsWith (prefix: String, ignoreCase: Boolean = false) 是否以 prefix 参数为该字符串的前缀开始
inline String	substring (startIndex: Int, endIndex: Int = length) 返回一个从 startIndex 参数开始到 endIndex 参数–1 结束的子字符串
actual inline CharArray	toCharArray () 转换为字符数组
inline String	toLowerCase () 根据默认语言环境的规则将字符都转换为英文小写
inline String	toUpperCase () 根据默认语言环境的规则将字符都转换为英文大写
inline String	trim () 返回字符串的副本，副本中会删除该字符串两端的空白字符
actual inline Boolean	toBoolean () 转换为 Boolean 类型数据
actual inline Byte	toByte () 转换为 Byte 类型数据
actual inline Short	toShort () 转换为 Short 类型数据

续表

类型和修饰符	方法及其描述
actual inline Int	toInt() 转换为 Int 类型数据
actual inline Long	toLong() 转换为 Long 类型数据
actual inline Float	toFloat(): 转换为 Float 类型数据
actual inline Double	toDouble() 转换为 Double 类型数据

在 C0208 工程中，演示了 String 类初始化字符串、字符串比较及处理字符串的方法。

C0208 工程：MainActivity.kt

```
val num = 123
val ch = charArrayOf('1', '2', '3')
val str1 = "123"
val str2 = String(ch)
val str3 = "$str1"
val str4 = String(ch)
var str5 = str2.intern()
var str6 = str4.intern()
val str7 = "Hello Hog"
val str8 = " hello hog "
val str9 = "hello hog"
println("字符串比较：")
println("str1 = = = str2: " + (str1 === str2)) //结果是 false
println("str1.equals(str2): " + str1.equals(str2)) //结果是 true
println("str1 = = = str3: " + (str1 === str3)) //结果是 true
println("str1.equals(str3): " + str1.equals(str3)) //结果是 true
println("str2 = = = str4: " + (str2 === str4)) //结果是 false
println("str2.equals(str4): " + str2.equals(str4)) //结果是 true
println("str5 = = = str6: " + (str5 === str6)) //结果是 true
println("str5 = = = str1: " + (str5 === str1)) //结果是 true
println("str5 = = = str2: " + (str5 === str2)) //结果是 false
println("str5 = = = str4: " + (str5 === str4)) //结果是 false
println("str5.equals(str6): " + str5.equals(str6)) //结果是 true
println("str7.equals(str9): " + str7.equals(str9)) //结果是 false
println("str7.equals(str9, ignoreCase = true): " + str7.equals(str9, ignoreCase = true)) //结果是 true
println("str7.compareTo(str9): " + str7.compareTo(str9)) //结果是-32
println("str7.compareTo(str9, ignoreCase = true): " + str7.compareTo(str9, ignoreCase = true)) //结果是 0
println("类型转换：")
println("Integer.parseInt(str1) = = num: " + (Integer.parseInt(str1) == num)) //结果是 true
println("num.toString() = = = str1: " + (num.toString() === str1)) //结果是 false
println("num.toString() = = = str2: " + (num.toString() === str2)) //结果是 false
println("字符串连接：")
str5 = str5 + "456"
println("str5+\"456\": $str5") //结果是 123456
println("字符串长度：")
println("str1.length: " + str1.length) //结果是 3
```

```
46 println("大小写转换: ")
47 println("str7.toLowerCase(): " + str7.toLowerCase()) //结果是hello hog
48 println("str8.toUpperCase(): " + str8.toUpperCase()) //结果是HELLO HOG
49 println("去除空格: ")
50 println("str8.trim():" + str8.trim()) //结果是hello hog
51 println("截取字符串: ")
52 println("str7.substring(6): " + str7.substring(6)) //结果是Hog
53 println("str7.substring(6, 7): " + str7.substring(6, 7)) //结果是H
54 println("分割字符串: ")
55 val arr1 = str7.split(" ")
56 for (i in arr1.indices) {
57     println("arr1[" + i + "]:" + arr1[i]) //结果是arr1[0]:Hello,arr1[1]:Hog
58 }
59 println("替换字符串: ")
60 println("str7.replace(\"W\",\"w\")): " + str7.replace("W", "w")) //结果是Hello Hog
61 println("str7.replaceFirst(\"h\",\"H\")): " + str8.replaceFirst("h", "H")) //结果是Hello hog
62 println("查找字符串: ")
63 println("str7.indexOf(\"o\"): " + str7.indexOf("o")) //结果是4
64 println("str7.indexOf(\"o\",5): " + str7.indexOf("o", 5)) //结果是7
65 println("str7.lastIndexOf(\"o\"): " + str7.lastIndexOf("o")) //结果是7
66 println("str7[3]: " + str7[3]) //结果是l
```

第 12 行直接使用字符串给变量赋值，变量通过赋值的数据自动判断数据类型。第 13、15 行使用 CharArray 类型数据实例化 String 类型变量。第 16、17 行使用 String 类型的 intern()方法获取规范化的表示形式，以节省内存开销。第 56 ~ 58 行使用 for 语句遍历 arr1 数组的元素下标，并使用数组名加下标的方式将存储的字符打印出来(此处如果无法理解数组的使用方法可参见 2.8 节)。

通过观察输出结果可以发现，str1 变量和 str2 变量虽然都是存储的"123"字符串，但是使用= =运算符可以判断出所使用的存储地址并不相同。而使用 equals()方法不判断存储地址，只判断字符串的内容是否相等，若相等则返回值为 true。

2.5.3　StringBuffer 类和 StringBuilder 类

如果希望存储的字符串可以调整大小，而不需要创建新的内存空间进行存储，可以使用 StringBuffer 类或 StringBuilder 类创建可变字符串。Kotlin 并没有直接提供这两个类，但是可以调用它们的 Java 类，它们的 Java 类方法名和参数是相同的，因此只列举 StringBuffer 类的方法，如表 2-8 所示。二者的区别在于，StringBuffer 类是线程安全的，采用了加锁机制。使用多线程时，当一个线程访问 StringBuffer 类的实例时，对其进行保护，其他线程不能进行访问，直到该线程读取完，其他线程才可进行访问，不会出现数据不一致问题。线程不安全就是不提供数据访问保护，可能出现多个线程先后更改数据，造成数据错乱的问题。

提示：Kotlin 调用 Java 类

Kotlin 可以直接调用 Java 类，但是需要注意以下几点。

- 由于两者数据的基本数据类型不同，会进行自动映射。
- Unit 类型映射为 void 类型。
- Java 的类名、接口名、方法名等是 Kotlin 的关键字时，需要使用反引号对关键字进行转义。

● Kotlin 提供了 ByteArray、ShortArray、IntArray、LongArray、CharArray、FloatArray、DoubleArray、BooleanArray 数组，用于代替 Java 的 byte[]、short[]等基本数据类型的数组。

● 泛型通配符语法不同，需要进行等价替换。

● 对参数个数可变的方法，Java 可以直接传入一个数组，Kotlin 要求只能传入多个参数值。可以通过使用*运算符解开数组的方式来传入多个数组元素作为参数值。

● Kotlin 没有提供 static 关键字，但是提供了伴生对象来实现静态成员，因此 Java 类中的静态成员都可以通过伴生对象的语法来调用。

表 2-8　StringBuffer 类的常用方法

修 饰 符	方法及其描述
	StringBuffer () 构造方法，构造一个其中不带字符的字符串缓冲区，其初始容量为 16 个字符
	StringBuffer (int capacity) 构造方法，构造一个不带字符且具有 capacity 参数指定的初始容量的字符串缓冲区
	StringBuffer (String str) 构造方法，构造一个字符串缓冲区，并将其内容初始化为指定的 str 参数字符串。初始容量为 str 参数的长度加 16 个字符。当存储的字符序列长度大于容量时，容量会自动增加，新容量是旧容量的两倍加 2
StringBuffer	append (char c) 将 c 参数的字符串表示形式追加到序列中
StringBuffer	append (char[] str) 将 str 参数的字符串表示形式追加到序列中
StringBuffer	append (char[] str, int offset, int len) 将 str 参数以 offset 参数为起始位置、len 参数为长度的子数组的字符串表示形式追加到序列中
StringBuffer	append (String str) 将 str 参数追加到字符序列中
StringBuffer	append (StringBuffer sb) 将 sb 参数追加到序列中
int	capacity () 返回当前容量
char	charAt (int index) 返回 index 参数指定的索引处的字符
StringBuffer	delete (int start, int end) 移除从 start 参数到 end 参数–1 之间的子字符串
StringBuffer	deleteCharAt (int index) 移除 index 参数指定位置的字符
void	ensureCapacity (int minimumCapacity) 确保容量至少等于 minimumCapacity 参数。如果当前容量小于 minimumCapacity 参数，则可分配一个具有更大容量的新内部数组。新容量应大于 minimumCapacity 参数，且是旧容量的两倍加 2
void	getChars (int srcBegin, int srcEnd, char[] dst, int dstBegin) 将字符从此序列复制到目标字符数组 dst 中。要复制的第一个字符在索引 srcBegin 参数处；要复制的最后一个字符在索引 srcEnd–1 处，复制的字符总数为 srcEnd–srcBegin。要复制到 dst 参数的字符从索引 dstBegin 参数处开始，结束于 dstBegin + (srcEnd-srcBegin) – 1 处
int	indexOf (String str) 返回第一次出现的 str 参数在该字符串中的索引。如果不存在，则返回–1
int	indexOf (String str, int fromIndex) 返回从 fromIndex 参数指定的索引处开始第一次出现的 str 参数在该字符串中的索引。如果不存在，则返回–1
StringBuffer	insert (int offset, char c) 从 offset 参数开始将 c 参数插入此序列中

续表

修 饰 符	方法及其描述
StringBuffer	insert (int offset, char[] str) 从 offset 参数开始将 str 参数的字符串表示形式插入此序列中
StringBuffer	insert (int index, char[] str, int offset, int len) 在 index 参数处将 str 参数从 offset 参数开始、以 len 参数为长度的子数组的字符串表示形式插入此序列中
StringBuffer	insert (int offset, String str) 从 offset 参数开始将 str 参数插入此字符序列中
int	lastIndexOf (String str) 返回最右边出现的指定 str 参数在此字符串中的索引
int	lastIndexOf (String str, int fromIndex) 返回从 fromIndex 参数开始最后一次出现的 str 参数在此字符串中的索引
int	length () 返回长度(字符数)
StringBuffer	replace (int start, int end, String str) 返回从 start 参数处开始到 end 参数-1 处结束的子字符串使用 str 参数替换后的字符
StringBuffer	reverse () 返回此字符序列的反转形式
void	setCharAt (int index, char ch) 将 index 参数处的字符设置为 ch 参数
void	setLength (int newLength) 序列将被更改为一个新的字符序列，新序列的长度由 newLength 参数指定。如果 newLength 参数小于当前长度，则长度将更改为指定长度。如果 newLength 参数大于或等于当前长度，则将追加有效的 null 字符 ('\u0000')，使长度满足 newLength 参数
String	substring (int start) 返回从 start 参数处开始的字符子序列
String	substring (int start, int end) 返回从 start 参数处开始、到 end 参数–1 处结束的字符子序列
String	toString () 返回此序列的字符串表示形式
void	trimToSize () 尝试减少用于字符序列的存储空间。如果缓冲区大于保存当前字符序列所需的存储空间，则将重新调整其大小，以便更好地利用存储空间

2.6 分 支 语 句

分支语句是通过逻辑判断确定执行某段代码的语句。if 语句是单分支结构语句，if…else 语句是双分支结构语句，if…else if…else 语句和 when 语句是多分支结构语句。

2.6.1 if 语句

if 语句是条件判断的单分支结构语句。程序运行到 if 语句时，对 condition 的逻辑值进行判断。当 condition 的逻辑值为 true 时，执行花括号内的 statement 语句体，否则直接执行花括号之后的语句。

语法格式：

```
if (condition) {
   statement
}
```

- 必选项：condition 是用于条件判断的 Boolean 类型变量或逻辑表达式。

在 C0209 工程中，演示了 if 语句的使用方法。由于 age<18 表达式结果的逻辑值为 false，所以没有执行 println("未成年")。

C0209 工程：MainActivity.kt

```
val age:Int = 20
if (age < 18) {
    println("未成年")
}
```

提示：语句体的简写格式

如果语句体中只有一条语句，可以将{}省略不写。

2.6.2 if…else 语句

if…else 语句是单一条件判断的双分支结构语句。程序运行到 if 语句时，对 condition 的逻辑值进行判断。当 condition 的逻辑值为 true 时，执行 statement1 语句体，否则执行 statement2 语句体。

语法格式：

```
if (condition) {
    statement1
} else {
    statement2
}
```

- 必选项：condition 是用于条件判断的 Boolean 类型变量或逻辑表达式。

在 C0210 工程中，演示了 if…else 语句的使用方法，根据 marriage 变量存储的数值判断婚姻状态并输出，输出结果如图 2-8 所示。

C0210 工程：MainActivity.kt

```
10  val marriage:Boolean = true
11  if (marriage) {
12      println("婚否：是")
13  } else {
14      println("婚否：否")
15  }
```

第 10 行直接将 marriage 变量值初始化为 Boolean 类型的值 true。第 11 行直接使用 marriage 变量判断逻辑值，并没有通过表达式计算逻辑值。如果 marriage 变量的值为 true，则执行第 12 行，否则执行第 14 行。

提示：为什么不直接存储婚姻状态的字符串？

如果婚姻状态只分为已婚和未婚，可以使用 Boolean 类型存储；如果包含丧偶或离异等婚姻状态，可以使用 Short 类型存储。当使用数据库存储数据时，如果数据量特别大，使用字符串存储会增加存储空间及查询时间。如果提供多语言版本，也不适合直接存储婚姻状态的字符串。

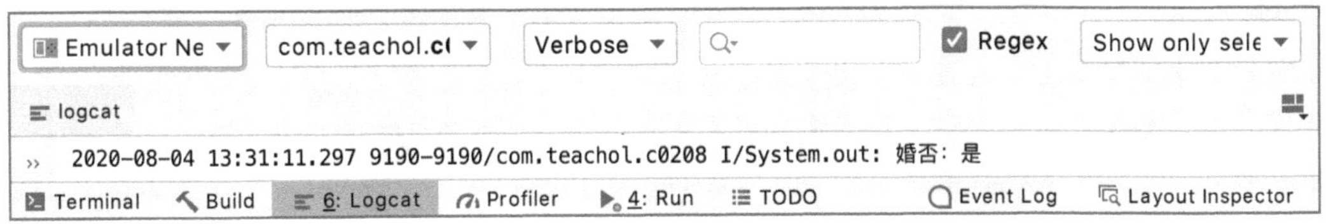

图 2-8　Logcat 窗口输出结果

2.6.3　if…else if…else 语句

if…else if…else 语句是多条件判断的分支结构语句。程序运行到 if 语句时，对 condition1 的逻辑值进行判断。当 condition1 的逻辑值为 true 时，执行完 statement1 语句，结束 if…else if…else 语句。否则，对 condition2 的逻辑值进行判断，当 condition2 的逻辑值为 true 时，执行 statement2 语句体，然后结束 if…else if…else 语句。else if 可以根据需要添加多组，当执行到 else 语句时，无须进行逻辑判断，直接执行 statementN 语句体。

语法格式：

```
if (condition1) {
    statement1
} else if (condition2) {
    statement2
}
...
else if (condition) {
    statementN
} else {
    statement(N+1)
}
```

- **必选项**：condition1，condition2，…，conditionN 是用于条件判断的 Boolean 类型变量或逻辑表达式。

在 C0211 工程中，演示了 if…else if…else 语句的使用方法，根据 sex 变量存储的数值来判断性别并输出，输出结果如图 2-9 所示。

C0211 工程：MainActivity.kt

```
10  val sex:Int = 1
11  if (sex == 1) {
12      println("性别：男")
13  } else if (sex == 0) {
14      println("性别：女")
15  } else {
16      println("性别：不详")
17  }
```

第 11 行计算 sex = =1 表达式的逻辑值。如果逻辑值为 true，则执行第 12 行语句；如果逻辑值为 false，则执行第 13 行语句，计算 sex = = 0 表达式的逻辑值。如果第 11 行和第 13 行表达式的逻辑值都为 false，则执行第 16 行语句。

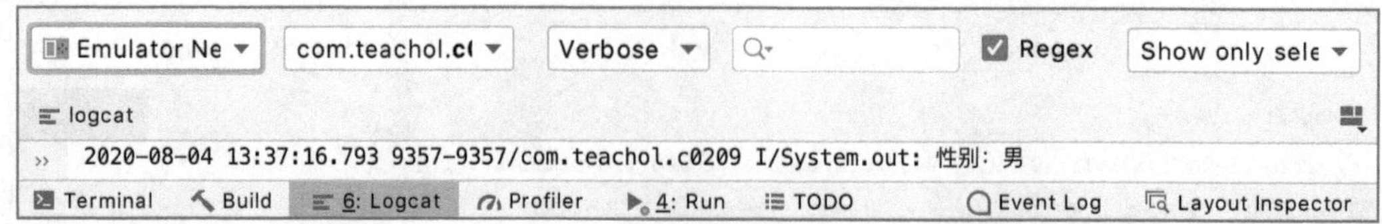

图 2-9 Logcat 窗口输出结果

2.6.4 when 语句

当分支情况较多时，使用 else if 语句会显得不够简洁，使用 when 语句可以解决这个问题。when 语句根据 expression 的值判断与哪个 value 值相等或符合 value 值的区间范围，然后执行“->”后的 statement 语句体。如果与 value 值都不相等或不符合 value 值的区间范围，则执行 else 语句后的 statementDefault 语句体。

语法格式：

```
when (expression) {
    [in|!in value1..value1N]|[value1[,…][,value1N]]|[is Type1]-> {statement1}
    [in|!in value2..value2N]|[value2[,…][,value2N]]|[is Type2]-> {statement2}
    …
    [in|!in valueN..valueNN]|[valueN[,…][,valueNN]]|[is TypeN]-> {statementN}
    [else -> {
        statementDefault
    }]
}
```

- **必选项**：expression 是条件判断的变量或表达式。
- **可选项**：[in|!in value1..value1N]表示在 value1 ~ value1N 区间内或外。例如，in 1..10 表示在 1 ~ 10 区间内，!in 1..10 表示在 1 ~ 10 区间外。
- **可选项**：[value1[,…][,value1N]]表示符合条件的并列关系值。例如，1,3,4 表示 expression 值等于 1、3 或 4。
- **可选项**：[is Type1]表示符合条件的数据类型。例如，is String 表示是 String 类型。
- **可选项**：[else -> {statementDefault}]表示以上条件均不符合时执行 StatementDefault 语句体。

在 C0212 工程中，演示了 when 语句的使用方法，根据 education 变量和 subject 变量存储的数值来显示相应的学历和学位名称，输出结果如图 2-10 所示。

C0212 工程：MainActivity.kt

```
val education:Int = 2
val subject:Int = 2
when (education) { //判断学历
    1 -> println("学历：专科")
    2 -> println("学历：本科")
    3 -> println("学历：研究生")
    else -> {
        println("学历：不详")
    }
}
```

```
22 when (subject) { //判断学位
23     1,2 -> println("学位：文学学士")
24     in 3..5 -> println("学位：理学学士")
25     6 -> println("学位：工学学士")
26     7 -> println("学位：农学学士")
27     else -> {
28         println("学位：其他学士")
29     }
30 }
```

第 13 ~ 20 行根据 education 变量的值判断学历，数值为 1 时是专科，数值为 2 时是本科，数值为 3 时是研究生，其余数值时为学历不详。第 22 ~ 30 行根据 subject 变量的值判断学位，数值为 1 或 2 时是文学学士，数值为 3 ~ 5 时是理学学士，数值为 6 时是工学学士，数值为 7 时是农学学士，其他数值时为其他学士。

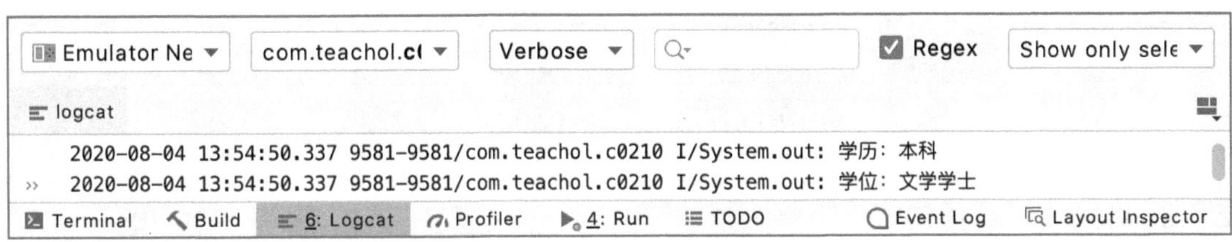

图 2-10　Logcat 窗口输出结果

2.7　循环语句

循环语句是通过条件判断是否多次执行某段代码的语句。while 语句和 do…while 语句通过表达式或者变量的逻辑值(true 或 false)进行循环条件判断，for 语句通过表达式的逻辑值进行循环条件判断，repeat 语句直接设定循环次数。

2.7.1　while 语句

while 语句是先进行条件判断后执行循环体代码的循环语句。程序运行到 while 语句时，对 condition 的逻辑值进行判断。当 condition 的逻辑值为 true 时，执行花括号包含的 statement 语句体，然后返回到 while 语句；当 condition 的逻辑值为 false 时，执行花括号之后的语句。

语法格式：

```
while (condition) {
    statement
}
```

- **必选项**：condition 是用于条件判断的 Boolean 类型变量或逻辑表达式。

在 C0213 工程中，演示了使用 while 语句进行 5 以内整数的累加过程，输出结果如图 2-11 所示。

C0213 工程：MainActivity.kt

```
var i:Int = 1
var s:Int = 0
while (i < 5) {
```

```
13      println("i=" + i)
14      s = s + i
15      i++
16  }
17  println("s=" + s)
```

第 12 行根据 i<5 的逻辑值判断是否结束循环。当逻辑值为 true 时，执行第 13 ~ 15 行，然后返回到第 12 行，重新进行循环条件判断；当逻辑值为 false 时，结束循环，执行第 17 行。

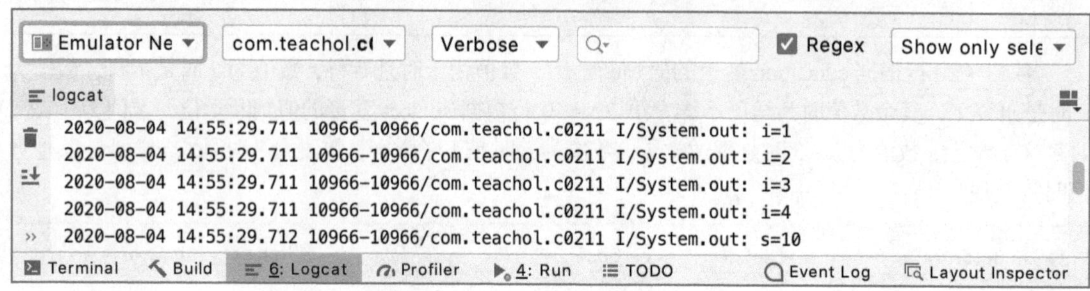

图 2-11　Logcat 窗口输出结果

2.7.2　do…while 语句

do…while 语句是先执行循环体代码后进行条件判断的循环语句。程序运行到 do 语句时，执行 statement 语句体。运行到 while 语句时，对 condition 的逻辑值进行判断。当 condition 的逻辑值为 true 时，返回到 do 语句；当 condition 的逻辑值为 false 时，结束循环。

语法格式：

```
do {
   statement
} while (condition)
```

- **必选项：** condition 是用于条件判断的 Boolean 类型变量或逻辑表达式。

在 C0214 工程中，演示了使用 do…while 语句进行 5 以内整数的累加过程，输出结果如图 2-12 所示。

C0214 工程：MainActivity.kt

```
10  var i:Int = 1
11  var s:Int = 0
12  do {
13      println("i=" + i)
14      s = s + i
15      i++
16  } while (i < 5)
17  println("s=" + s)
```

第 16 行判断 i<5 的逻辑值是否为 true，根据逻辑值决定执行代码的顺序。如果逻辑值为 true，则跳回到第 12 行再次执行；如果逻辑值为 false，则执行第 17 行语句。

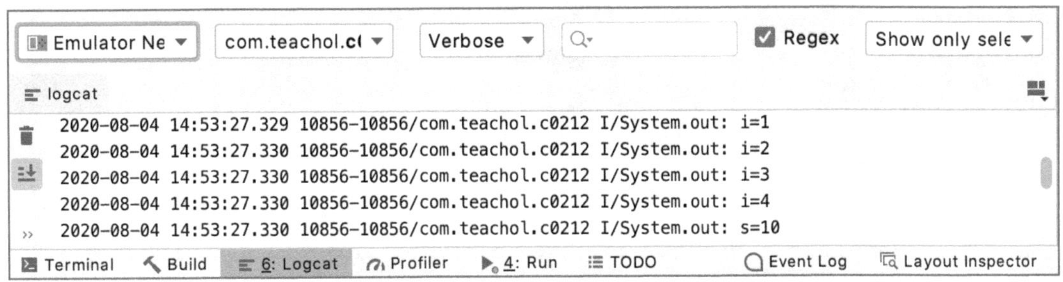

图 2-12　Logcat 窗口输出结果

2.7.3　for 语句

for 语句也是一种先进行条件判断后执行循环体代码的循环语句。for 语句使用 startInt 作为 variableName 循环变量的循环初始值，判断 variableName 循环变量是否达到终止循环的 endInt。如果达到或超过终止循环的 endInt，则终止循环，执行花括号后的语句；如果没有达到终止循环的 endInt，则执行花括号内的 statement 语句体。然后返回到 for 语句，将 variableName 循环变量与 stepInt 进行运算。使用 until 关键字时，variableName 的值加上 stepInt 循环增量；使用 downTo 关键字时，variableName 的值减去 stepInt 循环增量；使用..运算符代替 until 关键字和 downTo 关键字时，根据 startInt 和 endInt 的大小自动判断循环结束后加上或减去 stepInt 循环增量。stepInt 的默认值为 1。最后判断 variableName 循环变量是否达到终止循环的 endInt。

语法格式：

```
for (variableName in startInt until|downTo|.. endInt [step stepInt]) {
    statement
}
```

- **必选项**：variableName 是循环变量。
- **必选项**：startInt 是循环变量的初始值，endInt 是循环变量的结束值。
- **可选项**：[step stepInt]是循环增减量，默认值为 1。

在 C0215 工程中，演示了使用 for 语句进行 5 以内整数的累加过程，输出结果如图 2-13 所示。

C0215 工程：MainActivity.kt

```
10  var i:Int = 3
11  var s:Int = 0
12  for (i in 1 until 5) {
13      println("i=" + i)
14      s = s + i
15  }
16  println("s=$s")
```

第 12 行先对 i 变量进行赋值（为 1），然后根据 until 来判断 i 变量是否小于 5。如果小于 5，则执行第 13、14 行，然后返回到第 12 行，将 i 变量加 1，再判断 i 变量是否小于 5。如果还是小于 5，则再执行第 13、14 行，然后返回到第 12 行，判断是否循环，直至 i 变量大于等于 5 退出循环，执行第 16 行。

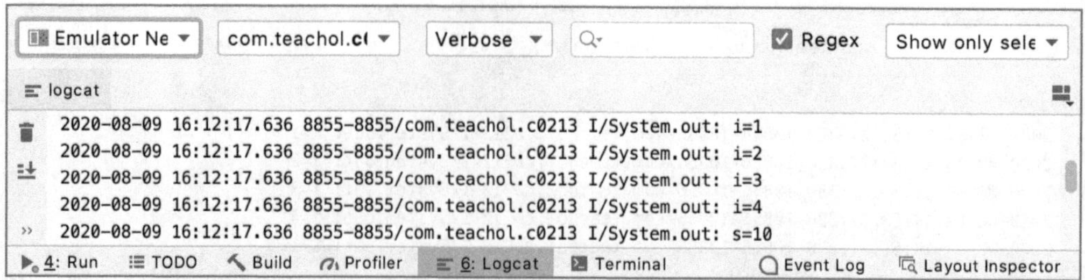

图 2-13　Logcat 窗口输出结果

2.7.4　repeat 语句

repeat 语句是指定循环次数的循环语句，同时默认使用 it 变量表示当前的循环次数(其本质是 Lambda 表达式作为方法的参数，实现原理参见 2.9.4 节)。程序运行到 repeat 语句时，根据 times 指定的次数执行 statement 语句体。

语法格式：

```
repeat (times) {
    statement
}
```

- 必选项：times 是循环次数，必须是整数。

在 C0216 工程中，演示了 repeat 语句的使用方法，输出结果如图 2-14 所示。

C0216 工程：MainActivity.kt

```
10  var s: Int = 0
11  repeat (4) {
12      println("it=" + it)
13      s = s + it
14  }
15  println("s=" + s)
```

第 11 行 repeat 语句的循环次数为 4 次，其语句体循环执行 4 次，退出循环再执行第 15 行。第 12 行 it 变量表示当前循环次数，数值从 0 开始。

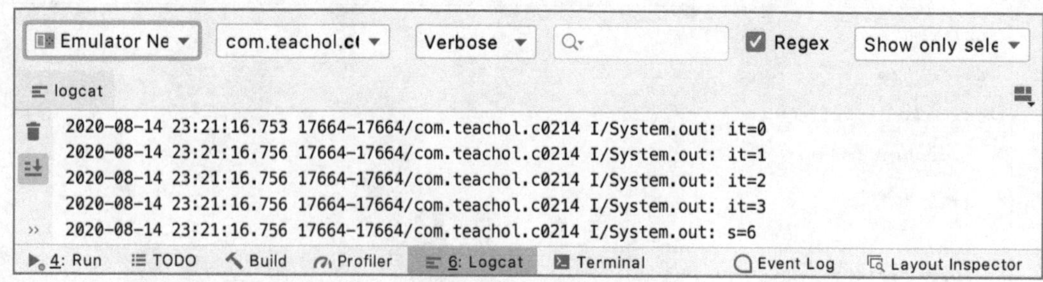

图 2-14　Logcat 窗口输出结果

2.7.5　break 语句

break 语句用于跳出当前循环语句或者 switch 语句，并且继续执行当前循环语句或者 switch 语句

之后的下一条语句。在循环语句中，break 语句通常嵌入 if 语句中使用，当满足一定条件时跳出循环。

在 C0217 工程中，演示了使用循环语句将 10 以内整数逐个累加过程，如果累加值不大于 8，输出累加值；如果大于 8，使用 break 语句结束循环，输出结果如图 2-15 所示。

C0217 工程：MainActivity.kt

```
10  var i: Int = 1
11  var s: Int = 0
12  while (i < 10) {
13      s = s + i
14      if (s > 8) break
15      i++
16      println("s=" + s)
17  }
```

第 14 行当 s 大于 8 时，执行 break 语句，结束循环，执行第 17 行以后的语句。当 s 小于等于 8 时，继续执行第 15、16 行语句。

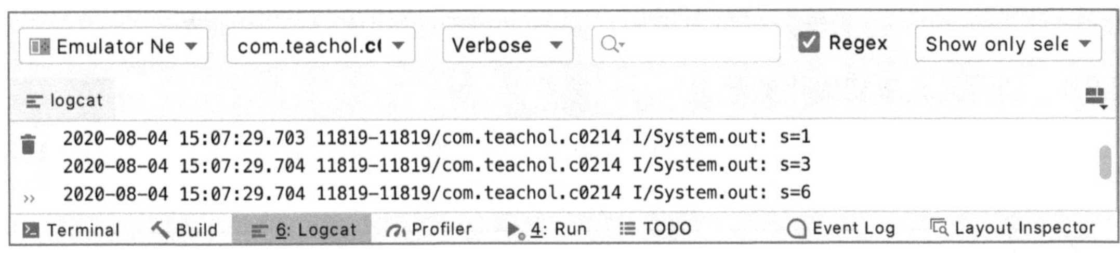

图 2-15　Logcat 窗口输出结果

2.7.6　continue 语句

continue 语句是在循环语句中能够立刻跳转到下一次循环的语句。在循环语句中，continue 语句通常嵌入 if 语句中使用，当满足一定条件时结束本次循环。

在 C0218 工程中，演示了使用循环语句累加 10 以内的奇数之和，当偶数时使用 continue 语句结束本次循环，达到只累加奇数的效果，输出结果如图 2-16 所示。

C0218 工程：MainActivity.kt

```
10  var i: Int = 1
11  var s: Int = 0
12  while (i < 10) {
13      i++
14      if (i % 2 == 0) {
15          continue
16      }
17      s = s + i
18      println("i=" + i)
19  }
20  println("s=" + s)
```

第 14 行 i%2 得到 i 除以 2 的余数，如果余数等于 0，就表示 i 是偶数。第 15 行只有当 i 是偶数时才会执行，执行后会结束本次循环，并跳转到第 12 行开始下一次循环的条件判断。

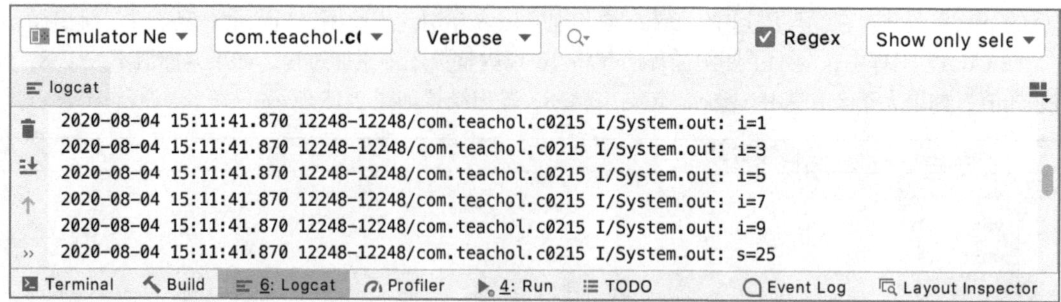

图 2-16　Logcat 窗口输出结果

2.8　数　　组

数组是用于存储多个相同类型数据的集合，可以提高数据的处理效率。在不使用数组的情况下，存储 30 个学生的姓名需要声明 30 个变量。如果使用数组的话，只需声明一个包含 30 个元素的 names 数组即可，names[0]表示数组中的第一个元素，names[29]表示数组中的最后一个元素。

2.8.1　创建元素未初始化的数组

在能够确定数组元素数量却无法确定元素值的情况下，可以创建指定长度的数组，然后根据需要再进行赋值。虽然可以定义数组长度，但是无法重新定义数组长度。

语法格式：

```
var|val arrayName = arrayOfNulls<dataType>(arraySize)
```

- **必选项**：arrayName 是数组的名称。
- **必选项**：dataType 是数组的类型。
- **可选项**：arraySize 是数组元素的数量。

在 C0219 工程中，演示了数组初始化类型和元素数量，通过数组下标对数组元素进行赋值，然后使用冒泡法查找数组元素的最大值，输出结果如图 2-17 所示。

C0219 工程：MainActivity.kt

```
10  var ages = arrayOfNulls<Int>(10)
11  ages[0] = 13
12  ages[1] = 15
13  ages[2] = 16
14  ages[3] = 11
15  var ageMax: Int? = 0
16  for (age in ages) {
17      age?: continue
18      if (age > ageMax!!) {
19          ageMax = age
20      }
21      println(age)
22  }
23  println("ageMax=" + ageMax)
```

第 10 行初始化 ages 数组，数据类型为 Int，元素数量为 10。第 11 ~ 14 行通过数组下标对 4 个数组元素进行赋值。第 15 行 ageMax 变量的 Int 类型后还有一个?符号，表示该变量是可空的 Int 类型变量。第 16～22 行使用 for 语句遍历 ages 数组的每一个元素，每次循环提取数组的一个元素存储在 age 变量内。第 17 行使用?:判断 age 变量内存储的数组元素值是否为 null，如果为 null 则执行 continue 语句结束本次循环。第 18 行 ageMax 变量后添加!!是因为 ageMax 变量为可空 Int 类型，而逻辑判断中不能有 null，所以需要添加!!，否则无法通过编译。

图 2-17　Logcat 窗口输出结果

提示：可空类型、?、!!和?:

可空类型是一种特殊的值类型，给变量设初值的时候给可以赋值为 null。例如，var ageMax:Int? = age[1]。如果 age[1]为空时，赋值为 null，而 Int 类型的值不能为 null，所以要使用 Int?类型。

?加在变量名后表示当前对象可以为 null，不会抛出空指针异常。

!!加在变量名后表示当前对象不为 null 时执行，如果为 null，抛出空指针异常。

?:是 Elvis 运算符，如果运算符左侧的结果为 null，则返回运算符右侧表达式的结果或执行右侧的语句，否则直接返回运算符左侧的结果。

2.8.2　创建元素初始化的数组

根据初始化元素的数据类型，分为两种语法格式：基本数据类型数组和其他类型数组，数组所包含元素的数据类型必须与数组的数据类型一致。

语法格式：

```
// 基本数据类型数组创建
var|val arrayName : typeArray = typeArrayOf([element0, element1,…, elementN])
// 其他类型数组创建
var|val arrayName[:Array<dataType>] = arrayOf([element0, element1,…, elementN])
```

- **必选项**：arrayName 是数组的名称。
- **必选项**：typeArray 是基本数据类型的数组类，包括 BooleanArray、CharArray、ByteArray、ShortArray、IntArray、LongArray、FloatArray、DoubleArray。
- **必选项**：typeArrayOf 需要与 typeArray 的类型相对应，如 IntArray 对应 intArrayOf。arrayOf 是固定的关键字。
- **可选项**：[:Array<dataType>]是数组的数据类型，默认时会根据元素的初始值自动判断数据类型。
- **可选项**：element0, element1,…, elementN 是每个元素的初始值，每个元素的数据类型必须一致。

在 C0220 工程中，演示了通过元素初始化创建 String 类型和 Int 类型数组，然后通过 for 语句遍历两个数组的所有元素并输出，输出结果如图 2-18 所示。

C0220 工程：MainActivity.kt

```
10 var names:Array<String> = arrayOf("张一","王二","李三","赵四")
11 var ages: IntArray = intArrayOf(10,14,12,45)
12 for (i in names.indices){
13     println("names[" + i + "] = " + names[i])
14     println("ages[" + i + "] = " + ages[i])
15 }
```

第 10 行通过元素初始化 String 类型的 names 数组。第 11 行通过元素初始化 Int 类型的 ages 数组。第 12 行通过数组的 indices 属性获取数组的所有下标，每次循环提取出一个下标赋给 i 变量。

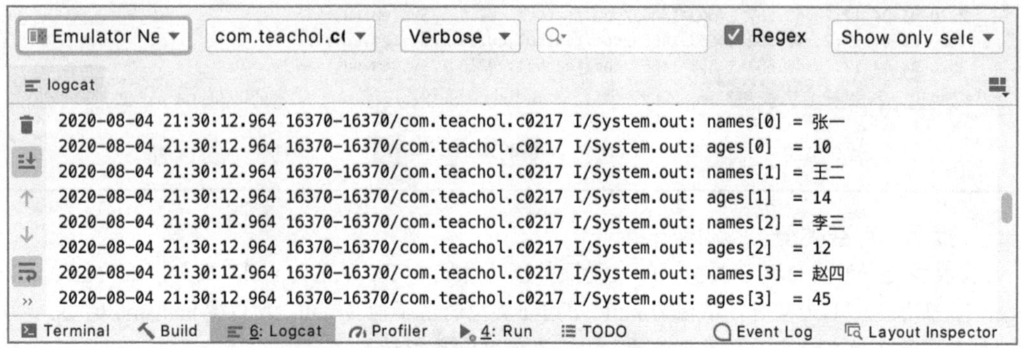

图 2-18 Logcat 窗口输出结果

2.8.3 空数组

可以使用 emptyArray<dataType>() 方法创建空数组。要使用数组时，需要使用 arrayOf 或 typeArrayOf 语句对数组进行初始化，或者赋值一个非空数组，才能对元素进行操作。

语法格式：

```
var|val arrayName = emptyArray<dataType>()
```

- 必选项：arrayName 是空数组的名称。
- 必选项：dataType 是空数组的类型。

在 C0221 工程中，演示了初始化空数组和通过 arrayOf 对空数组进行赋值的实例，输出结果如图 2-19 所示。

工程 C0221：MainActivity.kt

```
10 var height = emptyArray<Int>()
11 println("数组长度: " + height.size)
12 height = arrayOf(140,137,147,151)
13 println("数组长度: " + height.size)
```

第 10 行声明一个可变 Int 类型的空数组。第 11 行 height.size 获取 height 数组的元素数量，此时因为是空数组，所以元素数量为 0。第 12 行使用 arrayOf() 初始化数组元素。第 13 行 height.size 获取 height 数组的元素数量，此时数组已经包含 4 个元素。

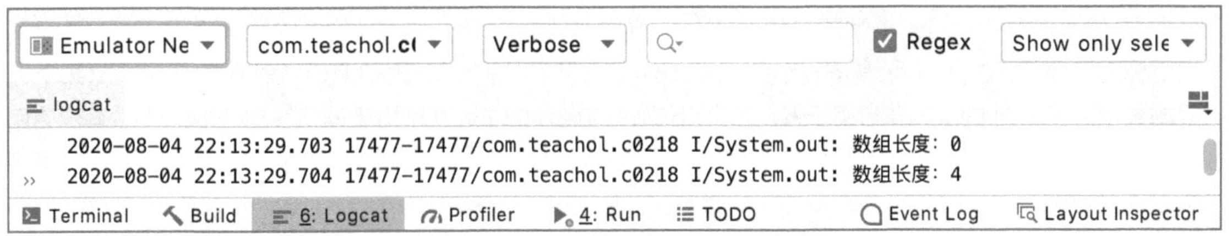

图 2-19　Logcat 窗口输出结果

2.8.4　二维数组

二维数组本质上是以数组作为元素的数组，即“数组的数组”，其中每一个元素都是一个一维数组。二维数组可以理解成一个表格，第一个下标表示行数，第二个下标表示列数。在创建二维数组时，可以嵌套或使用一维数组初始化的方式实现。

语法格式：

```
//使用嵌套的方式创建二维数组
val|var arrayName = Array(arrayLenght1){typeArray(arrayLenght2)}
或
//使用一维数组直接创建二维数组
val|var arrayName = arrayOf([array1, array2,…, arrayN])
```

- **必选项：** arrayName 是数组的名称。
- **必选项：** arrayLenght1 是包含的一维数组数量，arrayLenght2 是所包含一维数组的元素数量。
- **必选项：** typeArray 是基本数据类型的数组类，包括 BooleanArray、CharArray、ByteArray、ShortArray、IntArray、LongArray、FloatArray、DoubleArray。
- **可选项：** array1, array2,…, arrayN 是单独的一维数组，且数据类型必须一致。

在 C0222 工程中，演示了两种二维数组的实例化方式、二维数组元素的赋值及遍历二维数组，输出结果如图 2-20 所示。

C0222 工程：MainActivity.kt

```kotlin
val student = Array(3) { IntArray(2) } //实例化的方式创建二维数组
student[0][0] = 42
student[0][1] = 45
student[1][0] = 44
student[1][1] = 43
student[2][0] = 43
student[2][1] = 42
for (i in student.indices) { //遍历显示数组内容
    for (j in 0 until student[i].size) {
        println((i + 1).toString() + "年" + (j + 1) + "班学生人数: " + student[i][j])
    }
}
val cet4 = arrayOf(intArrayOf(345, 377, 491), intArrayOf(512), intArrayOf(423, 457)) //使用一维数组直接创建二维数组
for (i in cet4.indices) { //遍历显示数组内容
    for (j in 0 until cet4[i].size) {
        println("No." + (i + 1) + "的" + (j + 1) + "次 CET4 考试成绩: " + cet4[i][j])
    }
}
```

第 11 行声明一个 Int 类型的二维数组，一维包含 2 个元素，二维包含 3 个元素。第 12 ~ 17 行使用下标对二维数组的每个元素进行赋值。第 18 ~ 22 行通过嵌套 for 语句遍历二维数组，并输出每个元素的数值。外循环的 for 语句用于遍历二维下标，内循环的 for 语句用于遍历一维下标。第 24 行使用 3 个初始化元素的一维数组初始化二维数组。第 25 ~ 29 行通过嵌套 for 语句遍历二维数组，并输出每个元素的数值。

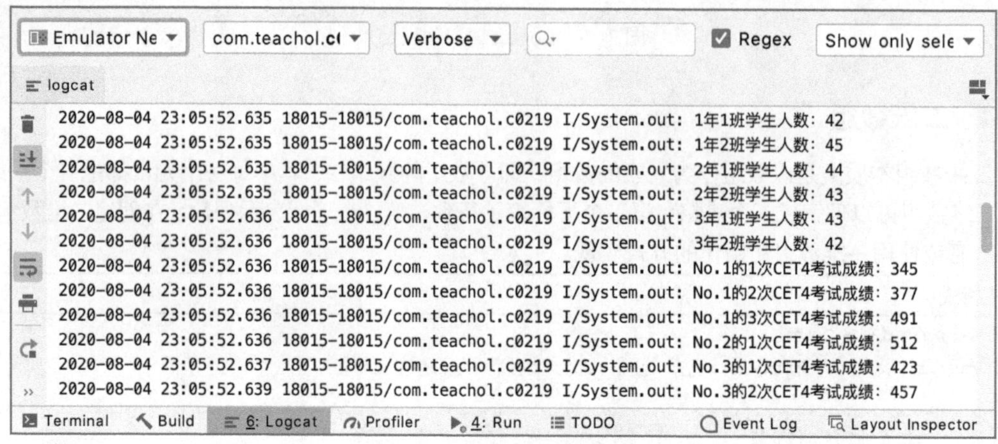

图 2-20　Logcat 窗口输出结果

2.9　函数和 Lambda 表达式

函数是可以直接被另一段程序或代码引用的代码片段。Lambda 表达式本质上是匿名函数，基于数学中的 λ 演算得名。引入 Lambda 表达式是为了简化语法结构，将开发者从原有烦琐的语法结构中解放出来。

2.9.1　函数和高阶函数

函数用于封装可重复使用的代码片段，使用 fun 关键字定义。函数可以包含参数和返回值，也可以不包含参数和返回值。函数参数的作用域是函数内部，没有返回值时的返回类型是 Unit 类型。调用函数时，使用函数名加括号包含参数值。

语法格式：

```
//声明函数
fun functionName([para0[:type0][,para1[:type1]][,…][,paraN[:typeN]]]) [:returnType]{
   statement
   [return returnValue]
}
//调用函数
functionName([para0[:type0][,para1[:type1]][,…][,paraN[:typeN]]])
```

- **必选项**：functionName 是函数的名称。
- **可选项**：para0、para1、…、paraN 是函数参数。
- **可选项**：type0、type1、…、typeN 是函数参数的数据类型。
- **可选项**：[:returnType]用于设置函数返回值的类型，需要与[return returnValue]一起使用，returnType 是返回值的类型。

● **可选项**：[return returnValue]用于设置函数的返回值，返回值的类型需要与[:returnType]设置的类型一致，returnValue 是返回值。

提示：函数参数的默认值

定义函数参数时，可以直接在参数类型后使用等号对参数设置默认值。在调用函数时，如果参数都具有默认值，则省略不写。如果想要单独设置某个参数的值，则在调用函数时，在括号内使用=运算符为参数赋值。

高阶函数是可以作为另一个函数参数或返回值的函数。每个函数都有一个类型，可以用于声明变量，也可以作为其他函数的参数或返回值。在 C0223 工程中，演示了高阶函数作为数据类型和返回值的使用方法，以及设置默认值的参数调用方法，输出结果如图 2-21 所示。

C0223 工程：MainActivity.kt

```
class MainActivity : AppCompatActivity() {
    override fun onCreate(savedInstanceState: Bundle?) {
        super.onCreate(savedInstanceState)
        setContentView(R.layout.activity_main)
        result("长方形", 5.1, 4.5, ::rectangleArea)
        resultByType("长方形", 5.1, 4.5)
        result("三角形", 2.3, 4.0, ::triangleArea)
        resultByType("三角形", 2.3, 4.0)
        var area = getCircularArea(5.0)
        area = getCircularArea(r=5.0)
        area = getCircularArea(5.0,3.1415926)
        println("圆形的面积为${area}平方米")
    }
    //函数类型为(Double,Double)->Double
    fun rectangleArea(width: Double, height: Double): Double {
        return width * height
    }
    //函数类型为(Double,Double)->Double
    fun triangleArea(width: Double, height: Double): Double = width * height / 2
    //高阶函数作为参数的数据类型，函数类型为()->Unit
    fun result(name: String, width: Double, height: Double, area: (Double, Double) -> Double) {
        println("${name}的面积为${area(width, height)}平方米")
    }
    //函数类型为()->Unit
    fun resultByType(type: String="长方形", width: Double, height: Double) {
        //高阶函数作为数据类型
        var area: (Double, Double) -> Double = getArea(type)
        println("${type}的面积为${area(width, height)}平方米")
    }
    //高阶函数作为返回值
    fun getArea(type: String): (Double, Double) -> Double {
        if (type == "长方形") {
            return ::rectangleArea
        } else {
            return ::triangleArea
        }
```

```
42     }
43     //带默认值的参数
44     fun getCircularArea(r: Double,pi:Double=3.14):Double {
45         return pi*r*r
46     }
47 }
```

第 10 行 result()方法的最后一个参数是::rectangleArea，方法作为参数时需要使用::运算符加方法名，此时该方法就是高阶函数。第 14 行调用函数时设置无默认值参数的数值。第 15 行调用函数时使用参数名设置无默认值参数的数值。第 16 行调用函数时设置所有的参数。第 20～22 行是计算长方形面积的方法，函数类型是(Double,Double)->Double，圆括号中的两个 Double 表示方法参数的类型，->后面的 Double 表示方法的返回值类型。第 24 行是计算三角形面积的方法，使用表达式返回值的形式，函数类型也是(Double,Double)->Double。第 26～28 行是打印面积计算结果的方法，最后一个参数类型是(Double,Double)->Double，只能接收高阶函数作为参数。第 30～34 行也是打印面积计算结果的方法，二者主要区别在于第 32 行使用(Double,Double)->Double 作为变量类型，通过 getArea(type)获取相应的高阶函数，第 33 行 area 变量实际上指代高阶函数名。第 36～42 行是获取计算面积的高阶函数的方法，使用(Double,Double)->Double 类型的高阶函数作为返回值。第 44～46 行是计算圆形面积的方法，其中 pi 参数设置了默认值。

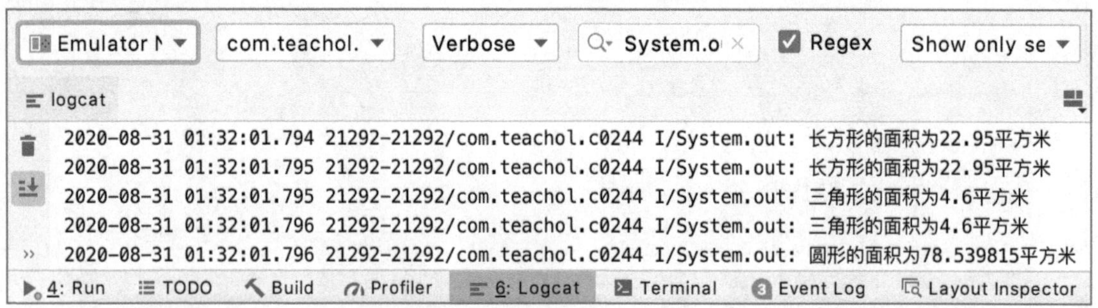

图 2-21　Logcat 窗口输出结果

2.9.2　匿名函数

匿名函数是没有函数名称的函数，可以直接赋给变量，然后通过变量名调用。在 C0224 工程中，演示了匿名函数的使用方法，输出结果如图 2-22 所示。

C0224 工程：MainActivity.kt

```
11 var circularArea = //使用匿名函数赋值
12     fun(r: Double, pi: Double): Double {
13         return pi * r * r
14     }
15 println("圆形的面积为${circularArea(5.0, 3.14)}平方米")
```

第 11～14 行定义了一个匿名函数，并赋给 circularArea 变量，此时 circularArea 变量可以理解为匿名函数的临时方法名。第 15 行使用存储匿名函数的 circularArea 变量调用匿名函数，circularArea 变量后面圆括号中参数数量和数据类型需要与匿名函数一致。

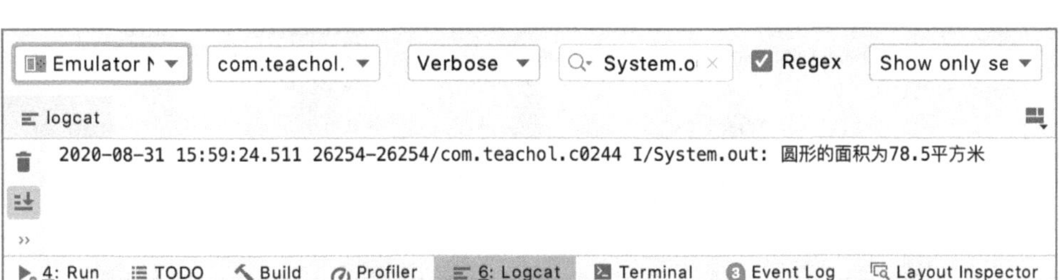

图 2-22　Logcat 窗口输出结果

2.9.3　Lambda 表达式的基本形式

Lambda 表达式的本质是匿名函数，多作为高阶函数使用，也可以直接使用。使用 Lambda 表达式可以减少代码量，使代码更加简洁，但是会增加理解的难度。在 Lambda 表达式中，->左侧的是匿名函数的参数，右侧的是匿名函数的语句体。语句体中可以有返回值，也可以没有返回值。返回值可以使用 return 语句进行设置，也可以不使用 return 语句而将最后一行的单个表达式或变量作为返回值。

语法格式：

```
{[para0[:type0][,para1[:type1]][,…][,paraN[:typeN]]] -> statement }
```

- **可选项**：para0、para1、…、paraN 是 Lambda 表达式的参数。
- **可选项**：type0、type1、…、typeN 是 Lambda 表达式参数的数据类型。

在 C0225 工程中，演示了 Lambda 表达式作为变量值、参数值和返回值及尾随形式的使用方法，输出结果如图 2-23 所示。

C0225 工程：MainActivity.kt

```
class MainActivity : AppCompatActivity() {
    override fun onCreate(savedInstanceState: Bundle?) {
        super.onCreate(savedInstanceState)
        setContentView(R.layout.activity_main)
        //标准形式
        var circularArea = {r:Double,pi:Double->pi*r*r}
        println("圆形的面积为${circularArea(5.0, 3.14)}平方米")
        //Lambda 表达式作为类型、参数和返回值
        resultByType("长方形", 5.1, 4.5)
        resultByLambda("长方形", 5.1, 4.5,{w:Double,h:Double->w*h})
        resultByLambda("长方形", 4.1, 2.5){w:Double,h:Double->w*h}//尾随形式
        resultByType("三角形", 2.3, 4.0)
        resultByLambda("三角形", 2.3, 4.0,{w,h->w*h/2})
        resultByLambda("三角形", 4.3, 2.0){w,h->w*h/2}//尾随形式
    }
    //根据类型计算面积，Lambda 表达式作为变量值
    fun resultByType(type: String, width: Double, height: Double) {
        var area: (Double, Double) -> Double = getArea(type)
        println("${type}的面积为${area(width, height)}平方米")
    }
    //根据 Lambda 表达式计算面积，Lambda 表达式为参数值
```

```
27     inline fun resultByLambda(type: String, width: Double, height: Double,lambda:(Double, Double)->Double) {
28         println("${type}的面积为${lambda(width, height)}平方米")
29     }
30     //根据类型获取计算面积的 Lambda 表达式，Lambda 表达式为返回值
31     fun getArea(type: String): (Double, Double) -> Double {
32         if (type == "长方形") {
33             return {width: Double, height: Double->width*height}
34         } else {
35             return {width: Double, height: Double->width*height/2}
36         }
37     }
38 }
```

第 11 行直接使用了 Lambda 表达式的标准形式，语句体是 pi*r*r，由于语句体只有一行且是表达式，所以这个表达式的值将作为 Lambda 表达式的返回值赋给 circularArea 变量。第 16 行是 Lambda 表达式作为参数时的尾随形式，当函数的最后一个参数是 Lambda 表达式时，可以放在括号的外面。第 19 行也是 Lambda 表达式作为参数时的尾随形式，还省略了 Lambda 表达式的参数类型。第 23 行 area 变量的类型定义为(Double, Double) -> Double，所以只能使用该类型的高阶函数进行赋值。getArea(type)根据传递的参数值会返回相应的 Lambda 表达式，再赋给 area 变量。第 24 行 area 变量已经被赋值为一个 Lambda 表达式，width 变量和 height 变量是该 Lambda 表达式的变量。第 27 行使用了 inline 修饰符将 resultByLambda()方法声明为内联方法，lambda 变量的类型定义为(Double, Double)->Double。第 28 行将 width 变量和 height 变量作为参数传递给 lambda 表达式，得到计算的面积。第 31 ~ 37 行根据 type 参数传递的参数值返回对应的(Double, Double) -> Double 类型的 Lambda 表达式。

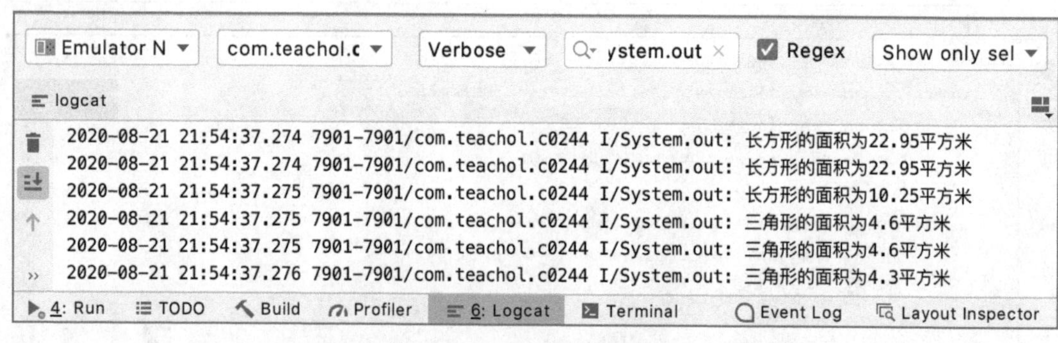

图 2-23　Logcat 窗口输出结果

提示：inline 关键字和 noinline 关键字

inline 关键字用于修饰方法(函数)，被修饰的方法称为内联方法。调用该关键字的方法时，会把方法中的所有代码移动到调用的地方，而不是通过方法之间压栈进栈的方式。不带参数或普通参数的方法，不建议使用 inline 关键字；带有 Lambda 表达式参数的函数，建议使用 inline 关键字。此外，使用 inline 关键字的方法还可以支持 return 语句退出方法。

内联方法的参数不允许作为参数传递给内联方法中调用的非内联方法，如果使用 noinline 关键字修饰内联方法的参数，该参数可以传递给内联方法中调用的非内联方法。

2.9.4　Lambda 表达式参数的省略形式

如果 Lambda 表达式只有一个参数，而且能够根据代码判断出来参数的数据类型，那么这个参数可以省略，并且可以在语句体中使用 it 关键字代替该参数。repeat 语句中也包含 it 关键字，所以 repeat 语句实际上就是一个使用 Lambda 表达式作为参数的函数。

在 C0226 工程中，演示了 repeatDown() 函数使用 Lambda 表达式实现递减功能，并使用了三种等价形式调用了该函数，输出结果如图 2-24 所示。

C0226 工程：MainActivity.kt

```
06 class MainActivity : AppCompatActivity() {
07     override fun onCreate(savedInstanceState: Bundle?) {
08         super.onCreate(savedInstanceState)
09         setContentView(R.layout.activity_main)
11         repeatDown(2, { s: Int -> println("a$s") }) //参数的标准形式
13         repeatDown(2) { s -> println("b$s") } //参数的尾随形式
15         repeatDown(2) { println("c$it") } //参数的尾随省略形式
16     }
17     //Lambda 表达式作为参数类型
18     inline fun repeatDown(r: Int, lambda: (Int) -> Unit) {
19         for (i in r downTo 0) {
20             lambda(i)
21         }
22     }
23 }
```

第 12 行使用标准形式调用 repeatDown() 函数。第 13 行使用尾随形式的 Lambda 表达式作为参数调用 repeatDown() 函数，s 参数省略了数据类型。第 15 行使用参数的尾随省略形式的 Lambda 表达式作为参数调用 repeatDown() 函数，it 关键字表示 Lambda 表达式的唯一参数。

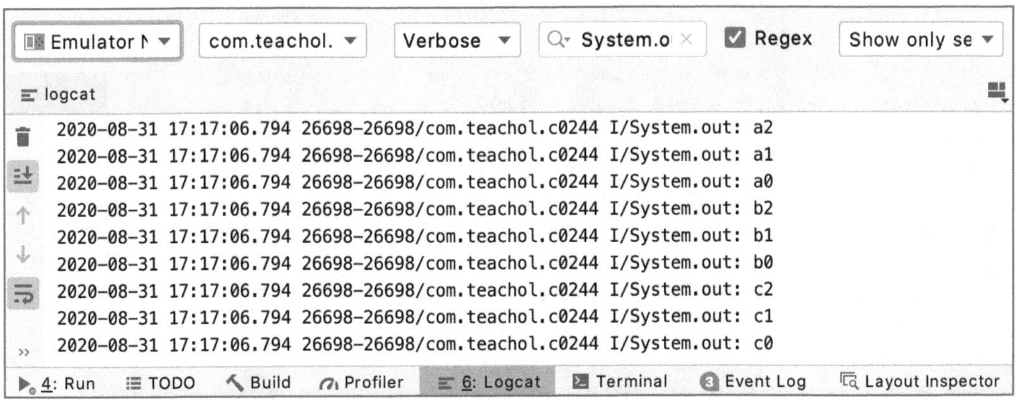

图 2-24　Logcat 窗口输出结果

2.9.5　let、also、apply、with 和 run 函数

let、also、apply、with 和 run 函数都是用于简化代码的函数，这些函数都使用了 Lambda 表达式的参数省略进行调用，可以查看这些函数的源代码来了解原理。这些函数的区别如表 2-10 所示，大家可以根据实际需求选择使用。

提示：查看源代码

在 Windows 操作系统中按住 Alt 键(在 Mac 操作系统中按住 Command 键)，单击类名、方法名或变量，可以查看公开的源代码。

表 2-10　let、also、apply、with 和 run 函数的区别

函 数 名	it 关键字	this 关键字	默认的返回值	带 返 回 值
let	调用对象本身	调用所属实例	Kotlin.Unit	返回返回值
also	调用对象本身	调用所属实例	调用对象本身	返回调用对象本身
apply	不支持	调用对象本身	调用对象本身	返回调用对象本身
with	不支持	调用对象本身	Kotlin.Unit	返回返回值
run	不支持	调用对象本身	Kotlin.Unit	返回返回值

在 C0227 工程中，演示了 let、also、apply、with 和 run 函数的使用方法，输出结果如图 2-25 所示。

C0227 工程：MainActivity.kt

```
var name = "tony"
println("MainActivity 哈希码: ${this.hashCode()}")
println("name 长度: ${name.length}, name 哈希码: ${name.hashCode()}")
var a = name.let {
    println("->执行 let 函数,   长度: ${it.length}, 哈希码: ${this.hashCode()}")
    "simba"
}
println("带返回值执行后 name=$name, a=$a") //name=tony, a=simba
println("不带返回值执行后, name=${name.let{}}") //name=kotlin.Unit
name = "tony"
var b = name.also {
    println("->执行 also 函数,   长度: ${it.length}, 哈希码: ${this.hashCode()}")
    "simba"
}
println("带返回值执行后 name=$name, b=$b") //name=tony, b=tony
println("不带返回值执行后, name=${name.also{}}") //name=tony
name = "tony"
var c = name.apply {
    println("->执行 apply 函数, 长度: ${length}, 哈希码: ${this.hashCode()}")
    "simba"
}
println("带返回值执行后 name=$name, c=$c") //name=tony, c=tony
println("不带返回值执行后, name=${name.apply{}}") //name=tony
name = "tony"
var d = with(name){
    println("->执行 with 函数,   长度: ${length}, 哈希码: ${this.hashCode()}")
    "simba"
}
println("带返回值执行后 name=$name, d=$d") //name=tony, d=simba
println("不带返回值执行后, name=${with(name){}}") //name=kotlin.Unit
name = "tony"
var e = name.run {
    println("->执行 run 函数,   长度: ${length}, 哈希码: ${this.hashCode()}")
```

```
43         "simba"
44     }
45     println("带返回值执行后 name=$name, e=$e") //name=tony, e=simba
46     println("不带返回值执行后, name=${name.run{}}") //name=kotlin.Unit
```

第 13 ~ 16 行使用了 name 变量调用 let() 函数。在函数内，it 关键字表示 name 变量，this 关键字表示当前类的实例。"samba"字符串作为 name.let{}的返回值会赋给 a 变量，name 变量的值不变。第 18 行函数中没有返回值的 name.let{}返回 kotlin.Unit。

第 20 ~ 23 行使用了 name 变量调用 also() 函数。在函数内，it 关键字表示 name 变量，this 关键字表示当前类的实例。"samba"字符串不会作为 name.also{}的返回值， b 变量和 name 变量的值不变。第 25 行函数中没有返回值的 name.also{}返回 name 变量。

第 27 ~ 30 行使用了 name 变量调用 apply() 函数，不支持 it 关键字，函数内可以省略 name 变量名而直接调用其 length 属性，this 关键字表示调用对象本身——name 变量。"simba"字符串不会作为 name.apply{}的返回值，c 变量和 name 变量的值不变。第 32 行函数中没有返回值的 name.apply{}返回 name 变量。

第 34 ~ 37 行 name 变量作为 with() 函数的参数，不支持 it 关键字，函数内可以省略 name 变量名而直接调用其 length 属性，this 关键字表示调用对象本身——name 变量。"simba"字符串作为 with(name){}的返回值会赋给 d 变量，name 变量的值不变。第 39 行函数中没有返回值的 with(name){}返回 kotlin.Unit。

第 41 ~ 44 行使用了 name 变量调用 run() 函数，不支持 it 关键字，函数内可以省略 name 变量名而直接调用其 length 属性，this 关键字表示调用对象本身——name 变量。"simba"字符串作为 name.run(){}的返回值会赋给 e 变量，name 变量的值不变。第 46 行函数中没有返回值的 name.run(){}返回 kotlin.Unit。

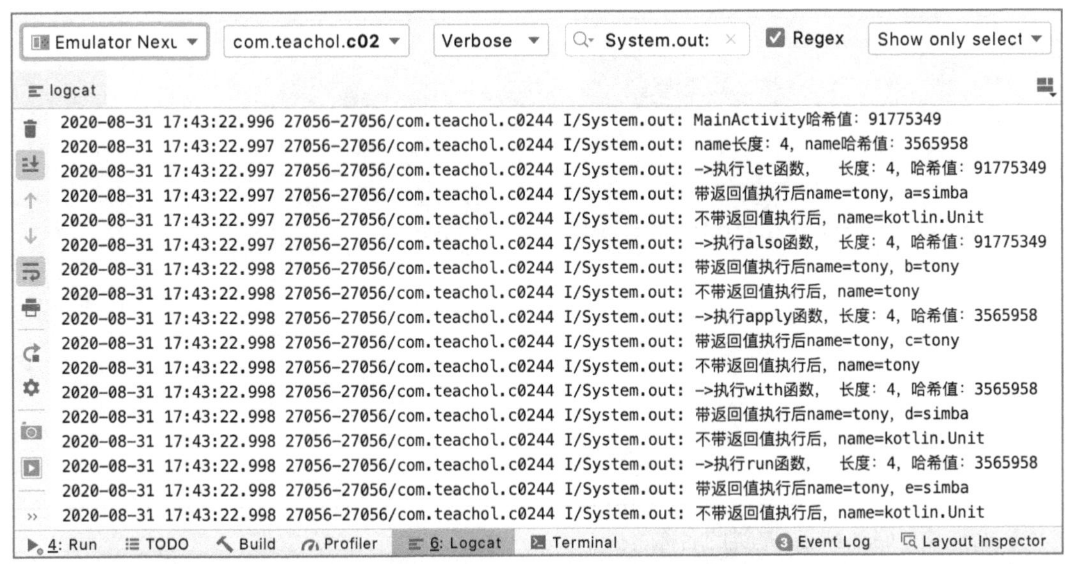

图 2-25　Logcat 窗口输出结果

第 3 章 Kotlin 的面向对象基础

在面向对象的编程思想中，万物皆对象。类是对一系列对象的抽象，对象是一个具体的事物。例如，动物就是一个类，熊猫、海鸥、青蛙和海豚就是对象。熊猫、海鸥、青蛙和海豚有各自的属性，但是它们共同的属性是以有机物为食，有神经系统，有血液循环系统，有感知觉，能运动。因此相对于熊猫、海鸥、青蛙和海豚的概念而言，动物就是类，熊猫、海鸥、青蛙和海豚就是对象，属于同一类的每一个对象都可以称为类的实例。

3.1 类

3.1.1 类的声明

Kotlin 提供了大量的内置类来实现各项基本功能，也可以使用 class 关键字声明自定义类。**类中一般包含主构造方法、初始化块、次构造方法、属性和方法**[①]。主构造方法在类名后使用 constructor 关键字定义，可以省略。初始化块使用 init 关键字，不能添加参数。次构造方法在类内部使用 constructor 关键字定义，可以添加参数。属性（即成员变量）使用 var 或 val 关键字在类内部的构造方法和方法外直接声明。方法是指类中的函数，方法和函数在定义时只有是否在类内部的区别。

语法格式：

```
class ClassName [constructor([[var|val] attrConst:Type])]{
    [var|val attribute[:Type][ = value] ]
    [init {
        statement
    }]
    [constructor([para:Type])[:this(attrConst)]{
        statement
    }]
    [fun functionName([para:Type]) [:ReturnType]{
        statement
        [return value]
    }]
}
```

- **必选项**：ClassName 是类的名称。
- **可选项**：[constructor([[var|val] attrConst:Type])]是主构造方法，如果主构造方法有注解或可见性修饰符，constructor 关键字可以省略。[[var|val] attrConst:Type]是主构造方法的可选参数，多个参数

① 主构造方法和次构造方法通常被直译为主构造器和次构造器，其他面向对象语言中的构造方法作用相同，编译成.class 文件也是构造方法。

之间使用逗号分隔。如果使用 var 或 val 声明参数，则会被自动转换为类的属性，否则只能使用在初始化块内。

- **可选项**：[var|val attribute[:Type][= value]]是类的属性，与变量的声明方法相同，一个类可以有多个属性。
- **可选项**：[init {statement}]是初始化块，在属性初始化后执行，没有参数，每个类只能有一个初始化块。
- **可选项**：[constructor([para:Type]) [:this(attribute)]{statement}]是次构造方法，在初始化块之后执行。[para:Type]是可选参数，多个参数之间用逗号分隔。可以有多个参数不同的次构造方法，即次构造方法的重载，在后续章节中会介绍重载的概念。如果主构造方法和次构造方法同时存在，则次构造方法需要使用[:this(attrConst)]调用主构造方法，先执行主构造方法再执行次构造方法。
- **可选项**：[fun functionName([para:Type]) [:ReturnType]{statement [return value] }]是类的方法。[para:Type]是可选参数，多个参数之间用逗号分隔。[:ReturnType]是方法的返回值类型，设置后需要使用[return value]提供返回值。

提示：成员变量和局部变量

成员变量即类的属性，是在类中的方法外定义的变量及主构造方法使用 var 或 val 关键字定义的参数，作用域是当前类；而局部变量是在方法中及次构造方法中定义的参数和变量，作用域是当前方法。

3.1.2　创建类的实例

类不能直接使用，需要通过类的实例来使用，使用 var 或 val 声明类实例的变量。声明类变量的同时还可以进行实例化，实例化后才能通过实例的变量调用类的属性和方法。

语法格式：

```
//声明类实例的变量
var|val instanceName:ClassName
//声明类实例的变量并进行实例化
var|val instanceName = ClassName([para])
```

- **必选项**：instanceName 是类的实例名。
- **必选项**：ClassName 是类名。
- **可选项**：[para]是调用主构造方法或次构造方法的参数，多个参数之间使用逗号分隔，ClassName([para])返回该类的实例。

类的实例可以使用.运算符调用类的属性和方法，var 声明的属性可以通过实例调用进行修改，val 声明的属性不可以修改。而主构造方法、初始化块和次构造方法无法直接调用，而是在实例化时自动调用的。

在 C0301 工程中，演示了通过自定义的 Animal 类实例化 panda 和 dolphin 变量，并调用属性和方法完成相应的操作。新建 C0301 工程后，在工具窗口中选择 Project 视图，将 c0301 文件夹展开。右键单击 c0301 文件夹，选择【New】→【Kotlin File/Class】命令，新建一个类文件，如图 3-1 所示。在“New Kotlin File/Class”对话框中输入类的名称“Animal”，然后双击“Class”类型完成类的创建，如图 3-2 所示。

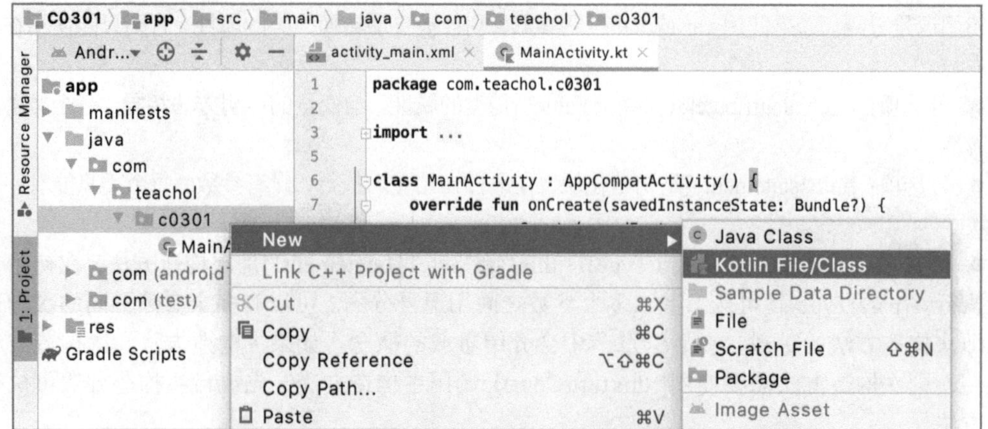

图 3-1　选择【New】→【Kotlin File/Class】命令

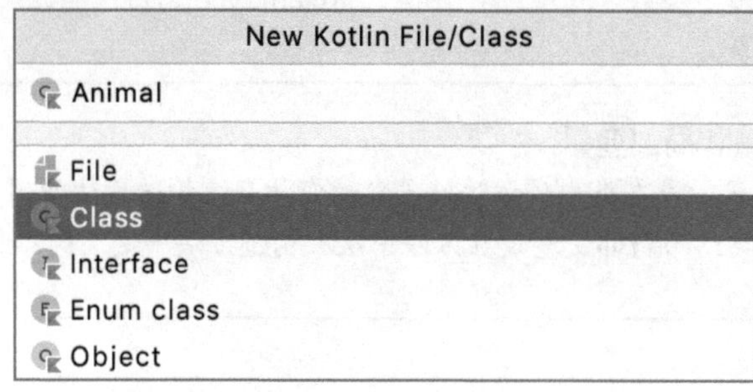

图 3-2　“New Kotlin File/Class”对话框

类的属性如果在声明时没有被初始化赋值，就必须在初始化块或构造方法内完成初始化，否则会提示报错。如果主构造方法没有任何注解和访问控制符，那么 constructor 关键字可以省略。

C0301 工程：Animal.kt

```
package com.teachol.c0301
/**
 * 动物类
 * @author 白喆
 * @version 1.0
 */
class Animal {
    var species: String//属性：物种名称，次构造方法内进行初始化
    var foot: Int//属性：行走足数，次构造方法内进行初始化
    var fly: Boolean = false//属性：是否可以飞行，声明时进行初始化
    var flying: Boolean = false//属性：飞行状态，声明时进行初始化
    /**
     * 初始化块
     */
    init {
        println("调用主构造方法")
    }
    /**
     * 次构造方法
```

```
    */
    constructor(species: String, foot: Int) {
        this.species = species
        this.foot = foot
        println("调用次构造方法")
    }
    /**
     * 开始进食
     */
    fun eat() {
        println(species + "开始进食")
    }
    /**
     * 行走
     * @param step
     */
    fun walk(step: Int) {
        if (foot > 0) {
            println(species + foot + "足行走" + step + "步")
        } else {
            println(species + "不会行走")
        }
    }
    /**
     * 开始飞行
     */
    fun startFlying() {
        if (fly) {
            flying = true
            println(species + "开始飞行")
        } else {
            println(species + "不会飞行")
        }
    }
    /**
     * 查询飞行状态
     * @return Boolean
     */
    fun isFlying(): Boolean {
        return if (fly) {
            flying
        } else {
            false
        }
    }
    /**
     * 结束飞行
     */
    fun endFlying() = if (fly) {
        flying = false
```

```
70            println(species + "停止飞行")
71        } else {
72            println(species + "不会飞行")
73        }
74    }
```

第 02～06 行使用多行备注该类的信息，@author 代表该类的作者，@version 代表该类的当前版本。第 08～12 行声明 4 个属性，并使用单行注释说明该属性的作用。第 15～17 行是初始化块。第 21～25 行是次构造方法，定义 2 个参数，这 2 个参数的变量名与类属性的变量名相同。第 22、23 行使用 this 关键字调用类属性，以便区分次构造方法的参数和类的属性。第 29～31 行是无参数的方法。第 36～42 行是带参数的方法。第 56 行@return 表示方法返回值的类型。第 68～73 行使用方法的表达式语句体格式，等价于将 endFlying()后的=替换成{，然后在第 73 行}后添加一个}的块形式。

提示：注释

注释用于说明某段代码的作用或者说明某个类的用途、某个方法的功能，以及该方法的参数和返回值的数据类型、意义等。书写注释的方式有以下 3 种。

- 单行注释：最常用的注释方式，注释内容从“//”开始到本行末尾。
- 多行注释：注释内容从“/*”开始，到“*/”结束。
- 文档注释：注释内容以“/**”开始，以“*/”结束。专门用于生成帮助文档的注释，可以包含以@开头的预设标签。

提示：this 关键字

this 表示当前实例本身，可以使用.运算符调用本类的属性和方法。

C0301 工程：MainActivity.kt

```
val panda = Animal("熊猫", 4)
println(panda.species + "的足数为" + panda.foot)
panda.fly = false
panda.eat()
panda.walk(10)
panda.startFlying()
if (panda.isFlying()) {
    panda.endFlying()
}
val dolphin: Animal
dolphin = Animal("海豚", 0)
println(dolphin.species + "的足数为" + dolphin.foot)
dolphin.fly = false
dolphin.eat()
dolphin.walk(15)
dolphin.startFlying()
if (dolphin.isFlying()) {
    dolphin.endFlying()
}
```

第 10 行 Animal("熊猫", 4) 调用初始化块和次后造函数后返回一个 Animal 类实例，然后赋给 panda 变量。第 11 行 panda 变量使用.运算符调用 Animal 类实例的 species 和 foot 属性。第 12 行对 Animal 类实例的 fly 属性重新赋值。第 13 行调用 Animal 类实例无参数的 eat() 方法。第 14 行调用 Animal 类实例带参数的 walk(step: Int) 方法。第 16 行调用 Animal 类实例带返回值的 isFlying() 方法，该方法的返回值作为 if 语句的条件判断逻辑值。第 19 行声明了一个 Animal 类型的 dolphin 变量，此时并未将 Animal 类实例赋给该变量，因此不能通过.运算符调用 Animal 类的属性和方法。第 20 行 Animal("海豚", 0) 实例化一个 Animal 类实例并赋给 dolphin 变量，这样 dolphin 变量就可以调用 Animal 类的属性和方法了。

工程运行后，输出结果如图 3-3 所示。两个实例都是先自动调用了初始化块，然后自动调用次构造方法。

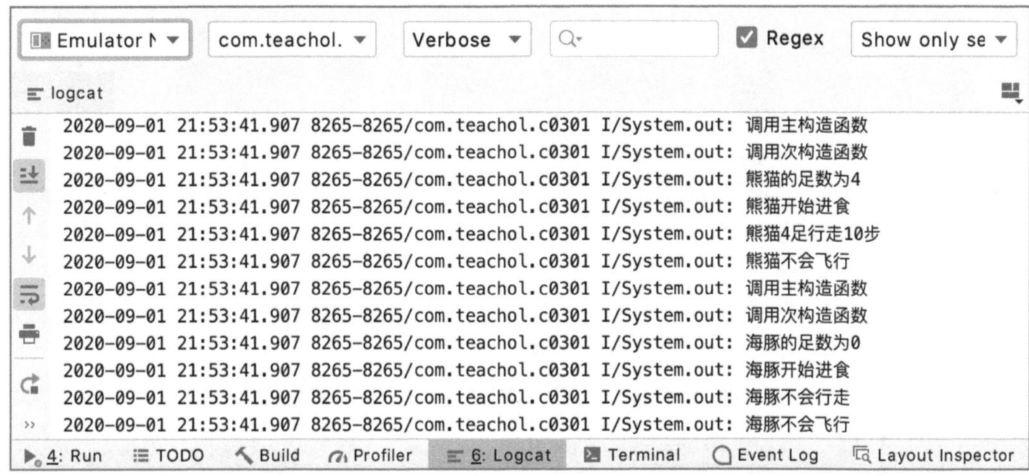

图 3-3　Logcat 窗口输出结果

3.1.3　属性的 get() 和 set() 方法

属性定义后可以使用 get() 和 set() 方法对属性的赋值和调用进行预处理。这两个方法只能在属性的声明后立即使用，无须使用 fun 关键字，方法内使用 field 关键字表示属性值。

当调用属性时，执行属性的 get() 方法。通常会在方法内对属性值进行预处理，通过 return 关键字返回属性值。如果块语句体可以简化为表达式语句体，也可以使用表达式语句体的形式。

语法格式：

```
get() {
   statement
   return value
}
或
get() = expression
```

- **必选项**：value 是 get() 方法的返回值，即调用属性的返回值。
- **必选项**：expression 是调用属性的返回值，可以直接使用单个变量作为返回值，也可以使用表达式。

提示：方法的块语句体形式和表达式语句体形式

方法的块语句体形式是在方法名和返回值类型后使用一对花括号将方法内的语句括起来。方法的表达式语句体形式是在方法名和返回值类型后使用等号直接连接一个表达式、分支语句或循环语句，方法的块语句体内的所有代码只有在能简化为一个表达式、一个单独的最外层分支语句或一个单独的最外层循环语句的情况下，才能使用方法的表达式语句体形式。

使用 val 声明属性时，无法设置 set() 方法。当属性被赋值时，执行属性的 set() 方法。该方法包含一个参数，该参数表示使用=运算符赋值时传递的数据。

语法格式：

```
set(value) {
    statement
    field = expression
}
或
set(value) = expression
```

- **必选项**：value 是存储属性使用=运算符赋值的数据的变量，可以使用任何合法的变量名。
- **必选项**：field 是使用 set() 方法预处理后的属性值。
- **必选项**：expression 是设置属性值的单个变量或表达式。

在 C0302 工程中，演示了使用属性的 get() 和 set() 方法对温度和湿度属性进行预处理的方法，输出结果如图 3-4 所示。新建 C0302 工程后，创建 Room 类的文件。

C0302 工程：Room.kt

```
class Room {
    var temperature: Float = 22.0f //属性：温度
        get() = field
        set(value) = if (field == value) {
            println("温度：temperature 属性数值无变化。")
        } else if (value > 30 || value < 18) {
            println("温度：temperature 属性超出设定范围。")
        } else {
            field = value
            println("温度：temperature 属性值赋值正常。")
            temperaturecontrolStartup(field)
        }
    var humidity: Float = 0.3f //属性：湿度
        get() {
            println("当前设定湿度：" + field)
            return field
        }
        set(value) {
            if (field == value) {
                println("湿度：humidity 属性数值无变化。")
            } else if (value > 0.6f || value < 0.2f) {
                field = value
                println("湿度：humidity 属性超出设定范围。")
            } else {
                field = value
                println("湿度：humidity 属性值赋值正常。")
```

```
29                  humidifierStartup(field)
30              }
31          }
32      constructor(temperature: Float, humidity: Float) {
33          this.temperature = temperature
34          this.humidity = humidity
35          println("调用次构造方法")
36      }
37      fun temperaturecontrolStartup(temperature: Float) {
38          println("启动温度调节设备。")
39      }
40      fun humidifierStartup(humidity: Float) {
41          println("启动湿度调节设备。")
42      }
43  }
```

第 05 行使用 get()方法直接将 field 作为调用 temperature 属性的返回值，实际上没有进行任何预处理，因此该行语句删除掉也不影响运行结果。第 06 ~ 14 行是 temperature 属性的 get()方法，赋值在合理范围内时执行第 11 行语句才真正对属性进行赋值。赋值成功后还会执行第 13 行调用 temperaturecontrolStartup()方法，根据设定温度启动空调。第 16 ~ 19 行是 humidity 属性的 get()方法。第 20 ~ 31 行是 humidity 属性的 set()方法。

```
C0302 工程: MainActivity.kt
11  var room:Room = Room(22.0f,0.3f)
12  room.humidity = 0.3f
13  room.humidity = 0.1f
14  room.humidity = 0.2f
16  var humidity = room.humidity
```

第 11 行实例化 Room 类型的 room 变量，此时会调用 Room 类的次构造方法对 temperature 属性和 humidity 属性进行赋值，属性被赋值时会调用属性的 set()方法。第 12、13 行对 humidity 属性的赋值被 humidity 属性的 set()方法分别判断为数值无变化和超出设定范围。第 14 行对 humidity 属性的赋值是 set()方法判断的正常范围，并在 set()方法内调用 temperaturecontrolStartup()方法启动湿度调节设备。第 16 行声明 humidity 变量，并将 room.humidity 属性赋给 humidity 变量。此时调用了 Room 类的 humidity 属性的 get()方法，返回属性值的同时还会输出设定的湿度。

工程运行后，输出结果如图 3-4 所示。当 humidity 属性赋值为 0.2f 时，在 set()方法内判断符合设置湿度的区间，然后调用 humidifierStartup()方法启动湿度调节设备。

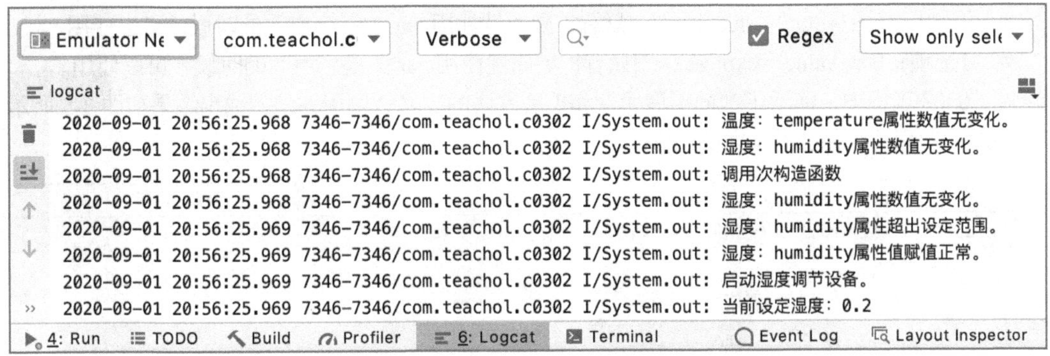

图 3-4　Logcat 窗口输出结果

3.1.4 扩展属性和扩展方法

可以在类的外部为该类扩展属性和方法，以达到增加功能的作用。声明扩展方法和扩展属性时，需要用被扩展的类名作为前缀。

语法格式：

```
fun ClassName.functionName([para:Type]) [:ReturnType]{
    statement
    [return value]
}
```

- 必选项：ClassName 是要扩展方法的类名。
- 必选项：functionName 是扩展的方法名。
- 可选项：[para:Type]是可选参数，多个参数之间用逗号分隔。
- 可选项：[:ReturnType]是方法的返回值类型，设置后需要使用[return value]提供返回值。

扩展属性与普通属性的区别是 get()方法和 set()方法无法使用 field 关键字获取属性值，且 get()方法的返回值是可选项。使用 val 声明扩展属性时，同样无法设置 set()方法。

语法格式：

```
var|val ClassName.attribute: Type
    get() {
        statement1
        [return returnValue]
    }
    [set(value) {
        statement2
    }]
```

或

```
var|val ClassName.attribute: Type
    get() = expression1
    [set(value) = expression2]
```

- 必选项：ClassName 是要扩展属性的类名。
- 必选项：attribute 是扩展属性的属性名。
- 可选项：[return returnValue]。
- 可选项：[set(value) {statement2}]只有扩展属性使用 var 关键字声明的时候才可以使用。
- 可选项：[set(value)=expression2]只有扩展属性使用 var 关键字声明的时候才可以使用。

在 C0303 工程中，演示了类的扩展属性和扩展方法的定义、调用方法，输出结果如图 3-5 所示。新建 C0303 工程后，创建 Car 类的文件。

C0303 工程：Car.kt

```
class Car(var name: String, var color: String) {
    fun stop() {
        println("$name(${color})停止行驶")
    }
}
```

第 06 ~ 10 行定义了 Car 类，该类通过主构造方法声明了两个属性——name 属性和 color 属性，还声明了 stop() 方法。

C0303 工程：MainActivity.kt

```
06  class MainActivity : AppCompatActivity() {
07      override fun onCreate(savedInstanceState: Bundle?) {
08          super.onCreate(savedInstanceState)
09          setContentView(R.layout.activity_main)
10          var car = Car("辽A520", "白色")
11          println("${car.info}")
12          car.info = "辽A1314|红色"
13          println("${car.info}")
14          car.forward()
15          car.stop()
16      }
17      //扩展属性
18      var Car.info: String
19          get() {
20              return this.name + "|" + this.color
21          }
22          set(value) {
23              var info = value.split("|")
24              if (info.size==2) {
25                  this.name = info[0]
26                  this.color = info[1]
27                  println("信息重新设置成功")
28              }
29          }
30      //扩展方法
31      fun Car.forward() {
32          println("$name(${color})向前行驶")
33      }
34  }
```

第 18 ~ 29 行定义了 Car 类的 info 扩展属性，并且添加了 get() 和 set() 方法。第 31 ~ 33 行定义了 Car 类的 forward() 扩展方法。

工程运行后，输出结果如图 3-5 所示。当修改 info 扩展属性时，会通过 set() 方法修改 name 属性和 color 属性，并输出“信息重新设置成功”信息。之后调用的 forward() 扩展方法和 stop() 方法会调用修改后的 color 属性值。

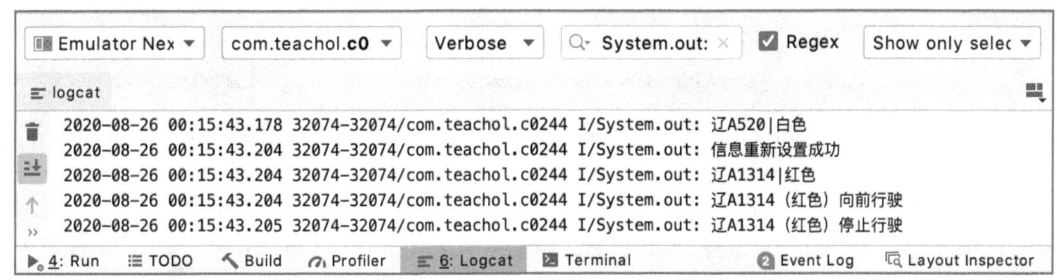

图 3-5　Logcat 窗口输出结果

3.1.5 自动生成 KDoc 文档

在多人合作开发或者开放 SDK 时，提供详细类库帮助文档有利于开发者了解类和接口的属性与方法。dokka 插件可以遵循 KDoc 注释规范自动生成帮助文档。KDoc 注释以/**开头、以*/结尾。注释的每一行以*开头，该*不会当成注释内容的一部分，@开头的标签(如表 3-1 所示)说明注释内容。按照 KDoc 标准格式进行注释时，可以使用 dokka 插件自动生成 html 格式的 KDoc 文档。

表 3-1　KDoc 支持的常用标签

标　　签	说　　明
@author	类或接口的作者
@param	参数
@constructor	构造方法
@property	属性
@return	返回值
@see	关联类或方法的链接
@exception/@throws	所抛出的异常类
@since	指定要编写文档的元素引入时的软件版本
@receiver	用于扩展函数的接收者
@suppress	从生成的文档中排除元素

在 C0304 工程中，演示了使用 dokka 插件自动生成 KDoc 文档的方法。

C0304 工程：build.gradle

```
03 buildscript {
04     ext.kotlin_version = "1.3.72"
05     ext.dokka_version = '0.10.1'
06     repositories {
07         google()
08         jcenter()
09         maven{url "http://maven.aliyun.com/nexus/content/groups/public/"}
10     }
11     dependencies {
12         classpath "com.android.tools.build:gradle:4.0.1"
13         classpath "org.jetbrains.kotlin:kotlin-gradle-plugin:$kotlin_version"
14         classpath "org.jetbrains.dokka:dokka-gradle-plugin:${dokka_version}"
15     }
16 }
```

第 05 行添加存储 dokka 插件版本号的变量。第 14 行添加了 dokka 插件的路径及版本号，此处的版本号使用 dokka_version 变量保存。添加完成后需要单击右上角的“Sync Now”按钮，如图 3-6 所示，下载 dokka 插件后同步 Gradle。

```
C0304 〉 build.gradle
activity_main.xml × MainActivity.kt × build.gradle (C0304) ×
Gradle files have changed since last project sync. A project sync may be neces...  Sync Now
// Top-level build file where you can add configuration options common to all sub
buildscript {
    ext.kotlin_version = "1.4.0"
    ext.dokka_version = '0.10.1'
    repositories {
        google()
        jcenter()
    }
    dependencies {
        classpath "com.android.tools.build:gradle:4.0.1"
        classpath "org.jetbrains.kotlin:kotlin-gradle-plugin:$kotlin_version"
        classpath "org.jetbrains.dokka:dokka-gradle-plugin:${dokka_version}"
        // NOTE: Do not place your application dependencies here; they belong
```

图 3-6　同步 Gradle

C0304 工程：build.gradle(:app)

```
04 apply plugin: 'org.jetbrains.dokka'
06 android {
27         dokka {
28         outputFormat = 'javadoc'
29         outputDirectory = "$buildDir/dokka"
31         configuration {
32             skipDeprecated = true //忽略过期的属性和方法
33             reportUndocumented = true //警告未记录属性
34             skipEmptyPackages = true //忽略空包
35             noJdkLink = true //不链接 JDK
36             noStdlibLink = true //不链接 Stdlib
37             noAndroidSdkLink = true //不链接 AndroidSDK
38         }
39     }
40 }
```

第 04 行添加应用 dokka 插件的代码。第 27～39 行添加 dokka 插件的配置信息。添加完成后需要再次单击右上角的“Sync Now”按钮，同步 Gradle 使 dokka 插件的配置信息生效。

提示：实例的代码删减

为了节省篇幅，本书中将实例中非核心代码进行了删减，因此程序中的行号是不连续的。读者可下载配套的资源文件，查看全部代码。

配置完成后，在 MainActivity.kt 和 Animal.kt 文件中添加代码注释，如图 3-7 所示。添加完所有注释后，选择右侧工具条上的“Gradle”选项卡，在展开的“Gradle”窗口中选择 app/Tasks 路径下的 dokka，如图 3-8 所示，双击 dokka 后自动生成 KDoc 文档，如图 3-9 所示。

```
MainActivity.kt ×   Animal.kt ×   build.gradle (C0304) ×   build.gradle (:app) ×

package com.teachol.c0304

/**
 * 动物类
 * @property species 物种名称，构造器内进行初始化。
 * @property foot 行走足数，构造器内进行初始化。
 * @property fly 是否可以飞行，声明时进行初始化。
 * @property flying 飞行状态，声明时进行初始化。
 */
class Animal {
    var species: String
    var foot: Int
    var fly: Boolean = false
    var flying: Boolean = false
    /**
     * 主构造器
     */
    init {
        println("调用主构造器")
    }
    /**
     * 次构造器
     * @param species 物种名称
     * @param foot 行走足数
     */
    constructor(species: String, foot: Int) {
        this.species = species
```

图 3-7　添加代码注释

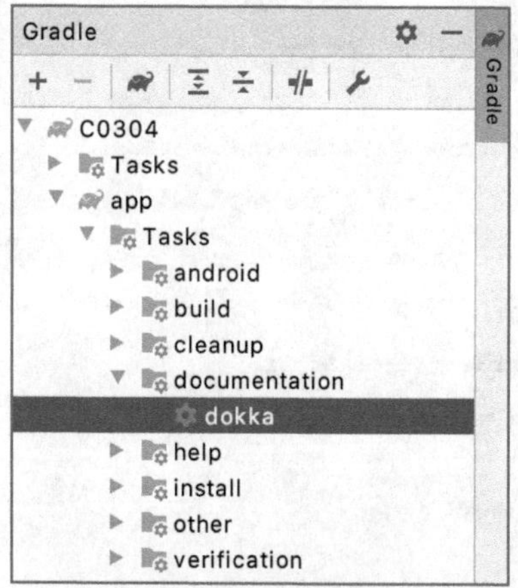

图 3-8　选择 dokka

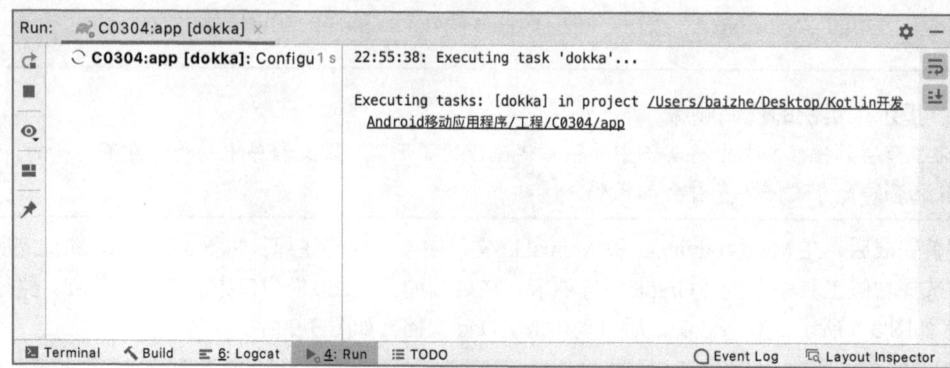

图 3-9　自动生成 KDoc 文档

自动生成 KDoc 文档会保存在工程文件夹的 app/build/dokka 文件夹中，双击打开 index.html 文件就可以浏览、查找该工程已经添加注释的类，如图 3-10 所示。

图 3-10　KDoc 文档

3.1.6　继承

类可以有继承的关系，子类可以继承父类的属性和方法。**Kotlin 中只允许单继承，即每个类最多只能继承一个父类，但是父类还可以再向上继承。**只要有继承关系，子类中无须包含父类的属性和方法就可以使用这些属性和方法。创建类时，在类名后使用:运算符连接父类名称，就可以继承父类所有的属性和方法。而父类则需要添加 open 关键字后才能允许其他类继承。

语法格式：

```
class ClassName : ParentClassName{
    statement
}
```

- **必选项**：ClassName 是类的名称。
- **必选项**：ParentClassName 是 ClassName 类继承的父类名称。

类的多态是指子类的实例可以称为父类的实例，但是父类的实例不能称为子类的实例。例如，Car 是一个父类，包含 color 属性和 forward()方法。Bus 类和 Truck 类定义为 Car 的子类，因为它们继承了 color 属性和 forward()方法，所以 Bus 类和 Truck 类的实例也可以称为 Car 类的实例。

在 C0305 工程中，演示了类的继承及通过子类调用父类的属性和方法，输出结果如图 3-11 所示。

C0305 工程：Car.kt

```
open class Car {
    var name: String? = null //属性：名称
    var color: String? = null //属性：外观颜色
    fun forward() {
        println(name!! + "向前行驶")
    }
}
```

第 03 行添加 open 关键字后该类可以被其他类继承。第 04、05 行添加两个非空字符串类型的属性。第 07 ~ 09 行添加 forward()方法模拟向前行驶。

C0305 工程：Bus.kt

```
03 class Bus : Car() {
04     var ridership: Int = 0 //属性：载客量
05     fun openDoor() {
06         println(name + "打开乘客车门")
07     }
08     fun closeDoor() {
09         println(name + "关闭乘客车门")
10     }
11 }
```

第 03 行 Bus 类使用:运算符继承 Car 类，继承后可以调用 Car 类的属性和方法。第 05 ~ 07 行添加 openDoor()方法模拟打开乘客车门。第 08 ~ 10 行添加 closeDoor()方法模拟关闭乘客车门。

C0305 工程：Truck.kt

```
03 class Truck:Car() {
04     var burden: Int = 0 //属性：载货量
05     fun loading() {
06         println(name + "装载货物")
07     }
08     fun unloading() {
09         println(name + "卸载货物")
10     }
11 }
```

第 03 行 Truck 类使用:运算符继承 Car 类，继承后可以调用 Car 类的属性和方法。第 05 ~ 07 行添加 loading()方法模拟装载货物。第 08 ~ 10 行添加 unloading()方法模拟卸载货物。

C0305 工程：MainActivity.kt

```
val bus = Bus() //Bus 类
bus.name = "客车" //父类属性
bus.color = "red" //父类属性
bus.ridership = 20 //Bus 类属性
bus.closeDoor() //Bus 类方法
bus.forward() //父类方法
bus.openDoor()
val truck = Truck() //Truck 类
truck.name = "货车" //父类属性
truck.color = "blue" //父类属性
truck.burden = 5 //Truck 类属性
truck.loading() //Truck 类方法
truck.forward() //父类方法
truck.unloading() //Truck 类方法
//类的多态
val cars = arrayOfNulls<Car>(3)
cars[0] = bus
cars[1] = truck
cars[2] = null
```

```
31 for (i in cars.indices) {
32     if (cars[i]?.javaClass === Bus::class.java) {
33         println(cars[i]?.name + "载客量: " + (cars[i] as Bus).ridership)
34     } else if (cars[i]?.javaClass === Truck::class.java) {
35         println(cars[i]?.name + "载货量: " + (cars[i] as Truck).burden)
36     } else {
37         println("无法识别" + cars[i] + "的类型")
38     }
39 }
```

第 11 行声明并实例化 Bus 类的 bus 变量。第 12、13 行对 Bus 父类的 name 属性和 color 属性进行了赋值。第 16 行调用 Bus 父类的 forward()方法。第 27 行声明一个 Car 类数组，数组长度为 3。第 28 行将 bus 变量赋给 cars[0]，虽然 bus 变量是 Bus 类的实例，但由于 Car 类是 Bus 类的父类，以及类的多态性，所以能够赋值成功。第 32 行判断 cars 数组元素是否是 Bus 类的实例。第 33 行(cars[i] as Bus)强制将 Car 类的 cars[i]元素转换成 Bus 类，再调用 Bus 类的 ridership 属性。

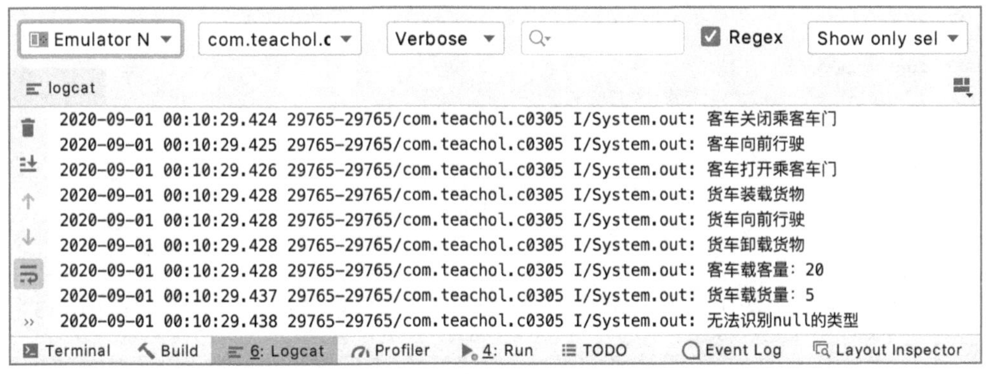

图 3-11　Logcat 窗口输出结果

提示：多态

多态是指同一个行为具有多种不同表现形式或同一对象属于不同的类。

多态的特性包含在继承、重载、接口、抽象类和抽象方法的运用中。例如，在继承中，父类的实例可以通过子类进行实例化；在重载中，同一个方法名可以有多个不同类型或数量的参数的方法；在接口中，有多个类实现同一个接口时，每个类中实现的抽象属性和方法都可能是不同的；在抽象类和抽象方法中，抽象类的继承和抽象方法与子类的实现不同。多态消除了类型之间的耦合关系，可以使程序有良好的扩展，并可以对所有类的对象进行通用处理。

3.2　重写与重载

重写(Override) 是在子类中对继承父类的方法重新编写的过程，方法的参数和返回值都不能改变。在子类中调用父类的被重写方法时，要使用 super 关键字。重载(Overload) 是一个类中方法名相同而参数不同的方法之间的关系，返回类型可以相同也可以不同。

在 C0306 工程中，演示了子类重写父类的方法及方法的重载，输出结果如图 3-12 所示。

C0306 工程：Car.kt

```
open class Car {
    var name: String? = null //名称
```

```
05      var color: String? = null //外观颜色
07      open fun forward() {
08          println(name!! + "向前行驶")
09      }
10  }
```

第 03 行 open 关键字表示 Car 类可以被继承。第 07 行 open 关键字表示 forward() 方法可以被重写。

C0306 工程：Bus.kt

```
03  class Bus : Car() {
04      var ridership: Int = 0 //载客量
06      override fun forward() { //重写方法
07          println(name + "装载" + ridership + "位乘客向前行驶")
08      }
10      fun forward(ridership: Int) { //重载方法
11          this.ridership = ridership
12          println(name + "装载" + ridership + "位乘客向前行驶")
13      }
14      fun openDoor() {
15          println(name + "打开乘客车门")
16      }
17      fun closeDoor() {
18          println(name + "关闭乘客车门")
19      }
20  }
```

第 06 ~ 08 行在 Bus 类中使用 override 关键字重写了 Car 类的 forward() 方法。第 10 ~ 13 行添加 forward(ridership: Int) 方法，该方法与 forward() 方法是重载的关系。

C0306 工程：Truck.kt

```
03  class Truck : Car() {
04      var burden: Int = 0 //属性：载货量
06      override fun forward() { //重写方法
07          super.forward() //调用父类的方法
08          println(name + "装载" + burden + "吨货物向前行驶")
09      }
10      fun loading() {
11          println(name + "装载货物")
12      }
13      fun unloading() {
14          println(name + "卸载货物")
15      }
16  }
```

第 06 行在 Truck 类中使用 override 关键字重写 Car 类的 forward() 方法。第 07 行 super.forward() 表示调用父类的 forward() 方法。

C0306 工程：MainActivity.kt

```
val bus = Bus() //Bus 类
bus.name = "客车" //父类属性
bus.ridership = 20 //Bus 类属性
bus.forward() //重写父类方法
```

```
15 bus.forward(10) //重写重载父类方法
17 val truck = Truck() //Truck 类
18 truck.name = "货车" //父类属性
19 truck.burden = 5 //Truck 类属性
20 truck.forward() //重写父类方法
```

第 14 行 bus.forward()调用的是 Bus 类重写的 forward()方法。第 15 行 bus.forward(10) 调用的是 Bus 类的 forward(Int)方法。第 14 行 truck.forward()调用的是 Truck 类重写的 forward()方法。

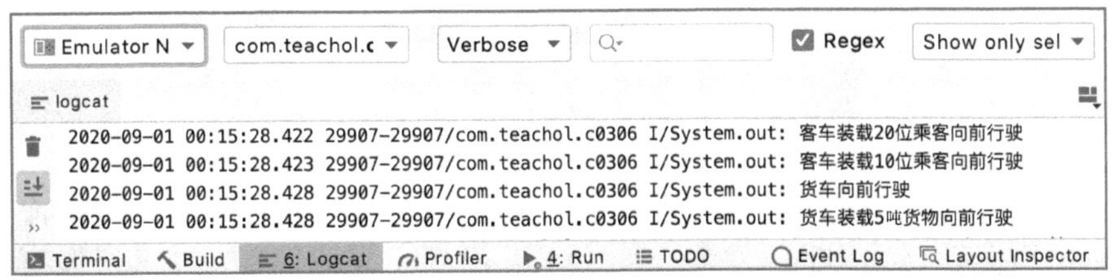

图 3-12　Logcat 窗口输出结果

3.3　抽象类和抽象方法

抽象类是一种特殊的类，声明了抽象方法，通常用于继承。抽象类不能实例化对象，而必须实例化其具体的子类。如果父类是一个高度抽象无须实例化的类，就可以设置为抽象类。抽象方法是一种特殊的方法，在父类中声明抽象方法，然后在子类中实现方法。父类中声明的抽象方法，在子类中必须要对该方法进行实现。**抽象类和抽象方法都使用 abstract 关键字。**

在 C0307 工程中，演示了抽象类和抽象方法的创建和使用方法，输出结果如图 3-13 所示。

```
C0307 工程: Car.kt
03 abstract class Car { //抽象类
04     abstract fun forward() //声明抽象方法
05 }
```

第 03 行使用 abstract 关键字声明 Car 类为抽象类。第 04 行使用 abstract 关键字声明 forward()方法为抽象方法，抽象方法需要在子类中实现。

```
C0307 工程: Bus.kt
03 class Bus : Car() {
05     override fun forward() { //实现抽象方法
06         println("向前行驶")
07     }
08 }
```

第 03 行 Bus 类通过:运算符继承了 Car 抽象类。第 05 行使用 override 关键字重写父类中的 forward()抽象方法。

```
C0307 工程: MainActivity.kt
10 val bus:Bus = Bus()
11 val car:Car = Bus()
12 bus.forward()
13 car.forward()
```

第 10 行将 Bus 类的实例赋给 bus 变量。第 11 行 car 变量虽然定义为 Car 类，但 Car 类是一个抽象类，抽象类无法被实例化。由于 Car 类是 Bus 类的父类，根据继承的多态性，所以可以将 Bus 类的实例赋给 Car 类的 car 变量。

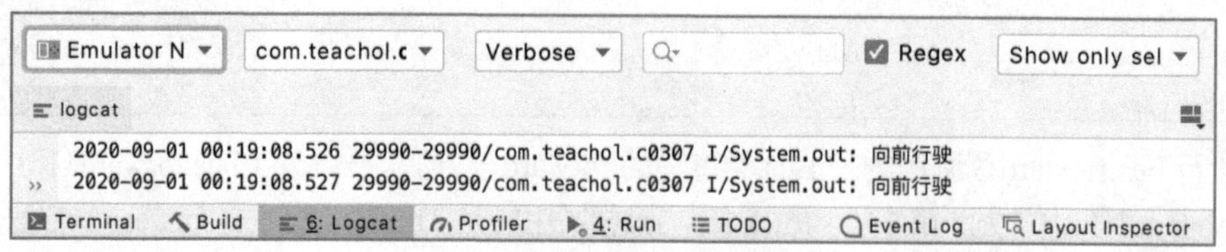

图 3-13 Logcat 窗口输出结果

3.4 包

包(package)对功能相似或相关的类和接口的组织形式，提供了访问保护和命名空间管理的功能。使用 package 关键字定义包，如同文件夹一样，包采用了树形目录的存储方式，每级包之间使用.运算符连接。同一个包中的类名是不能相同的，不同包中的类名是可以相同的，当同时调用两个不同包中相同类名的类时，需要加上包名来区别以避免名字冲突。

语法格式:
package packageName1[.packageName2][.…][.packageNameN]

- **必选项**：packageName1 是在工程内源代码根目录下包含该类的文件夹。
- **可选项**：[.packageName2][…][.packageN]是类在 packageName1 包内的文件夹路径(.运算符表示文件夹的层级关系)。

在不同包内的类不能直接调用，需要使用 import 关键字引入类所在的包后，才能使用。

语法格式:
import packageName1[.packageName2][.…] [.packageN].(className|*)

- **必选项**：.(className|*)引入某一个类或包内所有的类。className 表示要导入的类名，*表示导入包内所有的类。

在 C0308 工程中，演示了定义包和导入包的方法。新建工程 C0308 后，在工具窗口中选择 Project 视图，将 c0308 文件夹展开。右键单击 c0308 文件夹，选择【New】→【Package】命令(如图 3-14 所示)新建一个包。在“New Package”对话框中，输入包的名称(如图 3-15 所示)，回车完成创建。

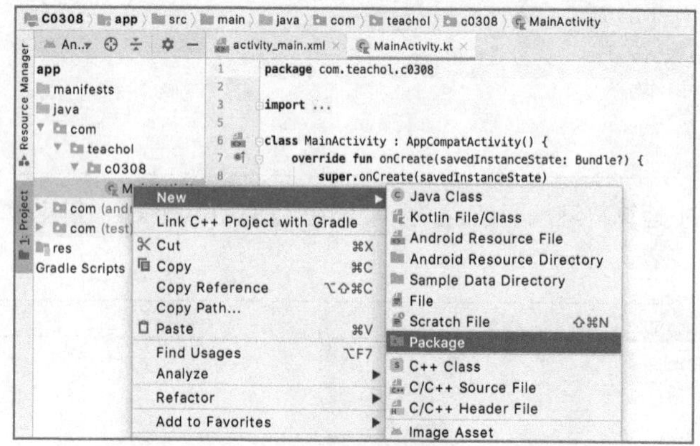

图 3-14 新建包

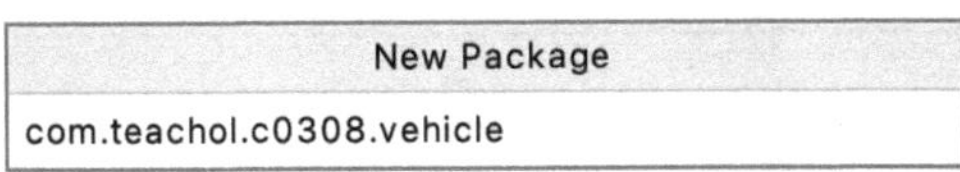

图 3-15　“New Package”对话框

右键单击 vehicle 文件，选择【New】→【New Kotlin File/ Class】命令，创建一个 Car 类。

C0308 工程：Car.kt

```
01 package com.teachol.c0308.vehicle
03 class Car {
04     var name: String? = null//名称
05     fun forward() {
06         println(name!! + "向前行驶")
07     }
08 }
```

第 01 行 package 关键字后面的 com.teachol.c0308.vehicle 表示该类所属的包位置，即该类所在的文件夹位置。

C0308 工程：MainActivity.kt

```
05 import com.teachol.c0308.vehicle.Car
07 class MainActivity : AppCompatActivity() {
08     override fun onCreate(savedInstanceState: Bundle?) {
09         super.onCreate(savedInstanceState)
10         setContentView(R.layout.activity_main)
11         val bus = Car()
12         bus.name = "公共汽车"
13         bus.forward()
14     }
15 }
```

第 05 行使用 import 关键字导入 vehicle 包的 Car 类，也可以将 Car 替换成*来导入 vehicle 包内的所有类。

运行工程后，可以在 Logcat 窗口中看到调用 Car 类的 forward()方法输出的信息，如图 3-16 所示。

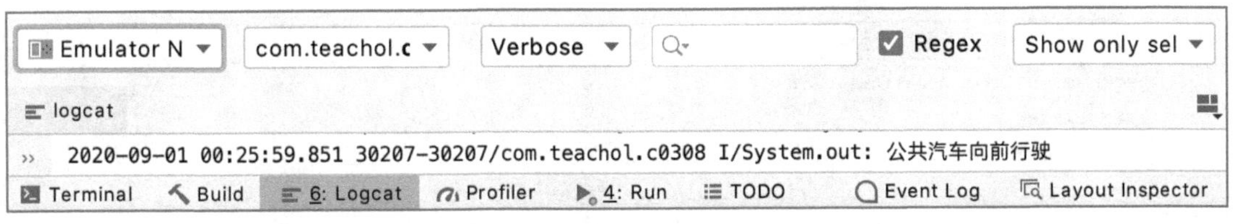

图 3-16　Logcat 窗口输出结果

3.5　封装和访问控制符

封装是将抽象性函数式接口的实现细节部分包装、隐藏起来的方法，让代码更容易理解与维护，也加强了代码的安全性。封装的主要优点包括：良好的封装能够减少耦合；封装内部的结构可以自由修改；可以对成员变量进行更精确的控制；隐藏信息实现细节。

提示：耦合

耦合性指软件系统结构中各模块间相互联系紧密程度的一种度量。在软件工程中经常提到的“高内聚低耦合”编程思想，是指一个完整系统的模块与模块之间尽可能使其独立存在。高内聚是指一个软件模块由相关性很强的代码组成，只负责一项任务，即单一责任原则。低耦合是指每个模块尽可能独立完成某个特定的子功能。模块与模块之间的接口，尽量少而简单。

类是一种典型的封装应用，可以使用访问控制符来保护对变量、方法和类的访问。Kotlin 的访问控制有 4 种类型：public、private、protected 和 internal，默认访问控制符是 public，访问控制权限的区别如表 3-2 所示。

表 3-2　访问控制权限的区别

访问控制符	同一包内	同包子类	同模块异包子类	同模块其他包	异模块
public	Y	Y	Y	Y	Y
private	N	N	N	N	N
protected	Y	Y	Y	N	N
internal	Y	Y	Y	Y	N

- public（公有）：对所有类可见。用于定义类、接口、变量和方法的访问控制权限。
- private（私有）：在同一类内可见。用于定义类、变量和方法的访问控制权限。
- protected（受保护）：对同一包内的类和所有子类可见。用于定义变量和方法的访问控制权限。
- internal（模块内）：对同一模块内的所有类可见。用于定义类、变量和方法的访问控制权限。

提示：模块

在 Kotlin 中，模块是指一个项目，包括以下几种：

- IDE 创建的 Module;
- Maven（软件仓库）或者 Gradle 项目;
- 通过一次调用 Ant 任务编译的一组文件。

在 C0309 工程中，演示了使用访问控制符的方法。新建工程后，在 c0309 文件夹内新建 car 包，在 car 包内新建 Car 类，为 Car 类添加两个属性和三个方法。

C0309 工程：Car.kt

```
package com.teachol.c0309.Car
open class Car {
    var name: String? = null //公有属性：名称
    private var drivingCondition: Boolean = false //私有属性：行驶状态
    //公有方法
    fun forward() {
        drivingCondition = true
        println(name + "向前行驶")
    }
    //公有方法
    fun stop() {
        drivingCondition = false
        println(name + "停止行驶")
```

```
15      }
16      //受保护方法
17      protected fun isDriving(): Boolean {
18          return drivingCondition
19      }
20  }
```

第 04 行 name 变量默认访问控制符，默认使用 public 权限供内部类和外部类使用。第 05 行 drivingCondition 变量使用 private 权限仅供 Car 类内部使用。第 07～15 行 forward()方法和 stop()方法均使用默认的 public 权限。第 17～19 行 isDriving()方法使用 protected 权限。

在 car 包内新建 Bus 类，使其继承 Car 类，并为其添加主构造方法的初始化块和三个方法。

C0309 工程：Bus.kt

```
01  package com.teachol.c0309.Car
03  class Bus(name: String): Car() {
04      //初始化块
05      init {
06          this.name = name
07      }
08      //公有权限方法
09      fun openDoor() {
10          if (isDriving()) {
11              println(name + "行驶中无法打开车门")
12          } else {
13              println(name + "打开车门")
14              passenger()
15          }
16      }
17      //公有权限方法
18      fun closeDoor() {
19          if (isDriving()) {
20              println(name + "行驶中已关闭车门")
21          } else {
22              println(name + "关闭车门")
23          }
24      }
25      //私有权限方法
26      private fun passenger() {
27          println(name + "乘客上下车")
28      }
29  }
```

第 09～24 行 openDoor()方法和 closeDoor()方法使用默认的 public 权限，都调用了父类的 isDriving()公有权限方法。第 26～28 行 passenger()方法使用 private 权限，只能在 Bus 类内被调用，openDoor()方法内的第 14 行调用该方法。

C0309 工程：MainActivity.kt

```
package com.teachol.c0309
import com.teachol.c0309.Car.Bus
class MainActivity : AppCompatActivity() {
```

```
08      override fun onCreate(savedInstanceState: Bundle?) {
09          super.onCreate(savedInstanceState)
10          setContentView(R.layout.activity_main)
12          val bus = Bus("公共汽车")
13          bus.forward()
14          bus.openDoor()
15          bus.closeDoor()
16          bus.stop()
17          bus.openDoor()
18          bus.closeDoor()
19      }
20  }
```

第 05 行使用 import 关键字引入 com.teachol.c0227.Car.Bus 包。第 17 行通过 openDoor()方法间接调用 Bus 类的 passenger()方法，而不能直接调用 Bus 类的 passenger()方法，因为 passenger()方法的权限是 private。

运行工程后，依次执行调用的方法并输出对应的字符串，Logcat 窗口输出结果如图 3-17 所示。

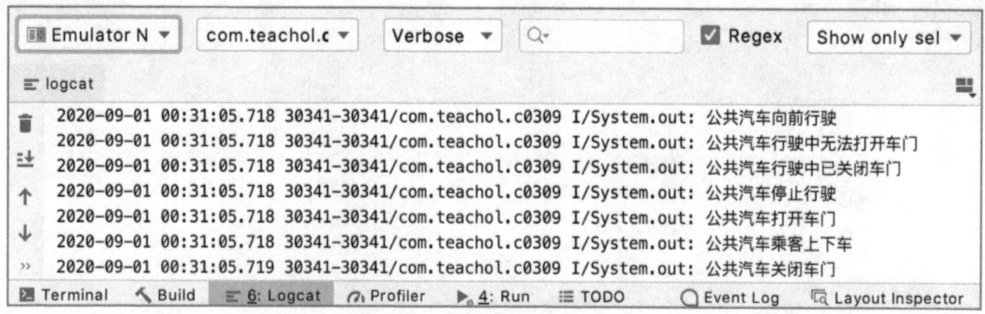

图 3-17　Logcat 窗口输出结果

3.6 接　　口

接口(Interface)定义了类必须包含的属性和方法，也可以直接包含已经实现的方法。接口并不是类，但可以继承另一个接口。接口不能用于实例化对象，但可以作为参数和返回值的类型。接口不能被类继承，但是类可以同时实现多个接口。接口中的抽象属性必须在类中重写并初始化，接口中的抽象方法必须在类中重写。

接口使用 interface 关键字进行定义。接口内的属性只能是抽象变量，且不能初始化赋值，默认修饰符是 public abstract。接口内只定义而未实现的方法是抽象方法，默认修饰符是 public abstract。

语法格式：

```
interface InterfaceName [: ParentInterfaceName]{
    statement
}
```

- **必选项**：InterfaceName 是接口的名称。
- **可选项**：[:ParentInterfaceName]是要继承的接口名称。

类实现接口时，需要使用:运算符，并且可以同时实现多个接口。实现接口的类必须重写抽象属性和抽象方法。

语法格式：

```
class ClassName :InterfaceName{
    statement
}
```

- 必选项：ClassName 是实现接口的类名称。
- 必选项：InterfaceName 是实现接口的名称，多个名称使用逗号分隔，这里的冒号表示实现接口。

提示：类声明时区分继承父类和实现接口的方法

继承父类和实现接口都使用冒号，而且可以不分先后顺序。接口只有名称没有花括号，而类名后需要花括号来初始化属性，即使没有属性也不能省略花括号。

在 C0310 工程中，演示了创建和使用接口的方法。新建工程后，在工具窗口中选择 Project 视图，将 c0310 文件夹展开。右键单击 c0310 文件夹，选择【New】→【New Kotlin File/Class】命令。在 “New Kotlin File/Class” 对话框中，输入 “Animal”，双击 “Interface” 类型完成接口的创建(如图 3-18 所示)。

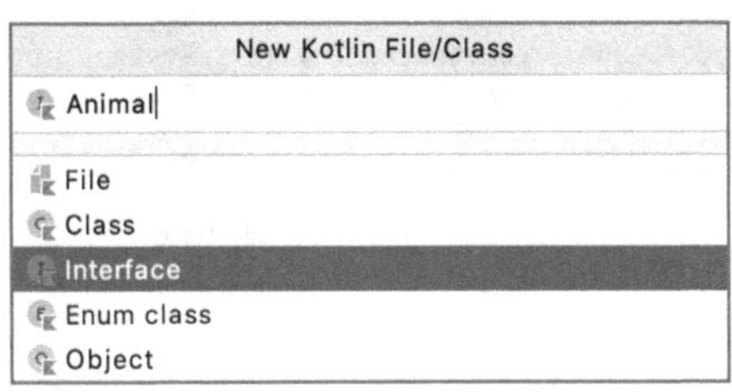

图 3-18　选择接口类型

C0310 工程：Animal.kt

```
03  interface Animal {
04      val name:String
05      fun eat(){
06          println(name + "开始进食")
07      }
08      fun walk()
09  }
```

第 03 行使用 interface 关键字定义 Animal 接口。第 04 行声明 name 属性，该属性是抽象属性，不能在接口中进行初始化。第 05～07 行声明并实现 eat()方法。第 08 行声明 walk()方法，因为在接口中并没有实现该方法，因此该方法是抽象方法。

新建 Panda 类，右键单击 c0310 文件夹，选择【New】→【New Kotlin File/Class】命令。在 “New Kotlin File/Class” 对话框中，输入 “Panda”，双击 “Class” 类型完成类的创建。

C0310 工程：Panda.kt

```
class Panda:Animal {
    override var name: String = "哺乳动物"

    override fun walk() {
        println(name + "开始行走")
    }
}
```

第 04 行使用 override 关键字重写了 name 属性，将 val 重写为 var，并进行初始化赋值。第 06～08 行使用 override 关键字重写并实现了 walk()方法。

C0310 工程：MainActivity.kt

```
10  val panda = Panda()
11  panda.name = "两栖动物"
12  panda.eat()
13  panda.walk()
```

第 10 行声明 panda 变量，并赋值了 Panda 类的实例。第 11 行对 Panda 类中重写 Animal 接口的 name 属性进行赋值。第 12 行 panda.eat()调用的是 Animal 接口中实现的 eat()方法。第 13 行 panda.walk()调用的是 Panda 类中重写并实现 Animal 接口的 walk()方法。

运行工程后，Logcat 窗口输出结果，如图 3-19 所示。

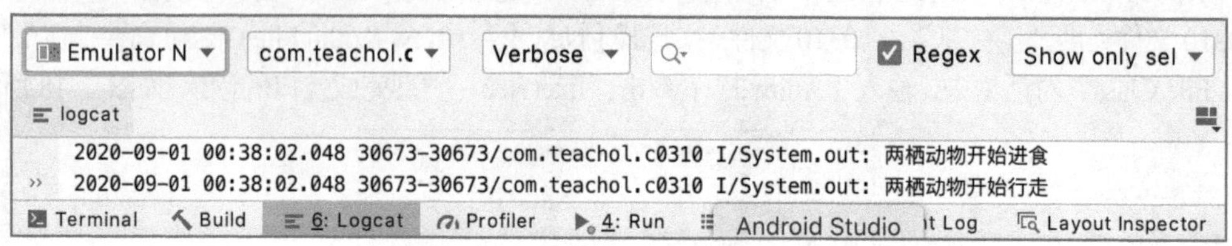

图 3-19 Logcat 窗口输出结果

3.7 委 托

委托模式是接收请求的对象将请求委托给另一个对象来处理的设计模式，Kotlin 通过关键字 by 直接支持委托模式。

提示：设计模式

设计模式(design pattern)是对面向对象程序设计中反复出现问题的解决方案，常用的设计模式包括工厂模式、抽象工厂模式、单例模式、适配器模式、代理模式、观察者模式、MVC 模式等。

3.7.1 类的委托

类的委托是将接口定义的方法通过实现该接口另一个类的对象实现的方式。委托类和被委托类声明时都定义了同一个接口，委托类实现了该接口的抽象方法。

语法格式：

```
class ClassName : InterfaceName by DelegateClassName(){
    statement
}
```

或

```
class ClassName (delegate:DelegateClassName) : InterfaceName by delegate{
    statement
}
```

- **必选项**：ClassName 是类的名称。
- **必选项**：InterfaceName 是接口的名称。
- **必选项**：DelegateClassName 是委托类的名称。

- **必选项**：delegate 是委托类实例的变量名。

在 C031 工程中，演示了委托类的使用方法，输出结果如图 3-20 所示。

C0311 工程：Produce.kt

```
01  package com.teachol.c0311
02  //接口类
03  interface Produce {
04      var brand: String
05      fun make(Num: Int)
06  }
07  //实现接口类
08  class CompanyA : Produce {
09      override var brand: String = "红星牌"
10
11      override fun make(Num: Int) {
12          println("生产" + brand + Num + "个杯子")
13      }
14  }
15  //直接设置委托类
16  class CompanyB : Produce by CompanyA() {
17  }
18  //通过主构造方法设置委托类
19  class CompanyC (delegate: CompanyA) : Produce by delegate
```

第 03 ~ 06 行定义 Produce 接口。第 08 ~ 14 行 CompanyA 类实现 Produce 接口的属性和方法。第 16 ~ 17 行 CompanyB 类通过 by 关键字设置 CompanyA 类为委托类。第 19 行 CompanyC 类通过 by 关键字主构造方法的 delegate 参数，将 CompanyA 类实例设置为委托。

C0311 工程：MainActivity.kt

```
10  var companyA = CompanyA()
11  companyA.make(20)
12  var companyB = CompanyB()
13  companyB.brand = "蓝星牌"
14  companyB.make(10)
15  var companyC = CompanyC(companyA)
16  companyA.brand = "紫星牌"
17  companyC.make(30)
```

第 11 行 companyA.make(20)调用的是 CompanyA 类重写 Produce 接口的 make(Int)方法。第 14 行 companyB.make(10)调用的是 CompanyA 类的委托类的 make(Int)方法。第 15 行将 companyA 变量作为参数传到 CompanyC 类主构造方法，间接地将其设置为委托类。第 17 行 companyC.make(30)调用的是通过 CompanyC 类主构造方法参数传递的委托类实例的 make(Int)方法。

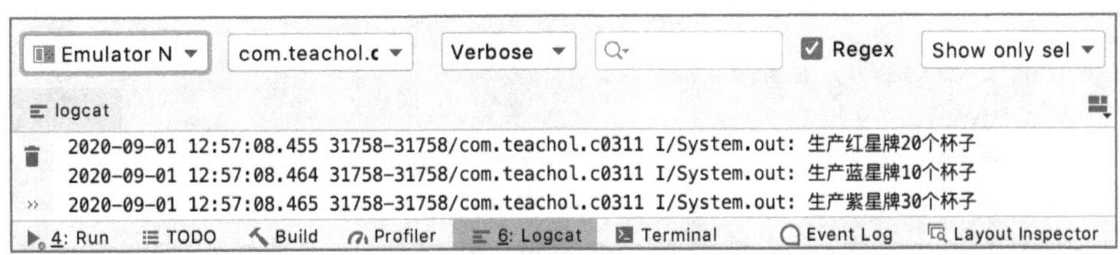

图 3-20　Logcat 窗口输出结果

3.7.2 变量的委托

变量的委托是指变量不使用 set()和 get()方法进行预处理，而将其委托给另一个类的方式。委托类需要导入 kotlin.reflect.KProperty 命名空间，并且需要添加 getValue()方法和 setValue()方法实现变量的调用与赋值。声明变量时使用 by 关键字设置委托类，设置委托类时变量不能进行初始化赋值。

语法格式：
`val|var identifier [: type] by DelegateClassName()`

- **必选项**：identifier 是变量的名称。
- **必选项**：DelegateClassName 是委托类的名称。
- **可选项**：[: type]是数据类型。

在 C0312 工程中，演示了创建属性的委托类及声明属性委托，输出结果如图 3-21 所示。

C0312 工程：Produce.kt

```
03  import kotlin.reflect.KProperty
05  class Delegate {
06      var field: Float = 0.0f
07      var num = 0
08      //获取委托属性值
09      operator fun getValue(thisRef: Any?, property: KProperty<*>): Float {
10          return field
11      }
12      //设置委托属性值
13      operator fun setValue(thisRef: Any?, property: KProperty<*>, value: Float) {
14          field = value
15          num ++
16          println("${property.name} 赋值为 $value")
17      }
18  }
```

第 03 行导入 kotlin.reflect.KProperty 命名空间。第 06 行 field 属性用于保存被委托属性的数据值。第 09 ~ 11 行是获取委托属性值的方法，thisRef 参数是进行委托的类对象，property 是被委托属性，返回值为 field。第 13 ~ 17 行是设置委托属性值的方法，value 参数是传入的被委托属性的赋值数据。

C0312 工程：MainActivity.kt

```
10  //通过委托类实例声明委托属性
11  var delegate = Delegate()
12  var pi: Float by delegate
13  pi = 3.14f
14  println("pi 当前数值: " + pi)
15  pi = 3.1415926f
16  println("pi 当前数值: " + pi)
17  println("pi 赋值次数: " + delegate.num)
18  //直接声明委托属性
19  var r: Float by Delegate()
20  r = 4f
```

第 11 行声明 Delegate 类实例并赋给 delegate 变量。第 12 行通过委托类实例的 delegate 变量设置

pi 变量的委托。第 13 行 pi 变量赋值时实际调用 Delegate 类的 setValue(Any?, KProperty<*>, Float) 方法进行预处理。第 19 行直接设置 pi 变量的委托类。

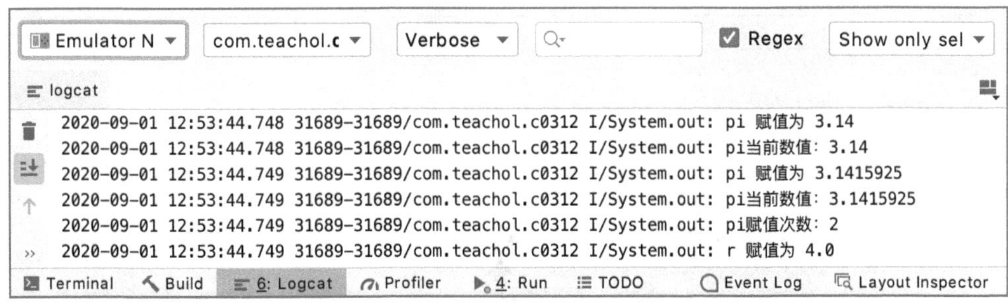

图 3-21 Logcat 窗口输出结果

3.8 属性的延迟初始化

属性需要在声明时或构造方法中初始化，否则会报错。如果不需要在声明时或构造方法中初始化就需要设置延迟初始化。延迟初始化的方式有 lateinit 和 by lazy 两种。

3.8.1 lateinit

lateinit 可以让编译器忽略对属性初始化的检查，未初始化的属性可以在方法中完成初始化。lateinit 只允许声明引用类型的变量，不允许声明基础类型的变量。

在 C0313 工程中，演示了使用 lateinit 延迟初始化属性，输出结果如图 3-22 所示。

C0313 工程: MainActivity.kt

```
06  class MainActivity : AppCompatActivity() {
07      lateinit var li:String
08      override fun onCreate(savedInstanceState: Bundle?) {
09          super.onCreate(savedInstanceState)
10          setContentView(R.layout.activity_main)
11          li = "延迟初始化变量"
12          println("$li")
13      }
14  }
```

第 07 行使用 lateinit 声明延迟初始化的 li 变量。第 11 行对 li 变量进行赋值完成初始化。

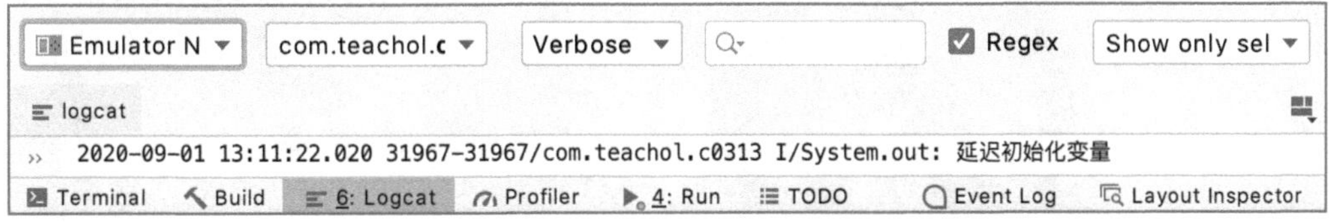

图 3-22 Logcat 窗口输出结果

3.8.2 by lazy

lazy() 函数是接收一个 Lambda 表达式并返回一个 Lazy<T>实例的函数。by lazy 本质上是变量的委托，委托后的属性可以在第一次被使用时自动初始化，这种方式延迟初始化会增加少量的系统资源开销。by lazy 不但允许声明引用类型的变量，还允许声明基础类型的变量。

在 C0314 工程中，演示了使用 by lazy 延迟初始化属性，输出结果如图 3-23 所示。

C0314 工程：MainActivity.kt

```
06 class MainActivity : AppCompatActivity() {
07     //声明延迟初始化的属性
08     private val car by lazy() {
09         Car("辽A1314","yellow")
10     }
11     override fun onCreate(savedInstanceState: Bundle?) {
12         super.onCreate(savedInstanceState)
13         setContentView(R.layout.activity_main)
14         car.forward()
15     }
17     class Car(var name: String,var color: String) { //嵌套类
18         fun forward() {
19             println(name!! + "向前行驶")
20         }
21     }
22 }
```

第 08～10 行声明属性时使用 lazy()方法委托实现初始化。第 14 行首次使用 car 属性，此时会调用 lazy()方法完成 car 属性的初始化。第 17～21 行使用嵌套类的方式声明了 Car 类。

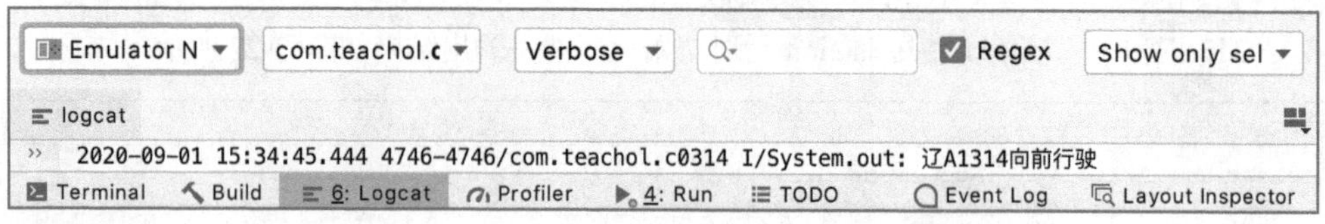

图 3-23　Logcat 窗口输出结果

提示：嵌套类和内部类

嵌套类是定义在另一个类内部的类。嵌套类提供了更好的封装，可以把嵌套类隐藏在外部类之内，不允许同一个包中的其他类访问该类。

内部类是使用 inner 修饰符定义在另一个类内部的类。内部类成员可以直接访问外部类的私有数据，因为内部类被当成其外部类成员，同一个类的成员之间可以互相访问。

第 4 章　Kotlin 的面向对象进阶

随着面向对象语言的发展，Kotlin 提供了更加丰富的编程功能。这些功能未必是使用面向对象语言的基础特性所无法实现的，但是一定会提高编程效率、减少代码量、增强代码可读性或强壮性，为程序设计提供了更多的选择。

4.1　数　据　类

数据类是专门用于快捷声明数据的类，在 class 关键字前添加 data 关键字声明数据类。数据类必须有主构造方法且至少有一个参数，不能继承和被继承，不能是抽象类、内部类或密封类。数据类可以自动声明与构造方法入参同名的属性字段，自动实现每个属性的 get()/set() 方法、equals() 方法、copy() 方法和 toString() 方法。

在 C0401 工程中，演示了定义数据类及数据类实例调用属性，输出结果如图 4-1 所示。

C0401 工程：User.kt

```
03  data class User(
04      var name: String,
05      var age: String,
06      var city: String
07  )
```

第 03 行使用 data 关键字声明 User 类为数据类，因此会自动添加隐含的 equals() 方法、copy() 方法和 toString() 方法。第 04 ~ 06 行通过主构造方法声明三个属性，由于是数据类的属性，所以会自动生成隐含的 get()/set() 方法。

C0401 工程：MainActivity.kt

```
10  val user = User("小白", "22", "沈阳")
11  val userCopy = user.copy()
12  if(user == userCopy){
13      println("$user")
14      userCopy.name = "小黑"
15      println("$userCopy")
16  }
```

第 10 行声明 user 变量，并赋值 User 类实例。第 11 行调用数据类的 copy() 方法复制一个该类实例并赋给 userCopy 变量。第 12 行由于 userCopy 变量赋值的是 user 变量复制的实例，所以两个变量存储的 User 类实例是相等的。

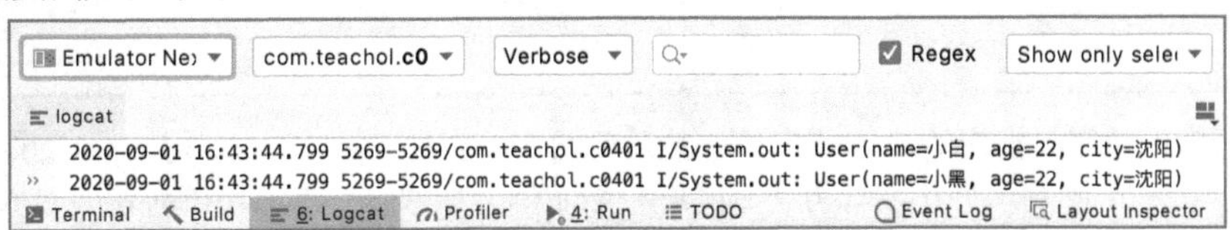

图 4-1　Logcat 窗口输出结果

4.2 密 封 类

密封类是所有子类都必须在与密封类自身相同的文件中声明的类，在 class 关键字前添加 sealed 关键字声明密封类。密封类的主构造方法和次构造方法默认访问权限为 private，且不能修改为其他权限。使用 when 语句时必须覆盖所有的情况，否则需要使用 else 语句。

在 C0402 工程中，演示了定义和使用密封类及 when 语句判断密封类中的子类类型，输出结果如图 4-2 所示。

C0402 工程：Car.kt

```
03 sealed class Car ( val name: String,  val color: String) {
04     fun forward() {
05         println(name + "向前行驶")
06     }
07     class Bus(name: String,color: String) :Car(name,color) {
08         var ridership: Int = 0//属性：载客量
09     }
10     class Truck(name: String,color: String) :Car(name,color) {
11         var burden: Int = 0//属性：载货量
12     }
13 }
```

第 03 行使用 sealed 关键字声明 Car 类为密封类。第 07～09 行声明 Bus 类为 Car 类的子类。第 10～12 行声明 Truck 类为 Car 类的子类。

C0402 工程：MainActivity.kt

```
06 class MainActivity : AppCompatActivity() {
07     override fun onCreate(savedInstanceState: Bundle?) {
08         super.onCreate(savedInstanceState)
09         setContentView(R.layout.activity_main)
10         val bus = Car.Bus("辽A21", "黄色")
11         val truck = Car.Truck("辽A22", "蓝色")
12         bus.forward()
13         truck.forward()
14         type(bus)
15         type(truck)
16     }
18     private fun type(car: Car) = when (car) { //判断车型
19         is Car.Bus -> println("${car.name}是大巴车")
20         is Car.Truck -> println("${car.name}是卡车")
21     }
22 }
```

第 14 行实例化密封类内的 Bus 类实例，然后赋给 bus 变量。第 18～21 行 type()方法通过 when 语句判断 car 参数的类型，使用这种方式判断密封类实例的类型时，需要提供所有的类型分支或者搭配 else 语句。

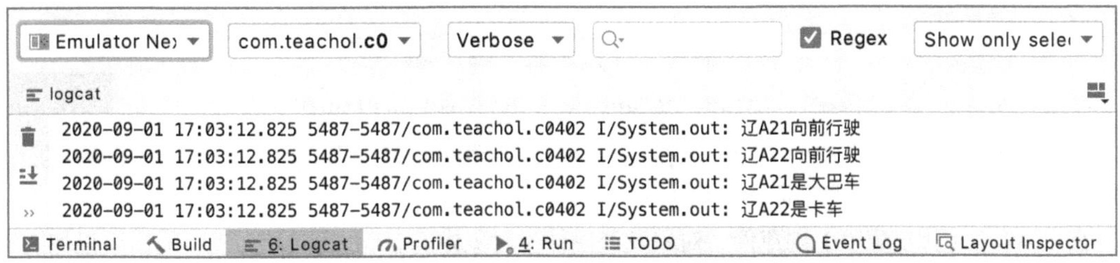

图 4-2　Logcat 窗口输出结果

4.3　对　象　类

对象类是单例模式的类，可以使用类名直接调用属性和方法。对象类不使用 class 关键字声明，而使用 object 关键字声明。由于对象类不能有主构造方法和次构造方法，所以也无法实例化，但是可以包含初始化块。内部的属性和方法等同于 Java 语言的静态属性和静态方法。静态属性只分配一次内存空间，程序结束后释放内存空间。静态方法内只能通过类名使用，不能被类实例使用。因此不能使用 this 关键字，且只能调用静态属性。

在 C0403 工程中，演示了定义和调用对象类，输出结果如图 4-3 所示。

C0403 工程：Office.kt

```
03  object Office{
04      var sendMailCount = 0
06      init{println("办公室准备就绪可以开始工作")} //初始化块
08      fun sendMail(num: Int) { //发送邮件
09          println("办公室发送了${num}封邮件")
10          sendMailCount += num
11      }
12  }
```

第 03 行使用 object 关键字声明 Office 对象类。第 04 行声明 sendMailCount 属性，该属性是静态属性。第 06 行初始化块内调用 println()方法打印“办公室准备就绪可以开始工作”。第 08～11 行声明 sendMail()方法用于处理发送邮件，该方法是静态方法。

C0403 工程：MainActivity.kt

```
10  Office.sendMail(2)
11  Office.sendMail(1)
12  Office.sendMail(5)
13  println("办公室发送的邮件总数为${Office.sendMailCount}封")
```

第 10～12 行连续三次使用 Office 对象类名直接调用 sendMail()方法，sendMail()中会将发送邮件的数量累加到 sendMailCount 属性，只有首次调用时会执行 Office 类初始化块中的代码。第 13 行使用 Office 对象类名直接调用 sendMailCount 属性。

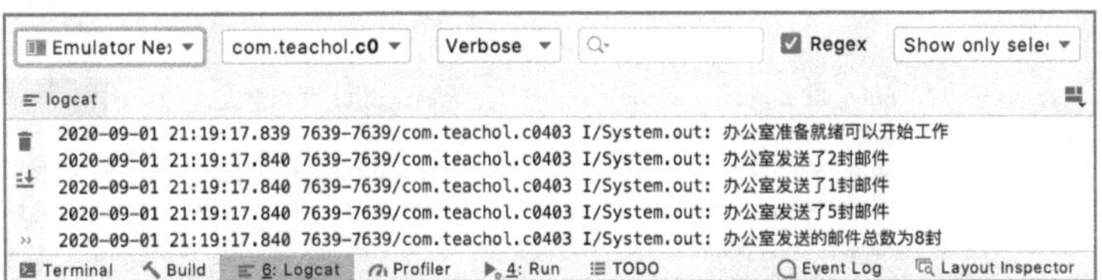

图 4-3　Logcat 窗口输出结果

4.4 伴 生 对 象

伴生对象用于在普通类中实现静态属性和静态方法的功能，可以使用 companion object{}语句创建伴生对象。伴生对象内，方法中不能使用在伴生对象外声明的属性，只能使用伴生对象内声明的属性，而伴生对象外声明的方法可以调用伴生对象内声明的属性。

在 C0404 工程中，演示了创建伴生对象及调用伴生对象内的态属性和静态方法，输出结果如图 4-4 所示。

C00404 工程：Office.kt

```
03 class Office {
05     companion object { //伴生对象
06         var sendMailCount = 0 //存储发送邮件的数量
07         fun mailCount(): Int {
08             println("所有办公室发送邮件数量为$sendMailCount")
09             return sendMailCount
10         }
11     }
12     fun sendMail(num: Int) {
13         println("办公室发送了${num}封邮件")
14         sendMailCount += num
15     }
16 }
```

第 05 ~ 11 行在 companion object{}语句内声明 sendMailCount 静态属性和 mailCount()静态方法，sendMailCount 静态属性用于存储发送邮件的数量，mailCount()静态方法用于获取发送邮件的数量。第 12 ~ 14 行 sendMail()方法调用 sendMailCount 静态属性用于累加发送邮件的数量。

C0404 工程：MainActivity.kt

```
10 val office1 = Office()
11 val office2 = Office()
12 val office3 = Office()
13 office1.sendMail(2)
14 office2.sendMail(5)
15 office3.sendMail(3)
16 //调用伴生对象内的方法
17 Office.mailCount()
```

第 10 ~ 12 行新建三个 Office 类实例，三个实例公用 sendMailCount 静态属性。第 13 ~ 15 行三个 Office 类实例通过 sendMail()方法发送邮件，并通过 sendMailCount 静态属性累加邮件数量。第 17 行 Office.mailCount()获取发送邮件的数量。

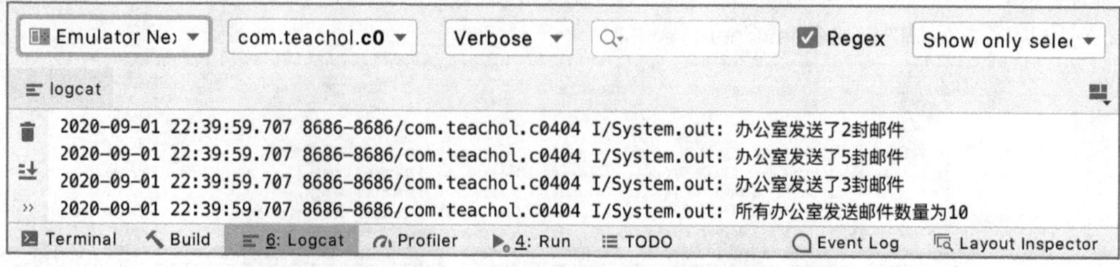

图 4-4 Logcat 窗口输出结果

4.5 枚　　举

枚举(Enum)是通过预定义列出所有值的标识符来定义的有序集合。枚举继承自 Enum 类，由于 Kotlin 不支持多继承，所以枚举对象不能再继承其他类，但是可以继承接口。枚举类提供了一些常用的属性和方法，如表 4-1 和表 4-2 所示。

语法格式：

```
enum class enumName[(para:Type)] {
    ELEMENT0[(para0)][,ELEMENT1[(para1)]][, …][,ELEMENTN[(paraN)]]
}
```

- 必选项：enumName 是枚举类的名称。
- 必选项：ELEMENT0 是枚举类的常量，一般采用大写英文字母，枚举类至少要包含一个常量。
- 可选项：[(para:Type)]定义了元素参数的名称和类型，多个 para:Type 之间使用逗号分隔。para 是元素参数的名称，Type 是元素参数的类型。
- 可选项：[(para0)]是元素的参数，其类型与[(para:Type)]定义的类型一致，多个 para0 之间使用逗号分隔。
- 可选项：[,ELEMENT1[(para1)]][,…][,ELEMENTN[(paraN)]]表示多个枚举类的常量。

表 4-1　枚举类的常用属性

类　型	属　性
Class<E>	declaringClass 返回与此枚举常量的枚举类型相对应的 Class 对象
String	Name 返回此枚举常量的名称，在其枚举声明中对其进行声明
Int	Ordinal 返回枚举常量的序数(它在枚举声明中的位置，其中初始常量序数为零)

表 4-2　枚举类的常用方法

类　型	方　法
Int	compareTo(other: E) 比较此枚举与指定对象的顺序
String	toString() 返回枚举常量的名称

4.5.1 枚举基础用法

新建 C0405 工程后，在工具窗口中选择 Project 视图，将 C0405 文件夹展开。右键单击 C0405 文件夹，选择【New】→【New Kotlin File/Class】命令。在“New Kotlin File/Class”对话框中输入名称为“Color”，双击“Enum class”类型完成枚举类的创建(如图 4-5 所示)。

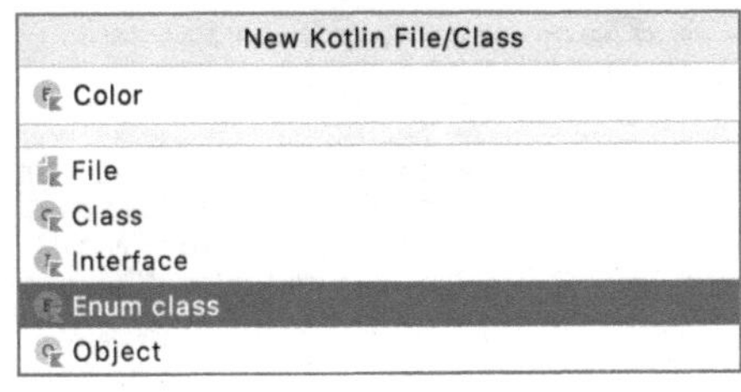

图 4-5　“New Kotlin File/Class”对话框

C0405 工程：Color.kt

```
03  enum class Color {
04      RED, YELLOW, BLUE
05  }
```

第 03 行使用 enum 关键字声明 Color 类为枚举类。第 04 行添加三个枚举常量，常量名使用大写英文字母。

C0405 工程：MainActivity.kt

```
10  //遍历枚举值
11  for (color in Color.values()) {
12      println("可选颜色：$color")
13  }
14  //判断枚举值
15  val colorSelect = Color.BLUE
16  when (colorSelect) {
17      Color.RED -> println("当前选择的颜色：红色")
18      Color.YELLOW -> println("当前选择的颜色：黄色")
19      Color.BLUE -> println("当前选择的颜色：蓝色")
20  }
21  //判断枚举值的位置
22  when (colorSelect.compareTo(Color.YELLOW)) {
23      -1 -> println("BLUE 在 YELLOW 之前")
24      1 -> println("BLUE 在 YELLOW 之后")
25      else -> println("BLUE 与 YELLOW 在同一位置")
26  }
27  //常用枚举类属性和方法
28  println("DeclaringClass 属性：" + colorSelect.declaringClass)
29  println("name 属性：" + colorSelect.name)
30  println("ordinal 属性：" + colorSelect.ordinal)
31  println("toString()方法：" + colorSelect.toString())
```

第 11～13 行遍历 Color 枚举类的常量。第 16～20 行判断 colorSelect 变量值与 Color 枚举类的哪个常量相等。第 22～26 行判断 colorSelect 变量值在 Color 枚举类常量中与 YELLOW 常量的位置关系。第 28 行 colorSelect.declaringClass 返回与此枚举常量的枚举类型相对应的 Class 对象。第 29 行 colorSelect.name 获取枚举常量名。第 30 行 colorSelect.ordinal 获取枚举常量在枚举类常量中的位置序号。第 31 行 colorSelect.toString() 与 colorSelect.name 的作用相同，即获取枚举常量名。

运行工程后，Logcat 窗口输出结果如图 4-6 所示。

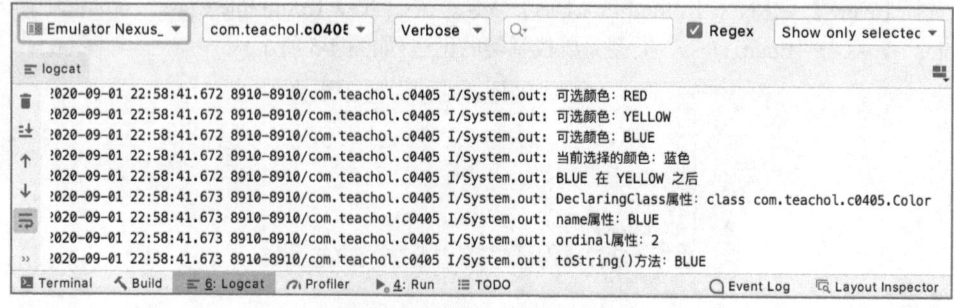

图 4-6　Logcat 窗口输出结果

4.5.2　枚举进阶用法

在 C0406 工程中，演示了带参数枚举类实例，输出结果如图 4-7 所示。

```
C0406 工程: Color.kt
03  enum class Color constructor(var RGB: String, var value: String) {
04      RED("#ff0000", "红色"), YELLOW("#ffff00", "黄色"), BLUE("#0000ff", "蓝色")
05  }
```

第 03 行主构造方法定义枚举常量的两个参数，第一个参数 RGB 用于存储颜色的 RGB 数值，第二个参数 value 用于存储颜色名称的字符串。第 04 行三个枚举常量根据主构造方法定义的参数类型设置相应的参数值。

```
C0406 工程: MainActivity.kt
10  val colorSelect = Color.BLUE
11  println("选取颜色的名称: " + colorSelect.value)
12  println("选取颜色的 RBG 值: " + colorSelect.RGB)
```

第 10 行声明 colorSelect 变量，并赋值枚举常量。第 11、12 行 colorSelect 变量通过.运算符直接获取 BLUE 常量的 RGB 和 value 的参数值。

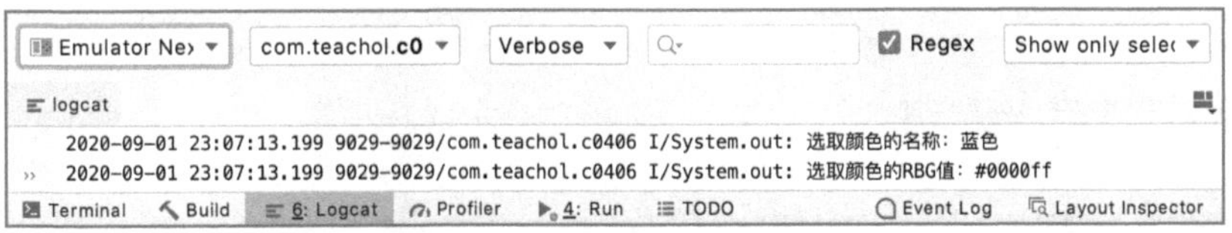

图 4-7　Logcat 窗口输出结果

4.6　集　　合

集合通常包含相同类型的对象，集合中的对象称为元素。集合分为可变集合和不可变集合，可变集合可以增减元素，而不可变集合不可以增加元素。

4.6.1　Set 集合

Set 集合用于存储不重复的同一类型数据，提供了 size 属性用于获取元素数量。如果向 Set 类实例内添加已有的元素，则自动忽略该元素，即不会重复添加元素。使用 setOf() 可以创建不可变的 Set 集合实例，Set 集合的常用方法如表 4-3 所示。使用 mutableSetOf() 可以创建可变的 MutableSet 类实例，MutableSet 集合的常用方法如表 4-4 所示。此外还包括 hashSetOf() 方法、linkedSetOf() 方法和 sortedSetOf() 方法，这里不再详细阐述。

表 4-3　Set 集合的常用方法

修饰符和类型	方　法
Boolean	isEmpty() 判断集合是否为空
Operator Boolean	contains(element: @UnsafeVariance E) 判断集合是否包含某元素
Boolean	containsAll(elements: Collection<@UnsafeVariance E>) 判断集合是否包含 elements 集合的所有元素
T	elementAt(index: Int) 获取指定索引位置的元素

表 4-4　MutableSet 集合的常用方法

修饰符和类型	方　法
Boolean	add (element: E) 添加元素
Boolean	remove (element: E) 删除元素
Boolean	addAll (elements: Collection<E>) 批量添加 elements 集合中的多个元素
Boolean	removeAll (elements: Collection<E>) 批量删除 elements 集合中的多个元素
Boolean	retainAll (elements: Collection<E>) 保留集合中与 elements 集合共有的元素
Unit	clear () 清空集合
T	elementAt (index: Int) 获取指定索引位置的元素

在 C0407 工程中，演示了 Set 集合和 MutableSet 集合的常用属性和方法，输出结果如图 4-8 所示。

C0407 工程：MainActivity.kt

```
11  val set = setOf(5, 2, 1, 1, 1) //不可变集合
12  println("set的元素数量: ${set.size}")
14  for (element in set) { //遍历集合
15      println("set的元素$element")
16  }
17  if (set.contains(2)) {println("set的元素中包含2")}
19  val mutableSet = mutableSetOf(5, 2, 1, 1, 1) //可变集合
20  println("mutableSet的元素数量: ${mutableSet.size}")
21  mutableSet.add(9) //添加单个元素
22  mutableSet.addAll(set) //添加集合内的所有元素
23  println("mutableSet的添加元素后的元素数量: ${mutableSet.size}")
25  for (i in mutableSet.indices) { //遍历集合
26      println("mutableSet的元素${mutableSet.elementAt(i)}")
27  }
28  if (mutableSet.containsAll(set)) {println("mutableSet包含set的所有元素")}
29  mutableSet.retainAll(set) //保留 set 集合内包含的元素
30  println("mutableSet保留与set共有元素后的元素数量: ${mutableSet.size}")
```

第 11 行使用 setOf() 方法创建不可变的 Set 集合，添加五个元素，但后三个元素是相同的，会自动忽略重复的元素，因此第 12 行 set.size 属性获取的元素数量为 3。第 14 ~ 16 行遍历 set 集合的所有元素。第 17 行判断 set 集合的元素中是否包含 2。第 19 行使用 mutableSetOf() 方法创建可变的 MutableSet 集合。第 21 行向 mutableSet 集合添加单个元素，由于之前集合内没有包含该元素，所以能够添加成功。第 22 行将 set 集合的所有元素添加到 mutableSet 集合，但是由于 mutableSet 集合已经包含了 set 集合的所有元素，所以第 23 行 mutableSet.size 属性获取的元素数量为 4。第 25 ~ 27 行遍历 mutableSet 集合的所有元素。第 28 行判断 mutableSet 集合是否包含 set 集合的所有元素。第 29 行 mutableSet 集合保留 set 集合内包含的元素，不包含的元素将被删除，所以第 30 行 mutableSet.size 属性获取的元素数量为 3。

图 4-8　Logcat 窗口输出结果

4.6.2　List 集合

List 集合用于存储不重复的同一类型数据，但是元素都有对应的顺序索引，因此允许使用重复元素，可以通过索引来访问指定位置的集合元素。使用 listOf() 方法创建不可变的 List 类实例，List 集合的常用方法如表 4-5 所示。使用 mutableListOf() 方法创建可变的 List 类实例，MutableList 集合的常用方法如表 4-6 所示。还包括 listOfNotNull() 方法和 arrayListOf() 方法，这里不再详细阐述。List 集合除包含 Set 集合的属性和方法外，还包含通过索引操作集合元素的方法。

表 4-5　List 集合的常用方法

类型和修饰符	方　法
Int	indexOf(element: @UnsafeVariance E) 查找该元素在 element 集合中首次出现的位置
Int	lastIndexOf(element: @UnsafeVariance E) 查找该元素在 element 集合中最后一次出现的位置
List<E>	subList(fromIndex: Int, toIndex: Int) 获取指定元素范围的子集合

表 4-6　MutableList 集合的常用方法

类型和修饰符	方　法
Unit	add(index: Int, element: E) 添加元素到指定位置
Boolean	addAll(elements: Collection<E>) 批量添加 elements 集合中的多个元素
Boolean	addAll(index: Int, elements: Collection<E>) 从指定位置批量添加 elements 集合中的多个元素
E	removeAt(index: Int) 查找该元素在集合中最后一次出现的位置
Boolean	removeAll(elements: Collection<E>) 批量删除 elements 集合中的多个元素
Boolean	retainAll(elements: Collection<E>) 保留集合中与 elements 集合共有的元素
Unit	clear() 清除所有元素
MutableList<E>	subList(fromIndex: Int, toIndex: Int) 获取指定元素范围的子集合

在 C0408 工程中，演示了 List 集合和 MutableList 集合的常用属性和方法，输出结果如图 4-9 所示。

```
C0408 工程：MainActivity.kt
11  val list = listOf(5, 2, 1, 1, 1) //不可变 List 集合
12  println("list 集合：$list")
13  println("list 集合的元素数量：${list.size}")
14  println("list 集合中元素值为 1 的第一个元素位置：${list.indexOf(1)}")
15  println("list 集合中元素值为 1 的最后一个元素位置：${list.lastIndexOf(1)}")
16  println("list 集合中从索引 1 到索引 3 的元素集合：${list.subList(1, 3)}")
18  val mutableList = mutableListOf(5, 2, 1, 1, 1) //可变 List 集合
19  println("mutableList 集合：$mutableList")
20  mutableList.add(3, 0) //在索引 3 处插入一个新元素
21  println("mutableList 集合插入元素后：$mutableList")
22  mutableList.removeAt(1) //删除索引 1 处的元素
23  println("mutableList 集合删除元素后：$mutableList")
24  mutableList[4] = 9 //将索引 4 处的元素替换为 9
25  println("mutableList 集合替换元素后：$mutableList")
26  mutableList.clear() //清空集合的所有元素
27  println("mutableList 集合清空后的元素数量：${mutableList.count()}")
```

第 13 行获取 list 集合的元素数量。第 14 行获取元素值为 1 的元素第一次出现在集合中的位置。第 15 行获取元素值为 1 的元素最后一次出现在集合中的位置。第 16 行将 list 集合中索引 1 到索引 3 的元素提取出来生成一个新的 List 集合。第 20 行在索引 3 处插入一个新元素，元素值为 0。第 22 行删除索引 1 处的元素，其后面的元素索引位置向前移动。第 24 行使用与数组相同的形式表示集合元素，将数组的下标替换成集合的元素索引，元素索引也是从 0 开始的。第 26 行清空 mutableList 集合的所有元素，所以第 27 行 count()方法获取 mutableList 集合元素的数量为 0。

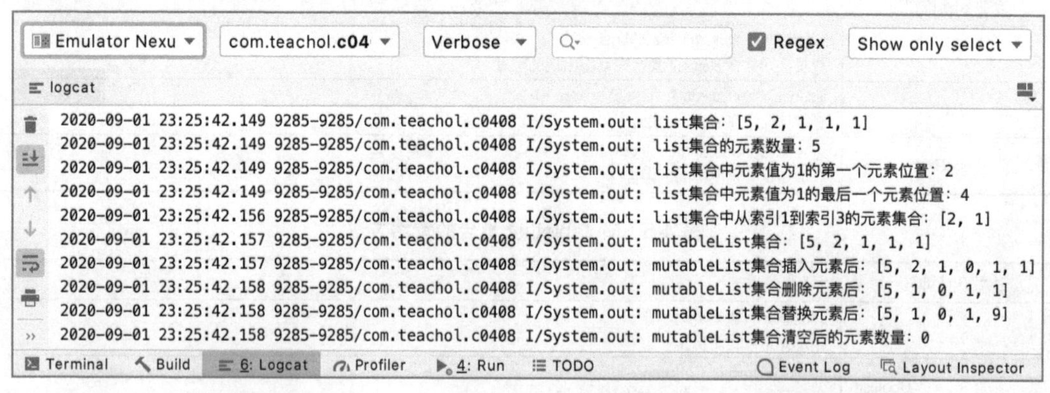

图 4-9　Logcat 窗口输出结果

4.6.3　Map 类

Map 类使用键名和键值保存同一类型数据，键名不能重复，键值可以重复。使用 mapOf()方法创建不可变的 Map 类实例，Map 类的常用属性和方法如表 4-7、表 4-8 所示。使用 mutableMapOf()方法创建可变的 MutableMap 类实例，MutableMap 类的常用属性和方法如表 4-9、表 4-10 所示。还包括 hashMapOf()方法、linkedMapOf()方法和 sortedMapOf()方法，这里不再详细阐述。Map 类除包含 Set 类的属性和方法外，还包含通过键名和键值操作集合的方法。

表 4-7 Map 类的常用属性

类型和修饰符	属 性
Set<K>	keys 键名的集合，只读属性
Collection<V>	values 键值的集合，只读属性
Set<Map.Entry<K, V>>	entries 键值对的集合，只读属性

表 4-8 Map 类的常用方法

类型和修饰符	方 法
Boolean	containsKey(key: K) 判断是否包含指定键名
Boolean	containsValue(value: @UnsafeVariance V) 判断是否包含指定键值
V?	get(key: K) 获取指定键名的键值

表 4-9 MutableMap 类的常用属性

类型和修饰符	属 性
MutableSet<K>	keys 键名的集合
MutableCollection<V>	values 键值的集合
MutableSet<MutableMap.MutableEntry<K, V>>	entries 键值对的集合

表 4-10 MutableMap 类的常用方法

类型和修饰符	方 法
V?	put(key: K, value: V) 添加一个键值对。如果该键名已存在，则修改其键值
Unit	putAll(from: Map<out K, V>) 批量添加键值对
V?	remove(key: K) 删除指定键名的键值对
Boolean	remove(key: K, value: V) 删除指定键名和键值的键值对
Unit	clear() 清空所有的键值对

在 C0409 工程中，演示了 Map 和 MutableMap 集合的常用属性和方法，输出结果如图 4-10 所示。

C0409 工程：MainActivity.kt

```
//不可变 Map 集合
val map = mapOf("a" to 5, "b" to 2, "c" to 1)
println("map 集合: $map")
//直接遍历 Map 集合获取键名和键值
for ((key, value) in map) println("key:${key}  value:${value}")
//遍历 Map 集合的键值对，再通过键值对获取键名和键值
for (en in map.entries) println("key:${en.key}  value:${en.value}")
//遍历 Map 集合的键名，再通过键名获取键值
for (key in map.keys) println("key:${key}  value:${map[key]}")
//可变 mutableMap 集合
val mutableMap = mutableMapOf("a" to 5, "b" to 2, "c" to 1, "d" to 9)
```

```
21 println("mutableMap 集合: $mutableMap")
22 mutableMap["e"] = 7//以方括号语法添加键值对
23 mutableMap.put("f", 3)//以 put 方法添加键值对
24 println("mutableMap 集合添加键值对后: $mutableMap")
25 mutableMap.remove("b")//删除键名为"b"的键值对
26 println("mutableMap 集合删除键名为 b 的键值对后: $mutableMap")
```

第 14 行直接遍历 Map 集合，获取键名和键值。第 16 行遍历 Map 集合的键值对，再通过键值对集合获取键值和键名。第 18 行遍历 Map 集合的键名，再通过键名获取键值。第 22、23 行使用两种方法添加或修改键值对，两种方式的效果是相同的。第 25 行根据键名删除键值对。

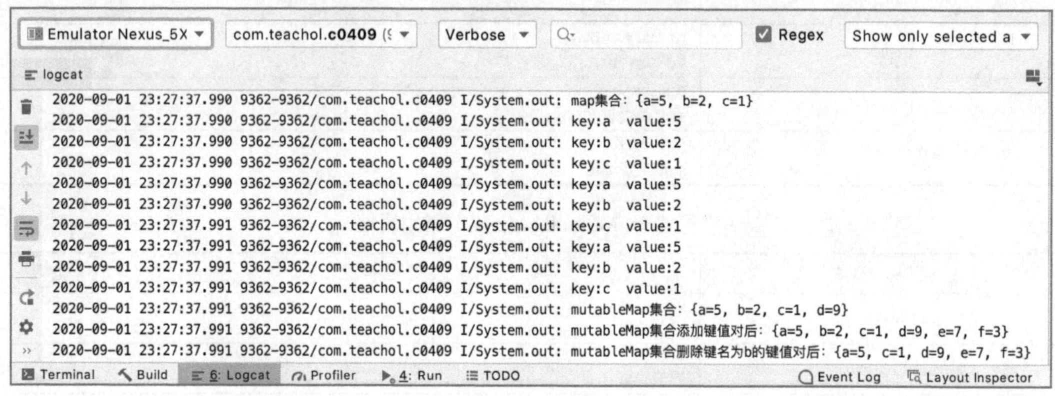

图 4-10　Logcat 窗口输出结果

4.7　泛　　型

泛型(generics)是一种特殊的类型，可以接收不同类型的数据。**泛型的本质是参数化类型，泛型并不是变量，所操作的数据类型被指定为一个参数**。例如，同样类型的箱子所存储的物品不一定是同一种类型的，可能是苹果，也可能是玩具。苹果和玩具是不同类型的物品，需要为每种类型的物品提供一个重载方法，每个方法提供一种不同类型的参数。如果使用泛型就可以只使用一个方法，通过泛型变量传入这个方法。

<>用于声明泛型类型。例如，<T>声明 T 为泛型，如果 T 不允许接收 null 类型，可以将:Any 放在 T 后，即 T:Any 声明 T 为非空泛型。

4.7.1　泛型类

泛型类是指声明类时同时声明泛型的类，声明类时在类名后添加<>声明泛型，<>内可以声明多个泛型，多个泛型之间使用逗号分隔。

语法格式

```
class ClassName <Type[:Any]> {
    statement
}
```

- **必选项**：ClassName 是泛型类的名称。
- **必选项**：Type 是泛型，可以作为类内部的数据类型。
- **可选项**：[:Any]表示非空泛型，即 T 可以是除 null 外的任何类型。

声明泛型类实例的变量时，<Type>中的 Type 需要替换成相应的类型。如果省略<Type>，会自动判断传入参数的数据类型。

语法格式：

```
var|val instanceName[:ClassName<Type>] = ClassName ([para])
```

- **必选项：**instanceName 是类的实例名。
- **必选项：**ClassName 是类名。
- **可选项：**[:ClassName<Type>]用于指定泛型的数据类型，Type 是数据类型。
- **可选项：**[para]是调用主构造方法或次构造方法的参数。

在 C0410 工程中，演示了声明和调用泛型类，输出结果如图 4-11 所示。

C0410 工程：Box.kt

```
03 class Box<T:Any>(private var collection: T) {
04     // 判断类型
05     fun type(): String {
06         val result: String
07         if (collection.javaClass == java.lang.Integer::class.java) {
08             result = "整数型"
09         } else if (collection.javaClass == java.lang.String::class.java) {
10             result = "字符型"
11         } else {
12             result = "无法识别类型"
13         }
14         return result
15     }
16 }
```

第 03 行<T:Any>定义 Box 类为非空泛型类，T 为非空泛型。collection 属性声明为 T 类型。第 07 行 collection.javaClass 用于获取 collection 属性的类型，该属性不能判断 null 的类型，因此之前 collection 属性被定义为非空泛型。

C0410 工程：MainActivity.kt

```
10 //初始化泛型类变量
11 val box1:Box<Int> = Box(123)//只能传入 Integer
12 val box2:Box<String> = Box("123")//只能传入 String
13 val box3 = Box(123)//可以传入任何类型
14 val box4 = Box("123")//可以传入任何类型
15 //输出类型
16 println("box1 存储的数据类型为: " + box1.type())
17 println("box2 存储的数据类型为: " + box2.type())
18 println("box3 存储的数据类型为: " + box3.type())
19 println("box4 存储的数据类型为: " + box4.type())
```

第 11、12 行声明两个指定类型的泛型类实例的变量，指定类型后传入的参数类型要与指定的类型一致。第 13、14 行声明变量时并未指定类型，而实例化时通过参数传入的类型进行自动识别 T 的类型。第 16～19 行使用 type()方法获取 Box 泛型类中 T 的类型。

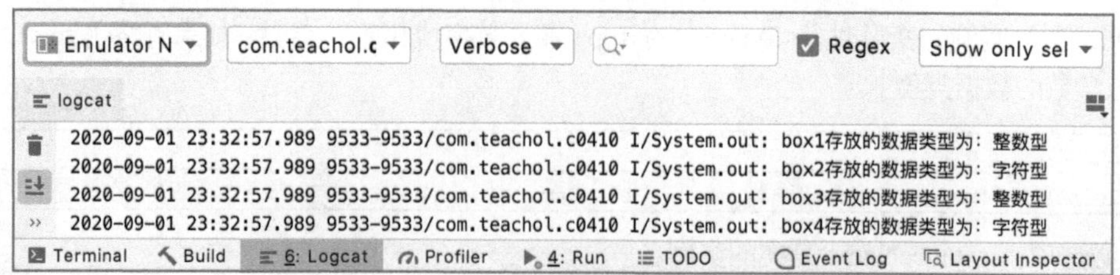

图 4-11　Logcat 窗口输出结果

4.7.2　泛型接口

泛型接口是指声明接口时同时声明泛型，声明接口时在接口名称后添加<>声明泛型，<>内可以声明多个泛型，多个泛型之间使用逗号分隔。

语法格式：

```
interface InterfaceName<Type[:Any]> [: ParentInterfaceName]{
    statement
}
```

- **必选项：**InterfaceName 是泛型接口的名称。
- **必选项：**Type 是泛型，可以作为接口内部的数据类型。
- **可选项：**[:Any]表示非空泛型，即 T 可以是除 null 外的任何类型。
- **可选项：**[:ParentInterfaceName]是要继承的接口名称。

类实现接口时，需要使用:运算符，可以同时实现多个接口。实现接口的类必须重写抽象属性和抽象方法。

语法格式：

```
class ClassName <Type [:Any]>:InterfaceName<Type>{
    statement
}
```

- **必选项：**ClassName 是实现泛型接口的类名称。
- **必选项：**Type 是泛型。
- **必选项：**InterfaceName 是实现泛型接口的名称。
- **可选项：**[:Any]表示非空泛型，即 T 可以是除 null 外的任何类型。

在 C0411 工程中，演示了声明和实现泛型接口，输出结果如图 4-12 所示。

C0411 工程：Box.kt

```
03  interface Box<T> {
04      fun type(): String
05  }
```

第 03 行声明 Box 接口为泛型接口，T 是泛型参数。第 04 行声明 type()抽象方法，该方法的返回值类型为 String，abstract 是接口中属性和方法的修饰符。

C0411 工程：RedBox.kt

```
class RedBox<T : Any>(var collection: T) : Box<T> {
    // 实现接口
    override fun type(): String {
```

```
06          val result: String
07          if (collection.javaClass == java.lang.Integer::class.java) {
08              result = "整数型"
09          } else if (collection.javaClass == java.lang.String::class.java) {
10              result = "字符型"
11          } else {
12              result = "无法识别类型"
13          }
14          return result
15      }
16  }
```

第 03 行声明 RedBox 类实现泛型接口，使用:Any 声明 T 是非空泛型。collection 属性声明为 T 类型。第 05 ~ 15 行重写 type()抽象方法，用于获取 collection 属性的类型。

C0411 工程: MainActivity.kt

```
10  //初始化泛型类变量
11  val box1:RedBox<Int> = RedBox(123)//只能传入 Integer
12  val box2:RedBox<String> = RedBox("123")//只能传入 String
13  val box3 = RedBox(123.0)
14  val box4 = RedBox("123")
15  //输出类型
16  println("存储的数据类型为: " + box1.type())
17  println("存储的数据类型为: " + box2.type())
18  println("存储的数据类型为: " + box3.type())
19  println("存储的数据类型为: " + box4.type())
```

第 11、12 行声明两个指定类型的实现泛型接口的泛型类实例的变量，指定类型后传入的参数类型要与指定的类型一致。第 13、14 行声明变量时并未指定类型，而实例化时通过参数传入的类型自动识别 T 类型。第 16 ~ 19 行使用 type()方法获取数据的类型。

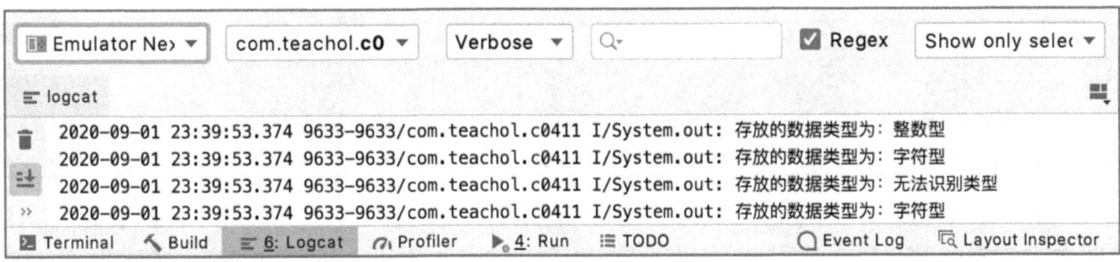

图 4-12　Logcat 窗口输出结果

4.7.3　泛型方法

泛型方法是指声明方法的同时声明泛型，声明方法时在 fun 关键字后添加<>声明泛型，<>内可以声明多个泛型，多个泛型之间使用逗号分隔。

语法格式：

```
fun<Type[:Any]> functionName ([para:ParaType])[:ReturnType] {
    statement
}
```

- **必选项：**Type 是泛型。

- **必选项**：functionName 是方法的名称。
- **可选项**：[:Any]表示非空泛型，即 T 可以是除 null 外的任何类型。
- **可选项**：[para:ParaType]是参数，多个参数之间使用逗号分隔。
- **可选项**：[:ReturnType]是返回值的类型。

在 C0412 工程中，演示了声明和调用泛型方法，输出结果如图 4-13 所示。

C0412 工程：Box.kt

```
03 class Box {
04     //获取数据类型
05     fun <T : Any> type(collection: T): String {
06         val result: String
07         if (collection.javaClass == java.lang.Integer::class.java) {
08             result = "整数型"
09         } else if (collection.javaClass == java.lang.String::class.java) {
10             result = "字符型"
11         } else {
12             result = "无法识别类型"
13         }
14         return result
15     }
16     //获取数据
17     fun <T> value(t: T): T {
18         return t
19     }
20 }
```

第 05 行声明 type(T)方法为非空泛型方法，<T:Any>声明 T 为非空泛型，collection: T 声明 collection 参数为 T 类型。第 17 行声明 value(T)方法为泛型方法，<T>声明 T 为泛型，t 参数和方法返回值的类型为 T。

C0412 工程：MainActivity.kt

```
10 val box = Box()
11 println(box.type(123) + ":" + box.value(123))
12 println(box.type("abc") + ":" + box.value("abc"))
```

第 10 行对 Box 类进行实例化。第 11、12 行分别调用 type(T)泛型方法和 value(T)泛型方法，传入整数型和字符型数据。

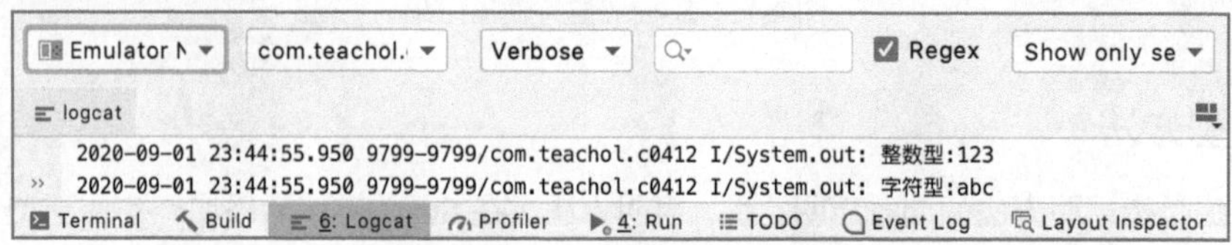

图 4-13 Logcat 窗口输出结果

4.8 异常处理

异常处理是指对程序运行中错误或问题的处理机制。例如，读取文件失败、网络通信中断、内存溢出、输出数据错误等。

当 try 语句内的 statement1 语句体出现异常时，则通过 catch 语句捕获相应类型异常后执行 statement2 语句体，无论是否存在异常最后都执行 finally 语句内的 statement3 语句体。try 语句和 catch 语句不能单独使用，一个 try 语句可以搭配多个 catch 语句以捕获多种不同类型的异常进行不同的处理。

语法格式：

```
try {
     statement1
} catch (exceptionName:Exception){
     statement2
}[ finally {
     statement3
}]
```

- **必选项**：Exception 是需要捕获的异常类型。
- **必选项**：exceptionName 是捕获的异常实例。
- **可选项**：[finally{statement3}]语句无论异常是否被捕获都会被执行。即使 catch 语句中执行了 return 语句或 break 语句，finally 语句内的语句依然会被执行。

内置异常类都继承自 Exception 类，如表 4-11 所示。直接使用 Exception 类可以捕获任意类型的异常，但是不便于分类处理。

表 4-11　内置异常类

异常类	描述
ArithmeticException	出现异常的运算条件，如被除数为 0
ArrayIndexOutOfBoundsException	超出数组下标范围
ArrayStoreException	试图将错误类型的对象存储到一个对象数组
ClassCastException	试图将对象强制转换为不是实例的子类
IllegalArgumentException	传递了一个不合法或不正确的参数
IllegalMonitorStateException	试图等待对象的监视器，或者试图通知其他正在等待对象的监视器而本身没有指定监视器的线程
IllegalStateException	在非法或不适当的时间调用方法
IllegalThreadStateException	线程没有处于请求操作所要求的适当状态
IndexOutOfBoundsException	指示某排序索引超出范围时抛出
NegativeArraySizeException	试图创建大小为负的数组
NullPointerException	在需要对象的地方为 null
NumberFormatException	字符串不能转换成数值类型格式
SecurityException	安全管理器抛出异常
StringIndexOutOfBoundsException	指示索引为负，或者超出字符串的大小
UnsupportedOperationException	不支持请求操作

4.8.1　异常处理基础用法

在 C0413 工程中，演示了基本形式的异常捕获，输出结果如图 4-14 所示。

C0413 工程：MainActivity.kt

```
val box = IntArray(2)
//任意类型异常处理(无异常)
try {
```

```
13      println("--异常检测 1 开始--")
14      box[1] = 0
15      println("无异常")
16  } catch (e: Exception) {
17      println("异常原因:" + e.message)
18  } finally {
19      println("--异常检测 1 结束--")
20  }
21  //任意类型异常处理(有异常)
22  try {
23      println("--异常检测 2 开始--")
24      box[3] = 0
25      println("无异常")
26  } catch (e: Exception) {
27      println("异常原因:${e.message}")
28  } finally {
29      println("--异常检测 2 结束--")
30  }
31  //固定类型异常处理(有异常)
32  try {
33      println("--异常检测 3 开始--")
34      box[3] = 0
35      println("无异常")
36  } catch (e: ArrayIndexOutOfBoundsException) {
37      println("异常原因:${e.message}")
38  } finally {
39      println("--异常检测 3 结束--")
40  }
41  //多种固定类型异常处理(有异常)
42  try {
43      println("--异常检测 4 开始--")
44      box[0] = 0 / 9
45      box[1] = 9 / 0
46      println("无异常")
47  } catch (e1: ArrayIndexOutOfBoundsException) {
48      println("异常原因:${e1.message}")
49  } catch (e2: ArithmeticException) {
50      println("异常原因:${e2.message}")
51  } finally {
52      println("--异常检测 4 结束--")
53  }
```

第 10 行声明 box 数组，该数组包含 2 个元素的 Int 型。第 12～20 行对 box[1]的赋值进行异常检测处理。第 22～30 行对 box[3]的赋值进行异常检测处理，由于该下标超出数组定义的长度，所以会捕获异常，使用字符串模板将异常信息输出。第 32～40 行对 box[3]赋值，进行 ArrayIndexOutOfBoundsException 异常检测，只有超出下标范围时才会捕获异常。第 42～53 行进行 ArrayIndexOutOfBoundsException 和 ArithmeticException 异常检测，被除数为 0 时会捕获 ArithmeticException 异常。

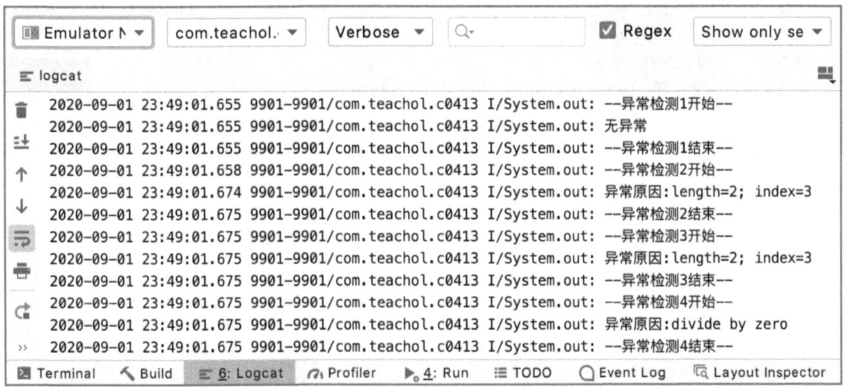

图 4-14　Logcat 窗口输出结果

提示：字符串模板

包含$和${}的字符串可以动态合成字符串，$和${}是字符串模板。$后直接连接变量名，可以将变量存储的数据合成到字符串中。${}的花括号中可以使用表达式，表达式的结果将被合成到字符串中。

4.8.2 异常处理进阶用法

在 C0414 工程中，演示了遇到 break 和 return 语句情况下的异常捕获，输出结果如图 4-15 所示。

C0414 工程：MainActivity.kt

```kotlin
class MainActivity : AppCompatActivity() {
    override fun onCreate(savedInstanceState: Bundle?) {
        super.onCreate(savedInstanceState)
        setContentView(R.layout.activity_main)
        println("+++++包含 return 的异常处理+++++")
        println("结果: " + division(10, 2))
        println("结果: " + division(10, 0))
        println("+++++包含 break 的异常处理+++++")
        for (i in 1 downTo -10) {
            try {
                println("--开始异常检测--")
                println("10/$i="+(10 / i))
            } catch (e: ArithmeticException) {
                println("异常原因:$e")
                break
            } finally {
                println("--结束异常检测--")
            }
        }
    }
}
private fun division(a: Int, b: Int): String {
    var error = false
    try {
        println("--开始异常检测--")
```

```
31              return a.toString() + "/" + b + "=" + a / b
32          } catch (e: ArithmeticException) {
33              println("异常原因:$e")
34              error = true
35              return a.toString() + "/" + b + "被除数不能为 0"
36          } finally {
37              if (!error) {
38                  println("无异常")
39              }
40              println("--结束异常检测--")
41          }
42      }
```

第 14 ~ 24 行使用 for 语句输出从 1 到-10 被 10 除的结果，并进行异常捕获。捕获到异常后，执行 break 语句结束循环。第 27 ~ 42 行的 division(Int, Int)方法对除法进行异常捕获。因为 try 语句和 catch 语句内都包含 return 语句，所以无论是否捕获异常都会为 division(Int, Int)方法提供返回值。执行了 return 语句后，finally 语句依然会执行。

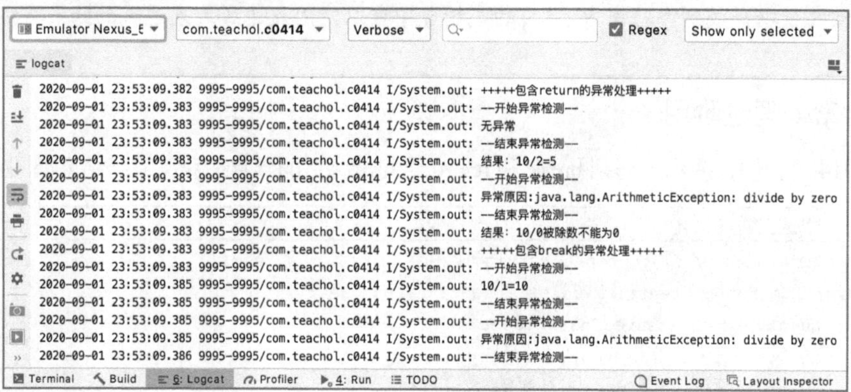

图 4-15　Logcat 窗口输出结果

4.8.3　自定义异常处理用法

自定义异常首先需要一个继承 Exception 类的自定义异常类，然后通过 throws 语句和 throw 语句抛出自定义异常，这样 catch 语句就可以对自定义异常进行捕获。

在 C0415 工程中，演示了抛出和捕获自定义异常类。

C0415 工程：BankCard.kt

```
class BankCard(private var balance: Double = 0.0) {
    // 方法：存钱
    fun deposit(amount: Double) {
        balance += amount
        println("存入：￥$balance")
    }
    // 方法：取钱
    @Throws(BankCardException::class)
    fun withdraw(amount: Double) {
        if (amount <= balance) {
```

```
13            balance -= amount
14            println("取出: ¥$balance")
15        } else {
16            throw BankCardException(amount)
17        }
18    }
19 }
20 class BankCardException(val amount: Double) : Exception()
```

第 03 行 BankCard 类通过主构造方法定义 number 属性和 balance 属性，分别存储余额和卡号，balance 属性的默认值为 0.0。第 05～08 行 deposit(Double)方法用于存钱。第 10 行使用@throws 注释了该方法可能抛出的异常类。第 11～18 行 withdraw(Double)方法用于取钱，当余额不足时通过 throw 关键字抛出 BankCardException 异常。第 20 行自定义的 BankCardException 类继承 Exception 类，定义 amount 属性。主构造方法没有任何注解和访问控制符的情况下，constructor 关键字可以省略。

```
C0415 工程: MainActivity.kt
06 class MainActivity : AppCompatActivity() {
07     override fun onCreate(savedInstanceState: Bundle?) {
08         super.onCreate(savedInstanceState)
09         setContentView(R.layout.activity_main)
10         val bankCard = BankCard("2103 8387 972")
11         bankCard.deposit(800.00)
12         try {
13             bankCard.withdraw(400.00)
14             bankCard.withdraw(600.00)
15         } catch (e: BankCardException) {
16             println("余额不足，取出¥" + e.amount + "失败。")
17             e.printStackTrace()
18         }
19     }
20 }
```

第 12～18 行用于捕获余额不足的异常，当执行到第 14 行 bankCard.withdraw(600.00)语句时银行卡内余额不足，此时 withdraw(Double)方法内抛出，然后第 15 行的 catch 语句捕获到 BankCardException 异常。第 16 行 e.amount 获取 bankCard 抛出异常时传递的余额数据。第 17 行 e.printStackTrace()方法在 Logcat 窗口中输出异常信息，即在程序中出错的位置及原因。

运行工程，Logcat 窗口中 BankCard.kt:16 和 MainActivity.kt:14 表示抛出异常的代码所属文件名称和行号，单击后可以直接跳转到代码的位置，如图 4-16 所示。

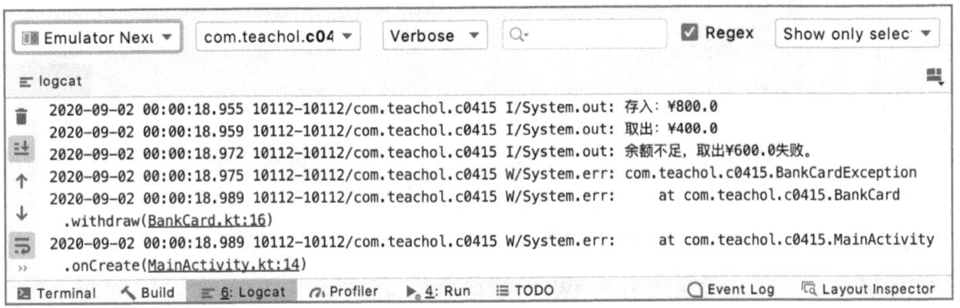

图 4-16　Logcat 窗口输出结果

4.9 多 线 程

理解多线程的概念，需要先理解进程和线程的概念。

进程是一段程序的执行过程，线程是程序执行流的最小单元。主流的操作系统都是支持多进程的，即在同一时间运行多个程序。进程在执行过程中拥有独立的内存单元，每个进程至少包含一个线程，线程不能脱离进程独立存在。如果进程中包含两个或两个以上的线程被称为多线程，多线程能更好地利用 CPU 资源，但是对内存的消耗也会更多。

每个线程都有独立的生命周期，包含新建状态、就绪状态、运行状态、阻塞状态和死亡状态等 5 种状态，如图 4-17 所示。

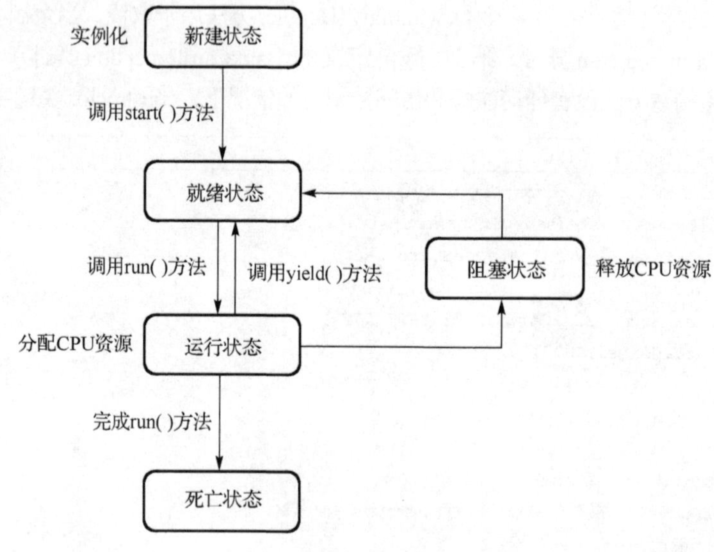

图 4-17 线程的生命周期

- **新建状态**：实例化线程后，进入新建状态。虽然已经分配内存空间，但是还没有开始执行。
- **就绪状态**：线程调用 start()方法后启动线程，在线程队列中等待被 CPU 调度执行，进入就绪状态。
- **运行状态**：被 CPU 调度执行后，调用 run()方法后，进入运行状态。在运行状态，调用 yield()方法暂停线程后，回到就绪状态。
- **阻塞状态**：在运行状态，如果进入阻塞状态，则释放 CPU 资源。阻塞结束后，回到就绪状态，重新进入线程队列等待，再次被 CPU 调度执行后继续执行。调用 suspend()方法、sleep()方法、wait()方法或线程处于 I/O 请求的等待时，进入阻塞状态。
- **死亡状态**：run()方法执行结束后，进入死亡状态。

4.9.1 Thread 类

Thread 类(java.lang.Thread)通常是作为父类使用的，子类通过继承 Thread 类实现多线程，重写 run()方法，然后通过子类的实例调用 start()方法开启线程，调用被重写的 run()方法。

Kotlin 实际上调用的是 Java 的 Thread 类，该类提供了很多构造方法和方法，如表 4-12 所示。在 **Kotlin 中调用 Java 的类时，需要使用 Kotlin 的语法规范进行调用。**

表 4-12　Java 中 Thread 类的构造方法和方法

类型和修饰符	方　法
	Thread () 构造方法
	Thread (String name) 构造方法。name 参数为线程名称
	Thread (Runnable target) 构造方法。Target 参数为实现 Runnable 接口的线程辅助类实例
	Thread (Runnable target, String name) 构造方法
void	checkAccess () 判断当前运行的线程是否有权修改该线程
static Thread	currentThread () 返回对当前正在执行的线程对象的引用
long	getId () 返回线程的标识符
String	getName () 返回线程的名称
int	getPriority () 返回线程的优先级
Thread.State	getState () 返回线程的状态
void	interrupt () 设置中断状态，但是并不会中断线程的执行。当线程被 Object.wait ()、Thread.join () 或 Thread.sleep () 方法阻塞时调用 interrupt () 方法，抛出 InterruptedException 异常，不会改变中断状态为 true。如果线程没有被阻塞，调用 interrupt () 方法将不起作用；直至执行到阻塞，才会抛出 InterruptedException 异常
static boolean	interrupted () 判断 Thread 对象的中断状态，并重置中断状态为 false
boolean	isAlive () 判断线程是否处于活动状态
boolean	isDaemon () 判断线程是否为守护线程
boolean	isInterrupted () 判断当前线程的中断状态，不会重置中断状态
void	join () 等待线程终止，阻塞调用它的线程。该线程结束后，继续执行调用它的线程
void	join (long millis) 等待线程终止的时间最长为 millis 毫秒。该方法会调用 Object.wait () 方法阻塞调用它的线程。线程结束或阻塞当前线程 millis 毫秒后，继续执行调用它的线程
void	join (long millis, int nanos) 等待线程终止的时间最长为 millis 毫秒+nanos 纳秒。该线程结束或阻塞当前线程 millis 毫秒+nanos 纳秒后，当前线程才能继续执行
void	run () 调用 start () 方法后，调度后执行该方法，分配 CPU 资源
void	setDaemon (boolean on) 将线程标记为守护线程或用户线程
void	setName (String name) 设置线程名称
void	setPriority (int newPriority) 设置线程的优先级
static void	sleep (long millis) 线程休眠 millis 毫秒，阻塞线程，不占用 CPU 资源
static void	sleep (long millis, int nanos) 线程休眠 millis 毫秒+nanos 纳秒，阻塞线程，不占用 CPU 资源
void	start () 开始执行线程，由 CPU 调度何时调用 run () 方法

续表

修饰符和类型	方　法
String	toString() 返回线程的字符串表示形式，包括线程名称、优先级和线程组
static void	yield() 当前线程从“运行状态”进入“就绪状态”，从而让其他等待线程获取执行权，但是不能保证在当前线程调用 yield() 方法后，其他线程一定能获得执行权，也可能是当前线程又回到“运行状态”继续运行

在 C0416 工程中，演示了自定义多线程类及启动多线程，输出结果如图 4-18 所示。

C0416 工程：MainActivity.kt

```
class MainActivity : AppCompatActivity() {
    override fun onCreate(savedInstanceState: Bundle?) {
        super.onCreate(savedInstanceState)
        setContentView(R.layout.activity_main)
        println("进入线程${Thread.currentThread().name}")
        val workerThread1 = WorkerThread1("a")
        val workerThread2 = WorkerThread2("b")
        workerThread1.start() //开始线程
        workerThread2.start() //开始线程
        Thread.sleep(100) //主线程休眠
        workerThread1.interrupt() //中断线程
        workerThread2.interrupt() //中断线程
        println("线程${Thread.currentThread().name}执行完毕")
    }
}
// 通过isInterrupted()结束的线程
class WorkerThread1(threadName:String) : Thread(threadName) {
    override fun run() {
        val threadName = currentThread().name
        var i = 0
        println("进入线程$threadName")
        while (!isInterrupted()) { //非中断状态时执行循环
            i ++
        }
        println("线程${threadName}生产了${i}个产品")
        println("线程${threadName}执行完毕")
    }
}
// 通过InterruptedException异常结束的线程
class WorkerThread2(threadName:String) : Thread(threadName) {
    override fun run() {
        val threadName = currentThread().name
        println("进入线程$threadName")
        var ii = 0
        while (true) {
            ii ++
            try {
                sleep(0,1)
            } catch (e:InterruptedException){
                e.printStackTrace() //Logcat窗口中打印异常的代码位置及原因
```

```
46                 break
47             }
48         }
49         println("线程${threadName}生产了${ii}个产品")
50         println("线程${threadName}执行完毕")
51     }
52 }
```

第 10 行 Thread.currentThread().name 用于获取当前线程的名称，默认情况下这个名称是系统自动分配的，也可以通过 setName() 方法自定义线程名称。第 11、12 行声明 WorkerThread1 类和 WorkerThread2 类的实例变量。第 13、14 行 workerThread1.start() 和 workerThread2.start() 开启各自的线程，执行 run() 方法。第 15 行主线程休眠 100 毫秒，为 workerThread1 线程和 workerThread2 线程提供足够的执行时间。第 16、17 行分别将 workerThread1 线程和 workerThread2 线程转为中断状态。第 22 ~ 33 行 WorkerThread1 类继承 Thread 类，重写 run() 方法，通过中断状态判断退出 while 循环，从而结束线程。第 35 ~ 52 行 WorkerThread2 类继承 Thread 类，重写 run() 方法，通过捕获 interrupt() 方法对 sleep() 方法触发的异常判断退出 while 循环，从而结束线程。

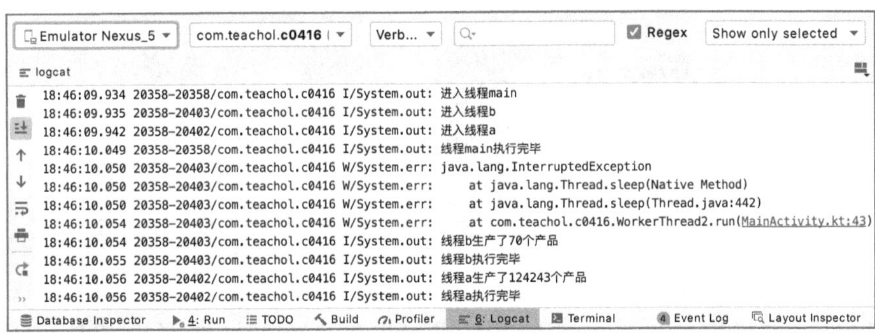

图 4-18　Logcat 窗口输出结果

在 Logcat 窗口输出结果中看到，线程 main(也就是主线程) 执行完成后，workerThread1 实例和 workerThread2 实例的线程才开始执行。线程的执行顺序是由 CPU 分配的，每次执行的顺序可能会有些差异。

4.9.2　Runnable 接口

由于类的单继承，有时无法再继承 Thread 类实现多线程，所以 Runnable 接口为实现多线程提供了另外一种方式。Runnable 接口只包含一个 run() 抽象方法，实现该方法后，将实现 Runnable 接口的类实例作为参数传递给 Thread 类的构造方法，然后通过 Thread 类的 start() 方法开启多线程。

在 C0417 工程中，演示了通过实现 Runnable 接口的类调用重写的 run() 抽象方法开启多线程，输出结果如图 4-19 所示。

C0417 工程：MainActivity.kt

```
class MainActivity : AppCompatActivity() {
    override fun onCreate(savedInstanceState: Bundle?) {
        super.onCreate(savedInstanceState)
        setContentView(R.layout.activity_main)
        println("进入线程" + Thread.currentThread().name)
        val worker1 = Worker("Thread-1", Thread.MIN_PRIORITY)
```

```
12         val worker2 = Worker("Thread-2", Thread.MAX_PRIORITY)
13         worker1.start()
14         worker2.start()
15         println("线程" + Thread.currentThread().name + "执行完毕")
16     }
17 }
18 class Worker(val threadName: String, val priority: Int) : Runnable {
19     var thread: Thread? = null
20     init {
21         println("新建线程 $threadName")
22     }
23     override fun run() {
24         val threadName = Thread.currentThread().name
25         println("进入线程$threadName")
26         for (i in 1..3) {
27             println(threadName + "生产第" + i + "个产品")
28             Thread.sleep(0, 1)
29         }
30         println("线程" + threadName + "执行完毕")
31     }
32     fun start() {
33         println("启动 $threadName")
34         if (thread == null) {
35             thread = Thread(this, threadName)
36             thread!!.start()
37             thread!!.priority = priority
38         }
39     }
40 }
```

第11、12行声明两个Worker类实例，Thread.MIN_PRIORITY和Thread.MAX_PRIORITY是Thread类的常量，分别表示最低优先级和最高优先级，也可以通过常量表示的 Int 型数值直接指定优先级。第 13、14 行使用 Worker 类的 start()方法分别开启 worker1 和 worker2 的线程。第 18 行定义 Worker 类，主构造方法定义了 threadName 属性和 priority 属性，并需要实现 Runnable 接口。第 23 ~ 31 行实现 Runnable 接口的 run()方法，由于 thread 属性是可空类型，所以第 36、37 行调用 start()方法和 priority 属性时，需要在 thread 属性后添加!!运算符。

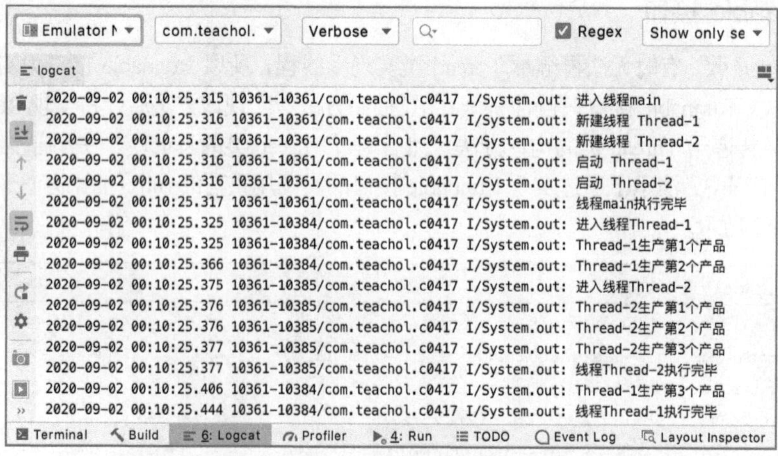

图 4-19 Logcat 窗口输出结果

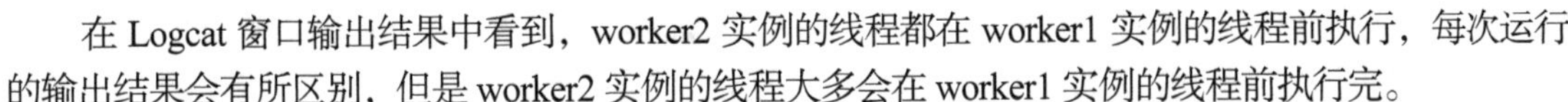

在 Logcat 窗口输出结果中看到，worker2 实例的线程都在 worker1 实例的线程前执行，每次运行的输出结果会有所区别，但是 worker2 实例的线程大多会在 worker1 实例的线程前执行完。

4.9.3 Callable 接口

Callable 接口使用泛型作为 call()方法返回值的类型，无须像 Runnable 接口那样单独创建一个类来实现接口，而是在当前类上实现接口；也可以像 Runnable 接口一样创建一个实现 Callable 接口的类，使用 FutureTask 类实现多线程。

使用 Callable 接口实现多线程时，首先创建一个类或使用当前类实现 Callable 接口，并实现 call()方法，该方法将作为线程的执行体且具有返回值。然后创建 FutureTask 类，将实现 Callable 接口的实例作为构造方法的参数，call()方法的返回值会传递给 FutureTask 类实例，使用 FutureTask 类的 get()方法可以获取到 call()方法的返回值。最后将 FutureTask 类的实例作为 Thread 类的构造方法参数，通过 Thread 类实现多线程。

在 C0418 工程中，演示了通过 Callable 接口重写 call()方法实现多线程，输出结果如图 4-20 所示。

C0418 工程：MainActivity.kt

```
05  import java.util.concurrent.Callable
06  import java.util.concurrent.FutureTask
07
08  class MainActivity : AppCompatActivity(), Callable<Int> {
09      override fun onCreate(savedInstanceState: Bundle?) {
10          super.onCreate(savedInstanceState)
11          setContentView(R.layout.activity_main)
12          println("进入线程" + Thread.currentThread().name)
13          val ft1 = FutureTask(this)
14          val ft2 = FutureTask(this)
15          Thread(ft1, "线程 1").start()
16          Thread(ft2, "线程 2").start()
17          try {
18              println("线程 1 已经生产了" + ft1.get() + "个产品")
19              println("线程 2 已经生产了" + ft2.get() + "个产品")
20          } catch (e: Exception) {
21              e.printStackTrace()
22          }
23          println("线程" + Thread.currentThread().name + "执行完毕")
24      }
25      override fun call(): Int? {
26          val threadName = Thread.currentThread().name
27          var i = 0
28          while (i < 4) {
29              i++
30              println(threadName + "生产第" + i + "个产品")
31              Thread.sleep(0, 1)
32          }
33          return i
34      }
35  }
```

第 05、06 行分别导入 Callable 和 FutureTask 命名空间。第 08 行 MainActivity 类声明 Callable 接

口，指定返回值的类型为 Integer。第 13、14 行声明两个 FutureTask 类实例的变量，并通过 this 关键字将 MainActivity 类实例传递给 FutureTask 类的构造方法。第 15、16 行分别将 ft1 变量和 ft2 变量作为 Thread 类的构造方法参数，同时调用 start()方法启动线程。第 18、19 行 ft1.get()和 ft2.get()用于获取 call()方法的返回值，call()方法返回值的类型必须与 Callable 接口定义的返回值类型一致。第 25 ~ 34 行重写实现 Callable 接口的 call()方法。

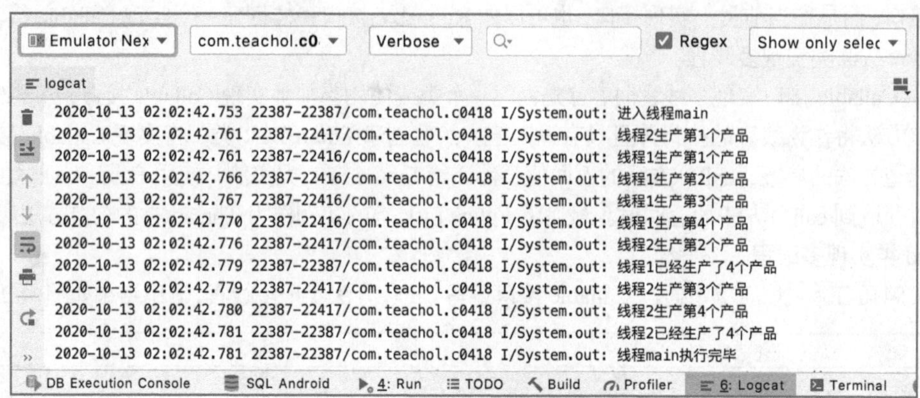

图 4-20 Logcat 窗口输出结果

在 Logcat 窗口输出结果中发现，通过 FutureTask 类实现的 ft1 和 ft2 开启的线程都执行完成后，才会执行 MainActivity 中的后续代码，与 Thread 类和 Runnable 接口的实例程序有很大区别。

4.9.4 Synchronized 注解和 synchronized 代码块

在多线程中，可能会出现多个线程同时访问同一个资源(如变量、对象、文件、数据库的表等)的情况，此时可能导致逻辑错误或异常，这就涉及线程安全问题。因此，线程中需要添加同步锁锁定该资源，对资源进行操作后再释放该资源，然后允许其他线程操作该资源。

Synchronized 注解和 synchronized 代码块都提供了同步机制，执行时添加同步锁，执行完成后解锁。因此，线程只能依次调用执行，当一个线程调用时，其他线程的调用进入等待状态，直到解锁后其他线程的调用才能执行。synchronized 注解用于方法，synchronized 代码块对需要同步的部分代码进行同步，因此 synchronized 代码块比 synchronized 注解更加灵活。

在 C0419 工程中，演示了多线程中 synchronized 注解和 synchronized 代码块的同步作用，输出结果如图 4-21 所示。

C0419 工程：MainActivity.kt

```
class MainActivity : AppCompatActivity() {
    override fun onCreate(savedInstanceState: Bundle?) {
        super.onCreate(savedInstanceState)
        setContentView(R.layout.activity_main)
        println("--线程调用普通方法--")
        val normalLog = NormalLog()
        object : Thread() {
            override fun run() {
                normalLog.insert("线程 1")
            }
        }.start()
```

```
        object : Thread() {
            override fun run() {
                normalLog.insert("线程 2")
            }
        }.start()
        Thread.sleep(200)
        println("--线程调用同步方法--")
        val syncMethodLog = SyncMethodLog()
        object : Thread() {
            override fun run() {
                syncMethodLog.insert("线程 3")
            }
        }.start()
        object : Thread() {
            override fun run() {
                syncMethodLog.insert("线程 4")
            }
        }.start()
        Thread.sleep(200)
        println("--线程调用方法同步代码块--")
        val syncBlockLog = SyncBlockLog()
        object : Thread() {
            override fun run() {
                syncBlockLog.insert("线程 5")
            }
        }.start()
        object : Thread() {
            override fun run() {
                syncBlockLog.insert("线程 6")
            }
        }.start()
    }
    //未同步
    internal class NormalLog {
        fun insert(thread: String) {
            println("$thread: 打开日志文件")
            println("$thread: 写入日志")
            Thread.sleep(50)
            println("$thread: 保存并关闭日志文件")
        }
    }
    //synchronized修饰符
    internal class SyncMethodLog {
        @Synchronized
        fun insert(thread: String) {
            println("$thread: 打开日志文件")
            println("$thread: 写入日志")
            Thread.sleep(50)
            println("$thread: 保存并关闭日志文件")
        }
```

```
67          }
68          //synchronized 代码块
69          internal class SyncBlockLog {
70              fun insert(thread: String) {
71                  synchronized(this) {
72                      println("$thread: 打开日志文件")
73                      println("$thread: 写入日志")
74                      Thread.sleep(50)
75                      println("$thread: 保存并关闭日志文件")
76                  }
77              }
78          }
79      }
```

第 12 ~ 16 行使用 object 关键字声明一个继承 Thread 类的匿名类，并重写 run()方法，通过 start()方法启动线程。第 50 ~ 57 行 NormalLog 类的 insert()方法没有使用同步，直接对日志文件进行操作。第 59 ~ 67 行 SyncMethodLog 类的 insert()方法使用 synchronized 注释定义为同步方法，一个线程调用该方法时会将该方法锁定，当该方法执行完成时解锁，再执行其他线程调用该方法。第 69 ~ 78 行 SyncBlockLog 类在方法中使用 synchronized 代码块为代码块内的代码设置同步锁。

提示：匿名类

匿名类是没有类名的类，不能被引用，使用 object 关键字代替匿名类。匿名类主要用于简化代码，可以直接获取它的实例并调用它的方法。

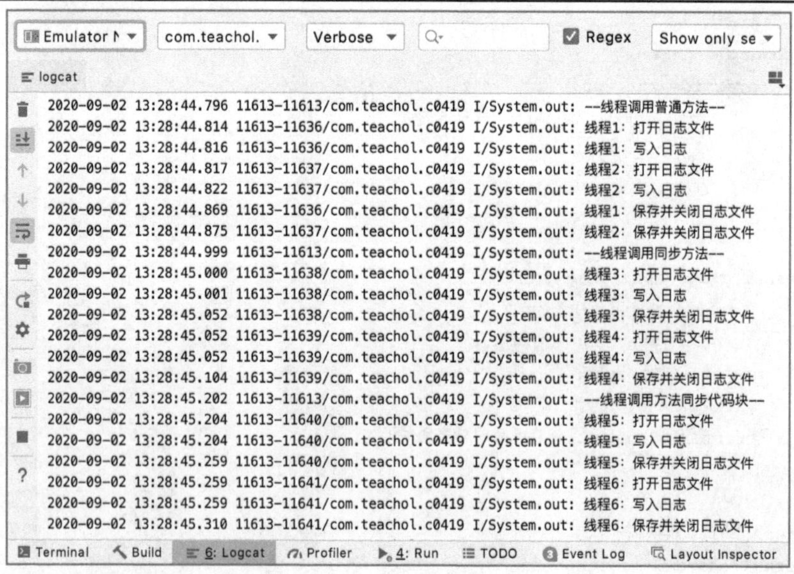

图 4-21　Logcat 窗口输出结果

在 Logcat 窗口中可以观察到，每个类的实例分别使用两个线程调用，每种类型的调用休眠 200 毫秒。线程调用普通方法会出现两个线程同时打开日志文件，只有一个线程能够将日志信息正确保存，另一个保存后被马上覆盖。这样无法保证数据的正确保存，会造成逻辑错误。调用同步方法和同步代码块的方法，会有同步锁，不同线程调用时依次执行。

4.9.5　volatile 注解

volatile 注解只能用于属性，提供一种轻量级的同步机制。volatile 注解的属性在每次被线程访问时，从共享内存中重读该属性的值，而不是从线程的本地内存中读取。当成员变量发生变化时，线程将变化值回写到共享内存，所以在任何时刻两个不同的线程总是看到某个属性的同一个值。

在 C0420 工程中，演示了模拟产品的生产过程，其中 Work 类的 count 属性使用@Volatile 进行注解，以避免多线程调用该属性时数据不同步，输出结果如图 4-22 所示。

C0420 工程：MainActivity.kt

```
06 class MainActivity : AppCompatActivity() {
07     override fun onCreate(savedInstanceState: Bundle?) {
08         super.onCreate(savedInstanceState)
09         setContentView(R.layout.activity_main)
10         val work = Work()
11         val worker1 = Thread(work, "worker1")
12         val worker2 = Thread(work, "worker2")
13         worker1.start()
14         worker2.start()
15     }
16 }
17 
18 class Work : Runnable {
19     @Volatile
20     var count = 1
21     override fun run() {
22         while (count < 10) {
23             println(Thread.currentThread().name + "生产第" + count + "件产品")
24             count++
25             try {
26                 Thread.sleep(1000)
27             } catch (e: InterruptedException) {
28                 e.printStackTrace()
29             }
30         }
31     }
32 }
```

第 10～14 行创建并开启两个线程，执行 Work 类的 run()方法。第 19、20 行使用@Volatile 注解 count 属性，将该属性声明为同步属性。

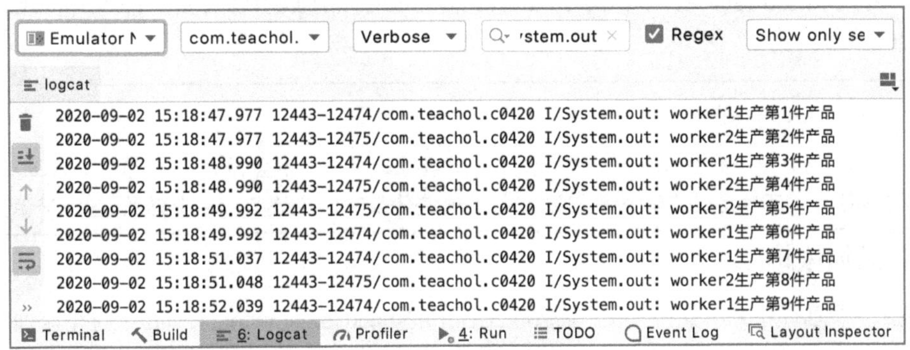

图 4-22　Logcat 窗口输出结果

在 Logcat 窗口中可以观察到，2 个线程依次生产产品。count 属性值改变后，其他线程会立即更新 count 属性值。

提示：Logcat 窗口的搜索历史功能

Logcat 窗口不仅会显示 println()语句输出的 System.out 信息，还会显示一些运行过程中其他信息。为了便于观察运行结果，可以在 Search History 文本框（带放大镜图标的输入框）中输入“System.out”，仅查看 println()语句输出的信息。

去掉@Volatile 注解后，再运行该工程。在 Logcat 窗口中可以观察到有重复的 count 属性值(每次的运行结果都可能不相同)，如图 4-23 所示。这种情况就是从本地内存读取数据造成的，而本地内存和共享内存的同步存在时间差。

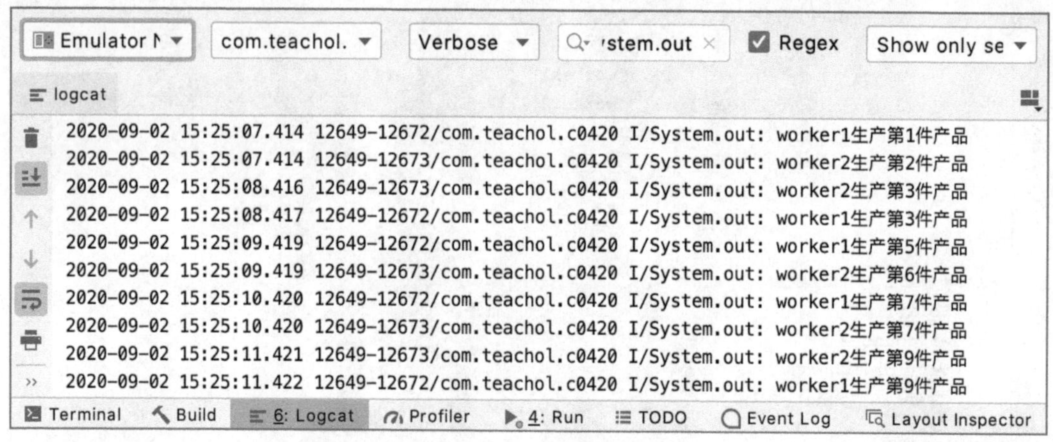

图 4-23　Logcat 窗口输出结果

4.10　协　　程

协程(Coroutine)是一种轻量级的异步解决方案。协程不是进程也不是线程，而是一个特殊的函数，这个函数可以在某个地方挂起，并且可以重新在挂起处外继续运行。协程依赖线程，一个线程中可以创建多个协程，但是协程挂起时不需要阻塞线程，消耗内存更少，几乎不会引起 OOM 异常。

协程需要主动释放使用权来切换到其他协程，同一时间其实只有一个协程拥有运行权，相当于单线程的能力。协程挂起能够将异步回调方式的代码简化成看似同步的代码，此外许多 Jetpack 库都包含支持协程的扩展。

提示：挂起和阻塞

挂起(suspend)是一种主动行为，因此恢复也需要主动完成，不会释放 CPU 资源。而阻塞(pend)是一种被动行为，不确定什么时候被阻塞，也不确定什么时候会恢复，但会释放 CPU 资源。

4.10.1　添加依赖库

Kotlin 的默认库中并不支持协程，需要在工程中的 build.gradle(Module.app)文件中添加依赖库，然后出现 Gradle 的同步提示，单击“Sync Now”图标下载并同步依赖库，如图 4-24 所示。

```
activity_main.xml ×   MainActivity.kt ×   build.gradle (:app) ×
Gradle files have changed since last project sync. A project sync may b...  Sync Now
26
27  dependencies {
28      implementation fileTree(dir: "libs", include: ["*.jar"])
29      implementation "org.jetbrains.kotlin:kotlin-stdlib:$kotlin_version"
30      implementation 'androidx.core:core-ktx:1.3.1'
31      implementation 'androidx.appcompat:appcompat:1.2.0'
32      implementation 'androidx.constraintlayout:constraintlayout:1.1.3'
33      implementation "org.jetbrains.kotlinx:kotlinx-coroutines-core:1.3.1"
34      testImplementation 'junit:junit:4.12'
```

图 4-24　添加依赖库

第 33 行 implementation "org.jetbrains.kotlinx:kotlinx-coroutines-core:1.3.1"用于添加协程的依赖库，然后同步 Gradle 才能使用协程。

4.10.2　协程作用域

协程需要在作用域内运行，协程的作用域使用了 CoroutineScope 接口。CoroutineScope 接口中 coroutineContext 属性是协程运行方式的规则和配置，用于实现作为作用域扩展的协同例程生成器。GlobalScope 是 CoroutineScope 的一个单例实现，是一个全局的作用域，适合创建与 App 生命周期一样的协程，不应和任何可被销毁的组件绑定使用。

CoroutineScope 接口的定义：

```
public interface CoroutineScope {
    public val coroutineContext: CoroutineContext
}
public object GlobalScope : CoroutineScope {
    override val coroutineContext: CoroutineContext
        get() = EmptyCoroutineContext
}
```

在 C0421 工程中，演示了调用 GlobalScope.launch()方法开启全局域协程，输出结果如图 4-25 所示。

C0421 工程：MainActivity.kt

```
import kotlinx.coroutines.*

class MainActivity : AppCompatActivity() {
    override fun onCreate(savedInstanceState: Bundle?) {
        super.onCreate(savedInstanceState)
        setContentView(R.layout.activity_main)
        println("主线程开始 线程 id: ${mainLooper.thread.id}")
        runCoroutine()
        println("主线程结束 线程 id: ${mainLooper.thread.id}")
    }
    //运行协程
    private fun runCoroutine() {
        GlobalScope.launch {
            println("全局域协程开始")
```

```
19            repeat(3) {
20                println("全局域协程执行repeat语句${it}次 线程id: ${Thread.currentThread().id}")
21                delay(100) //休眠100毫秒
22            }
23            println("全局域协程结束")
24        }
25    }
26 }
```

第 17 ~ 24 行使用 GlobalScope.launch() 方法启动一个全局域协程。第 20 行 Thread.currentThread().id 用于获取协程所在的线程 id 值。第 21 行 delay() 函数用于挂起协程，此时协程是等待状态，但是并不会影响线程，所以与 Thread 类的 sleep() 方法阻塞线程是不一样的。

在 Logcat 窗口中可以观察到，全局域协程没有在主线程上运行，也不会阻塞主线程，如图 4-25 所示。在默认情况下，全局域协程会随机分配主线程以外的线程运行全局域协程。

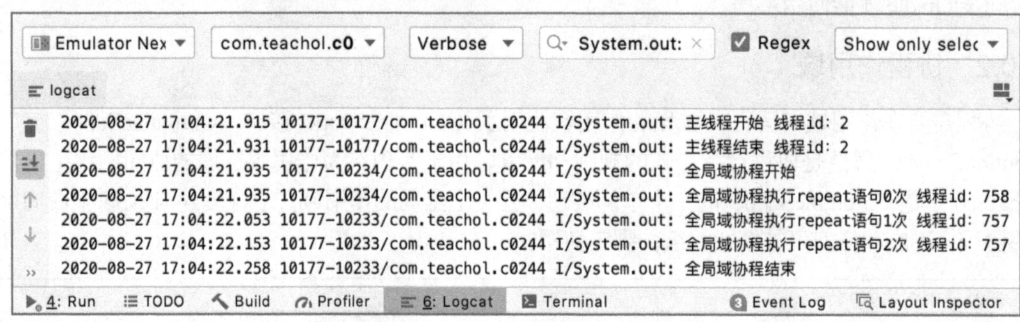

图 4-25　Logcat 窗口输出结果

4.10.3　启动协程

启动协程常用 CoroutineScope 接口的扩展方法和 runBlocking() 函数，这些方法中封装了启动协程的代码。

1. CoroutineScope 接口

CoroutineScope 接口定义了一些扩展方法用于启动协程，其中的 launch() 方法和 async() 方法是最常用的方法。launch() 方法启动的协程返回值类型为 Job 接口，async() 方法启动的协程返回值类型为 Deferred 接口。Deferred 接口继承了 Job 接口。Deferred 接口的 await() 方法是一个可中断方法，当需要获取 async() 方法的结果时，调用该方法等待结果。调用 await() 方法后中断当前协程，直至其返回结果。

```
CoroutineScope 接口的 launch()方法和 async()方法的定义:
public fun CoroutineScope.launch(
    context: CoroutineContext = EmptyCoroutineContext,
    start: CoroutineStart = CoroutineStart.DEFAULT,
    block: suspend CoroutineScope.() -> Unit
): Job {
    val newContext = newCoroutineContext(context)
    val coroutine = if (start.isLazy)
        LazyStandaloneCoroutine(newContext, block) else
        StandaloneCoroutine(newContext, active = true)
```

```
    coroutine.start(start, coroutine, block)
    return coroutine
}
public fun <T> CoroutineScope.async(
    context: CoroutineContext = EmptyCoroutineContext,
    start: CoroutineStart = CoroutineStart.DEFAULT,
    block: suspend CoroutineScope.() -> T
): Deferred<T> {
    val newContext = newCoroutineContext(context)
    val coroutine = if (start.isLazy)
        LazyDeferredCoroutine(newContext, block) else
        DeferredCoroutine<T>(newContext, active = true)
    coroutine.start(start, coroutine, block)
    return coroutine
}
```

通过 launch()方法和 async()方法的源代码可以看到参数都有默认值，因此调用方法时参数可以省略。但还是有必要了解这两个方法的三个参数，如下所述。

- context 参数：用于设置使用哪个线程运行协程，属性值包括 Dispatchers.Main（主线程运行）、Dispatchers.IO（主线程之外线程池运行磁盘或网络 I/O 操作）、Dispatchers.Default（主线程之外的线程池运行 CPU 使用密集型的工作）和 Dispatchers.Unconfined（不指定运行的线程），Dispatchers 实现了 CoroutineContext 接口。
- start 参数：用于设置启动模式，CoroutineStart 类型的属性值包括 CoroutineStart.DEFAULT（立即执行协程体）、CoroutineStart.LAZY（需要的情况下执行协程体）、CoroutineStart.ATOMIC（立即执行协程体，但在开始运行前无法取消）、CoroutineStart. UNDISPATCHED（立即在当前线程执行协程体，直至第一个 suspend 调用）。
- block 参数：用于设置带接收者的协程体，接收者是 CoroutineScope，suspend 关键字修饰的函数称为挂起函数，挂起函数会将整个协程挂起，协程中的挂起函数按顺序依次执行。

使用 launch()方法创建的协程都会返回 Job 实例，该实例唯一标识协程并管理其生命周期，Job 接口的常用方法如表 4-13 所示。Job 实例拥有三种状态：isActive（运行状态）、isCompleted（完成状态）、isCancelled（取消状态）。使用 async()方法创建的协程都会返回 Deferred 实例，Deferred 接口继承自 Job 接口。协程完成后，Job 实例没有返回值，Deferred 实例有返回值。

表 4-13　Job 接口的常用方法

类型和修饰符	方　法
Boolean	start() 当启动模式设置为 CoroutineStart.LAZY 时，用于启动协程，返回 true。如果协程已经启动，则返回 false
suspend	join() 挂起当前正在运行的协程，等待该 Job 执行完成
	cancel(cause: CancellationException? = null) 取消该 Job 实例运行
ChildHandle	attachChild(child: ChildJob) 附加一个子协程到当前协程上
	cancelChildren(cause: CancellationException? = null) 取消所有子 Job 实例运行
DisposableHandle	invokeOnCompletion(onCancelling: Boolean = false, invokeImmediately: Boolean = true, handler: CompletionHandler) 用于监听其完成或者其取消状态。onCancelling 参数用于判断是否监听取消事件，否则监听完成事件；invokeImmediately 参数用于监听已经完成协程时是否回调，当值为 false 时，如果在添加监听事件时协程已经完成或取消，将不会回调

在 C0422 工程中，演示了在 CoroutineScope.launch()方法开启协程内使用 launch()方法开启子协程，输出结果如图 4-26 所示。

C0422 工程：MainActivity.kt

```
import kotlinx.coroutines.*

class MainActivity : AppCompatActivity() {
    override fun onCreate(savedInstanceState: Bundle?) {
        super.onCreate(savedInstanceState)
        setContentView(R.layout.activity_main)
        println("主线程开始 线程 id: ${mainLooper.thread.id}")
        runCoroutine()
        println("主线程结束 线程 id: ${mainLooper.thread.id}")
    }
    //运行协程
    fun runCoroutine() {
        val job1 = Job()
        //launch()默认参数
        CoroutineScope(job1).launch() {
            println("job1 协程开始 线程 id: ${Thread.currentThread().id}")
            //启动子协程
            launch {
                repeat(30) {
                    println("job1 子协程执行 repeat 语句${it}次 线程 id: ${Thread.currentThread().id}")
                    delay(100)
                }
            }
            println("job1 协程结束 线程 id: ${Thread.currentThread().id}")
        }
        //launch()设置参数
        val scope = CoroutineScope(Job())
        val job2 = scope.launch(start=CoroutineStart.LAZY) {
            println("job2 协程开始 线程 id: ${Thread.currentThread().id}")
            //启动子协程
            var r = async() {
                repeat(3) {
                    println("job2 子协程执行 repeat 语句${it}次 线程 id: ${Thread.currentThread().id}")
                    delay(100)
                    if (it == 1) {
                        job1.join()
                    }
                }
            }
            println("job2 协程结束 线程 id: ${Thread.currentThread().id}")
        }
        job2.start()
        Thread.sleep(500)
        job1.cancel()
    }
}
```

第 19 ~ 28 行通过 CoroutineScope.launch() 方法开启默认协程，并在协程内使用 launch() 方法开启子协程。第 31 ~ 45 行间接使用 CoroutineScope.launch() 方法开启延迟协程，并在协程内使用 async() 方法开启子协程。第 38 ~ 40 行当 it==1 时将 job1 协程加入当前协程的进程中，并将当前协程挂起。第 45 行启动设置为延迟的 job2 协程。第 46 行当前线程阻塞 500 毫秒。第 47 行取消 job1 协程的执行。

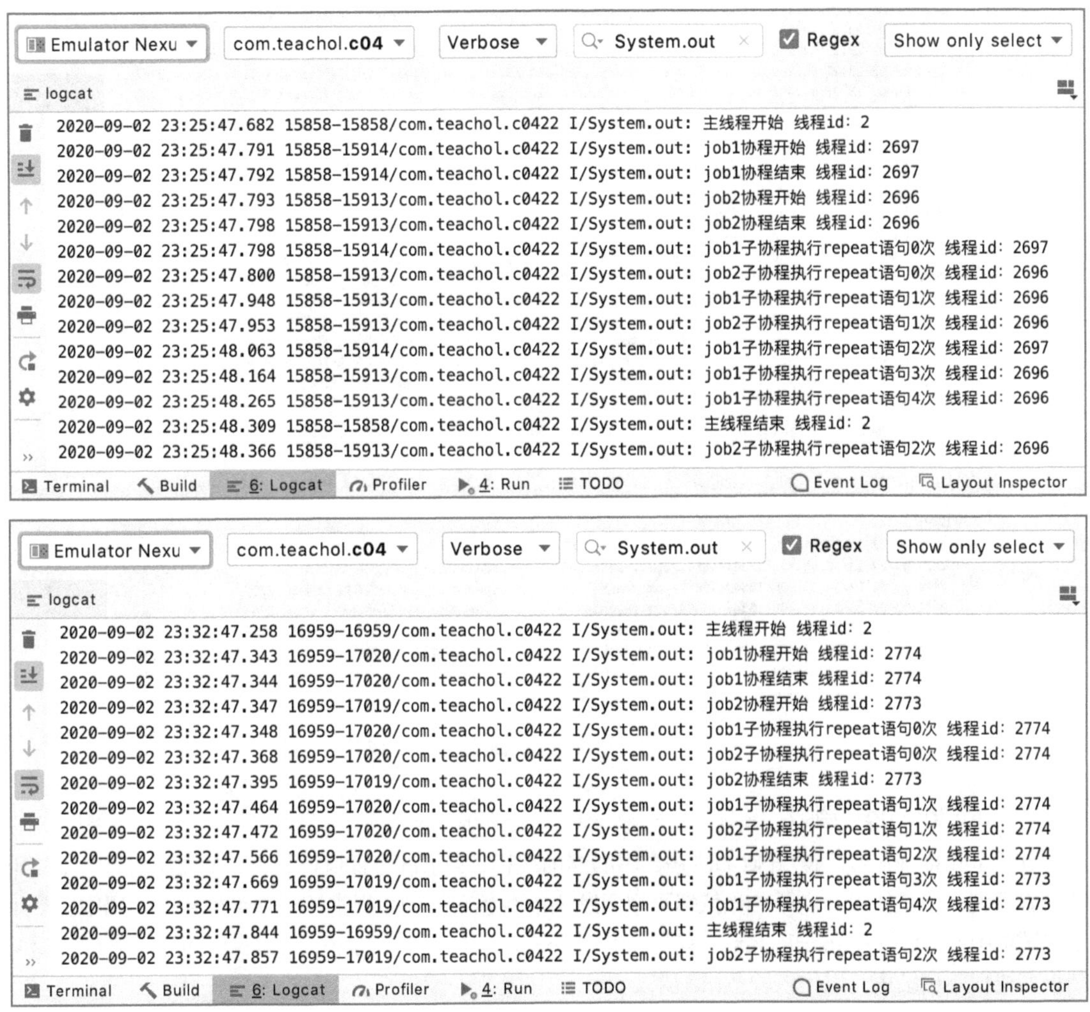

图 4-26　两次运行的 Logcat 窗口输出结果

多协程和多线程都是由底层调度分配资源的，因此每次执行的结果可能会不同。在 Logcat 窗口中可以观察到，job1 协程和 job2 协程及它们的子协程没有在主线程上运行，因此也不会阻塞主线程。但"主线程结束 线程 id:2"在倒数第二条输出，这是因为第 46 行的代码阻塞了主线程 500 毫秒，并不会阻塞其他线程，所以在其他线程上执行的协程不会受到影响。如果将第 46 行代码注释或删除，取消对主线程的阻塞，然后运行该工程三次，可以观察到有主线程和协程交替执行的情况、有主线程先执行完成再执行协程的情况，以及主线程先执行完成后 job1 协程还未开始就被取消的情况，输出结果如图 4-27 所示。

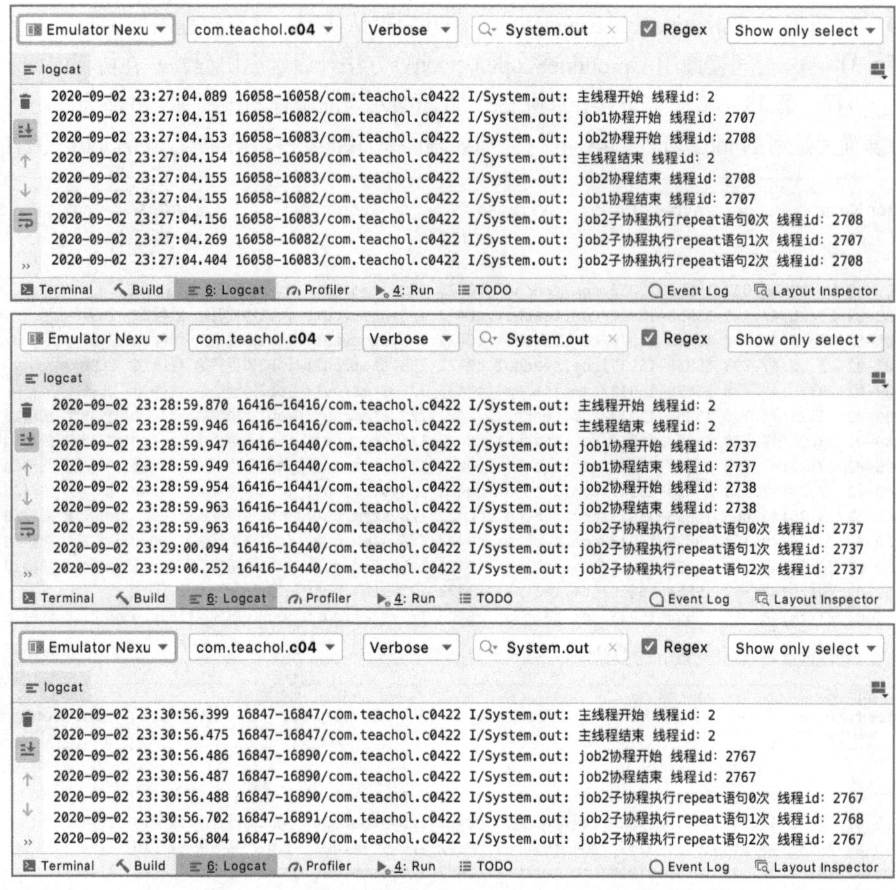

图 4-27　三次运行的 Logcat 窗口输出结果

2. runBlocking()函数

使用 runBlocking()函数开启的协程会阻断当前线程，直至该协程执行结束。单独使用时与不使用并没有区别，但是如果开启子协程依然可以阻断当前线程。该函数很少用于最终构建的代码中，多用于测试可中断方法。在 runBlocking()函数开启的协程内运行可中断方法，可以保证可中断方法的返回结果前线程不会结束，以便校验测试结果。

```
runBlocking()函数的定义:
public fun <T> runBlocking(context: CoroutineContext = EmptyCoroutineContext, block: suspend
CoroutineScope.() -> T): T {
    val currentThread = Thread.currentThread()
    val contextInterceptor = context[ContinuationInterceptor]
    val eventLoop: EventLoop?
    val newContext: CoroutineContext
    if (contextInterceptor == null) {
        eventLoop = ThreadLocalEventLoop.eventLoop
        newContext = GlobalScope.newCoroutineContext(context + eventLoop)
    } else {
        eventLoop = (contextInterceptor as? EventLoop)?.takeIf { it.shouldBeProcessedFromContext() }
            ?: ThreadLocalEventLoop.currentOrNull()
```

```
        newContext = GlobalScope.newCoroutineContext(context)
    }
    val coroutine = BlockingCoroutine<T>(newContext, currentThread, eventLoop)
    coroutine.start(CoroutineStart.DEFAULT, coroutine, block)
    return coroutine.joinBlocking()
}
```

在 C0423 工程中，演示了在主线程上使用 runBlocking() 函数开启协程及其子协程，输出结果如图 4-28 所示。

```
C0423 工程：MainActivity.kt
05  import kotlinx.coroutines.*
06
07  class MainActivity : AppCompatActivity() {
08      override fun onCreate(savedInstanceState: Bundle?) {
09          super.onCreate(savedInstanceState)
10          setContentView(R.layout.activity_main)
11          println("主线程开始 线程 id: ${mainLooper.thread.id}")
12          run()
13          println("主线程结束线程 id: ${mainLooper.thread.id}")
14      }
15
16      private fun run() {
17          runBlocking {
18              launch {
19                  repeat(3) {
20                      println("协程执行 repeat 语句${it}次 线程 id: ${Thread.currentThread().id}")
21                      delay(100)
22                  }
23              }
24              launch(context = Dispatchers.Default) {
25                  repeat(3) {
26                      println("协程执行 repeat 语句${it}次 线程 id: ${Thread.currentThread().id}")
27                      delay(100)
28                  }
29              }
30          }
31      }
32  }
```

第 16 ~ 30 行使用 runBlocking() 函数启动协程，阻断当前线程，并启动两个子协程。第 18 ~ 23 行子协程使用默认线程执行。第 24 ~ 29 行子协程使用主线程之外的线程池运行。

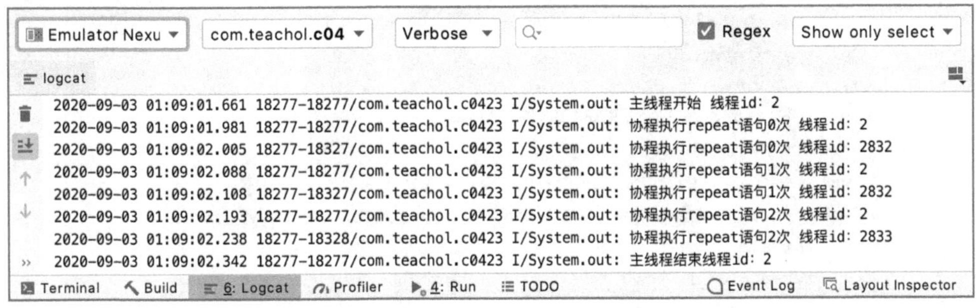

图 4-28　Logcat 窗口输出结果

4.10.4 挂起协程

挂起协程是指暂停当前协程，即暂时不在线程上运行该协程。使用 suspend 关键字修饰的函数称为挂起函数，该函数只能被协程调用，但是并不会直接挂起协程，而是通过能够挂起协程的函数挂起协程。挂起协程常用的方式包括 delay() 方法、await() 方法和 withContext() 函数。

delay() 方法用于设置当前协程挂起的时间，可以在协程内直接使用。如果需要在协程调用的方法或函数内调用该方法，则需要使用 suspend 关键字进行修饰。

await() 方法是 Deferred 接口的方法，会挂起协程直到得到返回值为止。

withContext() 函数是挂起函数，该函数不会创建新的协程，可以在指定的 CoroutineContext 上运行该函数，并挂起调用该函数的协程直至该函数运行完成。

提示：挂起和阻塞

协程可以被挂起而无须阻塞线程，协程挂起几乎是无代价的。而线程阻塞会浪费 CPU 资源，代价较高，特别是在高负载时，阻塞导致一些重要的任务被延迟。协程挂起不需要上下文切换或者操作系统的任何其他干预，可以由用户来控制。

在 C0424 工程中，演示了使用 withContext() 函数和使用 suspend 关键字自定义的挂起方法，输出结果如图 4-29 所示。

C0424 工程：MainActivity.kt

```
import kotlinx.coroutines.*

class MainActivity : AppCompatActivity() {
    override fun onCreate(savedInstanceState: Bundle?) {
        super.onCreate(savedInstanceState)
        setContentView(R.layout.activity_main)
        println("主线程开始 线程 id: ${Thread.currentThread().id}")
        runCoroutine()
        println("主线程结束 线程 id: ${Thread.currentThread().id}")
    }
    //运行协程
    private fun runCoroutine() {
        //启动协程
        CoroutineScope(Job()).launch {
            println("协程开始 线程 id: ${Thread.currentThread().id}")
            //启动子协程调用挂起方法
            val result1 = async {
                println("async()线程 id: ${Thread.currentThread().id}")
                asyncCount(10)
            }
            println("asyncCount()累加运算结果: ${result1.await()}")
            //withContext()函数
            val result2 = withContext(Dispatchers.IO) {
                println("withContext()线程 id: ${Thread.currentThread().id}")
                var sum = 0
                repeat(10) {
                    delay(10)
                    sum += it
```

```
33                }
34                sum
35            }
36            println("withContext()累加运算结果: ${result2}")
37            println("协程结束 线程 id: ${Thread.currentThread().id}")
38        }
39    }
40    //挂起方法
41    private suspend fun asyncCount(num: Int): Int {
42        println("asyncCount()线程 id: ${Thread.currentThread().id}")
43        var sum = 0
44        repeat(num) {
45            delay(10)
46            sum += it
47        }
48        return sum
49    }
50 }
```

第 21 ~ 24 行使用 async() 方法启动子协程运行 asyncCount() 挂起方法，将返回值赋给 result1 变量。第 25 行 result1 变量使用 await() 方法挂起当前协程直至接收到返回值。第 27 ~ 35 行使用 withContext() 函数执行与 asyncCount() 挂起方法内相同的代码，该函数会挂起调用该函数的线程，直至将该函数的返回值赋给 result2 变量。第 41 ~ 49 行声明 asyncCount() 挂起方法。

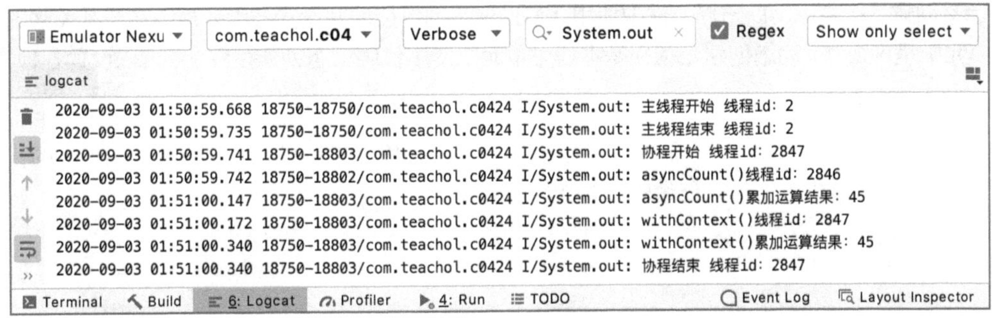

图 4-29　Logcat 窗口输出结果

第 5 章 Android 的基础控件

用户最直接接触的部分就是 App 的界面，界面的核心组成部分是控件。Android 提供了丰富的控件，不但能呈现丰富多彩的内容，还能进行人机交互。基础控件可以单独使用，无须依靠其他控件或适配器就可以实现核心功能。

5.1 控件基础

5.1.1 控件的创建方式

控件(Control)是具有用户界面(User Interface)功能的可视化组件。View 子类的控件，可以使用两种方式创建：XML 标签创建和动态创建。非 View 子类的控件只能通过动态创建。

提示：组件

组件(Component)是可重复使用且可以和其他对象进行交互的对象，控件是组件的子集。

标签创建是指在 XML 布局文件中使用 XML 标签创建控件并设置各种属性，在与其关联的基本程序单元中呈现出来；动态创建是指在基本程序单元中使用代码动态创建控件，并可以设置各种属性。

1. 标签创建控件

新建“Empty Activity”工程后，打开“activity_main.xml”文件，可以直接查看布局文件的代码视图(如图 5-1 所示)，能够通过代码编辑布局文件。

activity_main.xml × MainActivity.kt ×

Code Split Design

```
1  <?xml version="1.0" encoding="utf-8"?>
2  <androidx.constraintlayout.widget.ConstraintLayout xmlns:android="http://schemas.android.com/apk/res/android"
3      xmlns:app="http://schemas.android.com/apk/res-auto"
4      xmlns:tools="http://schemas.android.com/tools"
5      android:layout_width="match_parent"
6      android:layout_height="match_parent"
7      tools:context=".MainActivity">
8
9      <TextView
10         android:layout_width="wrap_content"
11         android:layout_height="wrap_content"
12         android:text="Hello World!"
13         app:layout_constraintBottom_toBottomOf="parent"
14         app:layout_constraintLeft_toLeftOf="parent"
15         app:layout_constraintRight_toRightOf="parent"
16         app:layout_constraintTop_toTopOf="parent" />
17
18 </androidx.constraintlayout.widget.ConstraintLayout>
```

androidx.constraintlayout.widget.ConstraintLayout › TextView

图 5-1 XML 格式布局文件的代码视图

标签由一个小于号(<)、一个大于号(>)及它们之间的文本组成。组件的标签由起始标签和结束标签组成，符号通常是成对出现的，结束标签的文本前需要添加一个“/”字符。如果不包含子标签，可以使用简写形式；如果不包含结束标签，则在起始标签的文本后添加一个“/”字符。第 2 ~ 18 行

<android.constraintlayout.widget.ConstrainLayout>标签由于包含<TextView>子标签，所以标签符号要成对出现。而第 9～16 行<TextView>标签没有包含子标签，可以使用简写形式。

提示：子标签和父标签

子标签和父标签的关系不同于子类和父类的关系。父标签是作为容器使用的，与子标签之间不是继承关系，而是包含关系。

单击“Design”按钮可以切换到设计视图。将左侧“Palette”中的控件拖曳到中间的预览视图中，选择控件后可以在右侧“Attributes”窗口中设置属性(如图 5-2 所示)。

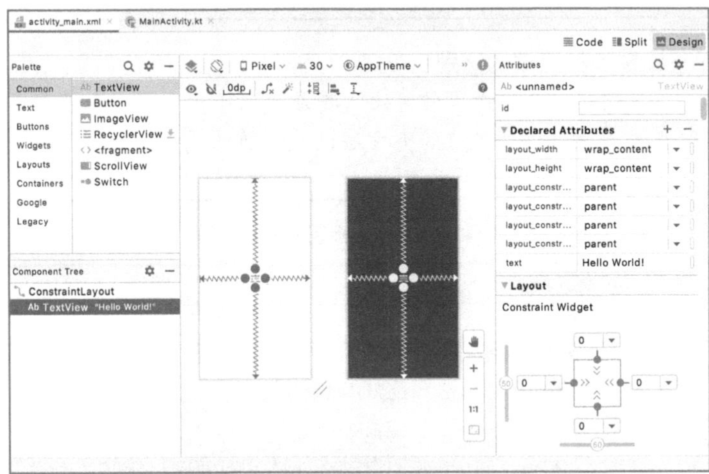

图 5-2　XML 格式布局文件的预览视图

2. 动态创建控件

新建“Empty Activity”工程后，打开“MainActivity.kt”文件，使用 Kotlin 代码创建一个 TextView 控件的实例，然后使用 setContentView()方法将其添加到视图中并显示出来(如图 5-3 所示)。

activity_main.xml　MainActivity.kt

```
package com.teachol.myapplication

import android.os.Bundle
import android.widget.TextView
import androidx.appcompat.app.AppCompatActivity

class MainActivity : AppCompatActivity() {
    override fun onCreate(savedInstanceState: Bundle?) {
        super.onCreate(savedInstanceState)
        setContentView(R.layout.activity_main)

        val textView = TextView( context: this)
        textView.text = "Hello Simba"
        textView.height = 70
        textView.width = 214
        textView.textSize = 36f

        setContentView(textView)//添加到视图覆盖原油XML布局文件的布局
    }
}
```

MainActivity › onCreate()

图 5-3　动态创建控件

5.1.2 View 类

View 子类的控件都可以使用继承自 View 类的属性和方法。View 子类的控件使用标签属性可以设置控件的部分属性和方法，设置方式较为便捷。但是 View 子类的控件使用类方法能够调用的属性和方法更多。

1. 标签的常用属性

在 XML 标签中，View 子类的控件使用“android:”作为属性的前缀(如表 5-1 所示)，使用=运算符设置属性。

提示：标签属性和类属性的区别

标签属性用于设置组件的属性或事件方法。类属性用于设置属性和获取属性。Kotlin 中可以直接设置和获取属性，也提供了对应的类方法设置和获取属性，官方推荐使用类方法设置和获取属性。

表 5-1 View 类标签的常用属性

共有属性	说明
android:id	设置标识符
android:layout_width	设置宽度，属性值包括： ● 具体的宽度值，长度单位推荐使用 dp ● wrap_content，自动匹配最小宽度 ● match_parent，自动匹配父标签组件的宽度，表示使当前组件和父标签组件的宽度一样 ● fill_parent，和 match_parent 的作用相同，已经不推荐使用
android:layout_height	设置高度，属性值同 android:layout_width
android:visibility	设置可见性，属性值包括： ● visible，可见 ● invisible，不可见，但占据原来的屏幕空间 ● gone，不可见，也不会占据屏幕空间
android:background	设置背景
android:foreground	设置前景
android:alpha	设置透明度(0～1 之间的数值，表示百分比)
android:padding	设置内边距(推荐使用 dp 作为单位)
android:layout_margin	设置外边距(推荐使用 dp 作为单位)
android:scrollbars	设置在滚动时是否显示滚动条
android:onClick	设置单击事件的方法

2. 类的常用方法

View 子类的控件能够继承 View 类的方法。View 类的方法不仅能够设置属性和获取属性，还提供了更多的设置监听器方法用于监听各类事件(如表 5-2 所示)。

表 5-2 View 类的常用方法

类型和修饰符	方法
open Int	getId() 获取标识符

续表

类型和修饰符	方　法
open Any!	getTag() 获取标记对象
open Unit	setId(id: Int) 设置标识符
open Unit	setTag(tag: Any!) 设置与视图关联的标记对象
open Unit	setLayoutParams(params: ViewGroup.LayoutParams!) 设置布局参数
open Unit	setVisibility(visibility: Int) 设置可见性
open Unit	setBackground(background: Drawable!) 设置背景
open Unit	setAlpha(alpha: Float) 设置透明度
open Unit	setPadding(left: Int, top: Int, right: Int, bottom: Int) 设置内边距，单位为像素
open Unit	setFocusable(focusable: Boolean) 设置是否可以获取焦点
Boolean	requestFocus() 获取焦点
open Unit	clearFocus() 清除焦点
open Unit	setOnClickListener(l: View.OnClickListener?) 设置单击监听器
open Unit	setOnLongClickListener(l: View.OnLongClickListener?) 设置长时间单击监听器
open Unit	setOnFocusChangeListener(l: View.OnFocusChangeListener!) 设置改变焦点监听器
open Unit	setOnScrollChangeListener(l: View.OnScrollChangeListener!) 设置滚动改变监听器
open Unit	setOnTouchListener(l: View.OnTouchListener!) 设置触碰监听器

提示：类方法设置和获取组件属性

组件属性多数可以使用 setXXX() 方法进行设置，使用 getXXX() 方法获取，布尔值的属性使用 isXXX() 方法获取。例如，android:id 属性可以使用 setId() 方法设置，使用 getId() 方法获取。

5.1.3　UI 控件的常用单位

常用的单位有 px、dp 和 sp，还可以使用 mm、dip、in、pt 等。Android 设备屏幕的 dpi(像素密度)不同(如表 5-3 所示)，dp 和 sp 单位可以解决不同 dpi 下显示的差异，在显示前会自动根据 dpi 转换成相应的 px 单位值。Android 项目中的控件大小主要使用 dp 作为单位，字体大小主要使用 sp 作为单位。

表 5-3　屏幕像素密度

密　度	LDPI	MDPI	HDPI	XHDPI	XXHPDI	XXXHDPI
像素密度	120dpi	160dpi	240dpi	320dpi	480dpi	640dpi
分辨率	240×320	320×480	480×800	720×1280	1080×1920	3840×2160
转换系数	0.75	1	1.5	2	3	4
转换结果	1dp=0.75px	1dp=1px	1dp=1.5px	1pd=2px	1dp=3px	1dp=4px

- px：像素(pixel)，每个像素单位代表屏幕上的一个显示点。100px 的图像在不同分辨率的手机上显示的大小是不同的，即使同样尺寸的屏幕，分辨率也可能不同，因此不建议使用该单位。
- dp：设备独立像素，等同于 dip(density-independent pixel)，转换公式：dp×dpi/160=px。160dpi 的中密度手机屏幕为基准屏幕，此时 1dp=1px。100dp 在 320×480(MDPI，160dpi)的屏幕上是 100px，那么 100dp 在 480×800(HDPI，240dpi)的屏幕上是 150px，但它们都是 100dp。不管屏幕的像素密度是多少，相同 dp 大小的元素在屏幕上显示的大小始终差不多，因此控件多使用该单位。
- sp：与缩放无关的抽象像素(Scale-independent pixel)，与 dp 类似，转换公式：sp×dpi/160=px。当用户通过手机设置修改手机字体时，以 sp 为单位的字体会随着改变，因此字号多使用该单位。

提示：分辨率、屏幕大小、dpi 和 ppi

- 分辨率：手机屏幕的像素点数，一般描述成屏幕的“宽×高”。常见的分辨率有 480×800、720×1280、1080×1920 等。
- 屏幕大小：屏幕大小是手机对角线的物理尺寸，以英寸(inch)为单位。5 英寸手机是指对角线的尺寸，5 英寸×2.54 厘米/英寸=12.7 厘米。
- dpi(dots per inch)：每英寸多少点，该值越高，图像越细腻。
- ppi(pixel per inch)：每英寸像素数，该值越高，屏幕显示越细腻。

5.2 文本视图

5.2.1 TextView 控件

TextView 控件(android.widget.TextView)是用于显示文本信息的控件，是 android.view.View 的子类，TextView 控件的常用标签属性如表 5-4 所示，TextView 类的常用方法如表 5-5 所示。

表 5-4 TextView 控件的常用标签属性

属性	说明
android:text	设置文本内容
android:gravity	设置文字的对齐方式，可选值有 top、bottom、left、right、center_vertical、center_horizontal、center 等，可以用 | 运算符连接使用多个值
android:textSize	设置文字的大小
android:textColor	设置文字的颜色

表 5-5 TextView 类的常用方法

类型和修饰符	方法
open Unit	setGravity(gravity: Int) 设置对齐方式
Unit	setText(text: CharSequence!) 设置文本内容
open Unit	setTextSize(size: Float) 设置文本大小，单位为 sp
open Unit	setTextColor(color: Int) 设置文本颜色
open CharSequence!	getText() 获取文本内容
open int	length() 获取文本长度
open int	addTextChangedListener(watcher: TextWatcher!) 添加文本改变监听器

提示：组件属性

在程序运行中，标签设置的组件属性只能通过类的实例进行关联后修改，而类不但能通过对应的方法设置组件属性，还提供了其他操作类的方法。由于篇幅限制，后续控件和布局组件内容中，只提供常用的标签属性表，在类的常用方法表中不提供设置和获取属性值的方法。

5.2.2　实例工程：显示文本

本实例演示了使用 TextView 控件显示文本的两种方法(如图 5-4 所示)。第一个 TextView 控件使用标签创建，使用 setText()方法修改文本内容；第二个 TextView 控件通过代码动态创建，使用代码设置内容、字号、布局等属性。

图 5-4　运行效果

1．新建工程

新建一个“Empty Activity”工程，工程名称为“C0501”。

2．主界面的布局

打开“activity_main.xml”文件，将原有标签删除，重新添加标签。

/res/layout/activity_main.xml

```
02  <LinearLayout xmlns:android = "http://schemas.android.com/apk/res/android"
03      android:id = "@+id/mainLinearLayout"
04      android:layout_width = "match_parent"
05      android:layout_height = "match_parent"
06      android:orientation = "vertical">
07      <TextView
08          android:id = "@+id/helloTextView"
09          android:layout_width = "match_parent"
10          android:layout_height = "match_parent"
11          android:textSize = "36sp"
12          android:text = "Hello World!" />
13  </LinearLayout>
```

第 02 ~ 06 行<LinearLayout>标签是用于线性布局组件的标签，xmlns:android 属性设置 XML 命名空间，用于识别标签的属性。android:id 属性值为“@+id/mainLinearLayout”，解析后等同于

“R.id.mainLinearLayout”，供代码调用。android:layout_width 和 android:layout_height 属性都设置为 match，直接匹配 Activity 的宽度和高度。android:orientation 属性设置为垂直布局。第 07 ~ 12 行 <TextView>标签用于显示文本，放在<LinearLayout>标签内，作为其子标签。android:text 属性设置显示的文本内容，高度和宽度都设置为 match_parent，匹配父标签的宽度和高度(如图 5-5 所示)。

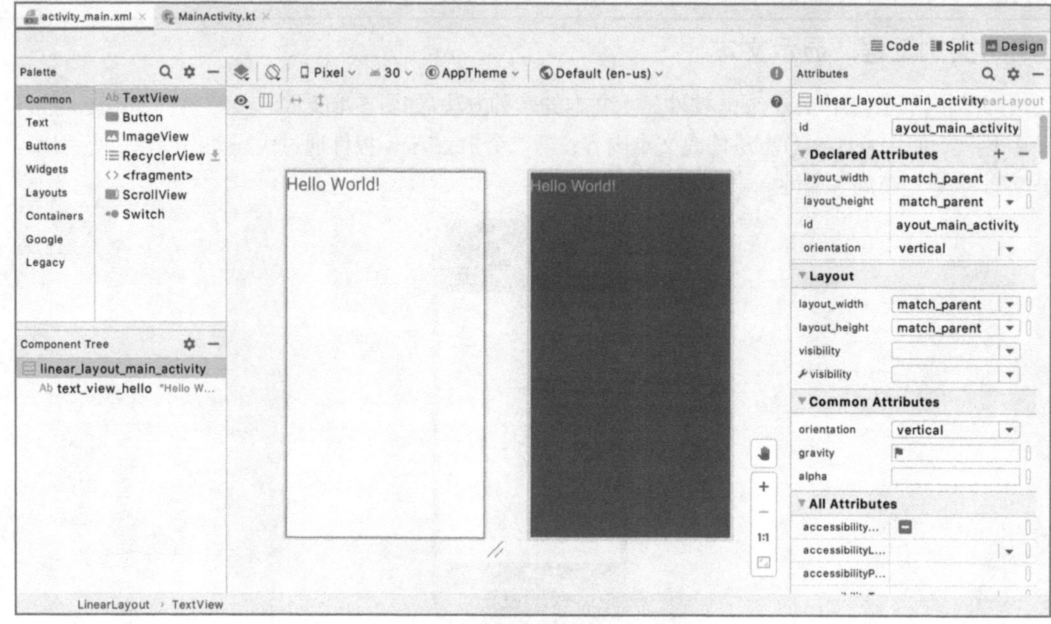

图 5-5　activity_main.xml 预览效果

3．主界面的 Activity

/java/com/teachol/c0501/MainActivity.kt

```kotlin
class MainActivity : AppCompatActivity() {
    override fun onCreate(savedInstanceState: Bundle?) {
        super.onCreate(savedInstanceState)
        setContentView(R.layout.activity_main)//设置布局文件
        // 设置 LayoutParams
        val helloLayoutParams = LinearLayout.LayoutParams(LinearLayout.LayoutParams.MATCH_PARENT, 500)
        // 通过 id 获取 TextView 对象并修改属性
        val helloTextView: TextView = findViewById(R.id.helloTextView)
        helloTextView.gravity = Gravity.CENTER_HORIZONTAL or Gravity.CENTER_VERTICAL//设置文本对齐方式
        helloTextView.text = "Hello Android!" //设置文本内容
        helloTextView.textSize = 36f //设置文本字号
        helloTextView.layoutParams = helloLayoutParams //设置文本布局参数
        // 通过 id 获取 XML 布局中的 linearLayout 布局
        val mainLinearLayout: LinearLayout = findViewById(R.id.mainLinearLayout)
        // 动态创建 TextView 对象
        val welcomeTextView = TextView(this)
        welcomeTextView.gravity = Gravity.CENTER //设置文本对齐方式
        welcomeTextView.text = "welcome!" //设置文本内容
        welcomeTextView.textSize = 36f //设置文本字号
        welcomeTextView.setTextColor(Color.rgb(0, 0, 0))//设置文本颜色
```

```
30          // 添加到 linearLayout 布局内
31          mainLinearLayout.addView(welcomeTextView)
32      }
33  }
```

第 13 行设置布局文件为“R.layout.activity_main”，解析后为“/res/layout/activity_main.xml”文件。第 15 行新建 LinearLayout.LayoutParams 类的 helloLayoutParams 对象，用于修改布局文件中文本控件的高度。由于标签中设置的高度匹配父标签，会填充满 Activity。如果不修改高度，后续代码动态添加的 TextView 控件不会显示出来。第 17 行新建 TextView 类的 helloTextView 对象，通过 findViewById(R.id.helloTextView)与布局文件中 id 为 helloTextView 的<TextView>标签关联。第 18 行设置对齐方式，Gravity.CENTER_HORIZONTAL 表示水平居中，Gravity.CENTER_VERTICAL 表示垂直居中，两者的值都是二进制的，使用 or 位运算符进行按位或运算，计算结果和 Gravity.CENTER 的值相等。第 19～21 行设置 helloTextView 对象的其他属性。第 23 行新建 LinearLayout 类的 mainLinearLayout 对象，与布局文件中 id 为 main LinearLayout 的<LinearLayout>标签进行关联。第 25 行新建 TextView 类的 welcomeTextView 对象，this 表示当前的 Activity。第 26～29 行设置 welcomeTextView 对象的属性。第 31 行将 welcomeTextView 对象添加到 mainLinearLayout 对象，此时才会显示在 Activity 视图中。

5.3 输　入　框

5.3.1 EditText 控件

EditText 控件(android.widget.EditText)是用于输入和编辑文本内容的控件，是 android.widget.TextView 的子类。EditText 控件的常用标签属性如表 5-6 所示，EditText 类的常用方法如表 5-7 所示。

表 5-6　EditText 控件的常用标签属性

属　性	说　明
android:hint	设置提示性的文字，输入内容后提示性的文字消失
android:ems	设置以 em 中单位的控件宽度。em 是一个在印刷排版中使用的单位，表示字宽。em 字面意思为 equal M(和 M 字符一样的宽度为一个单位)
android:focusable	设置是否可以获取焦点
android:maxEms	设置控件输入的最长字符数量，与 ems 同时使用时覆盖 ems 选项
android:maxLength	设置文本最大长度
android:minLines	设置最小显示行数
android:maxLines	设置最大显示行数
android:selectAllOnFocus	设置获得焦点后是否全选组件内所有文本内容
android:inputType	设置限制输入文本类型，常用的输入类型值包括： ● text，输入类型为普通文本 ● number，输入类型为数字文本 ● phone，输入类型为电话号码 ● numberDecimal，输入类型为十进制小数 ● numberPassword，输入类型为数字密码 ● textEmailAddress，输入一个电子邮件地址 ● textPassword，输入一个密码
android:singleLine	设置是否单行

表 5-7　EditText 类的常用方法

类型和修饰符	方　法
open Editable!	getText() 获取文本内容
open Unit	selectAll() 选取全部文本内容
open Unit	setSelection(start: Int, stop: Int) 选取指定范围内的文本内容
open Unit	setSelection(index: Int) 设置光标位置

5.3.2　实例工程：输入发送信息

本实例演示了三种不同效果的 EditText 控件的使用方式(如图 5-6 所示)。第一个 EditText 控件输入的字符超过 ems 属性设置的长度后，之前输入的字符会被推后隐藏起来；第二个 EditText 控件输入的字符超过 maxEms 属性设置的长度后，会在 maxLength 属性设置的范围内自动延长控件的长度；第三个 EditText 控件输入换行后，会在 maxLines 属性设置的范围内自动增加一行。

图 5-6　运行效果

1. 新建工程

新建一个“Empty Activity”工程，工程名称为“C0502”。

2. 输入框背景的资源文件

在 Project 视图中切换到 Android 选项，单击右键“drawable”文件夹，选择【New】→【Drawable resource file】命令(如图 5-7 所示)。

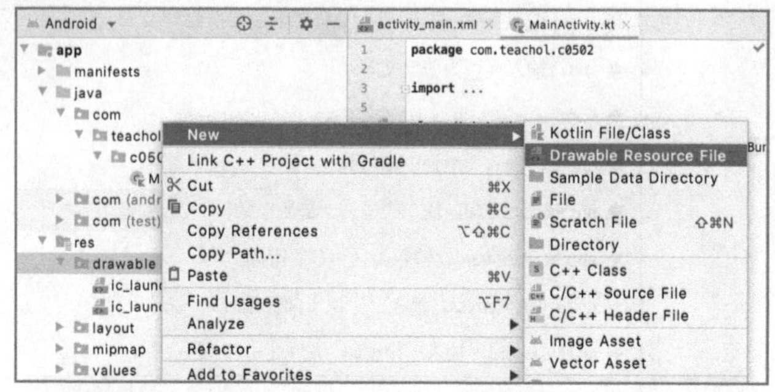

图 5-7　选择【New】→【Drawable Resource File】命令

在打开的“New Resource File”对话框中，输入文件名称和根标签名称（如图 5-8 所示），单击“OK”按钮，生成一个资源文件作为多行输入框的背景。

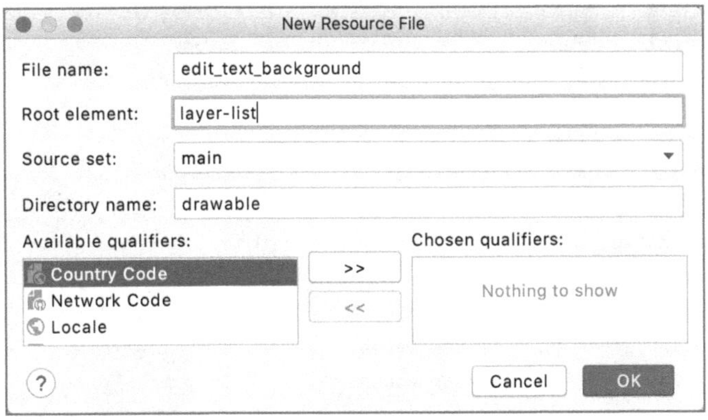

图 5-8　“New Resource File”对话框

/res/drawable/edit_text_background.xml

```
02 <layer-list xmlns:android = "http://schemas.android.com/apk/res/android">
03     <item>
04         <shape android:shape = "rectangle">
05             <solid android:color = "#EFEFEF" />
06             <corners android:radius = "3dp" />
07             <stroke android:width = "0.5dp" android:color = "#505050" />
08         </shape>
09     </item>
10 </layer-list>
```

第 02 行<layer-list>标签是图层列表标签，其原理是一层层叠加图层，后添加的会覆盖先添加的标签。第 04 行<shape>标签设置外形，可以设置为矩形（rectangle）、椭圆形（oval）、线性形状（line）、环形（ring）。第 05 行<solid>标签设置内部填充颜色。第 06 行<corners>标签设置圆角效果，radius 属性设置圆角的半径。第 07 行<stroke>标签设置描边效果，width 属性设置描边的宽度，color 属性设置描边的颜色。

3. 主界面的布局

/res/layout/activity_main.xml

```
<LinearLayout xmlns:android = "http://schemas.android.com/apk/res/android"
    android:id = "@+id/linearLayout"
    android:layout_width = "match_parent"
    android:layout_height = "match_parent"
    android:orientation = "vertical">
    <EditText
        android:layout_width = "wrap_content"
        android:layout_height = "wrap_content"
        android:layout_marginLeft = "10dp"
        android:layout_marginRight = "10dp"
        android:ems = "4"
        android:hint = "用户名"
        android:inputType = "textPersonName"
```

```
15         android:singleLine = "true" />
16     <EditText
17         android:layout_width = "wrap_content"
18         android:layout_height = "wrap_content"
19         android:layout_marginLeft = "10dp"
20         android:layout_marginRight = "10dp"
21         android:hint = "密码"
22         android:inputType = "textPassword"
23         android:maxEms = "4"
24         android:maxLength = "7"
25         android:singleLine = "true" />
26     <EditText
27         android:layout_width = "match_parent"
28         android:layout_height = "wrap_content"
29         android:layout_margin = "10dp"
30         android:background = "@drawable/edit_text_background"
31         android:hint = "需要发送的信息"
32         android:maxLines = "5"
33         android:padding = "5dp" />
34 </LinearLayout>
```

第 07 ~ 15 行<EditText>标签是第一个文本框，android:inputType="textPersonName"表示输入类型为个人名字，android:singleLine="true"表示单行。第 16 ~ 25 行<EditText>标签是第二个文本框，android:inputType="textPassword"表示输入类型为密码，输入后会以星号替代显示。第 26 ~ 33 行<EditText>标签是第三个文本框，android:background="@drawable/edit_text_background"设置背景，"@drawable/edit_text_background"解析后对应的文件为“/res/drawable/edit_text_background.xml”，并将该文件中的设置作为背景。android:maxLines="5"表示最多可以输入 5 行。

5.4 按　钮

5.4.1 Button 控件

Button 控件(android.widget.Button)是按钮的控件，是 android.widget.TextView 的子类，Button 控件的常用标签属性如表 5-8 所示。

表 5-8 Button 控件的常用标签属性

属　性	说　明
android:text	设置按钮上的文本
android:textColor	设置按钮上的文本颜色
android:textSize	设置按钮上的文本字号
android:enabled	设置按钮是否可用
android:gravity	设置按钮上文字的对齐方式

5.4.2 实例工程：单击按钮获取系统时间

本实例演示了 Button 控件及其单击监听事件的应用，单击“获取当前系统时间”按钮，在其上方的 TextView 控件中显示出当前的系统时间(如图 5-9 所示)。

图 5-9　运行效果

1. 新建工程

新建一个“Empty Activity”工程，工程名称为“C0503”。

2. 主界面的布局

/res/layout/activity_main.xml

```
07  <TextView
08      android:id = "@+id/text_view_time"
09      android:layout_width = "wrap_content"
10      android:layout_height = "wrap_content"
11      android:textSize = "28sp" />
12  <Button
13      android:id = "@+id/button_get_time"
14      android:layout_width = "wrap_content"
15      android:layout_height = "wrap_content"
16      android:text = "获取当前系统时间"
17      android:textSize = "32sp" />
```

第 07～11 行<TextView>标签用于显示获取的系统时间，第 12～17 行<Button>标签用于单击后获取当前系统时间并显示在<TextView>标签中。

3. 主界面的 Activity

/java/com/teachol/c0503/MainActivity.kt

```
14  val textView: TextView = findViewById(R.id.text_view_time)
15  val button: Button = findViewById(R.id.button_get_time)
16  // 添加监听器
17  button.setOnClickListener(View.OnClickListener {
18      val date = Date()
19      val dateFormat = SimpleDateFormat("yyyy 年 MM 月 dd 日 HH 时 mm 分 ss 秒")
20      textView.setText(dateFormat.format(date))
21  })
```

第 14 行声明 textView 对象与 activity_main.xml 中 id 为 text_view_time 的<TextView>标签相关联。第 15 行声明 button 对象与 activity_main.xml 中 id 为 button_get_time 的<Button>标签相关联。第 17～21 行 button 对象设置单击事件监听器，用于获取系统时间并显示在 textView 对象中。

5.5 图像视图

5.5.1 ImageView 控件

ImageView 控件(android.widget.ImageView)是显示图像的 UI 控件，是 android.view.View 的子类。图像通常存储在“drawable”资源文件夹中或下载后显示出来，ImageView 控件的常用标签属性如表 5-9 所示，ImageView 类的常用方法如表 5-10 所示。

表 5-9 ImageView 控件的常用标签属性

属 性	说 明
android:adjustViewBounds	设置是否保持比例调整视图边界，默认值是 false
android:maxHeight	设置最大高度
android:maxWidth	设置最大宽度
android:src	设置图像的路径
android:scaleType	设置图像显示的缩放匹配类型，可选类型包含：fitCenter(默认值)、center、centerCrop、centerInside、fitEnd、fitStart、fitXY 和 matrix
android:background	设置背景

表 5-10 ImageView 类的常用方法

类型和修饰符	方 法
open Unit	setImageAlpha(alpha: Int) 设置图像透明度
open Unit	setImageURI(uri: Uri?) 设置图像的 Uri 路径
open Unit	setImageResource(@DrawableRes int resId) 设置图像资源
open Unit	setSelected(selected: Boolean) 设置是否被选择

5.5.2 实例工程：显示图像

本实例演示了使用 ImageView 控件显示图像及三种缩放匹配效果(如图 5-10 所示)。第一幅图像是以等比例最大化居中显示的；第二幅图像是以原始尺寸在左上角显示的；第三幅图像是完全填充显示的，可能导致图像被拉伸变形。

图 5-10 运行效果

1. 新建工程并导入素材

新建一个“Empty Activity”工程，工程名称为“C0504”。然后选择“/res/mipmap”资源文件夹，将素材图像（素材文件夹路径为“/素材/C0504”）粘贴或拖曳到该文件夹中（如图 5-11 所示）。

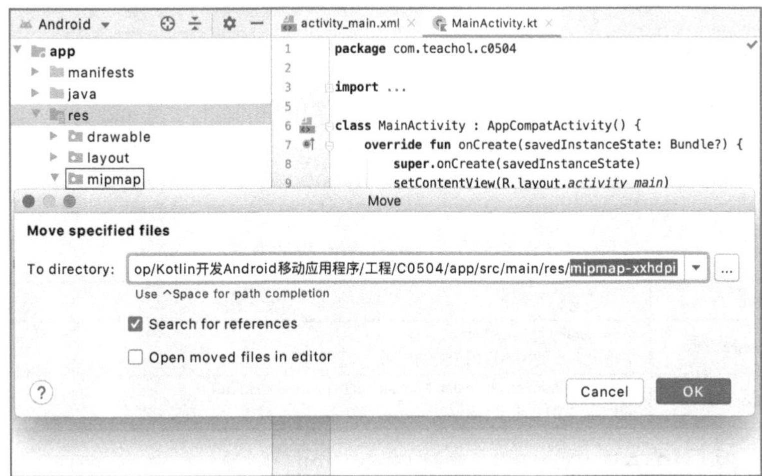

图 5-11　添加图像

2. 主界面的布局

/res/layout/activity_main.xml

```
<ImageView
    android:id = "@+id/image_view1"
    android:layout_width = "180dp"
    android:layout_height = "150dp"
    android:layout_margin = "5dp"
    android:background = "#00BCD4"
    android:padding = "3dp"
    android:src = "@mipmap/img01" />
<ImageView
    android:id = "@+id/image_view2"
    android:layout_width = "180dp"
    android:layout_height = "150dp"
    android:layout_margin = "5dp"
    android:background = "#00BCD4"
    android:padding = "3dp"
    android:scaleType = "matrix"
    android:src = "@mipmap/img02" />
<ImageView
    android:id = "@+id/image_view3"
    android:layout_width = "180dp"
    android:layout_height = "150dp"
    android:layout_margin = "5dp"
    android:background = "#00BCD4"
    android:padding = "3dp"
    android:scaleType = "fitXY"
    android:src = "@mipmap/img03" />
```

三个<ImageView>标签分别显示不同的图像，设置了相同的背景颜色和内间距。第一个没有使用 scaleType 属性，其他两个 scaleType 属性值分别为“matrix”和“fitXY”。

5.6 图像按钮

5.6.1 ImageButton 控件

ImageButton 控件(android.widget.ImageButton)是图像按钮的控件，是 android.widget.ImageView 的子类，不但能够显示图像，而且能够实现按钮的功能。除继承的方法外，还包含两个公有方法(如表 5-11 所示)。

表 5-11 ImageButton 类的方法

类型和修饰符	方 法
open CharSequence!	getAccessibilityClassName() 获取此对象的类名以用于辅助功能
open PointerIcon!	onResolvePointerIcon(event: MotionEvent!, pointerIndex: Int) 获取运动事件的指针图标

5.6.2 实例工程：提示广播信息状态的图像按钮

本实例演示了单击 ImageButton 控件后，图像中喇叭上方小圆点消失，再次单击小圆点出现的效果(如图 5-12 所示)。

图 5-12 运行效果

1. 新建工程并导入素材

新建一个“Empty Activity”工程，工程名称为“C0505”。然后选择“/res/drawable”资源文件夹，将素材图像(素材文件夹路径为“/素材/C0505”)粘贴或拖曳到该文件夹中。

2. 图像按钮选择状态的资源文件

在“/res/drawable”资源文件夹上单击右键，选择【New】→【Drawable Resource File】命令(如图 5-13 所示)。在“New Resource File”对话框的“File name”文本框中输入“image_button_msg”，单击“OK”按钮，创建“image_button_msg.xml”文件。该文件用于设置 ImageButton 控件不同状态时显示的图像资源。

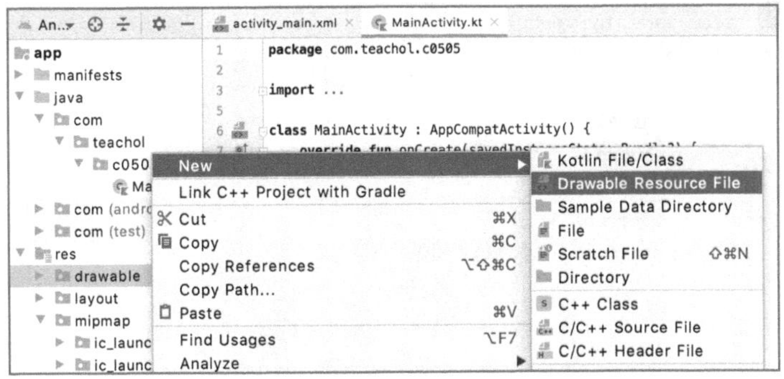

图 5-13　新建 Drawable 资源文件

/res/drawable/image_button_msg.xml

```
02  <selector xmlns:android = "http://schemas.android.com/apk/res/android">
03      <item android:drawable = "@mipmap/img_msg_new" android:state_selected="false" />
04      <item android:drawable = "@mipmap/img_msg" android:state_selected = "true" />
05  </selector>
```

第 03 行设置 ImageButton 控件未被选择时显示的图像，@mipmap/img_msg_new 表示 mipmap 文件夹中的 img_msg_new.png。第 04 行设置 ImageButton 控件被选择时显示的图像。

提示：StateListDrawable 资源状态列表

在资源文件的<selector>标签下的<item>标签中，可以设置资源的不同状态显示效果的属性，与“android:drawable”属性搭配使用。该属性值为布尔型，常用的状态属性如下。

- android:state_selected：表示控件是否被选择的状态。
- android:state_pressed：表示控件是否被单击或者触摸的状态。
- android:state_focused：表示控件是否获得焦点时的状态。
- android:state_checkable：表示控件是否可以被勾选的状态。
- android:state_checked：表示控件是否处于被勾选。
- android:state_enabled：表示控件是否处于可用状态。
- android:state_activated：表示控件是否被激活状态。

3. 主界面的布局

/res/layout/activity_main.xml

```
06  <ImageButton
07      android:id = "@+id/image_button"
08      android:layout_width = "100dp"
09      android:layout_height = "100dp"
10      android:background = "#E62F5A"
11      android:src = "@drawable/image_button_msg" />
```

第 10 行设置背景颜色，当图像有透明部分或者超出图像尺寸时，会显示为背景颜色。第 11 行未直接设置显示的图像资源，而是通过资源配置文件设置不同状态下显示的图像。

4. 主界面的 Activity

/java/com/teachol/c0505/MainActivity.kt

```
val button: ImageButton = findViewById(R.id.image_button)
```

```
14 button.setOnClickListener(object : View.OnClickListener{
15     override fun onClick(v: View) {
16         if (v.isSelected()) {
17             v.setSelected(false)
18             Log.i("ImageButton", "选中图像")
19         } else {
20             v.setSelected(true)
21             Log.i("ImageButton", "取消选中图像")
22         }
23     }
24 })
```

第 13 行将 button 对象与 activity_main.xml 中 id 为 image_button 的<ImageButton>标签相关联。第 14～24 行 button 对象设置单击事件监听器，用于切换图像按钮的显示状态。OnClick(v:View)方法的 v 参数表示被单击的 View 对象，由于 ImageButton 类是 View 类的子类，所以可以直接传递给 v 参数。第 18 行和第 20 行 Log.i()方法在控制台中输出测试信息，与 system.out.println()方法相比，主要优势在于能使用过滤器过滤日志。

5. 测试运行

选择【Run】→【Run'app'】命令，在 Logcat 窗口中单击过滤器的搜索输入框可以输入或选择<ImageButton>标签，用于筛选出 Log.i()方法输出的信息(如图 5-14 所示)。

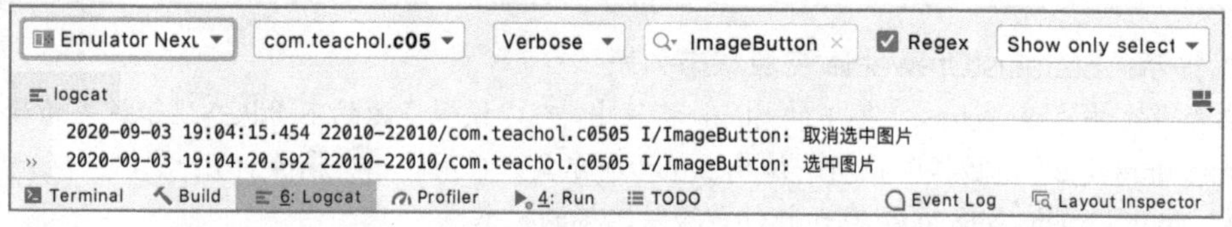

图 5-14　Logcat 窗口筛选信息

提示：Log 类

Log 类直接调用静态方法输出调试信息，在 Logcat 窗口中输出，并可以使用过滤器的级别选项进行筛选。Logcat 窗口中输出的调试信息共有以下 5 种级别。

- Verbose：任何信息都会输出，使用 Log.v()。
- Info：输出提示信息，使用 Log.i()。
- Error：输出错误信息，使用 Log.e()。
- Debug：输出调试信息，使用 Log.d()。
- Warning：输出警告信息，使用 Log.w()。

5.7 单选按钮

5.7.1 RadioButton 控件

RadioButton 控件(android.widget.RadioButton)是单选按钮的控件，是 android.widget.Compound Button 的子类。只有同一个 RadioGroup 中的 RadioButton 才能实现真正的单选功能。RadioButton 控件的常用标签属性如表 5-12 所示，RadioButton 类的常用方法如表 5-13 所示；RadioGroup 控件的常用标签属性如表 5-14 所示，RadioGroup 类的常用方法如表 5-15 所示。

表 5-12　RadioButton 控件的常用标签属性

属　性	说　明
android:button	设置文字前方的按钮形状，为@null 时不显示前面的圆形按钮
android:checked	设置是否被选中，同一个 RadioButtonGroup 中的 RadioButton 设置该属性后会取消之前被选中的 RadioButton 的选中状态
android:text	设置文本

表 5-13　RadioButton 类的常用方法

类　型	方　法
Boolean	isChecked() 判断单选按钮是否被选择
Unit	setChecked(checked: Boolean) 设置单选按钮是否被选择
Unit	setOnCheckedChangeListener(listener: CompoundButton.OnCheckedChangeListener?) 设置单选按钮选择状态改变的监听器

表 5-14　RadioGroup 控件的常用标签属性

属　性	说　明
android:checkedButton	设置默认选择的 RadioButton 选项 id
android:orientation	设置 RadioButton 的排列方式。horizontal 为水平排列，vertical 为垂直排列

表 5-15　RadioGroup 类的常用方法

类型和修饰符	方　法
open Unit	addView(child: View!, index: Int, params: ViewGroup.LayoutParams!) 添加子视图
open Unit	check(id: Int) 设置被选择的单选按钮
open Unit	clearCheck() 清除选择
int	getCheckedRadioButtonId() 获取被选择单选按钮的 id
open Unit	setOnCheckedChangeListener(listener: RadioGroup.OnCheckedChangeListener!) 设置单选按钮组选择状态改变的监听器

5.7.2　实例工程：选择性别的单选框

本实例演示了使用 RadioButton 控件选择性别(如图 5-15 所示)，单击改变选择的选项或提交按钮时，会在控制台中输出相应的信息。

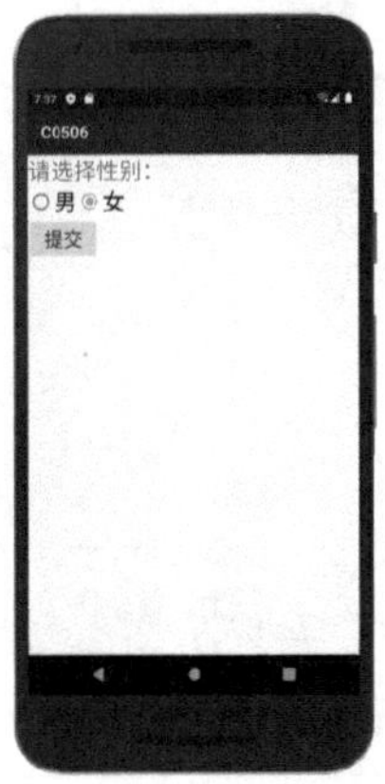

图 5-15　运行效果

1. 新建工程

新建一个“Empty Activity”工程，工程名称为“C0506”。

2. 字符串资源

/res/values/strings.xml

```
01 <resources>
02     <string name = "app_name">C0506</string>
03     <string name = "select_sex">请选择性别：</string>
04     <string name = "man">男</string>
05     <string name = "woman">女</string>
06     <string name = "submit">提交</string>
07 </resources>
```

第 03 ~ 06 行是添加的字符串资源，用于设置控件的 text 属性。与直接设置 text 属性相比，优势在于可以重复调用和国际化，便于统一修改和针对不同语言设置字符串。

3. 主界面的布局

/res/layout/activity_main.xml

```
06 <TextView
07     android:layout_width = "wrap_content"
08     android:layout_height = "wrap_content"
09     android:text = "@string/select_sex"
10     android:textSize = "28sp" />
11 <RadioGroup
12     android:id = "@+id/radio_group"
13     android:layout_width = "wrap_content"
14     android:layout_height = "wrap_content"
15     android:checkedButton = "@id/radio_button_women"
16     android:orientation = "horizontal">
17     <RadioButton
18         android:id = "@+id/radio_button_man"
19         android:layout_width = "wrap_content"
20         android:layout_height = "wrap_content"
21         android:text = "@string/man"
22         android:textSize = "28sp" />
23     <RadioButton
24         android:id = "@+id/radio_button_women"
25         android:layout_width = "wrap_content"
26         android:layout_height = "wrap_content"
27         android:text = "@string/woman"
28         android:textSize = "28sp" />
29 </RadioGroup>
30 <Button
31     android:id = "@+id/button"
32     android:layout_width = "wrap_content"
33     android:layout_height = "wrap_content"
34     android:text = "@string/submit"
35     android:textSize = "24sp" />
```

第 11 ~ 16 行的<RadioGroup>标签设置默认的 RadioButton 选项和 RadioButton 的排列方式。第 17 ~

28 行添加两个<RadioButton>标签，如果在<RadioButton>标签中设置了 checked 属性，将覆盖<RadioGroup>标签中设置的 checkedButton 属性及同组中之前<RadioButton>标签设置的 checked 属性。

4．主界面的 Activity

```
/java/com/teachol/c0506/MainActivity.kt
15 val radioGroup: RadioGroup = findViewById(R.id.radio_group)
16 radioGroup.setOnCheckedChangeListener { radioGroup, checkedId ->
17     val radioButton = findViewById<RadioButton>(checkedId)
18     Log.i("RadioButton", "您当前选择的选项: " + radioButton.text)
19 }
20 val button: Button = findViewById(R.id.button)
21 button.setOnClickListener(object : View.OnClickListener {
22     override fun onClick(v: View?) {
23         for (i in 0 until radioGroup.childCount) {
24             val radioButton = radioGroup.getChildAt(i) as RadioButton
25             if (radioButton.isChecked) {
26                 Log.v("RadioButton", "您提交的选项是:" + radioButton.text)
27                 break
28             }
29         }
30     }
31 })
```

第 16 ~ 19 行是 RadioGroup 对象改变选项的监听事件，用于在控制台中输出改变选项后的信息。第 21 ~ 31 行是 Button 对象的单击监听事件，用于获取选中的 RadioButton 对象，并在控制台中输出。

5.8 复 选 框

5.8.1 Checkbox 控件

Checkbox 控件(android.widget.Checkbox)是复选框的控件，是 android.widget.CompoundButton 的子类，Checkbox 控件的常用标签属性如表 5-16 所示，Checkbox 类的常用方法如表 5-17 所示。

表 5-16 Checkbox 控件的常用标签属性

属 性	说 明
android:button	设置文字前方的按钮形状，为@null 时不显示前面的方形按钮
android:checked	设置是否被选中
android:text	设置文本

表 5-17 Checkbox 类的常用方法

类 型	方 法
Boolean	isChecked() 判断复选框是否被选择
Unit	setChecked(checked: Boolean) 设置复选框是否被选择
Unit	setOnCheckedChangeListener(listener: CompoundButton.OnCheckedChangeListener?) 设置复选框选择状态改变的监听器

5.8.2 实例工程：兴趣爱好的复选框

本实例演示了使用 Checkbox 控件选择兴趣爱好（如图 5-16 所示）。未选择时，单击“确定”按钮提示未选择兴趣爱好；选择后，单击“确定”按钮显示所选的兴趣爱好。

图 5-16　运行效果

1．新建工程

新建一个“Empty Activity”工程，工程名称为“C0507”。

2．主界面的布局

/res/layout/activity_main.xml

```
<TextView
    android:layout_width = "match_parent"
    android:layout_height = "wrap_content"
    android:text = "请选择你的兴趣爱好："
    android:textSize = "28sp" />
<CheckBox
    android:id =" @+id/check_box1"
    android:layout_width = "wrap_content"
    android:layout_height = "wrap_content"
    android:text = "读书"
    android:textSize = "28sp" />
<CheckBox
    android:id = "@+id/check_box2"
    android:layout_width = "wrap_content"
    android:layout_height = "wrap_content"
    android:text = "旅行"
    android:textSize = "28sp" />
<CheckBox
    android:id = "@+id/check_box3"
    android:layout_width = "wrap_content"
    android:layout_height = "wrap_content"
    android:text = "摄影"
    android:textSize = "28sp" />
<CheckBox
```

```
30        android:id = "@+id/check_box4"
31        android:layout_width = "wrap_content"
32        android:layout_height = "wrap_content"
33        android:text = "绘画"
34        android:textSize = "28sp" />
35    <Button
36        android:id = "@+id/button_submit"
37        android:layout_width = "match_parent"
38        android:layout_height = "wrap_content"
39        android:text = "确定"
40        android:textSize = "24sp" />
41    <TextView
42        android:id = "@+id/text_view_result"
43        android:layout_width = "match_parent"
44        android:layout_height = "wrap_content"
45        android:textSize = "24sp" />
```

第 11 ~ 34 行添加 4 个<CheckBox>标签，用于选择兴趣爱好。第 35 ~ 40 行添加一个<Button>标签，用于确定选择。第 41 ~ 45 行添加一个<TextView>标签，用于显示选择结果。

3．主界面的 Activity

/java/com/teachol/c0507/MainActivity.kt

```kotlin
class MainActivity : AppCompatActivity(), CompoundButton.OnCheckedChangeListener {
    private val mHobbies: ArrayList<String> = ArrayList()
    override fun onCreate(savedInstanceState: Bundle?) {
        super.onCreate(savedInstanceState)
        setContentView(R.layout.activity_main)
        val resultTextView = findViewById<TextView>(R.id.text_view_result)
        val checkBox1 = findViewById<CheckBox>(R.id.check_box1)
        val checkBox2 = findViewById<CheckBox>(R.id.check_box2)
        val checkBox3 = findViewById<CheckBox>(R.id.check_box3)
        val checkBox4 = findViewById<CheckBox>(R.id.check_box4)
        val submitButton: Button = findViewById(R.id.button_submit)
        // CheckBox 设置监听器
        checkBox1.setOnCheckedChangeListener(this)
        checkBox2.setOnCheckedChangeListener(this)
        checkBox3.setOnCheckedChangeListener(this)
        checkBox4.setOnCheckedChangeListener(this)
        // Button 设置监听器
        submitButton.setOnClickListener(object : View.OnClickListener {
            override fun onClick(v: View?) {
                val sb = StringBuilder()
                for (i in 0 until mHobbies.size) {
                    // 选择的兴趣爱好添加到 StringBuilder 尾部
                    if (i == mHobbies.size - 1) {
                        sb.append(mHobbies[i])
                    } else {
                        sb.append(mHobbies[i] + "、")
                    }
                }
```

```
39                  // 显示选择结果
40                  if (sb.length == 0) {
41                      resultTextView.text = "您还没有进行选择了！"
42                  } else {
43                      resultTextView.text = "您选择了：$sb。"
44                  }
45              }
46          })
47      }
48      /**
49       * 实现改变选项的接口方法
50       * @param compoundButton
51       * @param isChecked
52       * @return Unit
53       */
54      override fun onCheckedChanged(compoundButton: CompoundButton, isChecked: Boolean) {
55          if (isChecked) {
57              mHobbies.add(compoundButton.text.toString().trim { it <= ' ' }) //添加到数组
58          } else {
60              mHobbies.remove(compoundButton.text.toString().trim { it <= ' ' }) //从数组中移除
61          }
62      }
63  }
```

第 11 行继承 CompoundButton.OnCheckedChangeListener 接口，用于实现 CheckBox 类的改变选项状态的监听事件接口。第 12 行定义一个 ArrayList 动态数组，用于保存选择的兴趣爱好。第 22 行是一行注释，用于说明下方的代码。第 48 ~ 53 行是多行注释，@param 表示方法的参数，@return 表示方法的返回值。第 54 ~ 62 行重写复选框的选取状态改变的接口方法，选中后添加到 mHobbies 数组中，取消选中后从 mHobbies 数组中移除。

5.9 开 关 按 钮

5.9.1 Switch 控件

Switch 控件(android.widget.Switch)是开关按钮的控件，是 android.widget.CompoundButton 的子类，Switch 控件的常用标签属性如表 5-18 所示，Switch 类的常用方法如表 5-19 所示。

表 5-18 Switch 控件的常用标签属性

属 性	说 明
android:checked	设置 on/off 的状态
android:showText	设置 on/off 时是否显示文本
android:switchMinWidth	设置开关的最小宽度
android:textOff	设置 off 状态时显示的文本
android:textOn	设置 on 状态时显示的文本
android:track	设置底部的图像
android:thumb	设置滑块的图像

表 5-19　Switch 类的常用方法

类型和修饰符	方　法
Boolean	isChecked() 判断开关按钮是否打开
open Unit	setChecked(checked: Boolean) 设置开关按钮是否打开
Unit	setOnCheckedChangeListener(listener: CompoundButton.OnCheckedChangeListener?) 设置开关按钮状态改变的监听器
open Unit	toggle() 改变选择状态

5.9.2　实例工程：房间灯的开关按钮

本实例演示了两种不同效果的 Switch 控件的方法（如图 5-17 所示）。单击 Switch 控件后，监听事件会判断被单击的控件及其状态，并在 Logcat 窗口中输出相应的信息。

图 5-17　运行效果

1. 新建工程

新建一个“Empty Activity”工程，工程名称为“C0508”。

2. 主界面的布局

/res/layout/activity_main.xml

```
<Switch
    android:id = "@+id/switch_livingroom"
    android:layout_width = "wrap_content"
    android:layout_height = "wrap_content"
    android:showText = "true"
    android:switchMinWidth = "50dp"
    android:text = "客厅灯"
    android:textSize = "28sp"
    android:textOff = "关"
    android:textOn = "开" />
<Switch
    android:id = "@+id/switch_bedroom"
    android:layout_width = "wrap_content"
    android:layout_height = "wrap_content"
```

```
21     android:checked = "true"
22     android:text = "卧室灯"
23     android:textSize = "28sp" />
```

第 07 ~ 23 行添加两个<Switch>标签，第一个标签设置显示开关的文本及不同状态的文本，第二个标签设置选中的状态。

3.主界面的 Activity

```
/java/com/teachol/c0508/MainActivity.kt
09 class MainActivity : AppCompatActivity(), CompoundButton.OnCheckedChangeListener {
10     override fun onCreate(savedInstanceState: Bundle?) {
11         super.onCreate(savedInstanceState)
12         setContentView(R.layout.activity_main)
13         val livingroomSwitch = findViewById<Switch>(R.id.switch_livingroom)
14         val bedroomSwitch = findViewById<Switch>(R.id.switch_bedroom)
15         livingroomSwitch.setOnCheckedChangeListener(this)
16         bedroomSwitch.setOnCheckedChangeListener(this)
17     }
18     // 重写开关状态改变的事件
19     override fun onCheckedChanged(compoundButton: CompoundButton, isChecked: Boolean) {
20         when (compoundButton.id) {
21             R.id.switch_livingroom -> if (isChecked) {
22                 Log.i("Switch", "打开客厅灯")
23             } else {
24                 Log.i("Switch", "关闭客厅灯")
25             }
26             R.id.switch_bedroom -> if (isChecked) {
27                 Log.i("Switch", "打开卧室灯")
28             } else {
29                 Log.i("Switch", "关闭卧室灯")
30             }
31         }
32     }
33 }
```

第 09 行 CompoundButton.OnCheckedChangeListener 设置需要实现的接口。第 19 ~ 32 行重写实现 CompoundButton.OnCheckedChangeListener 接口的 onCheckedChanged()方法，compoundButton 参数表示获取到被监听对象，compoundButton.id 用于获取到被监听对象的 id，isChecked 参数用于获取到被监听对象的开关状态。

5.10 提示信息

5.10.1 Toast 控件

Toast 控件(android.widget.Toast)是显示提示信息的控件，是 java.lang.Object 的子类。Toast 类的常量如表 5-20 所示常用方法如表 5-21 所示。

表 5-20　Toast 类的常量

类　型	常　量	说　明
int	LENGTH_LONG	长时间显示
int	LENGTH_SHORT	短时间显示

表 5-21　Toast 类的常用方法

类型和修饰符	方　法
	<init>(context: Context!) 主构造方法
open static Toast!	makeText(context: Context!, resId: Int, duration: Int) 创建指定文本资源的提示信息
open static Toast!	makeText(context: Context!, text: CharSequence!, duration: Int) 创建指定文本的提示信息
open Unit	setDuration(duration: Int) 设置持续时间
open Unit	setGravity(gravity: Int, xOffset: Int, yOffset: Int) 设置提示信息显示在屏幕上的位置，在 API Level 11 及以上版本中不产生效果
open Unit	setMargin(horizontalMargin: Float, verticalMargin: Float) 设置视图的边距
open Unit	setText(resId: Int) 设置提示信息的文本资源 id
open Unit	setText(s: CharSequence!) 设置提示信息的文本
open Unit	setView(view: View!) 设置提示信息的视图，用于自定义提示信息
open Unit	show() 根据指定的持续时间显示提示信息
open Unit	cancel() 如果提示信息正在显示，则取消显示

5.10.2　实例工程：不同位置显示的提示信息

本实例演示了使用两种方法显示 Toast 控件的效果(如图 5-18 所示)。单击“显示提示信息 1”按钮直接通过静态方法调用 show()方法显示 Toast 控件，单击“显示提示信息 2”按钮先使用静态方法进行赋值再调用 show()方法显示 Toast 控件。

图 5-18　运行效果

1. 新建工程

新建一个“Empty Activity”工程，工程名称为“C0509”。

2. 主界面的布局

/res/layout/activity_main.xml

```
06 <Button
07     android:id = "@+id/button_show_toast1"
08     android:layout_width = "wrap_content"
09     android:layout_height = "wrap_content"
10     android:text = "显示提示信息 1"
11     android:textSize = "24sp" />
12 <Button
13     android:id = "@+id/button_show_toast2"
14     android:layout_width = "wrap_content"
15     android:layout_height = "wrap_content"
16     android:text = "显示提示信息 2"
17     android:textSize = "24sp"  />
```

第 06 ~ 17 行添加两个<Button>标签，分别指定不同的 id 名称，用于单击后显示 Toast 提示信息。

3. 主界面的 Activity

/java/com/teachol/c0509/MainActivity.kt

```
14 val showToastButton1: Button = findViewById(R.id.button_show_toast1)
15 showToastButton1.setOnClickListener(object : View.OnClickListener {
16     override fun onClick(v: View?) {
17         // 直接通过静态方法调用 show()方法
18         Toast.makeText(applicationContext, "这是提示信息 1", Toast.LENGTH_LONG).show()
19     }
20 })
21 val showToastButton2: Button = findViewById(R.id.button_show_toast2)
22 showToastButton2.setOnClickListener(object : View.OnClickListener {
23     override fun onClick(v: View?) {
24         // 先使用静态方法进行赋值再调用 show()方法
25         val toast = Toast.makeText(this@MainActivity, "这是提示信息 1", Toast.LENGTH_LONG)
26         toast.setText("这是提示信息 2")
27         toast.duration = Toast.LENGTH_SHORT
28         toast.show()
29     }
30 })
```

第 18 行使用 Toast 类的 makeText()静态方法直接调用 show()方法显示提示信息，使用 applicationContext 属性或 this@MainActivity 获取 Context 对象。第 25 ~ 28 行使用 Toast 类的 makeText()静态方法为 toast 对象赋值，再通过方法和属性分别对文本内容、持续时间和显示位置进行修改，最后使用 show()方法显示提示信息。

提示：Context 类

Context 类是一个抽象类，有两个子类：ContextWrapper 类和 ContextImpl 类。ContextWrapper 类是上下文功能的封装类，而 ContextImpl 类是上下文功能的实现类。而 ContextWrapper 类的子类

包含 ContextThemeWrapper 类、Service 类和 Application 类。其中，ContextThemeWrapper 类是一个带主题的封装类，Activity 类是它的子类。

5.11 对 话 框

5.11.1 AlertDialog 控件

AlertDialog 控件（android.app.AlertDialog）是用于显示对话框的控件，是 android.app.Dialog 的子类。由于 AlertDialog 类的构造方法都是 protected 方法，所以需要通过 AlertDialog.Builder 类进行实例化。AlertDialog 类的常用方法如表 5-22 所示，AlertDialog.Builder 类的常用方法如表 5-23 所示。

表 5-22 AlertDialog 类的常用方法

类型和修饰符	方　法
Unit	show() 显示对话框
Unit	cancel() 关闭对话框
open Unit	setTitle(title: CharSequence?) 设置对话框标题
open Unit	setView(view: View!, viewSpacingLeft: Int, viewSpacingTop: Int, viewSpacingRight: Int, viewSpacingBottom: Int) 设置对话框中显示的视图，指定该视图周围显示的间距
Unit	setOnShowListener(listener: DialogInterface.OnShowListener?) 设置在对话框显示时要调用的监听器
Unit	setOnCancelListener(listener: DialogInterface.OnCancelListener?) 设置在对话框关闭时要调用的监听器
Unit	setCanceledOnTouchOutside(cancel: Boolean) 设置触摸到对话框以外的区域是否关闭对话框

表 5-23 AlertDialog.Builder 类的常用方法

类型和修饰符	方　法
	<init>(context: Context!) 主构造方法
open AlertDialog!	create() 根据对话框构建器设置的参数创建一个对话框
open AlertDialog.Builder!	setIcon(icon: Drawable!) 将 Drawable 设置为标题图标
open AlertDialog.Builder!	setIcon(iconId: Int) 根据资源 id 设置标题图标
open AlertDialog.Builder!	setMessage(message: CharSequence!) 设置要显示的消息
open AlertDialog.Builder!	setMessage(messageId: Int) 根据资源 id 设置要显示的消息
open AlertDialog.Builder!	setTitle(title: CharSequence!) 设置对话框标题
open AlertDialog.Builder!	setTitle(titleId: Int) 根据资源 id 设置标题
open AlertDialog.Builder!	setPositiveButton(text: CharSequence!, listener: DialogInterface.OnClickListener!) 设置肯定按钮的文字及其单击监听器
open AlertDialog.Builder!	setNegativeButton(text: CharSequence!, listener: DialogInterface.OnClickListener!) 设置否定按钮的文字及其单击监听器
open AlertDialog.Builder!	setNeutralButton(text: CharSequence!, listener: DialogInterface.OnClickListener!) 设置中性按钮的文字及其单击监听器
open AlertDialog.Builder!	setView(view: View!) 设置对话框的自定义视图

续表

类型和修饰符	方　法
open AlertDialog.Builder!	setSingleChoiceItems (items: Array<CharSequence!>!, checkedItem: Int, listener: DialogInterface.OnClickListener!) 设置在对话框中显示的单选按钮列表和被选项，以及单击单选按钮时的监听器
open AlertDialog.Builder!	setMultiChoiceItems (items: Array<CharSequence!>!, checkedItems: BooleanArray!, listener: DialogInterface.OnMultiChoiceClickListener!) 设置在对话框中显示的复选框列表和被选项，以及单击复选框时的监听器
open AlertDialog.Builder!	setItems (items: Array<CharSequence!>!, listener: DialogInterface.OnClickListener!) 设置在对话框中显示的列表，以及单击列表选项时的监听器
open AlertDialog!	show () 根据对话框构建器设置的参数创建一个对话框，并立即显示该对话框

提示：单选对话框、多选对话框和列表对话框

setSingleChoiceItems ()、setMultiChoiceItems () 和 setItems () 方法可以快速构建单选对话框、多选对话框和列表对话框，但是只能使用默认的样式。如果不想使用默认样式，需要使用自定义对话框。

5.11.2　实例工程：默认对话框和自定义对话框

本实例演示了默认 AlertDialog 控件和自定义 AlertDialog 控件的效果(如图 5-19 所示)。单击“默认对话框”按钮显示默认 AlertDialog 控件，单击 AlertDialog 控件以外的区域可以直接关闭对话框。单击“自定义对话框”按钮显示自定义 AlertDialog 控件，然后输入框自动获取焦点后弹出虚拟键盘，此时单击该控件以外的区域不会关闭对话框。

图 5-19　运行效果

1. 新建工程

新建一个“Empty Activity”工程，工程名称为“C0510”。

2. 对话框的布局

在“/res/layout”文件夹中，新建“dialog_setting_school.xml”布局文件，用于自定义对话框的布局。

```
/res/layout/dialog_setting_school.xml
<EditText
    android:id = "@+id/edit_text_school"
```

```
09      android:layout_width = "match_parent"
10      android:layout_height = "wrap_content"
11      android:layout_marginStart = "10dp"
12      android:layout_marginEnd = "10dp"
13      android:gravity = "center"
14      android:maxLength = "16" />
```

第 07 ~ 14 行添加<EditText>标签，左右外边距设置为 10dp，文字显示居中，输入文本的最大长度为 16。

3．主界面的布局

/res/layout/activity_main.xml

```
07  <Button
08      android:id = "@+id/button1"
09      android:layout_width = "match_parent"
10      android:layout_height = "wrap_content"
11      android:text = "默认对话框" />
12  <Button
13      android:id = "@+id/button2"
14      android:layout_width = "match_parent"
15      android:layout_height = "wrap_content"
16      android:text = "自定义对话框" />
```

第 07 ~ 16 行添加两个<Button>标签，指定不同的 id 名称，分别用于显示默认对话框和自定义对话框。

4．主界面的 Activity

/java/com/teachol/c0510/MainActivity.kt

```
class MainActivity : AppCompatActivity() {
    private var mContext: Context? = null
    override fun onCreate(savedInstanceState: Bundle?) {
        super.onCreate(savedInstanceState)
        setContentView(R.layout.activity_main)
        mContext = this@MainActivity
        val button1: Button = findViewById(R.id.button1)
        button1.setOnClickListener { showAlertDialog1() } //单击按钮显示默认对话框
        val button2: Button = findViewById(R.id.button2)
        button2.setOnClickListener { showAlertDialog2() } //单击按钮显示自定义对话框
    }
    private fun showAlertDialog1() {
        val adBuilder: AlertDialog.Builder = AlertDialog.Builder(mContext)
        adBuilder.setIcon(R.mipmap.ic_launcher) //设置默认对话框的图标
        adBuilder.setTitle("默认对话框") //设置默认对话框的标题
        adBuilder.setMessage("这是一个默认对话框吗？") //设置默认对话框的文本信息
        adBuilder.setPositiveButton("是", { dialog, which -> Toast.makeText(mContext, "单击了确认按钮", Toast.LENGTH_SHORT).show() }) //单击确认按钮事件
        adBuilder.setNegativeButton("否", { dialog, which -> Toast.makeText(mContext, "单击了取消按钮", Toast.LENGTH_SHORT).show() }) //单击取消按钮事件
        adBuilder.setNeutralButton("不确定", { dialog, which -> Toast.makeText(mContext, "单击了中性按钮", Toast.LENGTH_SHORT).show() }) //单击中性按钮事件
        adBuilder.show()
    }
```

```
45      private fun showAlertDialog2() {
46          //通过 LayoutInflater 加载 XML 布局文件作为一个 View 对象
47          val view: View = LayoutInflater.from(mContext).inflate(R.layout.dialog_setting_school, null)
48          // 设置自定义对话框的标题及布局文件
49          val adBuilder: AlertDialog.Builder =  AlertDialog.Builder(mContext)
50          adBuilder.setTitle("自定义对话框: 请输入院校的全称")
51          adBuilder.setView(view)
52          // 获取自定义布局中的 EditText 控件
53          val schoolEditText: EditText = view.findViewById(R.id.edit_text_school)
54          // 单击确定按钮事件: 对 EditText 输入的内容进行判断
55          adBuilder.setPositiveButton("确定") { dialog, which ->
56              val school = schoolEditText.text.toString().trim { it <= ' ' }
57              if (school != "") {
58                  if (school.endsWith("大学") || school.endsWith("学院") || school.endsWith("学校")) {
59                      Toast.makeText(mContext, school, Toast.LENGTH_LONG).show()
60                  } else {
61                      Toast.makeText(mContext, "请使用院校标准名称! ", Toast.LENGTH_LONG).show()
62                  }
63              } else {
64                  Toast.makeText(mContext, "院校未设置", Toast.LENGTH_LONG).show()
65              }
66          }
67          adBuilder.setNegativeButton("取消", null) //未设置单击监听器
69          val alertDialog: AlertDialog = adBuilder.create() //实例化对话框
70          alertDialog.setOnShowListener { //设置对话框显示的监听器
72              Handler(Looper.myLooper()!!).postDelayed({
73                  schoolEditText.requestFocus() //EditText 获取焦点
74                  showInputMethod() //弹出虚拟键盘
75              }, 100) //100 毫秒后 EditText 获取焦点并弹出虚拟键盘
76          }
77          alertDialog.setView(view, 0, 50, 0, 50) //设置对话框显示的位置
78          alertDialog.setCanceledOnTouchOutside(false) //单击对话框以外的区域不消失
79          alertDialog.show() //显示对话框
80      }
81      // 弹出虚拟键盘的方法
82      private fun showInputMethod() {
83          val inputMethodManager = mContext!!.getSystemService(INPUT_METHOD_SERVICE) as InputMethodManager
84          inputMethodManager.toggleSoftInput(0, InputMethodManager.HIDE_NOT_ALWAYS)
85      }
86  }
```

第 30 ~ 43 行添加 showAlertDialog1()方法用于显示默认对话框，使用 AlertDialog.Builder 类构建对话框，设置三个内置的按钮及其单击监听事件，使用 show()方法根据设置的参数创建对话框，并立即显示该对话框，没有直接使用 AlertDialog 类。第 45 ~ 80 行添加 showAlertDialog2()方法用于显示自定义对话框，通过 AlertDialog.Builder 类的 setView()方法设置对话框的自定义布局，再使用 AlertDialog.Builder 类的 create()方法初始化 AlertDialog 类实例，并使用 setOnShowListener()方法在对话框显示后使 EditText 控件获取后调用弹出虚拟键盘的 showInputMethod()方法，然后使用 AlertDialog 类的 setView()方法重新设置自定义布局，与 AlertDialog.Builder 类的 setView()方法相比可以多设置自定义布局周围的间距，最后使用 AlertDialog 类的 show()方法显示对话框。第 82 ~ 85 行添加 showInputMethod()方法，用于实现弹出虚拟键盘。

5.12　日期选择器

5.12.1　DatePicker 控件

DatePicker 控件(android.widget.DatePicker)是用于选择日期的控件，是 android.widget.FrameLayout 的子类。DatePicker 控件的常用标签属性如表 5-24 所示，DatePicker 类的常用方法如表 5-25 所示。

表 5-24　DatePicker 控件的常用标签属性

属　性	说　明
android:calendarTextColor	设置日历列表的文本颜色
android:calendarViewShown	设置是否显示日历视图
android:datePickerMode	设置外观样式，可选值为 spinner 和 calendar。calendar 是默认值，需要较大的显示面积，因此一般使用 spinner 选取日期
android:maxDate	设置最大日期，格式为 mm/dd/yyyy
android:minDate	设置最小日期，格式为 mm/dd/yyyy
android:spinnersShown	设置是否显示 spinner
android:startYear	设置起始年份
android:endYear	设置结束年份

表 5-25　DatePicker 类的常用方法

类型和修饰符	方　法
open Unit	init (year: Int, monthOfYear: Int, dayOfMonth: Int, onDateChangedListener: DatePicker.OnDateChangedListener!) 初始化默认日期及日期改变的监听器
open Unit	setOnDateChangedListener (onDateChangedListener: DatePicker.OnDateChangedListener!) 设置改变日期的监听器
open Unit	updateDate (year: Int, month: Int, dayOfMonth: Int) 更新日期，月份从 0 开始

5.12.2　实例工程：设置日期的日期选择器

本实例演示了 spinner 外观样式的 DatePicker 控件(如图 5-20 所示)。滑动 DatePicker 控件可以修改年月日。单击“重置”按钮，会将 DatePicker 控件的日期设置为“2020 年 1 月 1 日”。单击“确定”按钮，会获取 DatePicker 控件设置的日期。

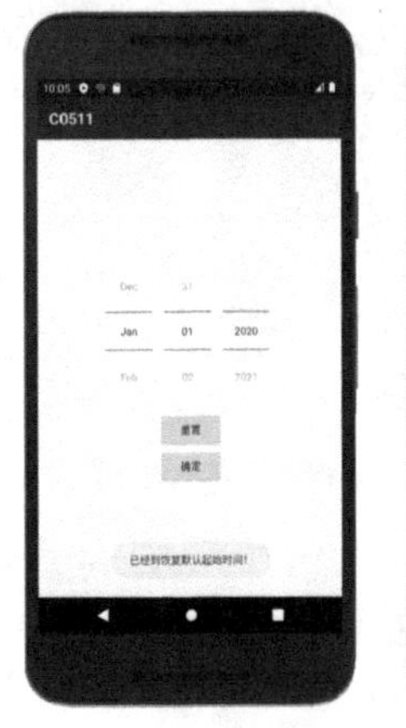

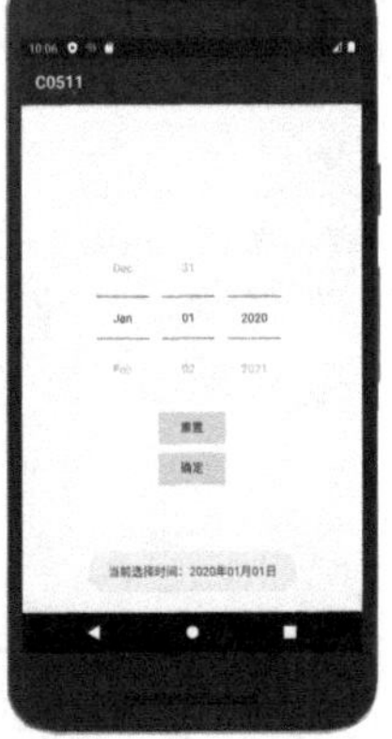

图 5-20　运行效果

1. 新建工程

新建一个“Empty Activity”工程，工程名称为“C0511”。

2. 主界面的布局

/res/layout/activity_main.xml

```
07 <DatePicker
08     android:id = "@+id/date_picker"
09     android:layout_width = "match_parent"
10     android:layout_height = "wrap_content"
11     android:calendarViewShown = "false"
12     android:datePickerMode = "spinner"
13     android:startYear = "2020" />
14 <Button
15     android:id = "@+id/button1"
16     android:layout_width = "wrap_content"
17     android:layout_height = "wrap_content"
18     android:text = "重置" />
19 <Button
20     android:id = "@+id/button2"
21     android:layout_width = "wrap_content"
22     android:layout_height = "wrap_content"
23     android:text = "确定" />
```

第 07 ~ 13 行添加<DatePicker>标签，设置为 spinner 模式、隐藏日历视图和起始年。第 14 ~ 23 行添加两个<Button>标签，分别用于重置 DatePicker 控件和显示 DatePicker 控件选择的日期。

3. 主界面的 Activity

/java/com/teachol/c0511/MainActivity.kt

```
val datePicker = findViewById<DatePicker>(R.id.date_picker)
datePicker.updateDate(2020, 0, 1) //更新日期选择器日期
datePicker.setOnDateChangedListener { view, year, monthOfYear, dayOfMonth ->
    val calendar: Calendar = Calendar.getInstance() //实例化日期
    calendar.set(year, monthOfYear, dayOfMonth) //获取日期
    val simpleDateFormat = SimpleDateFormat("yyyy 年 MM 月 dd 日") //简单日期格式
    Toast.makeText(applicationContext, "当前选择时间: " + simpleDateFormat.format(calendar.getTime()), Toast.LENGTH_SHORT).show()
}
val button1: Button = findViewById(R.id.button1)
button1.setOnClickListener(object : View.OnClickListener {
    override fun onClick(v: View?) {
        datePicker.updateDate(2020, 0, 1) //更新日期选择日期
        Toast.makeText(applicationContext, "已经到恢复默认起始时间! ", Toast.LENGTH_LONG).show()
    }
})
val button2: Button = findViewById(R.id.button2)
button2.setOnClickListener(object :  View.OnClickListener {
    override fun onClick(v: View?) {
        val calendar: Calendar = Calendar.getInstance() //实例化日期
        calendar.set(datePicker.year, datePicker.month, datePicker.dayOfMonth) //获取日期
```

```
36          val simpleDateFormat = SimpleDateFormat("yyyy年MM月dd日")
37          Toast.makeText(applicationContext, "当前设置时间: " + simpleDateFormat.format(calendar.getTime()),
    Toast.LENGTH_SHORT).show()
38      }
39  })
```

第 17 行设置 DatePicker 控件的日期，未设置时间时会以本地时间作为默认时间。第 18 ~ 23 行设置修改日期监听器，Calendar 类合成时间并通过 SimpleDateFormat 类进行时间格式化，然后使用提示信息将其显示出来。第 27 行重新设置 DatePicker 控件的日期。第 34 ~ 37 行获取 DatePicker 控件的日期，然后使用提示信息将其显示出来。

提示：DatePickerDialog 类

DatePickerDialog 类是 android.app.AlertDialog 的子类，相当于在 AlertDialog 控件上加载了 DatePicker 控件，与自定义对话框相比更加快捷。

5.13　时间选择器

5.13.1　TimePicker 控件

TimePicker 控件(android.widget.TimePicker)是用于选择时间的控件，是 android.widget.FrameLayout 的子类。TimePicker 控件的常用标签属性如表 5-26 所示，TimePicker 类的常用方法如表 5-27 所示。

表 5-26　TimePicker 控件的常用标签属性

属　性	说　明
android:timePickerMode	设置外观样式，可选值为 spinner 和 clock。clock 是默认值，以模拟表盘的方式选取时间，也可以直接输入数值

表 5-27　TimePicker 类的常用方法

类型和修饰符	方　法
open Int	getMinute() 获取选择的分钟
open Int	getHour() 获取选择的小时
open Boolean	is24HourView() 判断是否以 24 小时制显示时间
open Unit	setMinute(minute: Int) 设置当前选择的分钟
open Unit	setHour(hour: Int) 使用 24 小时制设置当前选择的小时
open Unit	setIs24HourView(is24HourView: Boolean) 设置是否 24 小时制显示时间
open Unit	setOnTimeChangedListener(onTimeChangedListener: TimePicker.OnTimeChangedListener!) 设置改变时间的监听器

5.13.2　实例工程：设置时间的时间选择器

本实例演示了 12 小时制和 24 小时制的 TimePicker 控件(如图 5-21 所示)。第一个 DatePicker 控件使用 24 小时制；第二个 DatePicker 控件使用 12 小时制，增加了“上午”和“下午”的选择。

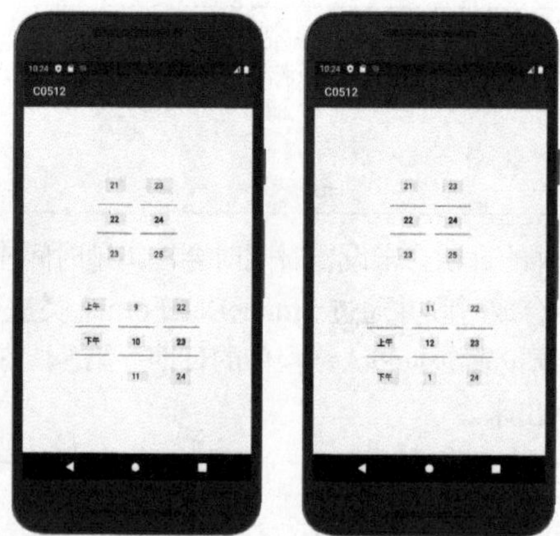

图 5-21　运行效果

1．新建工程

新建一个“Empty Activity”工程，工程名称为“C0512”。

2．主界面的布局

/res/layout/activity_main.xml

```
07 <TimePicker
08     android:id = "@+id/time_picker1"
09     android:layout_width = "match_parent"
10     android:layout_height = "wrap_content"
11     android:timePickerMode = "spinner" />
12 <TimePicker
13     android:id = "@+id/time_picker2"
14     android:layout_width = "match_parent"
15     android:layout_height = "wrap_content"
16     android:timePickerMode = "spinner" />
```

第 07 ~ 16 行添加两个<TimePicker>标签，除 id 外其他属性的设置都相同，用于对比 12 小时制和 24 小时制的区别。由于 24 小时制显示时间没有标签属性，所以需要通过类方法进行设置。

3．主界面的 Activity

/java/com/teachol/c0512/MainActivity.kt

```
// 24 小时制显示时间
val timePicker1: TimePicker = findViewById(R.id.time_picker1)
timePicker1.setIs24HourView(true)
timePicker1.setOnTimeChangedListener(object : TimePicker.OnTimeChangedListener {
    override fun onTimeChanged(view: TimePicker?, hourOfDay: Int, minute: Int) {
        Toast.makeText(this@MainActivity, "timePicker1 的时间是: " + hourOfDay + "时" + minute + "分!", Toast.LENGTH_SHORT).show()
    }
})
// 12 小时制显示时间
val timePicker2: TimePicker = findViewById(R.id.time_picker2)
timePicker2.setOnTimeChangedListener(object : TimePicker.OnTimeChangedListener {
    override fun onTimeChanged(view: TimePicker?, hourOfDay: Int, minute: Int) {
```

```
24             Toast.makeText(this@MainActivity, "timePicker2 的时间是: " + hourOfDay + "时" + minute +
    "分!", Toast.LENGTH_SHORT).show()
25         }
26     })}
```

第 13 ~ 19 行将 time_picker1 设置为 24 小时制显示，并设置改变时间的监听器用于显示改变后的时间。第 21 ~ 27 行 time_picker2 设置改变时间的监听器用于显示改变后的时间。

提示：TimePickerDialog 类

TimePickerDialog 类是 android.app.AlertDialog 的子类，相当于在 AlertDialog 控件上加载了 TimePicker 控件，与自定义对话框相比更加快捷。

5.14　滚动条视图

5.14.1　ScrollView 控件

ScrollView 控件 (android.widget.ScrollView) 是可滚动的用于显示其他控件的控件，是 android.widget.FrameLayout 的子类。ScrollView 控件只能包含一个子控件，否则会报错，但是可以在子控件内再添加控件。ScrollView 控件的常用标签属性如表 5-28 所示，ScrollView 类的常用方法如表 5-29 所示。

表 5-28　ScrollView 控件的常用标签属性

属　性	说　明
android:fillViewport	设置是否拉伸其内容以填充视图。嵌套的子控件高度达不到屏幕高度时，即使 ScrollView 高度设置了 match_parent，也无法充满整个屏幕，需要改属性值为 true，使 ScrollView 充满整个页面
android:scrollbars	设置在滚动时是否显示滚动条。可选值为 none、horizontal 和 vertical
android:scrollbarThumbHorizontal	设置水平方向的滑块图像
android:scrollbarThumbVertical	设置垂直方向的滑块图像
android:descendantFocusability	设置当获取焦点时和其子控件之间的关系。属性值有三种： ● beforeDescendants，优先其子控件而获取到焦点 ● afterDescendants，其子控件无须获取焦点时才获取焦点 ● blocksDescendants，覆盖子控件而直接获得焦点

表 5-29　ScrollView 类的常用方法

类型和修饰符	方　法
open Unit	addView (child: View!) 添加子控件
Int	getScrollX () 获取 mScrollX 属性的值
Int	getScrollY () 获取 mScrollY 属性的值
open Unit	scrollTo (x: Int, y: Int) 立即滚动到指定的坐标位置，不能超出子控件的范围
Unit	smoothScrollBy (dx: Int, dy: Int) 立即滚动到指定的相对坐标位置，不能超出子控件的范围
Unit	smoothScrollTo (x: Int, y: Int) 平滑滚动到指定的坐标位置，不能超出子控件的范围

续表

类型和修饰符	方　法
Unit	smoothScrollBy (dx: Int, dy: Int) 平滑滚动到指定的相对坐标位置，不能超出子控件的范围
open Unit	scrollToDescendant (child: View) 滚动显示出指定的子控件。如果该子控件已经显示出来，则无效果
Boolean	post (action: Runnable!) 将 Runnable 添加到消息队列中，在视图绘制完成后才会执行
open Unit	computeScroll () 父级调用其子控件时，更新其必要的 mScrollX 和 mScrollY 值

5.14.2 实例工程：滚动显示视图

本实例演示了 ScrollView 控件的子控件超出显示范围时通过滚动的方式进行显示（如图 5-22 所示）。ScrollView 控件先向下滑动到 1000 像素，然后单击其下方的 4 个按钮实现不同位置的滑动。

图 5-22　运行效果

1. 新建工程

新建一个“Empty Activity”工程，工程名称为“C0513”。

2. 主界面的布局

```
/res/layout/activity_main.xml
<LinearLayout xmlns:android = "http://schemas.android.com/apk/res/android"
    android:layout_width = "match_parent"
    android:layout_height = "match_parent"
    android:orientation = "vertical">
    <ScrollView
        android:id = "@+id/scroll_view"
        android:layout_width = "match_parent"
        android:layout_height = "520dp">
        <LinearLayout
            android:id = "@+id/linear_layout"
            android:layout_width = "match_parent"
            android:layout_height = "wrap_content"
            android:orientation = "vertical" />
    </ScrollView>
```

```
16     <LinearLayout
17         android:layout_width = "wrap_content"
18         android:layout_height = "wrap_content"
19         android:layout_gravity = "center"
20         android:orientation = "horizontal">
21         <Button
22             android:id = "@+id/button1"
23             android:layout_width = "wrap_content"
24             android:layout_height = "wrap_content"
25             android:text = "滑动到顶部" />
26         <Button
27             android:id = "@+id/button2"
28             android:layout_width = "wrap_content"
29             android:layout_height = "wrap_content"
30             android:text = "滑动到底部" />
31         <Button
32             android:id = "@+id/button3"
33             android:layout_width = "wrap_content"
34             android:layout_height = "wrap_content"
35             android:text = "滑动到指定控件" />
36         <Button
37             android:id = "@+id/button4"
38             android:layout_width = "wrap_content"
39             android:layout_height = "wrap_content"
40             android:text = "随机滑动" />
41     </LinearLayout>
42 </LinearLayout>
```

第 06 ~ 15 行添加一个<ScrollView>标签，其<LinearLayout>子标签用于添加其他控件。第 16 ~ 41 行在一个<LinearLayout>标签内添加 4 个<Button>子标签，用于实现不同的跳转滑动效果。

3. 主界面的 Activity

```
/java/com/teachol/c0513/MainActivity.kt
// scrollView 在视图上加载后执行
val scrollView = findViewById<ScrollView>(R.id.scroll_view)
scrollView.post { scrollView.smoothScrollTo(0, 1000) }
// scrollView 上添加 50 个 TextView 控件
val textView = arrayOfNulls<TextView>(50)
val linearLayout = findViewById<LinearLayout>(R.id.linear_layout)
for (i in 0..49) {
    // 实例化 TextView 数组的元素
    textView[i] = TextView(this)
    textView[i]!!.text = String.format("第%d 个 TextView 控件", 1 + i)
    textView[i]!!.textSize = 30f
    linearLayout.addView(textView[i])
}
// scrollView 滑动到顶部
val button1: Button = findViewById(R.id.button1)
button1.setOnClickListener(object : View.OnClickListener {
    override fun onClick(v: View?) {
```

```
32          scrollView.fullScroll(ScrollView.FOCUS_UP)
33      }
34  })
35  // scrollView滑动到底部
36  val button2: Button = findViewById(R.id.button2)
37  button2.setOnClickListener(object : View.OnClickListener {
38      override fun onClick(v: View?) {
39          scrollView.fullScroll(ScrollView.FOCUS_DOWN)
40      }
41  })
42  // scrollView滑动到指定控件
43  val button3: Button = findViewById(R.id.button3)
44  button3.setOnClickListener(object : View.OnClickListener  {
45      override fun onClick(v: View?) {
46          scrollView.scrollToDescendant(textView[20]!!)
47      }
48  })
49  // scrollView滑动到随机位置
50  val button4: Button = findViewById(R.id.button4)
51  button4.setOnClickListener(object : View.OnClickListener  {
52      override fun onClick(v: View?) {
53          scrollView.smoothScrollTo(0, (Math.random() * 5000).toInt())
54      }
55  })
```

第 17 行设置 ScrollView 控件平滑滚动到 y 坐标 1000 处，Lambda 表达式被识别成 Runnable 实例作为参数。第 19～27 行 TextView 数组的元素进行实例化，为其设置属性后添加到 LinearLayout 布局组件中。第 32 行 fullScroll()方法使用 ScrollView.FOCUS_UP 常量作为参数，移动焦点到顶部。第 39 行 fullScroll()方法使用 ScrollView.FOCUS_DWON 常量作为参数，移动焦点到底部。第 46 行 textView[20]是指 textView[]数组中的第 21 个 TextView 元素。第 53 行 Math.random()*5000 生成一个 5000 以内的随机数并强制转换成 Int 类型数值。

5.15 通　　知

5.15.1 Notification 控件

Notification 控件(android.app.Notification)是显示通知的控件，是 java.lang.Object 的子类。当 App 向系统发出通知时，先以图标的形式显示在通知栏中。当用户可以下拉通知栏后，可以查看通知的详细信息(如图 5-23 所示)。

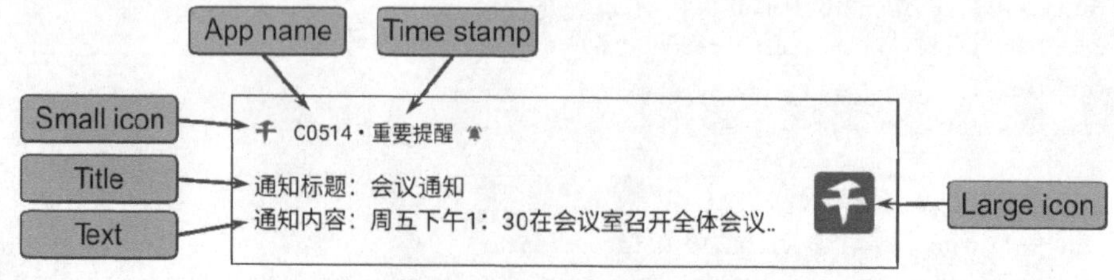

图 5-23　通知详细信息的组成

Notification类没有提供创建通知的方法，需要使用Notification.Builder类（常用方法如表5-30所示）构建通知，NotificationManager 类用于管理通知（常用方法如表 5-31 所示），NotificationChannel 类用于创建通知的通道（常用方法如表 5-32 所示）。

表 5-30　Notification.Builder 类的常用方法

类型和修饰符	方　法
	<init>(context: Context!, channelId: String!) 主构造方法
open Notification.Builder	setAutoCancel(autoCancel: Boolean) 设置用户触摸时状态栏通知是否自动取消
open Notification.Builder	setBadgeIconType(icon: Int) 设置状态栏通知的 badge 图标类型。可选类型包括： ● Notification.BADGE_ICON_NONE，无图标 ● Notification.BADGE_ICON_SMALL，小图标 ● Notification.BADGE_ICON_LARGE，大图标
open Notification.Builder	setChannelId(channelId: String!) 设置通知发送的通道 id
open Notification.Builder	setContentIntent(intent: PendingIntent!) 设置单击通知时要发送的 PendingIntent
open Notification.Builder	setContentText(text: CharSequence!) 设置通知的第二行文本
open Notification.Builder	setContentTitle(title: CharSequence!) 设置通知的第一行文本
open Notification.Builder	setCustomBigContentView(contentView: RemoteViews!) 设置大尺寸的自定义远程视图
open Notification.Builder	setCustomContentView(contentView: RemoteViews!) 设置自定义远程视图
open Notification.Builder	setCustomHeadsUpContentView(contentView: RemoteViews!) 设置自定义弹出式远程视图，自动消失后不在状态栏显示通知
open Notification.Builder	setExtras(extras: Bundle!) 设置附加数据。每次调用 build()时，Bundle 都会复制到通知中
open Notification.Builder	setFullScreenIntent(intent: PendingIntent!, highPriority: Boolean) 当 highPriority 为 true 时，在其他 App 全屏状态时直接 PendingIntent；当 highPriority 为 false 时，在状态栏显示通知
open Notification.Builder	setLargeIcon(b: Bitmap!) 设置通知的大图标
open Notification.Builder	setNumber(number: Int) 设置通知表示的项目数，支持 badge 时作为 badge 数量显示
open Notification.Builder	setOngoing(ongoing: Boolean) 设置是否为正在进行的通知。用户无法通过界面操作取消正在进行的通知，通常用于播放音乐、文件下载、同步操作、活动网络连接等
open Notification.Builder	setOnlyAlertOnce(onlyAlertOnce: Boolean) 设置通知是否只提示一次
open Notification.Builder	setSmallIcon(icon: Int) 设置通知的小图标
open Notification.Builder	setSubText(text: CharSequence!) 设置通知的附加信息
open Notification.Builder	setTicker(tickerText: CharSequence!) 设置辅助功能的文本
open Notification.Builder	setTimeoutAfter(durationMs: Long) 设置通知自动消失的时间

续表

类型和修饰符	方　法
open Notification.Builder	setVisibility (visibility: Int) 设置锁屏状态时通知的可见属性，可选值包括： ● Notification.VISIBILITY_PUBLIC，锁屏时显示通知 ● Notification.VISIBILITY_PRIVATE，锁屏时显示通知，但在安全锁屏时隐藏敏感或私有信息 ● Notification.VISIBILITY_SECRET，安全锁屏时不显示通知

表 5-31　NotificationManager 类的常用方法

类型和修饰符	方　法
open Unit	createNotificationChannel (channel: NotificationChannel) 创建通知通道
open Unit	cancel (id: Int) 取消显示指定 id 的通知
open Unit	cancel (tag: String?, id: Int) 取消显示指定 tag 和 id 的通知
open Unit	cancelAll () 取消所有显示的通知
open Unit	deleteNotificationChannel (channelId: String!) 删除指定 id 的通知通道
open Unit	notify (id: Int, notification: Notification!) 发送通知，为该通知指定 id
open Unit	notify (tag: String!, id: Int, notification: Notification!) 发送通知，为该通知指定 tag 和 id

表 5-32　NotificationChannel 类的常用方法

类型和修饰符	方　法
	<init> (id: String!, name: CharSequence!, importance: Int) 主构造方法
Unit	enableLights (lights: Boolean) 是否允许通知灯
Unit	enableVibration (vibration: Boolean) 是否允许通知震动
Unit	setAllowBubbles (allowBubbles: Boolean) 设置是否允许 Bubble
Unit	setImportance (importance: Int) 设置通知级别。可选值包括： ● NotificationManager.IMPORTANCE_UNSPECIFIED，未表示重要性的通知 ● NotificationManager.IMPORTANCE_NONE，不在状态栏中显示 ● NotificationManager.IMPORTANCE_MIN，最小化的通知，折叠在状态栏中 ● NotificationManager.IMPORTANCE_LOW，低级别的通知，显示在状态栏的折叠通知中，并可能显示在状态栏中，没有提示音 ● NotificationManager.IMPORTANCE_DEFAULT，普通重要通知，有提示音 ● NotificationManager.IMPORTANCE_HIGH，高级别重要通知，弹出式显示，可以在全屏下显示
Unit	setShowBadge (showBadge: Boolean) 设置是否在 App 启动图标上显示 badge
Unit	setSound (sound: Uri!, audioAttributes: AudioAttributes!) 设置通知的提示音
Unit	setVibrationPattern (vibrationPattern: LongArray!) 设置通知的震动间隔
Boolean	shouldShowLights () 获取是否触发通知灯
Boolean	shouldVibrate () 获取是否触发震动

5.15.2　实例工程：弹出式通知和自定义视图通知

本实例演示了弹出式状态栏通知和自定义视图状态栏通知(如图 5-24 所示)，单击“关闭状态栏通知”可以关闭通知。

图 5-24　运行效果

1. 打开基础工程

打开“基础工程”文件夹中的“C0514”工程，该工程已经包含 MainActivity 及资源文件。源代码部分都是之前讲过的内容，读者可自行查看源代码。

2. 主界面的 Activity

/java/com/teachol/c0514/MainActivity.kt

```
class MainActivity : AppCompatActivity(), View.OnClickListener {
    private val NOTIFY_TAG = "通知"
    private val CHANNL_ID1 = "2401"
    private val CHANNL_NAME1 = "通知"
    private val CHANNL_ID2 = "2402"
    private val CHANNL_NAME2 = "对话"
    private var mContext: Context? = null
    private var mNotificationManager: NotificationManager? = null
    private var mNotificationBuilder: Notification.Builder? = null
    private var mNotification: Notification? = null
    override fun onCreate(savedInstanceState: Bundle?) {
        super.onCreate(savedInstanceState)
        setContentView(R.layout.activity_main)
        mContext = this@MainActivity
        mNotificationManager = getSystemService(NOTIFICATION_SERVICE) as NotificationManager
        findViewById<View>(R.id.btn_show_notification).setOnClickListener(this)
        findViewById<View>(R.id.btn_show_custom_notification).setOnClickListener(this)
        findViewById<View>(R.id.btn_close_notification).setOnClickListener(this)
    }
    // 单击事件监听器
    override fun onClick(v: View) {
        when (v.getId()) {
```

```
            R.id.btn_show_notification -> {
                // 兼容Android 8.0(API Level 26)以后的版本
                if (Build.VERSION.SDK_INT >= Build.VERSION_CODES.O) {
                    val notificationChannel = NotificationChannel(CHANNL_ID1, CHANNL_NAME1, NotificationManager.IMPORTANCE_HIGH)
                    notificationChannel.setShowBadge(true)
                    notificationChannel.enableVibration(true)
                    notificationChannel.enableLights(true)
                    notificationChannel.setSound(RingtoneManager.getDefaultUri(RingtoneManager.TYPE_NOTIFICATION), Notification.AUDIO_ATTRIBUTES_DEFAULT)
                    notificationChannel.vibrationPattern = longArrayOf(100, 200, 300, 400, 500, 400, 300, 200, 400)
                    mNotificationManager!!.createNotificationChannel(notificationChannel)
                    mNotificationBuilder = Notification.Builder(mContext, CHANNL_ID1)
                } else {
                    mNotificationBuilder = Notification.Builder(mContext)
                }
                // 设置通知参数
                mNotificationBuilder!!.setContentTitle("通知标题：会议通知")
                        .setContentText("通知内容：周五下午1：30在会议室召开全体会议。")
                        .setTicker("收到一条会议信息")
                        .setSmallIcon(R.mipmap.notification_small)
                        .setLargeIcon(BitmapFactory.decodeResource(resources, R.mipmap.notification_icon))
                        .setBadgeIconType(Notification.BADGE_ICON_SMALL)
                        .setNumber(11)
                        .setSubText("重要提醒")
                        .setAutoCancel(true)
                mNotification = mNotificationBuilder!!.build()
                // 发出通知
                mNotificationManager!!.notify(NOTIFY_TAG, Integer.valueOf(CHANNL_ID1), mNotification)
            }
            R.id.btn_show_custom_notification -> {
                // 兼容Android 8.0(API Level 26)以后的版本
                if (Build.VERSION.SDK_INT >= Build.VERSION_CODES.O) {
                    val notificationChannel = NotificationChannel(CHANNL_ID2, CHANNL_NAME2, NotificationManager.IMPORTANCE_HIGH)
                    notificationChannel.setShowBadge(true)
                    notificationChannel.vibrationPattern = longArrayOf(100, 200, 300, 400, 500, 400, 300, 200, 400)
                    mNotificationManager!!.createNotificationChannel(notificationChannel)
                    mNotificationBuilder = Notification.Builder(mContext, CHANNL_ID2)
                } else {
                    mNotificationBuilder = Notification.Builder(mContext)
                }
                // 设置自定义视图
                val pendingIntent = PendingIntent.getActivity(mContext, 1, Intent(mContext, MainActivity::class.java), PendingIntent.FLAG_UPDATE_CURRENT)
                val remoteView = RemoteViews(packageName, R.layout.notification_custom)
```

```
81                  remoteView.setImageViewResource(R.id.user_face, R.mipmap.img_user_face)
82                  remoteView.setOnClickPendingIntent(R.id.btn_reply, pendingIntent)
83                  // 设置通知参数
84                  mNotificationBuilder!!.setContentTitle("通知标题: 会议通知")
85                          .setContentText("通知内容: 周五下午 1: 30 在会议室召开全体会议。")
86                          .setTicker("收到一条会议信息")
87                          .setSmallIcon(R.mipmap.notification_small)
88                          .setLargeIcon(BitmapFactory.decodeResource(resources,
R.mipmap.notification_icon))
89                          .setNumber(10)
90                          .setVisibility(Notification.VISIBILITY_PUBLIC)
91                          .setCustomContentView(remoteView)
92                  mNotification = mNotificationBuilder!!.build()
93                  // 发出通知
94                  mNotificationManager!!.notify(NOTIFY_TAG, Integer.valueOf(CHANNL_ID2),
mNotification)
95              }
96              R.id.btn_close_notification -> {
97                  // 除可以根据 ID 来取消 Notification 外，还可以调用 cancelAll();关闭该应用的所有通知
98                  mNotificationManager!!.cancel(NOTIFY_TAG, Integer.valueOf(CHANNL_ID1))
99                  mNotificationManager!!.cancel(NOTIFY_TAG, Integer.valueOf(CHANNL_ID2))
100                 mNotificationManager!!.deleteNotificationChannel(CHANNL_ID1)
101                 mNotificationManager!!.deleteNotificationChannel(CHANNL_ID2)
102             }
103         }
104     }
105 }
```

第 18～22 行定义通知需要使用的常量。第 31 行 NotificationManager 类没有主构造方法，需要将 getSystemService(NOTIFICATION_SERVICE) 转换为 NotificationManager，然后初始化 NotificationManager 实例。第 41 行 Build.VERSION.SDK_INT 表示生成 App 使用的 SDK 版本，Build.VERSION_CODES.O 表示使用 Android O 版本(API Level 26)的 SDK。第 42～47 行创建通知通道的实例，并设置相关属性。第 54～63 行设置通知所需的相关属性，然后使用 build()方法构建通知。第 79 行创建 PendingIntent 实例作为自定义视图中“回复”按钮单击时触发的对象。第 80～82 行创建 RemoteViews 实例，用于设置自定义视图，然后在第 91 行设置该实例为通知的自定义视图。

提示：RemoteViews

RemoteViews 只需要包名和待加载的资源文件 id，并不支持所有类型的 View，也不支持自定义的 View，支持的类型如下。

- Layout：FrameLyout、LinearLayout、RelativeLayout、GridLayout。
- View：Button、ImageView、ImageButton、ProgressBar、TextView、ListView、GridView、StackView、ViewStub、AdapterViewFlipper、ViewFlipper、AnalogClock、Chronometer。

第 6 章　Android 的布局组件

虽然控件可以不依靠布局组件单独使用，但是在实际使用中，几乎很少只使用一个控件。当多个控件同时使用时，需要通过布局组件组织其排列方式和位置。相同的界面效果可以使用不同的布局组件实现，因此需要熟悉各种布局组件的特点，选择最适合的使用。

6.1　线性布局

6.1.1　LinearLayout 组件

LinearLayout 组件(android.widget.LinearLayout)是进行水平或垂直排列布局的容器组件，是 android.view.ViewGroup 的子类。LinearLayout 控件的常用标签属性如表 6-1 所示，LinearLayout 子标签的标签属性如表 6-2 所示。

提示：Android 布局的归类

Android 的布局通常是不可见的，但是如果设置了分割线，就会部分可视化。本书采用主流观点，将布局归纳为组件。

表 6-1　LinearLayout 组件的常用标签属性

属　性	说　明
android:divider	设置分割线的图像资源
android:gravity	设置内部组件的对齐方式
android:measureWithLargestChild	设置是否以最大尺寸内部组件作为所有内部组件的尺寸
android:orientation	设置内部组件的排列方式，可选值包括 horizontal 和 vertical
android:weightSum	设置内部组件权重值的最大和

表 6-2　LinearLayout 子标签的标签属性

属　性	说　明
android:layout_gravity	设置布局组件内的定位方式
android:layout_weight	设置内部组件在布局组件内所占的权重。内部组件权重除以布局组件内所有内部组件权重之和的值是分配该内部组件剩余空间的比例

提示：gravity 和 layout_gravity

android:gravity 是针对子标签而言的，用来控制子标签的显示位置。当 Button 控件设置 android:gravity="left"时，Button 控件上的文字将会位于内部左侧。

android:layout_gravity 是针对自身而言的，用来控制所属布局组件中的位置。当 Button 控件设置 android:layout_gravity="left"时，Button 控件将位于所属布局组件内的左侧。

提示：RTL 布局

从 Android 4.2 开始，Android SDK 支持一种从右到左(Right-to-Left，RTL)的 UI 布局方式，以右边上角作为原点，从右到左、从上到下进行排列。经常使用在阿拉伯语、希伯来语等环境中。如果使用 RTL 布局，需要在 AndroidManifest.xml 文件中将<application>标签的 android:supportsRtl 属性值设为"true"，然后将控件标签的 android:layoutDirection 属性值设为"rtl"即可。

6.1.2　实例工程：动态视图的线性布局

本实例演示了水平和垂直两种线性布局及嵌套布局的效果(如图 6-1 所示)。整体使用垂直线性布局，局部使用了多重嵌套的线性布局。

图 6-1　运行效果

1. 新建工程并导入素材

新建一个“Empty Activity”工程，工程名称为“C0601”。选择“/res/drawable”资源文件夹，将素材图像(素材文件夹路径为“/素材/C0601”)粘贴或拖曳到该文件夹中。

2. 主界面的布局

/res/layout/activity_main.xml

```
<!--垂直线性布局-->
<LinearLayout xmlns:android = "http://schemas.android.com/apk/res/android"
    android:layout_width = "match_parent"
    android:layout_height="wrap_content"
    android:orientation = "vertical">
    <!--水平线性布局-->
    <LinearLayout
        android:layout_width = "match_parent"
        android:layout_height = "50dp"
        android:layout_margin = "5dp"
        android:orientation = "horizontal">
        <ImageView
            android:layout_width = "50dp"
            android:layout_height = "50dp"
            android:scaleType = "centerCrop"
            android:src = "@mipmap/img_face" />
        <!--垂直布局-->
```

```
19         <LinearLayout
20             android:layout_width = "0dp"
21             android:layout_height = "match_parent"
22             android:layout_marginStart = "5dp"
23             android:layout_weight = "1"
24             android:orientation = "vertical">
25             <TextView
26                 android:layout_width = "wrap_content"
27                 android:layout_height = "wrap_content"
28                 android:text = "沉睡的海螺"
29                 android:textColor = "#000"
30                 android:textSize = "18sp" />
31             <TextView
32                 android:layout_width = "wrap_content"
33                 android:layout_height = "0dp"
34                 android:layout_weight = "1"
35                 android:gravity = "bottom"
36                 android:text = "2020 年 4 月 7 日" />
37         </LinearLayout>
38         <Button
39             android:layout_width = "0dp"
40             android:layout_height = "wrap_content"
41             android:layout_gravity = "center"
42             android:layout_weight = "0.3"
43             android:text = "关注" />
44     </LinearLayout>
45     <ImageView
46         android:layout_width = "411dp"
47         android:layout_height = "411dp"
48         android:scaleType = "centerCrop"
49         android:src = "@mipmap/img_pic" />
50     <TextView
51         android:layout_width = "wrap_content"
52         android:layout_height = "wrap_content"
53         android:layout_gravity = "end"
54         android:layout_margin = "5dp"
55         android:text = "举报" />
56 </LinearLayout>
```

第 06 行 android:orientation="vertical"设置所有子标签以垂直线性布局排列。第 12 行 android:orientation="horizontal"设置所有子标签以水平线性布局排列。第 23 行 android:layout_weight="1"设置 LinearLayout 布局方向上的剩余空间分配给<TextView>标签的比例。第 42 行 android:layout_weight="0.3"设置父标签布局方向上的剩余空间分配给<Button>标签的比例。第 53 行 android:layout_gravity="end"设置该标签在父标签内右侧对齐。

6.2 相对布局

6.2.1 RelativeLayout 组件

RelativeLayout 组件(android.widget.RelativeLayout)是相对排列布局的容器组件，是 android.view.ViewGroup 的子类。除继承的标签属性外只有 2 个标签属性(如表 6-3 所示)。子标签在 RelativeLayout 标签内的定位通过子标签的属性进行设置(如表 6-4 所示)。

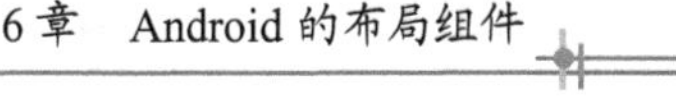

表 6-3 RelativeLayout 组件的 XML 标签属性

属 性	说 明
android:gravity	设置子标签的对齐方式
android:ignoreGravity	设置是否忽略 android:gravity 属性的设置

表 6-4 RelativeLayout 子标签的常用 XML 标签属性

属 性	描 述
android:layout_above	设置该标签位于指定 id 的标签上方，下边与指定 id 的标签上边对齐
android:layout_below	设置该标签位于指定 id 的标签下方，上边与指定 id 的标签下边对齐
android:layout_toStartOf	设置该标签位于指定 id 的标签左侧，右边与指定 id 的标签左边对齐
android:layout_toEndOf	设置该标签位于指定 id 的标签右侧，左边与指定 id 的标签右边对齐
android:layout_alignTop	设置该标签的上边和指定 id 的标签的上边对齐
android:layout_alignBottom	设置该标签的下边和指定 id 的标签的下边对齐
android:layout_alignLeft	设置该标签的左边和指定 id 的标签的左边对齐
android:layout_alignRight	设置该标签的右边和指定 id 的标签的右边对齐
android:layout_alignParentStart	设置该标签的左边和 RelativeLayout 标签的左边对齐
android:layout_alignParentTop	设置该标签的上边和 RelativeLayout 标签的上边对齐
android:layout_alignParentEnd	设置该标签的右边和 RelativeLayout 标签的右边对齐
android:layout_alignParentBottom	设置该标签的下边和 RelativeLayout 标签的下边对齐
android:layout_centerInParent	设置该标签位于 RelativeLayout 标签的中心位置
android:layout_centerHorizontal	设置该标签位于水平方向的中心位置
android:layout_centerVertical	设置该标签位于垂直方向的中心位置

6.2.2 实例工程：显示方位的相对布局

本实例演示了 9 种相对布局的效果，“中”、“顶部”、“底部”、“左侧”和“右侧”是相对于父标签的相对布局，“上”、“下”、“左”和“右”是相对于“中”的相对布局(如图 6-2 所示)。

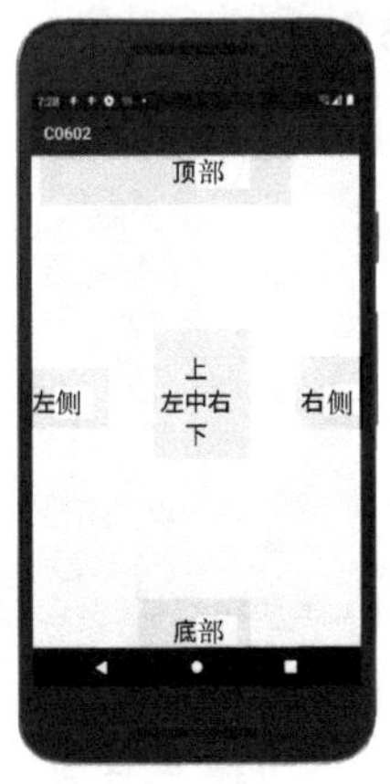

图 6-2 运行效果

1. 新建工程

新建一个“Empty Activity”工程，工程名称为“C0602”。

2. 主界面的布局

/res/layout/activity_main.xml

```
<RelativeLayout xmlns:android = "http://schemas.android.com/apk/res/android"
    android:layout_width = "match_parent"
    android:layout_height = "match_parent">
    <!--在RelativeLayout控件内顶部显示-->
    <TextView
        android:layout_alignParentTop = "true"
        android:layout_centerHorizontal = "true"
        android:text = "顶部"
        android:textSize = "30sp" />
    <!--在RelativeLayout控件内底部显示-->
    <TextView
        android:layout_alignParentBottom = "true"
        android:layout_centerHorizontal = "true"
        android:text = "底部"
        android:textSize = "30sp" />
    <!--在RelativeLayout控件内左侧显示-->
    <TextView
        android:layout_alignParentStart = "true"
        android:layout_centerVertical = "true"
        android:text = "左侧"
        android:textSize = "30sp" />
    <!--在RelativeLayout控件内右侧显示-->
    <TextView
        android:layout_alignParentEnd = "true"
        android:layout_centerVertical = "true"
        android:text = "右侧"
        android:textSize = "30sp" />
    <!--在RelativeLayout控件内居中显示-->
    <TextView
        android:id = "@+id/text_view"
        android:layout_centerInParent = "true"
        android:text = "中"
        android:textSize = "30sp" />
    <!--显示在text_view上方-->
    <TextView
        android:layout_above = "@id/text_view"
        android:layout_centerHorizontal = "true"
        android:text = "上"
        android:textSize = "30sp" />
    <!--显示在text_view下方-->
    <TextView
```

```
57            android:layout_below = "@id/text_view"
58            android:layout_centerHorizontal = "true"
59            android:text = "下"
60            android:textSize = "30sp" />
61        <!--显示在 text_view 左侧-->
62        <TextView
65            android:layout_centerVertical = "true"
66            android:layout_toStartOf = "@id/text_view"
67            android:text = "左"
68            android:textSize = "30sp" />
69        <!--显示在 text_view 右侧-->
70        <TextView
73            android:layout_centerVertical = "true"
74            android:layout_toEndOf = "@id/text_view"
75            android:text = "右"
76            android:textSize = "30sp" />
77    </RelativeLayout>
```

第 06～44 行的 5 个<TextView>标签通过属性设置 RelativeLayout 组件内部的 5 种定位方式，android:layout_centerVertical="true"设置左侧和右侧的垂直居中，android:layout_centerHorizontal="true"设置顶部和底部的垂直居中。第 46～76 行的 4 个标签通过属性设置与 text_view 的相对位置。

6.3　表格布局

6.3.1　TableLayout 组件

TableLayout 组件(android.widget.TableLayout)是使用表格形式布局的容器组件，是 android.widget.LinearLayout 的子类，除继承的标签属性外还有 3 个标签属性(如表 6-5 所示)。TableLayout 组件还需要与 TableRow 组件配合使用，TableRow 组件是表格行的容器组件，其每一个子标签放置于一个单元格内，并可以通过子标签设置单元格的相关属性(如表 6-6 所示)。

表 6-5　TableLayout 组件的标签属性

属　性	说　明
android:collapseColumns	设置需要被隐藏的列序号
android:shrinkColumns	设置允许被收缩的列序号
android:stretchColumns	设置运行被拉伸的列序号

表 6-6　TableRow 子组件的标签属性

属　性	说　明
android:layout_column	设置跳过的单元格数量
android:layout_span	设置与当前单元格合并的单元格数量

6.3.2　实例工程：登录界面的表格视图

本实例演示了使用表格布局显示登录界面，“登录”按钮占用两个单元格，其余控件各占用一个单元格(如图 6-3 所示)。

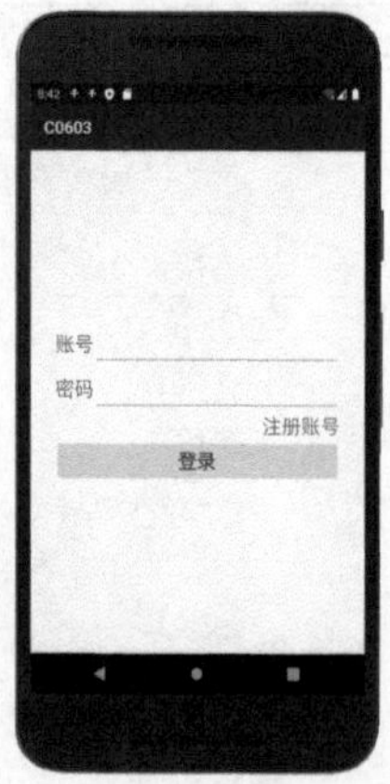

图 6-3　运行效果

1. 新建工程

新建一个“Empty Activity”工程，工程名称为“C0603”。

2. 主界面的布局

/res/layout/activity_main.xml

```
<TableLayout xmlns:android = "http://schemas.android.com/apk/res/android"
    android:layout_margin = "30dp"
    android:layout_gravity = "center"
    android:collapseColumns = "2"
    android:stretchColumns = "1">
    <!--第1行-->
    <TableRow>
        <!--第1列-->
        <TextView
            android:text = "账号"
            android:textSize = "24sp" />
        <!--第2列-->
        <EditText
            android:ems = "10"
            android:inputType = "textPersonName"
            android:textSize = "24sp" />
    </TableRow>
    <!--第2行-->
    <TableRow>
        <!--第1列-->
        <TextView
            android:text = "密码"
            android:textSize = "24sp" />
        <!--第2列-->
        <EditText
            android:ems = "10"
            android:inputType = "textPassword"
            android:textSize = "24sp" />
        <!--第3列-->
```

```
41          <TextView
44              android:text = "忘记密码"
45              android:textSize = "18sp" />
46      </TableRow>
47      <!--第 3 行-->
48      <TableRow>
49          <!--第 2 列-->
50          <TextView
51              android:id = "@+id/textView4"
54              android:layout_column = "1"
55              android:gravity = "end"
56              android:text = "注册账号"
57              android:textSize = "24sp" />
58      </TableRow>
59      <!--第 4 行-->
60      <TableRow>
61          <!--第 1 列和第 2 列合并-->
62          <Button
65              android:layout_span = "2"
66              android:text = "登录"
67              android:textSize = "24sp" />
68      </TableRow>
69  </TableLayout>
```

第 07 行 android:collapseColumns="2"设置表格布局的第 3 列隐藏不显示。第 08 行 android:stretchColumns="1"设置表格布局的第 2 列可以被拉伸。第 54 行 android:layout_column="1"设置跳过当前行前一个单元格。第 65 行 android:layout_span="2"设置表格布局的第 4 行第 1 列和第 2 列的单元格合并为同一个单元格。

6.4　网格布局

6.4.1　GridLayout 组件

GridLayout 组件（android.widget.GridLayout）是网格形式排列布局的容器组件，是 android.view.ViewGroup 的子类，GridLayout 组件的常用标签属性如表 6-7 所示。

表 6-7　GridLayout 组件的常用标签属性

属　性	说　明
android:alignmentMode	设置对齐模式，可选值为 alignBounds 和 alignMargins
android:columnCount	设置最大的列数
android:orientation	设置子标签排列顺序的方向，可选值包括 horizontal 和 vertical
android:rowCount	设置最大的行数
android:useDefaultMargins	设置是否使用默认边距

6.4.2　实例工程：模仿计算器界面的网格布局

本实例演示了使用网格布局显示计算器的界面，第一行控件占用 4 列网格，第二行控件各占用 2 列网格（如图 6-4 所示）。

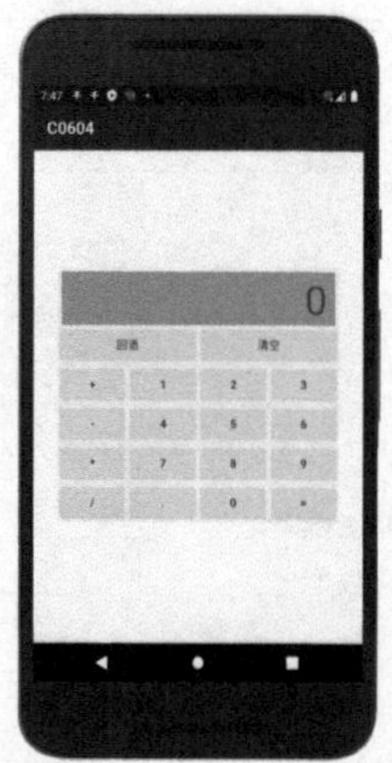

图 6-4　运行效果

1．新建工程

新建一个“Empty Activity”工程，工程名称为“C0604”。

2．主界面的布局

/res/layout/activity_main.xml

```
<GridLayout xmlns:android = "http://schemas.android.com/apk/res/android"
    android:layout_gravity = "center"
    android:background = "#eeeeee"
    android:columnCount = "4"
    android:rowCount = "6">
    <TextView
        android:layout_columnSpan = "4"
        android:layout_gravity = "fill"
        android:gravity = "right"
        android:layout_marginLeft = "5dp"
        android:layout_marginRight = "5dp"
        android:paddingRight = "10dp"
        android:background = "#999999"
        android:text = "0"
        android:textSize = "50sp" />
    <Button
        android:layout_columnSpan = "2"
        android:layout_gravity = "fill"
        android:text = "回退" />
    <Button
        android:layout_columnSpan = "2"
        android:layout_gravity = "fill"
        android:text = "清空" />
    <Button android:text = "+" />
    <Button android:text = "1" />
    <Button android:text = "2" />
    <Button android:text = "3" />
    <Button android:text = "-" />
    <Button android:text = "4" />
```

```
33      <Button android:text = "5" />
34      <Button android:text = "6" />
35      <Button android:text = "*" />
36      <Button android:text = "7" />
37      <Button android:text = "8" />
38      <Button android:text = "9" />
39      <Button android:text = "/" />
40      <Button android:text = "." />
41      <Button android:text = "0" />
42      <Button android:text = "=" />
43  </GridLayout>
```

第 07 行 android:columnCount="4"设置 GridLayout 组件分为 4 列。第 08 行 android:rowCount="6"设置 GridLayout 组件分为 6 行。第 10 行 android:layout_columnSpan="4"设置 TextView 控件占用 4 列网格。第 19 ~ 26 行 android:layout_columnSpan="2" 设置 Button 控件占用 2 列网格，android:layout_gravity="fill"设置完全充满 GridLayout 组件水平和垂直方向的空间。

6.5　帧　布　局

6.5.1　FrameLayout 组件

FrameLayout 组件(android.widget.FrameLayout)是依次堆叠形式排列布局的容器组件，是 android.view.ViewGroup 的子类，后添加的子标签置于上层，除继承的标签属性外还有 2 个标签属性(如表 6-8 所示)。

表 6-8　GridLayout 组件的标签属性

属　性	说　明
android:foregroundGravity	设置前景图像显示的位置
android:measureAllChildren	设置是否对子标签进行测量，默认值为 true

6.5.2　实例工程：分层显示图像的帧布局

本实例演示了使用帧布局叠加显示图像的效果，文字和汽车图像包含透明区域，透明区域直接显示出背景图像(如图 6-5 所示)。

图 6-5　运行效果

1. 新建工程并导入素材

新建一个“Empty Activity”工程，工程名称为“C0605”。选择“/res/mipmap”资源文件夹，将素材图像(素材文件夹路径为“/素材/C0605”)粘贴或拖曳到该文件夹中。

2. 主界面的布局

```
/res/layout/activity_main.xml
02  <FrameLayout xmlns:android = "http://schemas.android.com/apk/res/android"
03     android:layout_width = "wrap_content"
04     android:layout_height = "wrap_content">
05     <!--背景-->
06     <ImageView
09        android:scaleType = "fitXY"
10        android:src = "@mipmap/img_bg" />
11     <!--汽车-->
12     <ImageView
15        android:src = "@mipmap/img_car" />
16     <!--标题-->
17     <ImageView
20        android:src = "@mipmap/img_title" />
21  </FrameLayout>
```

第 06 ~ 20 行添加三个<ImageView>标签，自下而上进行排列。图像都使用了带透明度通道的 PNG 格式，汽车和标题的图像透明区域直接显示出背景图像。

6.6 约 束 布 局

6.6.1 ConstraintLayout 组件

ConstraintLayout 组件(androidx.constraintlayout.widget.ConstraintLayout)是依次堆叠形式排列布局的容器组件，其子标签通过属性进行约束(如表 6-9 所示)。ConstraintLayout 组件与 RelativeLayout 组件类似，但嵌套层级更少且更加灵活。定位标签的属性值为“parent”时，是指子标签相对于 ConstraintLayout 组件的位置。定位标签的属性值为其他子标签的 id 时，是指子标签相对于其他子标签的位置。

表 6-9　ConstraintLayout 子标签的常用标签属性

属　性	说　明
app:layout_constraintStart_toStartOf	设置左侧与指定标签的左侧约束
app:layout_constraintStart_toEndOf	设置左侧与指定标签的右侧约束
app:layout_constraintEnd_toStartOf	设置右侧与指定标签的左侧约束
app:layout_constraintEnd_toEndOf	设置右侧与指定标签的右侧约束
app:layout_constraintTop_toTopOf	设置顶部与指定标签的顶部约束
app:layout_constraintTop_toBottomOf	设置顶部与指定标签的底部约束
app:layout_constraintBottom_toTopOf	设置底部与指定标签的顶部约束
app:layout_constraintBottom_toBottomOf	设置底部与指定标签的底部约束

续表

属　性	说　明
app:layout_constraintDimensionRatio	设置宽度和高度的比例。设置该属性时高度和宽度至少有一个值设置为 0dp，然后在运行时根据该属性值自动计算设置为 0dp 的尺寸；如果都是 0dp，在运行时根据所有约束条件和该属性值自动计算最大尺寸
app:layout_constraintCircleAngle	设置约束角度
app:layout_constraintCircleRadius	设置约束角度的半径

提示：AndroidX

AndroidX 是 Google 从 API Level 28（Android 9.0）开始推荐使用的新支持库，具有实时更新、无须经常修改版本的优点。例如，老版本的 android.support.v4.app 支持库的 v4 表示 Android API 版本号，提供的 API 会向下兼容到 Android 1.6（API Level 4.0）。原有的支持库虽然被保留可以继续使用，但从 Android Studio 3.4.2 开始，新建项目默认使用 AndroidX 架构。原有项目可以使用菜单【Refactor】→【Migrate to AndroidX】命令进行迁移。

6.6.2　实例工程：模仿朋友圈顶部的约束布局

本实例演示了使用约束布局模仿朋友圈顶部的效果（如图 6-6 所示）。背景图像与其父控件进行约束，头像图像与背景图像进行约束，昵称文本与头像图像进行约束。

图 6-6　运行效果

1．新建工程并导入素材

新建一个“Empty Activity”工程，工程名称为“C0606”。选择“/res/mipmap”资源文件夹，将素材图像（素材文件夹路径为“/素材/C0606”）粘贴或拖曳到该文件夹中。

2．主界面的布局

/res/layout/activity_main.xml

```
<androidx.constraintlayout.widget.ConstraintLayout
    android:layout_width = "match_parent"
    android:layout_height = "match_parent">
    <!--背景图像-->
```

```
08    <ImageView
09        android:id = "@+id/image_view_bg"
10        android:layout_width = "wrap_content"
11        android:layout_height = "0dp"
12        app:layout_constraintDimensionRatio = "1.5"
13        app:layout_constraintStart_toStartOf = "parent"
14        app:layout_constraintTop_toTopOf = "parent"
15        app:srcCompat = "@mipmap/img_bg" />
16    <!--头像图像-->
17    <ImageView
18        android:id = "@+id/image_view_face"
19        android:layout_width = "80dp"
20        android:layout_height = "80dp"
21        android:layout_marginTop = "212dp"
22        android:layout_marginEnd = "30dp"
23        app:layout_constraintEnd_toEndOf = "@+id/image_view_bg"
24        app:layout_constraintTop_toTopOf = "@+id/image_view_bg"
25        app:srcCompat = "@mipmap/img_face" />
26    <!--昵称-->
27    <TextView
28        android:layout_width = "wrap_content"
29        android:layout_height = "wrap_content"
30        android:layout_marginTop = "25dp"
31        android:layout_marginEnd = "20dp"
32        android:text = "沉睡的海螺"
33        android:textColor = "#FFFFFF"
34        android:textSize = "18sp"
35        app:layout_constraintEnd_toStartOf = "@+id/image_view_face"
36        app:layout_constraintTop_toTopOf = "@+id/image_view_face" />
37 </androidx.constraintlayout.widget.ConstraintLayout>
```

第 12 ~ 14 行 app:layout_constraintDimensionRatio="1.5"设置<ImageView>标签宽度和高度的约束比例，app:layout_constraintEnd_toEndOf="parent"设置<ImageView>标签右侧与父标签的右侧进行约束，app:layout_constraintTop_toTopOf="parent"设置<ImageView>标签顶部与父标签的顶部进行约束。第 21 ~ 24 行设置<ImageView>标签与 id 为“image_view_bg”的<ImageView>标签的约束关系，通过 android:layout_marginTop="212dp"设置与 id 为“image_view_bg”的<ImageView>标签顶部的距离，通过 android:layout_marginEnd="30dp"设置与 id 为“image_view_bg”的<ImageView>标签顶部的距离。第 35、36 行设置了<TextView>标签与 id 为“image_view_bg”的<ImageView>标签的约束关系。

提示：AbsoluteLayout（绝对布局）

AbsoluteLayout 在 API Level 3 的时候就已经不推荐使用了，其子标签使用 android:layout_x 和 android:layout_y 设置坐标。但是，对于屏幕适配的情况，绝对布局还需要进行转换，可以考虑使用 ConstraintLayout 组件。

第 7 章　Android 的进阶控件与适配绑定

有一类控件单独使用无法实现其应有的功能，需要适配器或其他控件配合使用。这类控件中，有些通过适配器会将数据匹配后显示出来，有些可以作为其他控件的容器。这类控件的主要特点是功能和内容相分离，降低了耦合性，特别适合可变长度的内容呈现。

7.1　数据适配原理

有一类控件视图(View)的内部包含子视图(ItemView)，每个子视图对应一组子数据(ItemData)。但是，不直接通过控件标签属性或类方法设置所需的数据，而是通过适配器(Adapter)将子视图与对应的子数据关联起来(如图 7-1 所示)。当子数据包含的数据项较多时，如果使用数组存储和传递子数据，就需要几个甚至十几个数组，不利于管理和使用。这种情况下，推荐将子数据存储在模型(Model)中，然后使用模型(Model)通过适配器(Adapter)与子视图(ItemView)进行关联。

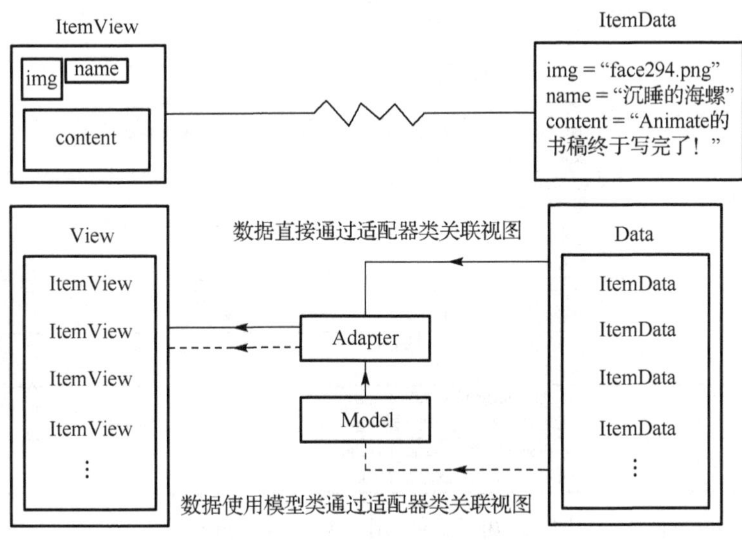

图 7-1　数据适配原理

7.2　列 表 视 图

7.2.1　ListView 控件

ListView 控件(android.widget.ListView)是滚动显示列表信息的控件，是 android.widget.AbsListView 的子类，通过适配器将数据与列表项相匹配。除继承的标签属性外有 5 个标签属性(如表 7-1 所示)，而类方法可以提供设置适配器、顶部视图、底部视图、滚动等更多功能(如表 7-2 所示)。

表 7-1 ListView 控件的常用标签属性

属 性	说 明
android:divider	设置分割线的图像资源
android:dividerHeight	设置分割线的高度
android:entries	设置填充控件的数组资源引用
android:footerDividersEnabled	设置是否显示与底部视图之间的分割线
android:headerDividersEnabled	设置是否显示与顶部视图之间的分割线

表 7-2 ListView 类的常用方法

类型和修饰符	方 法
open Unit	addFooterView (v: View!) 添加底部视图
open Unit	addHeaderView (v: View!) 添加顶部视图
open ListAdapter!	getAdapter () 获取适配器对象
View!	getChildAt (index: Int) 获取指定位置的视图
open Int	getMaxScrollAmount () 获取响应箭头事件滚动的最大值
open Boolean	removeFooterView (v: View!) 移除底部视图
open Boolean	removeHeaderView (v: View!) 移除顶部视图
Unit	setOnItemClickListener (listener: AdapterView.OnItemClickListener?) 设置单击列表项的监听器
open Unit	setAdapter (adapter: ListAdapter!) 设置适配器对象
Unit	setSelector (resID: Int) 设置被选中项的背景
open Unit	setSelection (position: Int) 设置被选择列表项的位置
open Unit	setSelectionAfterHeaderView () 选中顶部视图下方的第一个列表项
open Unit	smoothScrollByOffset (offset: Int) 平滑地滚动到指定的适配器偏移位置，offset 是与当前位置的偏移量
open Unit	smoothScrollToPosition (position: Int) 平滑地滚动到指定的适配器位置

简单数据的列表项适配可以使用 ArrayAdapter 类或 SimpleAdapter 类。自定义列表项适配需要创建 BaseAdpater 类的子类和自定义列表视图。

7.2.2 实例工程：简单数据的列表视图

本实例演示了从资源文件中获取字符串数组显示到 ListView 控件中，并为 ListView 控件添加顶部视图，单击列表项后显示相应的提示信息（如图 7-2 所示）。

1. 新建工程

新建一个“Empty Activity”工程，工程名称为“C0701”。

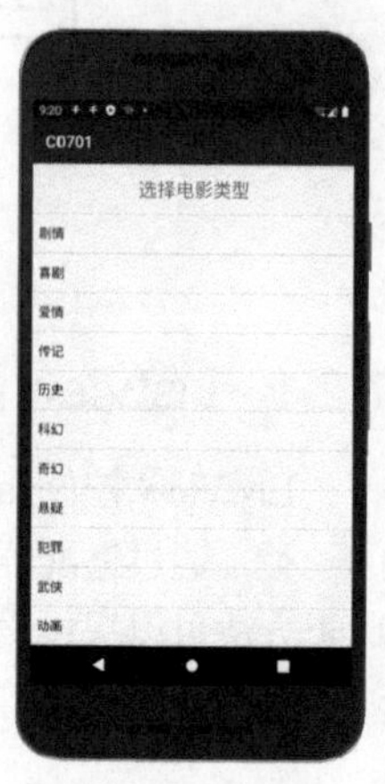

图 7-2 运行效果

2. 列表项的布局

在“layout”文件夹中，新建“list_view_header.xml”文件，用于保存 ListView 控件的顶部视图。

/res/layout/list_view_header.xml

```
02 <LinearLayout xmlns:android = "http://schemas.android.com/apk/res/android"
05     android:orientation = "vertical">
06     <TextView
10         android:gravity = "center"
11         android:text = "选择电影类型"
12         android:textSize = "24dp" />
13 </LinearLayout>
```

第 06 ~ 12 行<TextView>标签用于显示 ListView 控件顶部视图中的标题文字，居中显示，外边距设置为 15dp。

3. 主界面的布局

/res/layout/activity_main.xml

```
02 <LinearLayout xmlns:android = "http://schemas.android.com/apk/res/android"
05     android:orientation = "vertical">
06     <ListView
07         android:id = "@+id/list_view"
08         android:layout_width = "match_parent"
09         android:layout_height = "match_parent" />
10 </LinearLayout>
```

第 06 ~ 09 行使用<ListView>标签的属性并不能设置顶部视图，需要通过类方法进行设置。

4. 字符串资源

/res/values/strings.xml

```
01 <resources>
02     <string name = "app_name">C0701</string>
03     <string-array name = "movie_type_array">
04         <item>剧情</item>
05         <item>喜剧</item>
06         <item>爱情</item>
07         <item>传记</item>
08         <item>历史</item>
09         <item>科幻</item>
10         <item>奇幻</item>
11         <item>悬疑</item>
12         <item>犯罪</item>
13         <item>武侠</item>
14         <item>动画</item>
15     </string-array>
16 </resources>
```

第 03 行<string-array>标签用于保存字符串数组，数组名称属性设置为“movie_type_array”，该标签的资源 id 为 R.array.movie_type_array。第 04 ~ 14 行<item>标签表示数组项，保存电影类型的名称。

5. 主界面的 Activity

/java/com/teachol/c0701/MainActivity.kt

```
12 class MainActivity : AppCompatActivity() {
13     private lateinit var mListView: ListView
14     override fun onCreate(savedInstanceState: Bundle?) {
17         val  resAdapter  =  ArrayAdapter.createFromResource(this,  R.array.movie_type_array,
   R.layout.support_simple_spinner_dropdown_item) //movie_type_array 数组资源适配
18         val headerView: View =  View.inflate(this,R.layout.list_view_header, null)//顶部视图
19         mListView = findViewById(R.id.list_view) //获取 ListView 标签
20         mListView.adapter = resAdapter //设置数据适配器
21         mListView.addHeaderView(headerView) //添加顶部视图
22         // 添加单击列表项事件监听器
23         mListView.onItemClickListener = OnItemClickListener { parent, view, position, id ->
   Toast.makeText(this@MainActivity, (mListView.getChildAt(position) as TextView).text, Toast.
   LENGTH_SHORT).show() }
24     }
25 }
```

第 17 行使用 ArrayAdapter.createFromResource() 静态方法初始化一个 ArrayAdapter 实例。android.R.layout.support_simple_spinner_dropdown_item 是 API 自带的列表项视图，内部只包含一个 TextView 控件。第 18 行通过 View.inflate 类的静态方法获取顶部视图。第 19 ~ 23 行从布局文件中获取 ListView 控件，设置适配器并添加顶部视图，然后为列表项添加单击事件监听器。

7.2.3 实例工程：带缓存的自定义视图列表

本实例演示了最常用的自定义 ListView 控件(如图 7-3 所示)，列表项的布局保存在布局文件中，列表项数据通过数据类传递给适配器，为了提升运行效率还进行了缓存处理。

图 7-3 运行效果

1. 新建工程并导入素材

新建一个“Empty Activity”工程，工程名称为“C0702”。选择“/res/mipmap”资源文件夹，将素材图像(素材文件夹路径为“/素材/C0702”)粘贴或拖曳到该文件夹中。

2．图像模型

在“/java/com/teachol/c0702”文件夹中，新建“PhotoModel.kt”文件，用于构建图像及其相关数据的数据类。

```
/java/com/teachol/c0702/PhotoModel.kt
03  data class PhotoModel(
04      var photoResId: Int, //图像资源 id
05      var hits: Int//图像单击次数
06  )
```

第 03 ~ 06 行声明 PhotoModel 数据类，包含 photoResId 属性和 hits 属性用于存储图像的资源 id 和图像单击的次数。

3．列表项的布局

在“/res/layout”文件夹中，新建“list_view_item.xml”文件，用于保存列表项的布局，将 PhotoModel 类实例提供的数据处理后呈现在 ListView 控件中。

```
/res/layout/list_view_item.xml
02  <LinearLayout xmlns:android = "http://schemas.android.com/apk/res/android"
06      android:orientation = "vertical"
07      android:showDividers = "middle">
08      <TextView
09          android:id = "@+id/text_view_order"
15          tools:text = "NO.1" />
16      <ImageView
17          android:id = "@+id/image_view_photo"
20          android:adjustViewBounds = "true"
21          tools:srcCompat = "@tools:sample/avatars" />
22      <TextView
23          android:id = "@+id/text_view_hits"
28          tools:text = "单击量: 201" />
29  </LinearLayout>
```

第 16 ~ 21 行<ImageView>标签用于显示图像，srcCompat 属性与 src 属性相比可以保证版本的兼容性。命名空间使用 tools 替换 android，用于在 IDE 中的预览渲染。@tools:sample/avatars 是自带的样例图像，便于在 IDE 中预览效果。

4．列表视图的适配器

在“/java/com/teachol/c0702”文件夹中，新建“PhotoListAdapter.kt”，用于创建 ListView 控件的自定义适配器类——PhotoListAdapter 类，该类继承了 BaseAdapter 类。

```
/java/com/teachol/c0702/PhotoListAdapter.kt
class PhotoListAdapter(private val mContext: Context, private var mPhotoModel: List<PhotoModel>) :
    BaseAdapter() {
    private lateinit var mViewHolder: ViewHolder
    private var mConvertViewCount = 0
    override fun getCount(): Int { //获取列表项总数
        return mPhotoModel.size
    }
    override fun getItem(position: Int): Any { //获取列表项对象
        return mPhotoModel
    }
```

```
26        override fun getItemId(position: Int): Long { //获取列表项 id
27            return position.toLong()
28        }
29        // 获取列表项视图
30        override fun getView(position: Int, convertView: View?, parent: ViewGroup?): View? {
31            var cacheView: View? = convertView
32            if (cacheView == null) { //convertView 未实例化时
33                cacheView =
34                    LayoutInflater.from(mContext).inflate(R.layout.list_view_item, parent, false)
35                mConvertViewCount++
36                // 实例化 ViewHolder 缓存类对象
37                mViewHolder = ViewHolder(cacheView)
38                cacheView.tag = mViewHolder //holder 缓存到 convertView
39                cacheView.id = mConvertViewCount
40                Log.i("getView", "当前位置position: $position，初始化convertView的id: ${cacheView.id}。")
41            } else { //convertView 实例化时
42                mViewHolder = cacheView.tag as ViewHolder //从 convertView 获取 holder 缓存
43                Log.i("getView", "当前位置position: $position，重用convertView的id: ${cacheView.id}。")
44            }
45            with (mViewHolder) {
46                orderTextView.text = String.format("No.%1s", position + 1)
47                photoImageView.setImageResource(mPhotoModel[position].photoResId)
48                hitsTextView.text = String.format("单击量:%1s", mPhotoModel[position].hits)
49                photoImageView.id = position
50                photoImageView.setOnClickListener { v ->
51                    mPhotoModel[v.id].hits++
52                    notifyDataSetChanged() //更新视图数据
53                    Toast.makeText(mContext, "单击 No." + (v.id + 1), Toast.LENGTH_SHORT).show()
54                }
55            }
56            return cacheView
57        }
59        class ViewHolder(view: View) { //ViewHolder 缓存类
60            var orderTextView: TextView = view.findViewById(R.id.text_view_order)
61            var photoImageView: ImageView = view.findViewById(R.id.image_view_photo)
62            var hitsTextView: TextView = view.findViewById(R.id.text_view_hits)
63        }
64    }
```

第 30 行实现的 getView(position: Int, convertView: View?, parent: ViewGroup?)方法是 Adapter 接口的抽象方法，初始化、滚动父视图再调用列表项或执行 notifyDataSetChanged()时被调用。position 参数是列表项的位置，convertView 参数是列表项的视图，parent 参数是列表项的父视图。第 32～44 行判断 convertView 是否为空，等于 null 时对其进行初始化并缓存到 convertView 对象中，不等于 null 时直接从 convertView 对象中获取。第 50～54 行设置 photoImageView 对象的单击事件监听器，notifyDataSetChanged()语句能够立即根据数据的改变更新 ListView 控件的列表项视图。第 59～63 行声明用于缓存列表项视图中控件的 ViewHolder 类，该类包含与列表项视图中控件相对应类型的三个属性。

5. 主界面的布局

```
/res/layout/activity_main.xml
<ListView
    android:id = "@+id/list_view_photo"
    android:layout_width = "match_parent"
    android:layout_height = "match_parent" />
```

第 07～10 行<ListView>标签显示图像序号、图像和图像单击次数的列表。

6. 主界面的 Activity

/java/com/teachol/c0702/MainActivity.kt

```
08  class MainActivity : AppCompatActivity() {
09      private var mContext: Context = this@MainActivity
10      private var mPhotoModel = ArrayList<PhotoModel>()
11      private lateinit var mAdapter: PhotoListAdapter
12      private lateinit var photoListView: ListView
13      override fun onCreate(savedInstanceState: Bundle?) {
16          mPhotoModel.add(PhotoModel(R.mipmap.img_pic01, 203))
17          mPhotoModel.add(PhotoModel(R.mipmap.img_pic02, 44))
18          mPhotoModel.add(PhotoModel(R.mipmap.img_pic03, 311))
19          mPhotoModel.add(PhotoModel(R.mipmap.img_pic04, 97))
20          mPhotoModel.add(PhotoModel(R.mipmap.img_pic05, 462))
21          mPhotoModel.add(PhotoModel(R.mipmap.img_pic06, 117))
22          mPhotoModel.add(PhotoModel(R.mipmap.img_pic07, 187))
23          mPhotoModel.add(PhotoModel(R.mipmap.img_pic08, 87))
24          mAdapter = PhotoListAdapter(mContext, mPhotoModel)
25          photoListView = findViewById(R.id.list_view_photo)
26          photoListView.setAdapter(mAdapter)
27      }
28  }
```

第 16～23 行向 ArrayList<PhotoModel>类型的 mPhotoModel 变量添加 8 个 PhotoModel 实例，PhotoModel 实例保存图像的资源 id 和单击次数。第 24 行将保存上下文的 mContext 变量和保存 PhotoModel 实例的 mPhotoModel 变量作为参数，实例化自定义的适配器——PhotoListAdapter。第 26 行 ListView 控件设置适配器。

7. 测试运行

运行程序后，拖动 ListView 控件直至显示出第 8 个列表项视图，ListView 控件回到顶部，可以在 Logcat 窗口中观察到输出的日志信息(如图 7-4 所示)。在 ListView 控件中列表项同时最多可以显示出 3 个，因此初始化的 convertView 对象只有 3 个，滚动 ListView 控件时 3 个 convertView 对象被循环重用。

图 7-4　Logcat 窗口输出的日志信息

7.3 网格视图

7.3.1 GridView 控件

GirdView 控件（android.widget.GridView）是以滚动网格形式显示信息的控件，是 android.widget.AbsListView 的子类，通过适配器将数据与网格项相匹配。除继承的标签属性外有 6 个标签属性(如表 7-3 所示)，而类方法可以提供设置适配器、滚动等更多功能(如表 7-4 所示)。

表 7-3 GridView 控件的标签属性

属性	说明
android:columnWidth	设置列宽度
android:gravity	设置对齐方式
android:horizontalSpacing	设置水平方向单元格的间距
android:numColumns	设置列数
android:stretchMode	设置拉伸模式，可选值包括 none(不拉伸)、spacingWidth(拉伸单元格的间距)、columnWidth(拉伸列宽度)和 spacingWidthUniform(均匀拉伸单元格的间距)
android:verticalSpacing	设置垂直方向单元格的间距

表 7-4 GridView 类的常用方法

类型和修饰符	方法
open Unit	setAdapter(adapter: ListAdapter!) 设置适配器对象
Unit	setOnItemClickListener(listener: AdapterView.OnItemClickListener?) 设置单击列表项的监听器
Unit	setSelector(sel: Drawable!) 设置被选中项的背景
open Unit	setSelection(position: Int) 设置被选择列表项的位置
open Unit	smoothScrollByOffset(offset: Int) 平滑地滚动到指定的适配器偏移位置，offset 是与当前位置的偏移量
open Unit	smoothScrollToPosition(position: Int) 平滑地滚动到指定的适配器位置

7.3.2 实例工程：显示商品类别的网格视图

本实例演示了使用 GirdView 控件分两行排列商品类别，每个类别都配有文字和图像，单击后显示相应的提示信息(如图 7-5 所示)。

图 7-5 运行效果

1. 新建工程并导入素材

新建一个“Empty Activity”工程，工程名称为“C0703”。选择“/res/mipmap”资源文件夹，将素材图像(素材文件夹路径为“/素材/C0703”)粘贴或拖曳到该文件夹中。

2. 商品分类的模型

在“/java/com/teachol/c0703”文件夹中，新建“CategoryModel.kt”文件，用于创建商品分类的模型。

/java/com/teachol/c0703/CategoryModel.kt

```
03 data class CategoryModel(var ico: Int, var name: String)
```

第 03 行声明 CategoryModel 数据类，包含 ico 属性和 name 属性，分别用于存储商品类别图标的资源 id 和商品类别名称。

3. 网格项的视图

在“/res/layout”文件夹中，新建“grid_view_item.xml”，用于创建 GridView 控件的网格项视图。

/res/layout/grid_view_item.xml

```
02 <RelativeLayout xmlns:android = "http://schemas.android.com/apk/res/android"
04     android:layout_width = "wrap_content"
05     android:layout_height = "wrap_content">
06     <ImageView
07         android:id = "@+id/image_view_ico"
10         android:layout_centerHorizontal = "true"
11         android:adjustViewBounds = "true"
12         tools:srcCompat = "@tools:sample/avatars" />
13     <TextView
14         android:id = "@+id/text_view_name"
17         android:layout_below = "@id/image_view_ico"
18         android:layout_centerHorizontal = "true"
19         android:textSize = "18sp"
20         tools:text = "类别名称" />
21 </RelativeLayout>
```

第 6 ~ 12 行<ImageView>标签用于显示商品类别的图像，使用自带的样例图像进行占位。命名空间使用 tools 替换 android 可以在预览中渲染，srcCompat 属性替换 src 属性可以保证版本的兼容性。

4. 网格视图的适配器

在“/java/com/teachol/c0703”文件夹中，新建“CategoryAdapter.kt”文件，用于创建 GridView 控件的自定义适配器类——CategoryAdapter 类，该类继承了 BaseAdapter 类。

/java/com/teachol/c0703/CategoryAdapter.kt

```
class CategoryAdapter (private var mContext: Context, private var mCategoryModel: List<CategoryModel>) : BaseAdapter() {
    private lateinit var mViewHolder: ViewHolder
    private var mConvertViewCount = 0
    override fun getCount(): Int { //获取网格项总数
        return mCategoryModel.size
    }
```

```
20      override fun getItem(position: Int): Any { //获取网格项对象
21          return mCategoryModel
22      }
24      override fun getItemId(position: Int): Long { //获取网格项 id
25          return position.toLong()
26      }
27      // 获取网格项视图
28      override fun getView(position: Int, convertView: View?, parent: ViewGroup?): View? {
29          var cacheView: View? = convertView
30          if (cacheView == null) { //convertView 未实例化时
31              cacheView = LayoutInflater.from(mContext).inflate(R.layout.grid_view_item, parent, false)
32              mConvertViewCount++
33              // 实例化 ViewHolder 缓存类对象
34              mViewHolder = ViewHolder(cacheView)
35              cacheView.tag = mViewHolder //holder 缓存到 convertView
36              cacheView.id = mConvertViewCount
37              Log.i("getView", "当前位置 position: $position，初始化 convertView 的 id: ${cacheView.id}。")
38          } else { //convertView 实例化时
39              mViewHolder = cacheView.tag as ViewHolder //从 convertView 获取 holder 缓存
40              Log.i("getView", "当前位置 position: $position，重用 convertView 的 id: ${cacheView.id}。")
41          }
42          mViewHolder.icoImageView.setImageResource(mCategoryModel[position].ico)
43          mViewHolder.nameTextView.text = mCategoryModel[position].name
44          return cacheView
45      }
47      internal class ViewHolder(view: View) { //ViewHolder 缓存类
48          var icoImageView: ImageView = view.findViewById(R.id.image_view_ico)
49          var nameTextView: TextView = view.findViewById(R.id.text_view_name)
50      }
51  }
```

第 28 行实现的 getView(int, View, ViewGroup)方法是 Adapter 接口的抽象方法，用于返回网格项显示的视图。第 30 ~ 41 行判断 convertView 是否为空，等于 null 时对其进行初始化并缓存到 convertView 对象中，不等于 null 时直接从 convertView 对象中获取。第 47 ~ 50 行声明用于缓存网格项视图中控件的 ViewHolder 类，该类包含与网格项视图中控件相对应类型的两个属性。

5．主视图的布局

```
/res/layout/activity_main.xml
05  <GridView
06      android:id = "@+id/grid_view"
09      android:columnWidth = "80dp"
10      android:gravity = "center_horizontal"
11      android:numColumns = "4"
12      android:stretchMode = "spacingWidthUniform"
13      android:verticalSpacing = "10dp" />
```

第 05 ~ 13 行<GridView>标签用于显示商品类别，android:columnWidth="80dp"设置列的宽度为 80dp，android:numColumns="4"设置网格视图为 4 列，android:stretchMode="spacingWidthUniform"设置均匀拉伸单元格的间距。

6．主视图的 Activity

```
/java/com/teachol/c0703/MainActivity.kt
12  class MainActivity : AppCompatActivity() {
13      private var mContext: Context = this@MainActivity
14      private var mCategoryModel = ArrayList<CategoryModel>()
15      private lateinit var mAdapter: CategoryAdapter
16      private lateinit var mGridView: GridView
17      override fun onCreate(savedInstanceState: Bundle?) {
20          mCategoryModel.add(CategoryModel(R.mipmap.img_category1, "图书音像"))
21          mCategoryModel.add(CategoryModel(R.mipmap.img_category2, "日化护肤"))
22          mCategoryModel.add(CategoryModel(R.mipmap.img_category3, "美食外卖"))
23          mCategoryModel.add(CategoryModel(R.mipmap.img_category4, "鞋靴箱包"))
24          mCategoryModel.add(CategoryModel(R.mipmap.img_category5, "食品饮料"))
25          mCategoryModel.add(CategoryModel(R.mipmap.img_category6, "运动户外"))
26          mCategoryModel.add(CategoryModel(R.mipmap.img_category7, "手机数码"))
27          mCategoryModel.add(CategoryModel(R.mipmap.img_category8, "日用百货"))
28          mAdapter = CategoryAdapter(mContext, mCategoryModel) //实例化适配器
29          mGridView = findViewById(R.id.grid_view)
30          mGridView.setAdapter(mAdapter) //设置网格视图的适配器
31          mGridView.setSelector(ColorDrawable(Color.TRANSPARENT)) //设置被选择网格项的背景颜色
32          mGridView.setOnItemClickListener(OnItemClickListener { parent, view, position, id ->
    Toast.makeText(mContext, "单击了第" + position + "项", Toast.LENGTH_SHORT).show() })
33      }
34  }
```

第 20～27 行 List<CategoryModel>类的 mCategoryModel 变量添加 8 个 CategoryModel 类实例，CategoryModel 类实例保存商品类别的图像资源 id 和商品类别的名称。第 34 行将保存上下文的 mContext 变量和保存 CategoryModel 实例的 mCategoryModel 变量作为参数，实例化自定义的适配器——CategoryAdapter。第 29～32 行 mGridView 变量获取布局文件中的 GridView 控件，然后设置适配器和被选择网格项的背景颜色为透明，最后添加单击网格项的监听事件。

7．测试运行

运行程序后，可以在 Logcat 窗口中观察到输出的日志信息(如图 7-6 所示)。在 GridView 控件中所有的网格项都同时显示出来，因此需要初始化 8 个 convertView 对象。

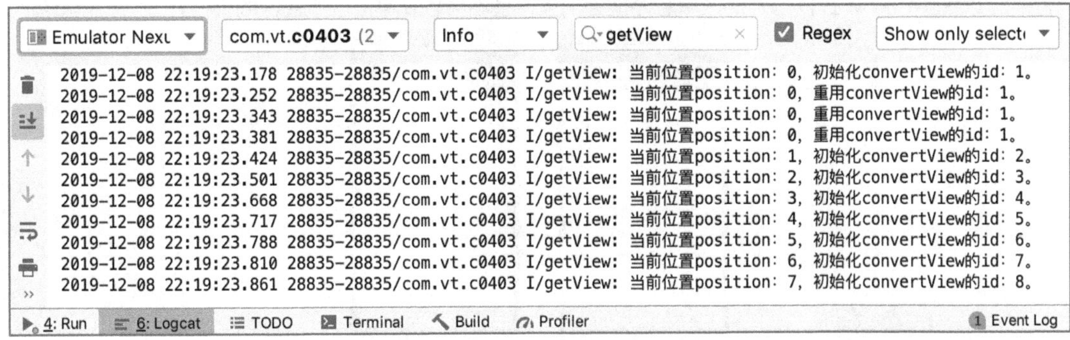

图 7-6　Logcat 窗口输出的日志信息

7.4 悬 浮 框

7.4.1 PopupWindow 控件

PopupWindow 控件(android.widget.PopupWindow)是悬浮显示的控件，是 java.lang.Object 的子类。除继承的标签属性外还有 6 个标签属性(如表 7-5 所示)，而类方法还提供了根据位置显示悬浮框、关闭悬浮框等更多功能(如表 7-6 所示)。

表 7-5 PopupWindow 控件的标签属性

属 性	说 明
android:overlapAnchor	设置悬浮框是否应重叠其锚定视图
android:popupAnimationStyle	设置悬浮框的动画样式
android:popupBackground	设置悬浮框的背景
android:popupElevation	设置悬浮框的高度(*z* 轴的高度)，默认值为 0dp
android:popupEnterTransition	设置悬浮框的显示过渡动画
android:popupExitTransition	设置悬浮框的消失过渡动画

表 7-6 PopupWindow 类的常用方法

类型和修饰符	方 法
open Unit	dismiss () 关闭悬浮框的视图
open View!	getContentView () 获取悬浮框的视图
open Unit	setContentView (contentView: View!) 设置悬浮框的视图
open Unit	showAsDropDown (anchor: View!) 在锚点视图左下角弹出悬浮框的视图
open Unit	showAsDropDown (anchor: View!, xoff: Int, yoff: Int) 根据偏移值(像素)在锚点视图左下角弹出悬浮框的视图
open Unit	showAtLocation (parent: View!, gravity: Int, x: Int, y: Int) 在指定视图位置(像素)弹出悬浮框的视图

7.4.2 实例工程：单击按钮显示自定义悬浮框

本实例演示了单击按钮显示 PopupWindow 控件，单击悬浮框内的按钮后显示提示信息且 PopupWindow 控件消失(如图 7-7 所示)，PopupWindow 控件显示和消失的过程使用动画效果。当 PopupWindow 控件显示时，单击 PopupWindow 控件以外的区域悬浮框会消失，且不会触发单击区域控件的事件。

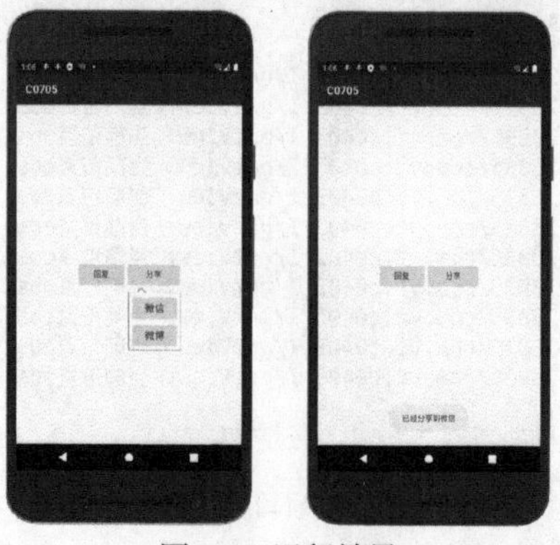

图 7-7 运行效果

1. 打开基础工程

打开“基础工程”文件夹中的“C0705”工程，该工程已经包含 MainActivity 及资源文件。

2. 悬浮框背景图像

双击打开“/res/mipmap”资源文件夹中的“img_popup_bg.9.png”文件，在 9-Patch 图中，将图像四周的 1 像素宽作为边界的设置区域，左侧和上方设置成黑色表示可以拉伸的区域，右侧和下方设置成黑色表示显示内容的区域。按住 Ctrl 键拖曳出黑色线条，按住 Shift 键拖曳可以删除黑色线条区域。勾选“Show content”和“Show patches”复选框后，中间区域是可拉伸区域，右侧预览图显示了三种拉伸效果(如图 7-8 所示)。

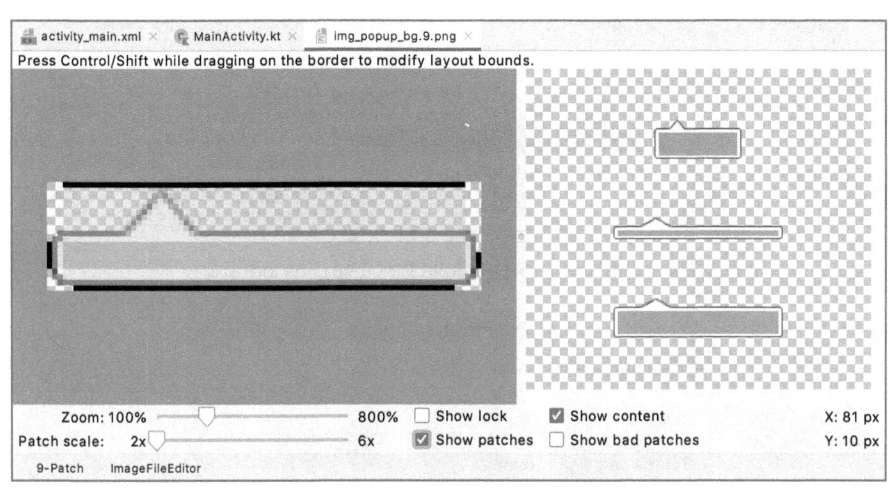

图 7-8　9-Patch 图的可拉伸区域

> **提示：9-Patch 图**
>
> 9-Patch 图是可拉伸的位图，会自动调整大小使图像在充当背景时可以在界面中自适应。9-Patch 图是标准 PNG 格式的图像，并且将四周 1 像素宽作为边界的设置区域，程序运行时不会被显示出来。保存时文件以“.9.png”为扩展名，放置在项目的“/res/drawable”文件夹中。

3. 悬浮框的视图

在“/res/layout”文件夹中，新建“popup_share.xml”文件，用于创建弹出的悬浮框视图。

```
/res/layout/popup_share.xml
<LinearLayout xmlns:android = "http://schemas.android.com/apk/res/android"
   android:background = "@mipmap/img_popup_bg"
   android:orientation = "vertical">
   <Button
      android:id = "@+id/btn_weixin"
      android:padding = "5dp"
      android:text = "微信"
      android:textSize = "18sp" />
   <Button
      android:id = "@+id/btn_weibo"
      android:padding = "5dp"
      android:text = "微博"
```

```
20          android:textSize = "18sp" />
21  </LinearLayout>
```

第 05 行设置“img_popup_bg.9.png”图像作为<LinearLayout>标签的背景图。第 07～20 行两个<Button>标签作为悬浮框的菜单按钮。

4．悬浮框出现的动画

选择“/res”文件夹，单击右键，选择【New】→【Android Resource Directory】命令（如图 7-9 所示）。在“New Resource Directory”对话框中，设置“Directory name”为“anim”，“Resource type”为“anim”（如图 7-10 所示），单击“OK”按钮完成资源文件夹的创建。

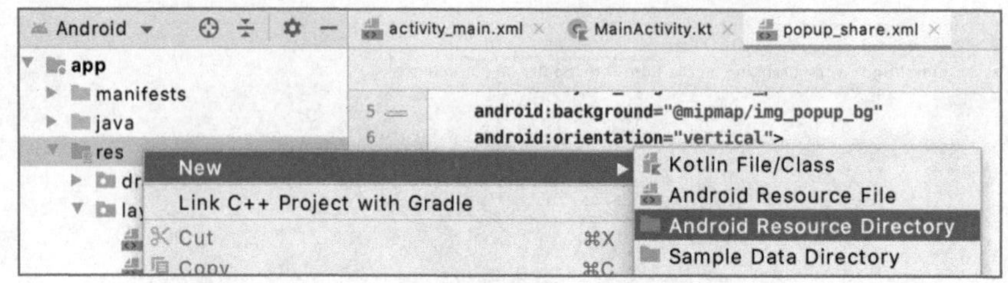

图 7-9 【New】→【Android Resource Directory】命令

图 7-10 “New Resource Directory”对话框

选择“/res/anim”文件夹，单击右键，选择【New】→【Animation Resource File】命令（如图 7-11 所示）。在“New Resource File”对话框中，设置“File name”为“popup_enter”（如图 7-12 所示），单击“OK”按钮，完成新建“popup_enter.xml”文件，用于添加悬浮框出现的动画效果。

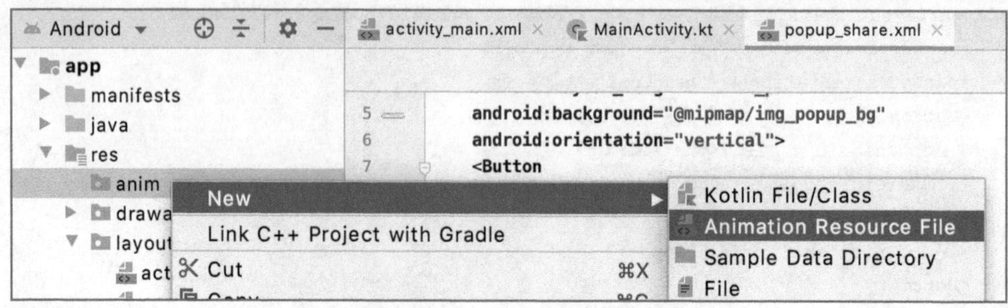

图 7-11 【New】→【Animation Resource File】命令

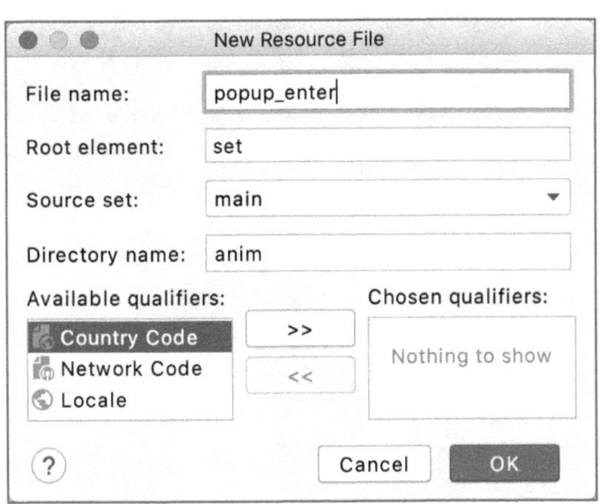

图 7-12　“New Resource File”对话框

/res/anim/popup_enter.xml

```
01  <?xml version = "1.0" encoding = "utf-8"?>
02  <set xmlns:android = "http://schemas.android.com/apk/res/android">
03      <scale
04          android:duration = "100"
05          android:fromXScale = "0.6"
06          android:fromYScale = "0.6"
07          android:pivotX = "50%"
08          android:pivotY = "50%"
09          android:toXScale = "1.0"
10          android:toYScale = "1.0" />
11      <translate
12          android:duration = "200"
13          android:fromYDelta = "10%"
14          android:toYDelta = "0" />
15      <alpha
16          android:duration = "100"
17          android:fromAlpha = "0.0"
18          android:interpolator = "@android:anim/decelerate_interpolator"
19          android:toAlpha = "1.0" />
20  </set>
```

第 03 ~ 10 行<scale>标签用于设置缩放动画的参数，pivotX 属性和 pivotY 属性设置缩放中心点的 *x*、*y* 轴位置，属性值可以使用整数值、百分数（或者小数）或百分数 p 的形式，如 50（表示相对于控件左上角的像素）、50% / 0.5（表示相对于控件左上角的百分比）、50%p（表示相对于父控件左上角的百分比）。第 11 ~ 14 行<translate>标签用于设置位移动画的参数，fromYDelta 属性和 toYDelta 属性用于设置 *y* 轴位移动画的起始位置，也可以使用整数值、百分数（或者小数）或百分数 p 的形式。第 15 ~ 19 行<alpha>标签用于设置透明度动画的参数，interpolator 属性用于设置动画补间的差值器，@android:anim/decelerate_interpolator 是系统自带的减速差值器。

5. 悬浮框消失的动画

在“/res/anim”文件夹中，新建“popup_exit.xml”动画资源文件，用于添加悬浮框消失的动画效果。

```
/res/anim/popup_exit.xml
01 <?xml version = "1.0" encoding="utf-8"?>
02 <set xmlns:android = "http://schemas.android.com/apk/res/android">
03     <scale
04         android:duration = "500"
05         android:fromXScale = "1.0"
06         android:fromYScale = "1.0"
07         android:pivotX = "50%"
08         android:pivotY = "50%"
09         android:toXScale = "0.5"
10         android:toYScale = "0.5" />
11     <translate
12         android:duration = "200"
13         android:fromYDelta = "0"
14         android:toYDelta = "10%" />
15     <alpha
16         android:duration = "500"
17         android:fromAlpha = "1.0"
18         android:interpolator = "@android:anim/accelerate_interpolator"
19         android:toAlpha = "0.0" />
20 </set>
```

第 03～10 行<scale>标签用于设置缩放动画的参数。第 11～14 行<translate>标签用于设置位移动画的参数。第 15～19 行<alpha>标签用于设置透明度动画的参数。

6．悬浮框的动画

```
/res/values/styles.xml
10 <style name = "PopupAnimation" parent = "android:Animation">
11     <item name = "android:windowEnterAnimation">@anim/popup_enter</item>
12     <item name = "android:windowExitAnimation">@anim/popup_exit</item>
13 </style>
```

第 10 行<style>标签设置悬浮框的动画效果文件，name 属性表示动画效果的名称，使用 R.style.PopupAnimation 调用，parent 属性表示该样式所继承的样式。第 11、12 行两个<item>标签分别设置出现和消失的动画资源。

7．主界面的 Activity

```
/java/com/teachol/c0705/MainActivity.kt
class MainActivity : AppCompatActivity() {
    private lateinit var mContext: Context
    override fun onCreate(savedInstanceState: Bundle?) {
        mContext = this@MainActivity
        val replyBtn: Button = findViewById(R.id.btn_reply)
        replyBtn.setOnClickListener { Toast.makeText(mContext, "回复信息", Toast.LENGTH_SHORT).show() }
        val shareBtn: Button = findViewById(R.id.btn_share)
        shareBtn.setOnClickListener(object : View.OnClickListener {
            override fun onClick(v: View) {
                popWindow(v)
            }
        })
```

```
27      }
29      private fun popWindow(v: View) {
30          // 实例化悬浮框视图
31          val popupView: View = LayoutInflater.from(mContext).inflate(R.layout.popup_share, null, false)
32          // 实例化悬浮框
33          val popupWindow = PopupWindow(popupView, ViewGroup.LayoutParams.WRAP_CONTENT, 
    ViewGroup.LayoutParams.WRAP_CONTENT, true)
34          popupWindow.animationStyle = R.style.PopupAnimation //设置动画
35          popupWindow.showAsDropDown(v, 10, 0) //显示悬浮框
36          // 设置悬浮框视图中的按钮事件
37          val weixinBtn: Button = popupView.findViewById(R.id.btn_weixin)
38          val weiboBtn: Button = popupView.findViewById(R.id.btn_weibo)
39          weixinBtn.setOnClickListener {
40              Toast.makeText(mContext, "已经分享到微信", Toast.LENGTH_SHORT).show()
41              popupWindow.dismiss() //关闭悬浮框
42          }
43          weiboBtn.setOnClickListener(object : View.OnClickListener {
44              override fun onClick(v: View?) {
45                  Toast.makeText(mContext, "已经分享到微信", Toast.LENGTH_SHORT).show()
46                  popupWindow.dismiss() //关闭悬浮框
47              }
48          })
49      }
50  }
```

第 20 行设置 replyBtn 的单击监听事件监听器，使用 Lambda 表达式实现匿名类。第 22～26 行设置 shareBtn 的单击监听事件监听器，使用常规方式实现监听器的匿名类，调用 popWindow(v:View)方法，并将 onClick(View)方法中的 v 参数作为参数继续传递下去。第 33～35 行设置并显示悬浮框，PopupWindow 构造方法的第 4 个参数设置是否可以获取焦点，设置为 true 时，单击悬浮框以外的区域悬浮框会消失，且不会触发单击区域的控件。第 39～42 行设置 weixinBtn 的单击监听事件监听器，使用 Lambda 表达式实现匿名类。第 43～48 行设置 weiboBtn 的单击监听事件监听器，使用常规方式实现监听器的匿名类。

7.5　翻 转 视 图

7.5.1　ViewFlipper 控件

ViewFlipper 控件(android.widget.ViewFlipper)是多视图轮转的控件，是 android.widget.ViewAnimator 的子类。ViewFlipper 控件的常用标签属性如表 7-7 所示，ViewFlipper 类的常用方法如表 7-8 所示。

表 7-7　ViewFlipper 控件的常用标签属性

属　性	说　明
android:autoStart	设置是否自动开始翻转子视图
android:flipInterval	设置翻转子视图的间隔时间
android:inAnimation	设置子视图显示的动画
android:outAnimation	设置子视图消失的动画

表 7-8　ViewFlipper 类的常用方法

类型和修饰符	方　法
open Int	getFlipInterval() 获取翻转子视图的间隔时间
open Boolean	isAutoStart() 判断是否自动开始翻转子视图
open Boolean	isFlipping() 判断是否正在翻转子视图
open Unit	startFlipping() 开始翻转子视图
open Unit	stopFlipping() 停止翻转子视图
Unit	addView(child: View!) 添加翻转子视图
Unit	addView(child: View!, index: Int) 添加翻转子视图到指定的索引号位置，子视图的索引号从 0 开始
View!	getCurrentView() 获取当前翻转子视图
Int	getDisplayedChild() 获取当前翻转子视图的索引号，子视图的索引号从 0 开始
Unit	removeAllViews() 移除所有翻转子视图
Unit	removeViewAt(index: Int) 移除指定索引号的翻转子视图，子视图的索引号从 0 开始
Unit	setOnClickListener(l: View.OnClickListener?) 设置单击事件监听器
Unit	showNext() 显示下一个翻转子视图
Unit	showPrevious() 显示上一个翻转子视图

7.5.2　实例工程：轮流显示图像的翻转视图

本实例演示了使用 ViewFlipper 控件轮转显示图像，单击图像后通过提示信息显示当前子视图的索引号(如图 7-13 所示)。由于 ViewFlipper 控件设置了显示动画，所以当子视图开始执行显示动画时当前子视图索引号会改变。

图 7-13　运行效果

1. 打开基础工程

打开“基础工程”文件夹中的“C0706”工程，该工程已经包含 ViewFlipper 子视图布局、MainActivity 及资源文件。

2. 图像滚动出现的动画

在“/res”文件夹中，新建“anim”文件夹。在“/res/anim”文件夹中，新建“view_flipper_right_in.xml”文件，设置子视图显示的动画效果。

/res/anim/view_flipper_right_in.xml

```
02 <set xmlns:android = "http://schemas.android.com/apk/res/android">
03     <translate
04         android:duration = "2000"
05         android:fromXDelta = "100%p"
06         android:toXDelta = "0" />
07 </set>
```

第 03 ~ 06 行<translate>标签设置动画持续时间为 2000 毫秒，沿着 x 轴从 0 坐标点处移动到距离 0 坐标点左侧为父控件 100%宽度处。

3. 图像滚动离开的动画

在“/res/anim”文件夹中，新建“view_flipper_left_out.xml”文件，设置子视图消失的动画效果。

/res/anim/view_flipper_left_out.xml

```
02 <set xmlns:android = "http://schemas.android.com/apk/res/android">
03     <translate
04         android:duration = "2000"
05         android:fromXDelta = "0"
06         android:toXDelta = "-100%p" />
07 </set>
```

第 03 ~ 06 行<translate>标签设置动画持续时间为 2000 毫秒，沿着 x 轴从距离 0 坐标点左侧父控件 100%宽度处移动到 0 坐标点处。

4. 主界面的布局

/res/layout/activity_main.xml

```
10 <ViewFlipper
11     android:id = "@+id/view_flipper"
14     android:layout_columnSpan = "4"
15     android:flipInterval = "3000"
16     android:inAnimation = "@anim/view_flipper_right_in"
17     android:outAnimation = "@anim/view_flipper_left_out">
18     <include layout = "@layout/view_flipper_page1" />
19     <include layout = "@layout/view_flipper_page2" />
20 </ViewFlipper>
```

第 15 行设置翻转子视图的间隔时间。第 16、17 行设置子视图显示和消失的动画。第 18、19 行两个<include>标签设置子视图。

5. 主界面的 Activity

/java/com/teachol/c0706/MainActivity.kt

```
11 class MainActivity : AppCompatActivity() {
12     private lateinit var mViewFlipper: ViewFlipper
13     override fun onCreate(savedInstanceState: Bundle?) {
16         // 通过布局文件获取子视图
17         val view3: View = View.inflate(applicationContext, R.layout.view_flipper_page3, null)
18         // 创建图像视图作为子视图
19         val imageView4 = ImageView(this)
20         imageView4.adjustViewBounds = true
21         imageView4.setImageResource(R.mipmap.img_9546)
22         // 设置翻转视图
23         mViewFlipper = findViewById(R.id.view_flipper)
24         mViewFlipper.addView(imageView4)
25         mViewFlipper.addView(view3, 2)
26         mViewFlipper.startFlipping()
27         mViewFlipper.setOnClickListener {
28             Toast.makeText(applicationContext, "当前显示的子视图索引号为" + mViewFlipper.
displayedChild, Toast.LENGTH_SHORT).show()
29         }
31         val startBtn: Button = findViewById(R.id.button_start) //开始按钮
32         startBtn.setOnClickListener { mViewFlipper.startFlipping() }
34         val stopBtn: Button = findViewById(R.id.button_stop) //停止按钮
35         stopBtn.setOnClickListener { mViewFlipper.stopFlipping() }
37         val previousBtn: Button = findViewById(R.id.button_previous) //上一页按钮
38         previousBtn.setOnClickListener { mViewFlipper.showPrevious() }
40         val nextBtn: Button = findViewById(R.id.button_next) //下一页按钮
41         nextBtn.setOnClickListener { mViewFlipper.showNext() }
42     }
43 }
```

第 24 行 mViewFlipper 添加 imageView4 作为子视图放在其他子视图的后面，mViewFlipper 已有两个子视图，因此 imageView4 作为第三个子视图的索引号为 2。第 25 行 mViewFlipper 添加 view3 作为子视图放在 2 号索引位置，imageView4 的索引号顺位后移后变为 3。第 28 行 mViewFlipper.displayedChild 获取当前显示子视图的索引号，由于 mViewFlipper 通过标签设置 android:inAnimation 属性，所以开始执行显示动画时当前显示子视图的索引号变为该子视图的索引号。

7.6 分页视图

7.6.1 ViewPager 控件

ViewPager 控件（androidx.viewpager.widget.ViewPager）是分页滚动的控件，是 android.view.ViewGroup 的子类。该控件没有独有的标签属性，标签属性都继承自父类。ViewPager 类的常用方法如表 7-9 所示。

表 7-9 ViewPager 类的常用方法

类型和修饰符	方法
open Unit	addOnAdapterChangeListener(@NonNull listener: ViewPager.OnAdapterChangeListener) 添加更改适配器的监听器

续表

类型和修饰符	方　法
open Unit	addOnPageChangeListener (@NonNull listener: ViewPager.OnPageChangeListener) 添加页面改变的监听器
open PagerAdapter?	getAdapter () 获取适配器
open Int	getCurrentItem () 获取当前分页的索引号
open Unit	setAdapter (@Nullable adapter: PagerAdapter?) 设置适配器
open Unit	setCurrentItem (item: Int) 跳转到指定索引号的分页
open Unit	setCurrentItem (item: Int, smoothScroll: Boolean) 是否平滑地滚动到指定索引号的分页

PagerAdapter 类为 ViewPager 控件提供数据适配的功能（常用方法如表 7-10 所示），PagerAdapter 的子类需要重写 4 个方法：instantiateItem (@NonNull container: ViewGroup, position: Int)、destroyItem (@NonNull container: ViewGroup, position: Int, @NonNull object: Any)、getCount () 和 isViewFromObject (@NonNull view: View, @NonNull object: Any)。

表 7-10　PagerAdapter 类的常用方法

类型和修饰符	方　法
open Any	instantiateItem (@NonNull container: ViewGroup, position: Int) 为指定索引位置创建分页视图。适配器负责将视图添加到此处给定的容器中，从 finishUpdate (@NonNull container: ViewGroup) 返回时调用该方法
open Unit	destroyItem (@NonNull container: ViewGroup, position: Int, @NonNull object: Any) 删除指定索引位置的分页视图
open Unit	finishUpdate (@NonNull container: ViewGroup) 当前分页视图更新完成时调用
abstract Int	getCount () 获取分页视图的数量
abstract Boolean	isViewFromObject (@NonNull view: View, @NonNull object: Any) 判断视图是否与对象相关联
open Unit	notifyDataSetChanged () 数据改变时更新视图
open Unit	startUpdate (@NonNull container: ViewGroup) 显示的分页视图更新开始时调用此方法

ViewPager.OnPageChangeListener 接口用于监听 ViewPager 控件的事件，包含三个抽象方法（如表 7-11 所示）。

表 7-11　ViewPager.OnPageChangeListener 接口的方法

类型和修饰符	方　法
abstract void	onPageScrollStateChanged (state: Int) 页面滚动状态改变事件，state 参数是事件状态，包括 SCROLL_STATE_IDLE（空闲状态）、SCROLL_STATE_DRAGGING (滚动状态)、SCROLL_STATE_SETTLING (最终位置)
abstract Unit	onPageScrolled (position: Int, positionOffset: Float, @Px positionOffsetPixels: Int) 页面滚动事件，position 参数是分页视图的索引号，positionOffset 参数是分页视图的偏移百分比，positionOffsetPixels 参数是分页视图的偏移像素
abstract Unit	onPageSelected (position: Int) 页面选择事件，position 参数是分页视图的索引号

7.6.2　实例工程：欢迎引导页

本实例演示了使用 ViewPager 控件制作的欢迎引导页，该页面包含 4 个分页，左右滑动、单击左右箭头或底部 4 个圆点可以切换分页。切换分页时，底部的 4 个圆点标识显示的分页位置(如图 7-14 所示)。为了便于演示监听事件，添加了两个 TextView 控件放置于顶部和底部，用于显示监听信息。

图 7-14　运行效果

1．打开基础工程

打开“基础工程”文件夹中的“C0707”工程，该工程已经包含 MainActivity 及资源文件。

2．主题的样式

/res/values/styles.xml

```
03 <style name = "AppTheme" parent = "Theme.AppCompat.Light.NoActionBar">
04     <item name = "colorPrimary">@color/colorPrimary</item>
05     <item name = "colorPrimaryDark">@color/colorPrimaryDark</item>
06     <item name = "colorAccent">@color/colorAccent</item>
07     <item name = "android:windowBackground">@mipmap/img_loding_bg</item>
08     <item name = "android:windowTranslucentStatus">true</item>
09 </style>
```

第 03 行将 Theme.AppCompat.Light.DarkActionBar 修改为 Theme.AppCompat.Light.NoActionBar，用于取消显示页面顶部的 ActionBar。第 04 行设置主题 tabar 背景颜色。第 05 行设置主题状态栏背景颜色。第 06 行设置控件重点外观元素的颜色，如单选按钮的圆点颜色、复选框的方块颜色、对话框的默认按钮文字颜色。第 07 行添加 android:windowBackground 属性，为所有的 Activity 设置背景图像，主要作用是避免在启动 App 的加载期间显示空白背景，增强用户体验。第 08 行添加了 android:windowTranslucentStatus 属性，属性值为 true，表示状态栏的颜色为透明，这样就可以将背景图像全屏显示出来。

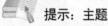

提示：主题

主题(theme)是 App 界面元素显示效果的规则，Android Studio 3.6 创建的 Empty Activity 项目使用的默认主题继承自 Theme.AppCompat.Light.DarkActionBar。主题通过“AndroidManifest.xml”文件中<application>标签的 android:theme 属性设置。

3．状态栏的背景颜色

/res/values/colors.xml

```
02 <resources>
```

```
03      <color name = "colorPrimary">#008577</color>
04      <color name = "colorPrimaryDark">#dd4661</color>
05      <color name = "colorAccent">#D81B60</color>
06  </resources>
```

第 04 行修改<color>标签的 colorPrimaryDark 属性值为#dd4661，该颜色与背景图像@mipmap/img_loding_bg 的背景颜色相同，消除状态栏与页面之间的分割感。

4．分页视图的适配器

在“/java/com/teachol/c0707”文件夹中，新建“WelcomePagerAdapter.kt”文件。WelcomePagerAdapter 类继承自 PagerAdapter 类，用于自定义分页视图的适配器。

```
/java/com/teachol/c0707/WelcomePagerAdapter.kt
08  class WelcomePagerAdapter (
09      private val mPagerViews: ArrayList<View>//分页视图的数组列表
10  ) : PagerAdapter() {
11      //获取分页视图的数量
12      override fun getCount(): Int {
13          return mPagerViews.size
14      }
15      //判断是否由对象生成界面
16      override fun isViewFromObject(view: View, 'object': Any): Boolean {
17          return view === 'object'
18      }
19      //显示分页视图或缓存分页时进行布局的初始化
20      override fun instantiateItem(container: ViewGroup, position: Int): Any {
21          container.addView(mPagerViews[position])
22          return mPagerViews[position]
23      }
24      //销毁分页视图时移除相应的分页
25      override fun destroyItem(container: ViewGroup, position: Int, `object`: Any) {
26          container.removeView(mPagerViews[position])
27      }
28  }
```

第 09 行将分页视图的数组列表作为构造方法的参数传递进来。第 17 行使用官方建议使用的返回值表达式 view= = =object，如果直接使用 true 或 false 作为返回值会出现显示错误。第 20 行在初始化或缓存分页视图时将相应的分页视图添加到使用该适配器的控件内，ViewPager 控件作为 ViewGroup 类的子类可以自动向上转型为 ViewGroup 类实例。第 22 行将初始化或缓存分页视图作为返回对象。第 25 行在销毁分页面时将该分页面从 container 对象中移除。

5．主界面的布局

```
/res/layout/activity_main.xml
<!-- 分页视图 -->
<androidx.viewpager.widget.ViewPager
    android:id = "@+id/viewPager"
    android:layout_gravity = "center"
    android:persistentDrawingCache = "animation">
</androidx.viewpager.widget.ViewPager>
<!-- 显示滚动事件信息 -->
```

```
<TextView
    android:id = "@+id/pagePositionTextView"
    android:layout_gravity = "bottom|center_horizontal"
    android:layout_marginBottom = "250dp"
    android:textColor = "#ffffff" />
<!-- 前进按钮 -->
<ImageView
    android:id="@+id/leftImageView"
    android:layout_gravity = "bottom|start"
    android:layout_marginStart = "10dp"
    android:layout_marginBottom = "60dp"
    android:src = "@mipmap/img_left" />
<!-- 后退按钮 -->
<ImageView
    android:id = "@+id/rightImageView"
    android:layout_gravity = "bottom|end"
    android:layout_marginEnd = "10dp"
    android:layout_marginBottom = "60dp"
    android:src = "@mipmap/img_right" />
<!-- 分页标记提示 -->
<LinearLayout
    android:layout_gravity = "bottom"
    android:layout_marginBottom = "70dp"
    android:gravity = "center">
    <ImageView
        android:id = "@+id/pageImageView0"
        android:scaleType = "fitXY"
        android:src = "@mipmap/img_page_now" />
    <ImageView
        android:id = "@+id/pageImageView1"
        android:layout_marginStart = "10dp"
        android:scaleType = "fitXY"
        android:src = "@mipmap/img_page" />
    <ImageView
        android:id = "@+id/pageImageView2"
        android:layout_marginStart = "10dp"
        android:scaleType = "fitXY"
        android:src = "@mipmap/img_page" />
    <ImageView
        android:id = "@+id/pageImageView3"
        android:layout_marginStart = "10dp"
        android:scaleType = "fitXY"
        android:src = "@mipmap/img_page" />
</LinearLayout>
<!-- 显示滚动状态改变事件信息 -->
<TextView
    android:id = "@+id/pageStateTextView"
    android:layout_gravity = "bottom|center_horizontal"
    android:layout_marginBottom = "20dp"
    android:textColor = "#ffffff" />
```

第 08 ~ 14 行<ViewPager>标签用于显示分页视图，android:persistentDrawingCache="animation"表示在布局动画之后进行缓存。第 20 行使用“|”连接两个属性值，top|center_horizontal 属性值表示顶部对齐且水平居中。第 42 ~ 75 行在<LinearLayout>标签内添加 4 个<ImageView>标签，用于标识当前显示的分页。第 77 ~ 83 行<TextView>标签用于显示分面状态。

6. 主界面的 Activity

/java/com/teachol/c0707/MainActivity.kt

```
class MainActivity : AppCompatActivity(), View.OnClickListener {
    private lateinit var mViewPager: ViewPager
    private lateinit var mPagePositionTextView: TextView
    private lateinit var mPageStateTextView: TextView
    private lateinit var mPage: MutableList<ImageView>
    private var mCurrentPosition = 0
    override fun onCreate(savedInstanceState: Bundle?) {
        mPagePositionTextView = findViewById(R.id.text_view_page_position)
        mPageStateTextView = findViewById(R.id.text_view_page_state)
        mPage = ArrayList()
        mPage.add(findViewById(R.id.image_view_page0))
        mPage.add(findViewById(R.id.image_view_page1))
        mPage.add(findViewById(R.id.image_view_page2))
        mPage.add(findViewById(R.id.image_view_page3))
        for (i in mPage) {
            i.setOnClickListener(this)
        }
        val leftImageView = findViewById<ImageView>(R.id.image_button_left)
        val rightImageView = findViewById<ImageView>(R.id.image_button_right)
        leftImageView.setOnClickListener(this)
        rightImageView.setOnClickListener(this)
        mViewPager = findViewById(R.id.view_pager)
        mViewPager.addOnPageChangeListener(PageChangeListener())
        init()
    }
    private fun init() { // 初始化
        val pagerViews = ArrayList<View>()
        pagerViews.add(View.inflate(this, R.layout.view_pager_welcome0, null))
        pagerViews.add(View.inflate(this, R.layout.view_pager_welcome1, null))
        pagerViews.add(View.inflate(this, R.layout.view_pager_welcome2, null))
        pagerViews.add(View.inflate(this, R.layout.view_pager_welcome3, null))
        val welcomePagerAdapter = WelcomePagerAdapter(pagerViews)
        mViewPager.adapter = welcomePagerAdapter
        mPagePositionTextView.text = String.format("页面索引号:%1d  偏移百分比:%2f  偏移像素:%3d:", 0, 0f, 0)
        mPageStateTextView.text = "空闲状态"
    }
    override fun onClick(v: View) { //单击监听事件
        when (v.id) {
            R.id.image_button_left -> { if (mCurrentPosition > 0) mCurrentPosition-- } //向左移动一页
            R.id.image_button_right -> { if (mCurrentPosition < 3) mCurrentPosition++ } //向右移动一页
            R.id.image_view_page0 -> { mCurrentPosition = 0 } //第一页
```

```
56                R.id.image_view_page1 -> { mCurrentPosition = 1 } //第二页
57                R.id.image_view_page2 -> { mCurrentPosition = 2 } //第三页
58                R.id.image_view_page3 -> { mCurrentPosition = 3 } //第四页
59            }
60            mViewPager.setCurrentItem(mCurrentPosition, true)
61        }
63        inner class PageChangeListener : ViewPager.OnPageChangeListener { //页面改变监听器
65            override fun onPageSelected(position: Int) { //页面选择事件
66                //翻页时当前page，改变当前状态圆点图像
67                mCurrentPosition = position
68                for(i in mPage.indices) {
69                    if (i == position) {
70                        mPage[i].setImageResource(R.mipmap.img_page_now)
71                    } else {
72                        mPage[i].setImageResource(R.mipmap.img_page)
73                    }
74                }
75            }
77            override fun onPageScrolled(position: Int, positionOffset: Float, positionOffsetPixels: Int) { //页面滚动事件
78                mPagePositionTextView.text = String.format("页面索引号:%1d  偏移百分比:%2f  偏移像素:%3d:", position, positionOffset, positionOffsetPixels)
79            }
81            override fun onPageScrollStateChanged(state: Int) { //页面滚动状态改变事件
82                when (state) {
83                    ViewPager.SCROLL_STATE_IDLE -> mPageStateTextView.text = "空闲状态"
84                    ViewPager.SCROLL_STATE_DRAGGING -> mPageStateTextView.text = "拖动状态"
85                    ViewPager.SCROLL_STATE_SETTLING -> mPageStateTextView.text = "结束状态"
86                }
87            }
88        }
89    }
```

第 35 行实例化一个 PageChangeListener 类对象作为 mViewPager 控件的分页改变监听器。第 39 ~ 49 行初始化并设置适配器。第 51 ~ 61 行重写单击监听事件，根据单击的图像按钮滚动到相应的页面。第 63 ~ 88 行声明 PageChangeListener 内部类实现了三个 ViewPager.OnPageChangeListener 接口方法，重写 onPageSelected()方法实现翻页时底部圆点标识当前页面的位置，重写 onPageScrolled()方法实现在顶部的 TextView 控件上显示页面索引号、偏移百分比和偏移像素信息，重写 onPageScrollStateChanged()方法在底部的 TextView 控件上显示 ViewPager 控件的状态。

7.7 视图绑定

7.7.1 ViewBinding

ViewBinding 是通过为 XML 布局自动生成一个绑定类，将控件的 id 转换为绑定类的属性，通过绑定类的实例直接调用控件的操作方式。启用 ViewBinding 后，能够省去 findById()方法，避免控件 id 无效出现的空指针问题。例如，activity_main.xml 文件会自动生成 ActivityMainBinding 类用于绑定

布局文件中的控件，该类的文件保存在“/app/build/generated/data_binding_base_class_source_out”的子文件夹中，如图 7-15 所示。自动生成的绑定类无法在 Android Studio 中进行编辑，避免程序混乱。

提示：ViewBinding 的效率

ViewBinding 可以有效提高开发效率，但是不可避免地会增加编译时间，降低执行效率，增加安装文件的大小。

启用 ViewBinding 后，在布局文件的根标签中添加 tools:viewBindingIgnore="true"属性，可以不生成该布局文件的绑定类。

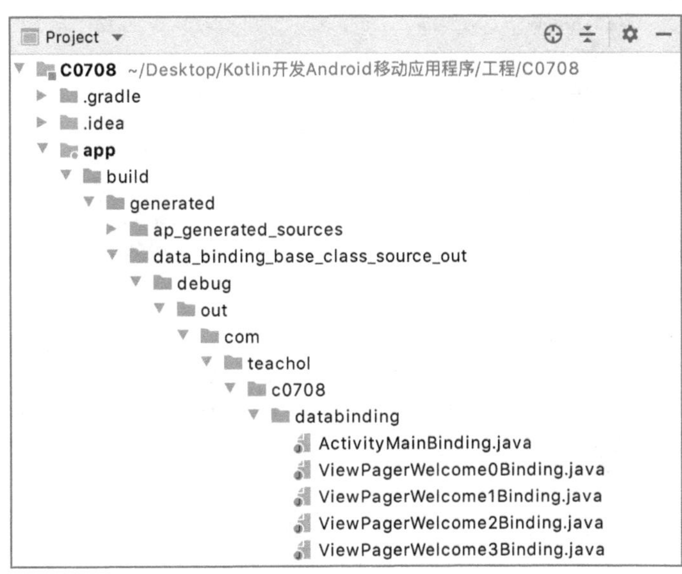

图 7-15　自动生成的绑定类存储位置

7.7.2　实例工程：使用视图绑定改造欢迎引导页

本实例演示了使用视图绑定的方式改造“欢迎引导页”实例工程(C0707 工程)，从而精简代码，但是运行效果并没有改变。

1. 打开基础工程

打开“基础工程”文件夹中的“C0708”工程，该工程与“工程”文件夹中的“C0707”工程相同。

2. 启用视图绑定

build.gradle(Module:C0708.app)

```
19  buildFeatures{
20      viewBinding = true
21  }
```

第 19 ~ 21 行添加启用 ViewBinding 功能的代码，然后单击“Sycn Now”按钮对工程进行同步。

3. 主界面的布局

将控件的 id 名称以小驼峰命名法的规则重新进行命名。例如，将 view_pager 修改为 viewPager、image_button_left 修改为 leftImageButton、image_view_page0 修改为 pageImageView0。

4. 主界面的 Activity

/java/com/teachol/c0708/MainActivity.kt

```
08  import com.teachol.c0708.databinding.ActivityMainBinding
10  class MainActivity : AppCompatActivity(), View.OnClickListener {
11      private lateinit var activityMainBinding: ActivityMainBinding
12      private var currentPosition = 0
13      override fun onCreate(savedInstanceState: Bundle?) {
14          super.onCreate(savedInstanceState)
15          activityMainBinding = ActivityMainBinding.inflate(layoutInflater)
16          setContentView(activityMainBinding.root)
17          activityMainBinding.pageImageView0.setOnClickListener(this)
18          activityMainBinding.pageImageView1.setOnClickListener(this)
19          activityMainBinding.pageImageView2.setOnClickListener(this)
20          activityMainBinding.pageImageView3.setOnClickListener(this)
21          activityMainBinding.leftImageView.setOnClickListener(this)
22          activityMainBinding.rightImageView.setOnClickListener(this)
23          activityMainBinding.viewPager.addOnPageChangeListener(PageChangeListener())
24          init()
25      }
39      override fun onClick(v: View) { //重写单击事件
40          when (v) {
41              activityMainBinding.leftImageView -> { if (currentPosition > 0) currentPosition-- }
42              activityMainBinding.rightImageView -> { if (currentPosition < 3) currentPosition ++ }
47          }
48          activityMainBinding.viewPager.setCurrentItem(currentPosition, true)
49      }
114 }
```

第 08 行导入 ActivityMainBinding 绑定类。第 11 行声明 ActivityMainBinding 绑定类的实例。第 15 行调用 ActivityMainBinding.inflate(layoutInflater)进行实例化，并绑定 activity_main.xml 布局文件。第 16 行 activityMainBinding.root 获取布局文件的视图。第 17 行 activityMainBinding.pageImageView0 通过绑定类的实例获取布局文件中的控件，pageImageView0 是布局文件中设置的控件 id。第 39～49 行通过 v 参数获取单击的控件实例，与 activityMainBinding 获取的控件实例直接进行分支条件匹配。

7.8 数据绑定

7.8.1 DataBinding

DataBinding 是通过为 XML 布局自动生成一个绑定类(和 ViewBinding 公用一个绑定类)，将布局文件中控件的属性与数据类的属性或方法进行关联，对数据类的实例间接修改控件属性或处理控件事件的操作方式。启用 DataBinding 后，可以提高应用性能，并且有助于防止内存泄漏及避免发生空指针异常。

使用 DataBinding 的布局文件的根节点必须是<layout>标签，通过<data>标签设置用于数据绑定的数据类，同时<layout>标签只能包含一个子布局标签。数据类的属性可以与控件的属性值进行单向数据绑定或双向数据绑定，数据类的方法可以与控件的事件进行绑定。

数据类中，用于进行数据绑定的属性需要继承 BaseObservable 类的子类，用于绑定事件的方法需要使用一个 View 参数。

提示：DataBinding 的报错

目前，Android Studio 对 XML 标签布局文件中的错误没有直接提示定位，而是提示自动生成的绑定类中的错误位置，所以要及时调试、修改错误的绑定代码。

7.8.2　BaseObservable 类

BaseObservable 类继承 Object 类，实现 Observable 接口，其方法如表 7-12 所示。继承 BaseObservable 类的子类，通过 notifyPropertyChanged(int) 和 notifyChange() 方法通知 UI 控件进行刷新。

表 7-12　BaseObservable 类的方法

类　型	方　法
Unit	addOnPropertyChangedCallback(callback:Observable.OnPropertyChangedCallback) 添加监听属性变化的回调
Unit	notifyChange() 通知监听器此实例的所有属性都已更改
Unit	notifyPropertyChanged(fieldId: Int) 通知监听器特定属性已更改
Unit	removeOnPropertyChangedCallback(callback:Observable.OnPropertyChangedCallback) 删除监听属性变化的回调

7.8.3　ObservableField 类

数据类的属性通常声明为 ObservableField 类的实例，ObservableField 类是 BaseObservable 类的子类，其方法如表 7-13 所示。ObservableField 类的实例不能使用=运算符进行赋值，而是使用 set(T) 方法，获取变量值时使用 get() 方法。

表 7-13　ObservableField 类的方法

类　型	方　法
T	get() 获取属性值
Unit	set(value: T) 设置属性值

Kotlin 还提供了一些常用数据类型对应的可观察类，包括 ObservableBoolean 类、ObservableByte 类、ObservableChar 类、ObservableShort 类、ObservableInt 类、ObservableLong 类、ObservableFloat 类、ObservableDouble 类、ObservableParcelable 类、ObservableArrayList 类、ObservableArrayMap 类等 BaseObservable 类的子类。其中没有包含 String 类型对应的可观察类，因此 String 类型需要使用 ObservableField 类的实例进行数据绑定。

7.8.4　实例工程：使用数据绑定改造欢迎引导页

本实例演示了使用数据绑定的方式改造 “欢迎引导页” 实例工程(C0707 工程)，实现对控件属性的单向数据绑定和双向数据绑定。单击第 4 个分页的“开始”按钮，通过双向绑定获取控件的 text 属性值，并使用通知显示出来，如图 7-16 所示。

图 7-16　运行效果

1. 打开基础工程

打开“基础工程”文件夹中的“C0709”工程，该工程与“工程”文件夹下的“C0708”工程相同。

2. 启用数据绑定

build.gradle(Module:C0709.app)

```
19 buildFeatures{
20     viewBinding = true
21     dataBinding = true
22 }
```

第 21 行添加启用 DataBinding 功能的代码，单击“Sycn Now”按钮对工程进行同步。

3. 数据类

右键单击“/java/com/teachol/c0709”文件夹，选择【New】→【New Kotlin Class/File】命令。在弹出的“New Kotlin Class/File”对话框中输入名称为“WelcomePagerModel”，双击“Class”类型完成类的创建(如图 7-17 所示)。

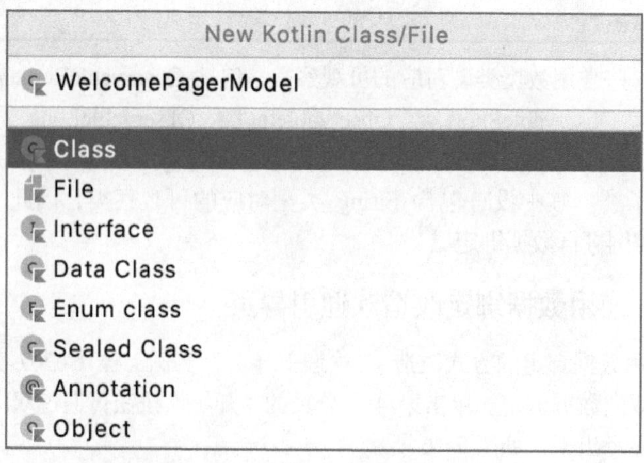

图 7-17　“New Kotlin Class/File”对话框

/java/com/teachol/c0709/WelcomePagerModel.kt

```
04 import androidx.databinding.ObservableField
06 class WelcomePagerModel(var context: MainActivity) {
07     var pageState = ObservableField<String>() //当前分页状态
08     var pagePosition = ObservableField<String>() //当前分页位置及偏移数据
09     var currentPosition = 0 //当前分页位置
11     fun onClick(v: View) { //单击事件
12         when (v) {
13             context.activityMainBinding.leftImageView -> { if (currentPosition > 0) currentPosition-- } //向左移动一页
14             context.activityMainBinding.rightImageView -> { if (currentPosition < 3) currentPosition++ } //向右移动一页
15             context.activityMainBinding.pageImageView0 -> { currentPosition = 0 } //第一页
16             context.activityMainBinding.pageImageView1 -> { currentPosition = 1 } //第二页
17             context.activityMainBinding.pageImageView2 -> { currentPosition = 2 } //第三页
18             context.activityMainBinding.pageImageView3 -> { currentPosition = 3 } //第四页
19         }
20         context.activityMainBinding.viewPager.setCurrentItem(currentPosition, true)
21     }
22 }
```

第 04 行导入 androidx.databinding.ObservableField 命名空间。第 06 行将 MainActivity 作为 context 参数的数据类型，以控制 MainActivity 的控件。第 07、08 行 pageState 和 pagePosition 变量声明为 ObservableField<String>类型，该类型作为字符型的可观察字段，用于更新绑定的控件属性。第 11 ~ 21 行声明处理控件单击事件的方法，context.activityMainBinding 通过 MainActivity 实例调用布局的绑定类实例，再通过绑定类实例调用布局中的控件。

4. 主界面的布局

/res/layout/activity_main.xml

```
<layout xmlns:android = "http://schemas.android.com/apk/res/android"
    xmlns:tools = "http://schemas.android.com/tools">
    <!-- 绑定数据类 -->
    <data>
        <variable
            Name = "welcomePagerModel"
            Type = "com.teachol.c0709.WelcomePagerModel" />
    </data>
    <!-- 帧布局 -->
    <FrameLayout
        tools:context = ".MainActivity">
        <!-- 分页视图 -->
        <androidx.viewpager.widget.ViewPager
            android:id = "@+id/viewPager"
            android:layout_gravity = "center"
            android:persistentDrawingCache = "animation" />
        <!-- 显示滚动事件信息 -->
        <TextView
            android:id = "@+id/pagePositionTextView"
            android:text = "@={welcomePagerModel.pagePosition}"
            android:textColor = "#ffffff" />
```

```
31        <!-- 前进按钮 -->
32        <ImageView
33           android:id = "@+id/leftImageView"
39           android:onClick = "@{welcomePagerModel::onClick}"
40           android:src = "@mipmap/img_left" />
41        <!-- 后退按钮 -->
42        <ImageView
43           android:id = "@+id/rightImageView"
49           android:onClick = "@{welcomePagerModel::onClick}"
50           android:src = "@mipmap/img_right" />
94        <!-- 显示滚动状态改变事件信息 -->
95        <TextView
96           android:id = "@+id/pageStateTextView"
101          android:text = "@{welcomePagerModel.pageState}"
102          android:textColor = "#ffffff" />
103    </FrameLayout>
104 </layout>
```

第 02 行添加<layout>标签，将原有根节点的<FrameLayout>标签包含起来，作为<layout>标签唯一的子布局标签。第 05 ~ 09 行添加设置数据绑定的<data>标签，<variable>标签的 name 属性设置数据绑定的变量名，type 属性设置数据绑定的数据类。第 29 行对 text 属性值进行双向数据绑定。第 39 行将 welcomePagerModel 实例的 onClick(View)方法与 ImageView 控件的单击事件绑定。第 101 行将 welcomePagerModel 实例的 pageState 属性与 TextView 控件的 text 属性进行单向数据绑定。

提示：单向数据绑定和双向数据绑定

在布局文件中绑定数据时，单向数据绑定使用@运算符连接包含可观察属性的花括号，双向数据绑定与单向数据绑定区别在于左侧花括号前增加了=运算符。

5. 主界面的 Activity

/java/com/teachol/c0709/MainActivity.kt

```
import com.teachol.c0709.databinding.ActivityMainBinding
class MainActivity : AppCompatActivity(){
   lateinit var activityMainBinding: ActivityMainBinding
   private var welcomePagerModel = WelcomePagerModel(this)
   override fun onCreate(savedInstanceState: Bundle?) {
      super.onCreate(savedInstanceState)
      activityMainBinding = ActivityMainBinding.inflate(layoutInflater)
      setContentView(activityMainBinding.root)
      init()
   }
   private fun init() { //初始化
      val view3 = View.inflate(this, R.layout.view_pager_welcome3, null)
      view3.findViewById<Button>(R.id.startButton).setOnClickListener { //最后一个分页的开始按钮
         Toast.makeText(this, welcomePagerModel.pagePosition.get(),Toast.LENGTH_LONG).show()
      }
      activityMainBinding.welcomePagerModel= welcomePagerModel //设置绑定的数据类实例
      // 添加 Fragment
      val pagerViews = ArrayList<View>()
      pagerViews.add(View.inflate(this, R.layout.view_pager_welcome0, null))
```

```
31         pagerViews.add(View.inflate(this, R.layout.view_pager_welcome1, null))
32         pagerViews.add(View.inflate(this, R.layout.view_pager_welcome2, null))
33         pagerViews.add(view3)
35         val welcomePagerAdapter = WelcomePagerAdapter(pagerViews) //实例化适配器
36         activityMainBinding.viewPager.adapter = welcomePagerAdapter //设置适配器
37         activityMainBinding.viewPager.addOnPageChangeListener(PageChangeListener())//设置监听器
39         welcomePagerModel.pagePosition.set(String.format("页面索引号:%1d  偏移百分比:%2f  偏移像素:%3d", 0, 0f, 0)) //设置初始显示的信息
40         welcomePagerModel.pageState.set("空闲状态")
41     }
90 }
```

第 08 行导入 ActivityMainBinding 绑定类。第 11 行声明 ActivityMainBinding 绑定类的实例。第 12 行声明并实例化数据类。第 15 行调用 ActivityMainBinding.inflate(layoutInflater)进行实例化，并绑定 activity_main.xml 布局文件。第 16 行 activityMainBinding.root 获取布局文件的视图。第 27 行 activityMainBinding.welcomePagerModel 对应布局文件中<variable>标签声明的 welcomePagerModel 变量。第 39、40 行 pagePosition 和 pageState 可观察字段变量，调用 set()方法设置数值，同时更新绑定该变量的控件属性。

第 8 章 Android 的基本程序单元

控件无法单独使用呈现给用户，无论是使用标签形式的布局文件还是通过代码动态创建的控件，都需要通过基本程序单元作为载体进行呈现。Activity 和 Fragment 类似于页面的功能，即把控件呈现给用户，并与用户进行交互。

8.1 活　　动

8.1.1 Activity 组件

Activity(android.app.Activity)是以窗口的形式显示的组件(常用方法如表 8-1 所示), App 通常由多个彼此松散绑定的 Activity 组成。

表 8-1 Activity 类的常用方法

类型和修饰符	方　法
open Unit	addContentView(view: View!, params: ViewGroup.LayoutParams!) 添加内容视图
open Unit	finish() 关闭 Activity
open Unit	finishActivity(request Code: Int) 关闭使用 startActivityForResult(Intent!, Int)启动的 Activity
open Intent!	getIntent() 获取启动 Activity 的 Intent 对象
open Unit	setContentView(layout Res ID: Int) 设置 Activity 的布局资源
open Unit	startActivity(intent: Intent!) 启动 Activity
open Unit	startActivity(intent: Intent!, options: Bundle?) 启动 Activity 并传递数据
open Unit	startActivityForResult(intent: Intent!, request Code: Int) 启动 Activity 并传递请求码
open Unit	startActivityForResult(intent: Intent!, request Code: Int, options: Bundle?) 启动 Activity，在关闭该 Activity 时能够获取返回的数据
open Unit	onNewIntent(intent: Intent!) 重新调用栈内 Activity 到前台时调用该方法
open Unit	onCreate(savedInstanceState: Bundle?) 创建时调用该方法
open Unit	onRestart() 即将重新启动时调用该方法
open Unit	onStart() 即将启动时调用该方法
open Unit	onResume() 即将进入前台时调用该方法
open Unit	onPause() 即将进入后台时调用该方法

续表

类型和修饰符	方　法
open Unit	onStop () 进入后台后调用该方法
open Unit	onDestroy () 即将销毁时调用该方法
open Unit	onActivityResult (requestCode: Int, resultCode: Int, data: Intent?) 当使用 startActivityForResult () 方法启动的 Activity 退出时调用该方法，并将请求码、返回码及附加数据回传

Activity 需要在 AndroidManifest.xml 中添加<activity>标签后才能被调用，<activity>标签属性（如表 8-2 所示）大多无法通过 Activity 类的方法进行设置和获取。

表 8-2　<activity>标签属性

属　性	说　明
android:alwaysRetainTaskState	设置是否始终保持 Activity 所在任务的状态而不被系统回收，该属性只对任务的根 Activity 设置有效
android:directBootAware	是否支持直接启动，即是否可以在用户解锁设备之前运行
android:enabled	是否可实例化 Activity
android:excludeFromRecents	是否从最近任务列表中删除该 Activity 启动的任务
android:exported	设置是否可由其他 App 组件启动
android:icon	设置图标
android:immersive	设置是否使用沉浸模式
android:label	设置标签
android:launchMode	设置启动模式。可选值包括：standard、singleTop、singleTask 和 singleInstance
android:mul 提示 rocess	设置是否可将 Activity 实例启动到该实例的组件进程中，默认值为 false
android:name	设置实现 Activity 的类名
android:process	设置运行 Activity 的进程名称。如果名称以“:”开头，则系统会在需要时创建专用新进程运行 Activity；如果名称以小写字母开头，则 Activity 将在使用该名称的全局进程中运行
android:resizeableActivity	设置是否支持多窗口显示。设置为 true 时，在分屏和自由窗口模式下启动；设置为 false 时，在多窗口模式下启动将全屏显示
android:screenOrientation	设置非多窗口模式下屏幕上显示的方向。 ● unspecified：默认值，系统决定方向 ● behind：与栈中其后的 Activity 方向相同 ● landscape：横向显示 ● portrait：纵向显示 ● reverseLandscape：与正常横向方向相反的横向 ● reversePortrait：与正常纵向方向相反的纵向 ● sensorLandscape：横向显示，根据传感器调整为正常或反向的横向 ● sensorPortrait：纵向显示，根据传感器调整为正常或反向的纵向 ● userLandscape：横向显示，根据传感器和用户首选项调整为正常或反向的横向 ● userPortrait：纵向显示，根据传感器和用户首选项调整为正常或反向的纵向 ● sensor：屏幕方向由设备方向传感器决定，一些设备在正常情况下不使用反向纵向或反向横向 ● fullSensor：屏幕方向由设备方向传感器决定，支持所有 4 种可能的屏幕方向 ● nosensor：忽略传感器

续表

属　性	说　明
android:screenOrientation	● user：用户当前的首选方向 ● fullUser：如果用户锁定基于传感器的旋转，则与 user 属性值相同；否则，与 fullSensor 属性值相同 ● locked：锁定为其当前的方向
android:supportsPictureInPicture	设置是否支持画中画显示功能。如果 android:resizeableActivity 为 false，则系统会忽略该属性

提示：AppCompatActivity

AppCompatActivity 类间接继承自 Activity 类，与 Activity 类相比，在视图顶部增加了 ActionBar，可以设置 Material 风格及直接使用 Material 风格，如 toolBar、Snackbar 和 AlertDialog 等。使用 AppCompatActivity 或直接使用 AppCompateDelegate，都必须使用 Theme.AppCompat 样式。

8.1.2 Activity 的创建和删除

1. 创建 Activity

新建一个"Empty Activity"工程，工程名称为"C0801"。在 AndroidManifest.xml 文件中，已经注册了该 Activity 的配置信息，并在<intent-filter>标签中设置该 Activity 为默认启动项。

/Manifests/AndroidManifest.xml

```
<activity android:name = ".MainActivity">
    <intent-filter>
        <action android:name = "android.intent.action.MAIN" />
        <category android:name = "android.intent.category.LAUNCHER" />
    </intent-filter>
</activity>
```

在 MainActivity 所在的文件夹上单击右键，选择【New】→【Activity】子菜单中的相应 Activity 预设可以快速创建 Activity(如图 8-1 所示)。

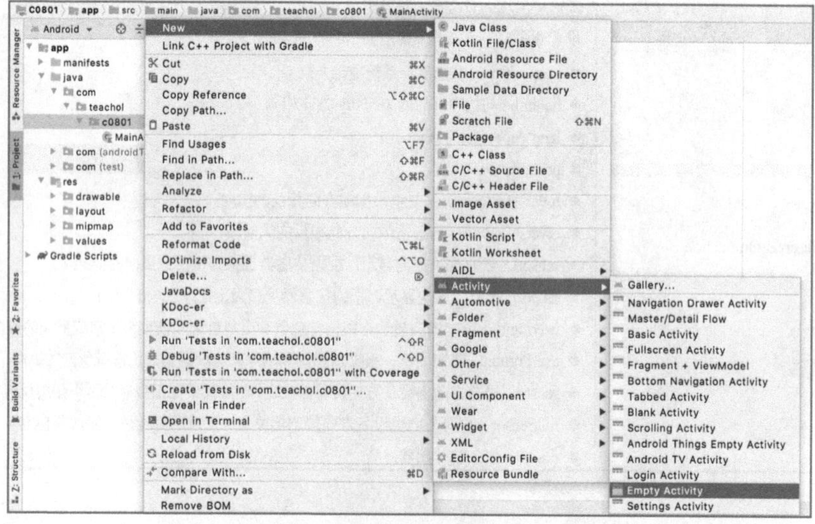

图 8-1　创建 Activity 的子菜单

在打开的“New Android Activity”对话框中，默认勾选“Generate a Layout File”复选框，用于同时创建一个布局文件(如图 8-2 所示)。“Launcher Activity”复选框用于设置工程的默认启动项，勾选后会替代之前的启动项作为默认启动的 Activity。

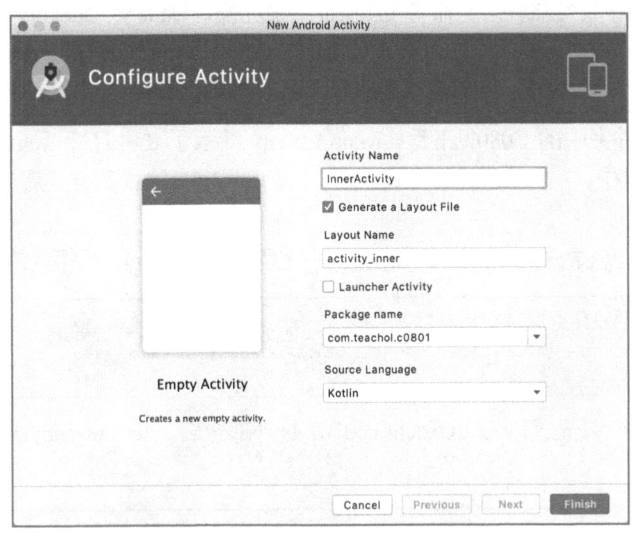

图 8-2　“New Android Activity”对话框

创建完成后，会在 AndroidManifest.xml 文件中自动添加<activity android:name=".InnerActivity"></activity>代码，用于注册 Activity。

2. 删除 Activity

在 Project 视图的 InnerActivity 上单击右键，选择【Delete】命令。在打开的“Delete”对话框(如图 8-3 所示)中，单击“OK”按钮进行删除。勾选所有复选框后会将与该 Activity 唯一相关的配置信息和文本资源删除，否则需要手动删除相应选项的内容。

删除 Activity 后，并不会删除创建该 Activity 时的布局文件。在 Project 视图的 layout 文件夹中的 activity_inner.xml 上单击右键，选择【Delete】命令，打开的“Delete”对话框(如图 8-4 所示)。勾选两个复选框后会将与该布局文件唯一相关的内容删除，否则需要手动删除相应选项的内容。

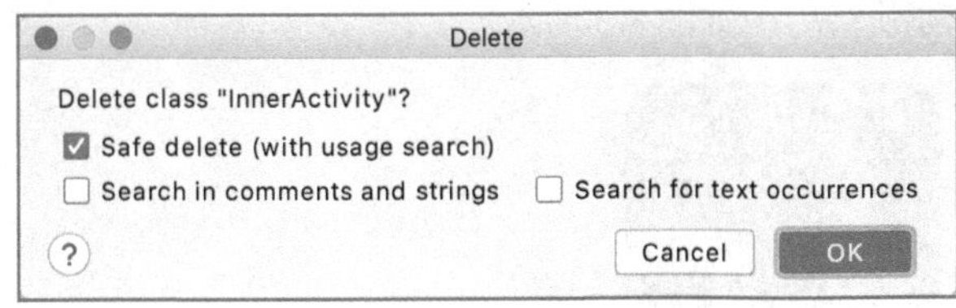

图 8-3　“Delete”对话框(1)

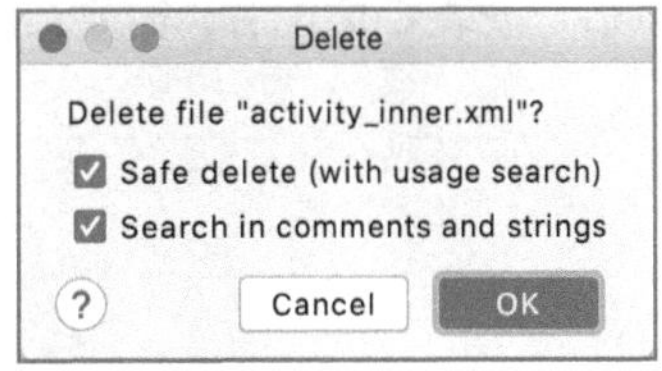

图 8-4　“Delete”对话框(2)

8.1.3 Activity 的启动和关闭

1. 启动 Activity

通过启动项的 Activity 启动其他 Activity 时，最简单的方法是使用 startActivity(Intent intent)方法，该方法需要使用一个 Intent 对象。Intent 是一个消息传递对象，用来从其他应用组件请求操作。Intent 可以通过多种方式促进组件之间的通信，主要功能是启动 Activity、启动 Service 和传递广播。

提示：显式 Intent 和隐式 Intent

显式 Intent 直接通过组件名指定启动的组件，会明确指定启动的组件路径。

隐式 Intent 不直接指定组件名，而指定 Intent 的 Action、Data 或 Category 多用于调用系统默认提供的功能。当启动组件时，自动匹配系统中所有安装 App 的 AndroidManifest.xml 中的<intent-filter>标签，匹配出满足条件的组件。当不止一个组件满足条件时，会弹出一个选择组件的对话框。

打开“工程”文件夹中的 C0802 工程，MainActivity 演示了 8 种打开 Activity 的方式，可分为显式和隐式两种类型。其中，2 个以显式的方式启动，1 个以自定义隐式的方式启动和 5 个以使用预设隐式的方式启动。

- 启动内部 Activity（显式）：InnerActivity 是同一工程中的 Activity，后缀名使用 class.java 表示类名。

/java/com/teachol/c0802/MainActivity.kt

```kotlin
val intent = Intent(applicationContext, InnerActivity::class.java)
startActivity(intent)
```

- 启动外部 Activity（显式）：com.teachol.c0707 是外部包名，com.teachol.c0707.MainActivity 是外部包内的类名。

/java/com/teachol/c0802/MainActivity.kt

```kotlin
val intent = Intent("android.intent.action.MAIN")
intent.setClassName("com.teachol.c0707", "com.teachol.c0707.MainActivity")
startActivity(intent)
```

- 启动外部 Activity（预设隐式）：MediaStore.INTENT_ACTION_STILL_IMAGE_CAMERA 设置打开拍摄照片的组件。如果没有安装第三方拍摄照片 App，则直接调用系统相机 App。如果安装了第三方拍摄照片 App，则需要进行选择。

/java/com/teachol/c0802/MainActivity.kt

```kotlin
val intent = Intent()
intent.action = MediaStore.INTENT_ACTION_STILL_IMAGE_CAMERA
startActivity(intent)
```

打开 C0801 工程 manifest 文件夹中的 AndroidManifest.xml 文件，设置 Sub1Activity 为有拍照功能的 Activity。安装 C0801 后，需要选择启动哪个应用（如图 8-5 所示）。

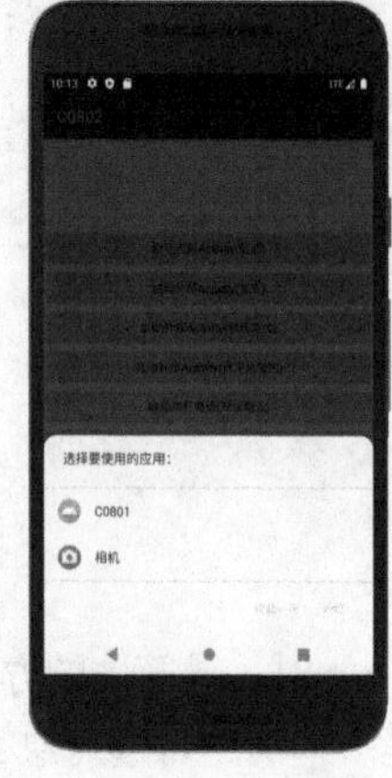

图 8-5　选择启动的应用

```
/manifests/AndroidManifest.xml
<activity android:name = ".Sub1Activity">
    <intent-filter>
        <action android:name = "android.media.action.STILL_IMAGE_CAMERA" />
        <category android:name = "android.intent.category.DEFAULT" />
    </intent-filter>
</activity>
```

● 启动外部 Activity（自定义隐式）：intent.action="custom_action"设置外部 Activity 的<action>标签名称，intent.addCategory("custom_category")设置外部 Activity 的<category>标签名称，这两个标签属性相对应的 Activity 是 C0801 工程中的 Sub2Activity。安装 C0801 后，才能启动 C0801 的 Sub2Activity。

```
/java/com/teachol/c0802/MainActivity.kt
val intent = Intent()
intent.action = "custom_action"
intent.addCategory("custom_category")
startActivity(intent)
```

在 C0801 工程的 AndroidManifest.xml 文件中，设置 Sub2Activity 的<action>和<category>标签。

```
/manifests/AndroidManifest.xml
<activity android:name = ".Sub2Activity">
    <intent-filter>
        <action android:name = "custom_action" />
        <category android:name = "custom_category" />
        <category android:name = "android.intent.category.DEFAULT" />
    </intent-filter>
</activity>
```

● 启动拨打电话（预设隐式）：Intent.ACTION_DIAL 设置启动拨号组件，uri 设置拨打的电话号码。

```
/java/com/teachol/c0802/MainActivity.kt
val uri = Uri.parse("tel:10086")
val intent = Intent(Intent.ACTION_DIAL, uri)
startActivity(intent)
```

● 启动发送短信（预设隐式）：Intent.ACTION_SENDTO 设置启动发送指定目标的组件，uri 设置接收短信的电话号码，putExtra("sms_body", "短信内容")设置发送短信的内容为“测试短信”。系统根据"sms_body"确定启动发送短信。

```
/java/com/teachol/c0802/MainActivity.kt
val uri = Uri.parse("smsto:10086")
val intent = Intent(Intent.ACTION_SENDTO, uri)
intent.putExtra("sms_body", "测试短信")
startActivity(intent)
```

● 启动发送邮件（预设隐式）：Intent.ACTION_SEND 设置启动发送的组件，putExtra(Intent.EXTRA_EMAIL, "baizhe_22@qq.com")设置接收邮箱的地址，putExtra(Intent. EXTRA_SUBJECT, "邮件标题")设置邮件的标题，putExtra(Intent.EXTRA_TEXT, "邮件内容")设置邮件的内容，setType("text/plain")设置纯文本的邮件且未包含附件。

```
/java/com/teachol/c0802/MainActivity.kt
```

```
val intent = Intent(Intent.ACTION_SEND)
intent.putExtra(Intent.EXTRA_EMAIL, "baizhe_22@qq.com")
intent.putExtra(Intent.EXTRA_SUBJECT, "邮件标题")
intent.putExtra(Intent.EXTRA_TEXT, "邮件内容")
intent.type = "text/plain"
startActivity(intent)
```

- 启动浏览器(预设隐式)：Intent.ACTION_VIEW 设置启动显示数据的组件，uri 设置访问的网址。

```
/java/com/teachol/c0802/MainActivity.kt
val uri = Uri.parse("http://www.weiju2014.com")
val intent = Intent(Intent.ACTION_VIEW, uri)
startActivity(intent)
```

2. 关闭 Activity

在 C0802 工程的 InnerActivity 中，调用 Activity 的 finish()方法关闭 Activity。

```
/java/com/teachol/c0802/InnerActivity.kt
val button = findViewById<Button>(R.id.button)
button.setOnClickListener { finish() }
```

8.1.4 Activity 的生命周期

Activity 的生命周期(如图 8-6 所示)主要包含 7 个方法和 6 个状态(Created、Started、Resumed、Paused、Stopped、Destroyed)。完整生命周期从回调 onCreate()方法开始，到回调 onDestroy()方法结束。可见生命周期从回调 onStart()方法开始，到回调 onStop()方法结束。前台生命周期从回调 onResume()方法开始，到回调 onPause()方法结束。

打开“工程”文件夹中的 C0803 工程，MainActivity 重写 7 个生命周期方法，用于测试 Activity 的生命周期执行顺序。

```
/java/com/teachol/c0803/MainActivity.kt
class MainActivity : AppCompatActivity() {
    val TAG = "生命周期"
    override fun onCreate(savedInstanceState: Bundle?) {
        Log.d(TAG, "MainActivity.onCreate()")
        val button: Button = findViewById(R.id.button)
        button.setOnClickListener(object : View.OnClickListener {
            override fun onClick(v: View?) {
                val intent = Intent(applicationContext, SubActivity::class.java)
                startActivity(intent)
                Log.d(TAG, "打开 SubActivity")
            }
        })
    }
    override fun onStart() { //Activity 即将启动时调用
        super.onStart()
        Log.d(TAG, "MainActivity.onStart()")
    }
    override fun onRestart() { //Activity 即将重新启动时调用
```

```
        super.onRestart()
        Log.d(TAG, "MainActivity.onRestart()")
    }
    override fun onResume() { //Activity 即将进入后台时调用
        super.onResume()
        Log.d(TAG, "MainActivity.onResume()")
    }
    override fun onPause() { //Activity 即将进入后台时调用
        super.onPause()
        Log.d(TAG, "MainActivity.onPause()")
    }
    override fun onStop() { //Activity 进入后台后调用
        super.onStop()
        Log.d(TAG, "MainActivity.onStop()")
    }
    override fun onDestroy() { //Activity 即将被销毁时调用
        super.onDestroy()
        Log.d(TAG, "MainActivity.onDestroy()")
    }
}
```

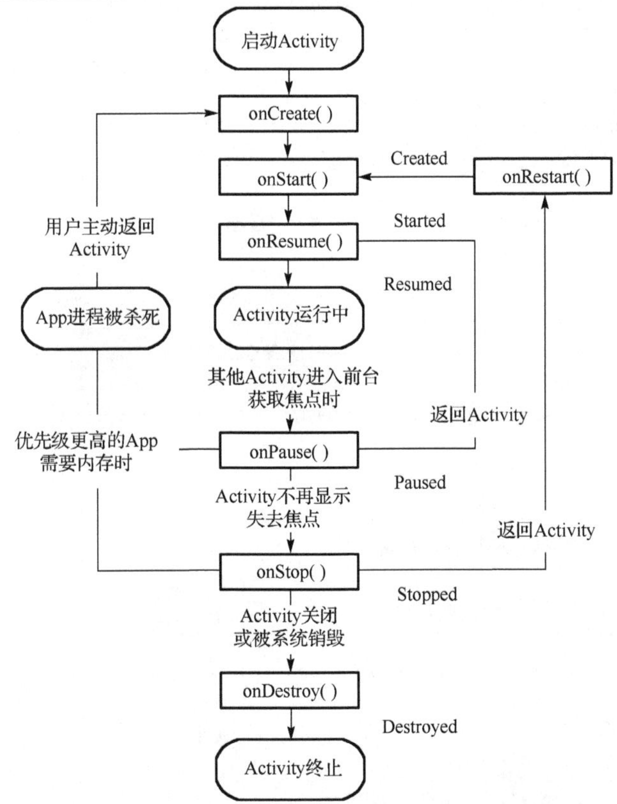

图 8-6　Activity 的生命周期

运行工程后，在 Logcat 窗口的过滤器搜索输入框中输入“生命周期”，过滤输出结果，可以查看启动后执行的生命周期方法名称(如图 8-7 所示)。

图 8-7　Logcat 窗口输出结果(1)

SubActivity 主要将 Log.d(String, String)方法中的“MainActivity”替换成了“SubActivity”，onClick(View)方法中的启动 SubActivity 替换成了关闭 SubActivity。

```
/java/com/teachol/c0803/SubActivity.kt
class SubActivity : AppCompatActivity() {
    val TAG = "生命周期"
    override fun onCreate(savedInstanceState: Bundle?) {
        Log.d(TAG, "SubActivity.onCreate()")
        val button = findViewById<Button>(R.id.button)
        button.setOnClickListener {
            finish()
            Log.d(TAG, "关闭 SubActivity")
        }
    }
    override fun onStart() { //Activity 即将启动时调用
        super.onStart()
        Log.d(TAG, "SubActivity.onStart()")
    }
    override fun onRestart() { //Activity 即将重新启动时调用
        super.onRestart()
        Log.d(TAG, "SubActivity.onRestart()")
    }
    override fun onResume() { //Activity 即将进入后台时调用
        super.onResume()
        Log.d(TAG, "SubActivity.onResume()")
    }
    override fun onPause() { //Activity 即将进入后台时调用
        super.onPause()
        Log.d(TAG, "SubActivity.onPause()")
    }
    override fun onStop() { //Activity 进入后台后调用
        super.onStop()
        Log.d(TAG, "SubActivity.onStop()")
    }
    override fun onDestroy() { //Activity 即将被销毁时调用
        super.onDestroy()
        Log.d(TAG, "SubActivity.onDestroy()")
    }
}
```

在 Logcat 窗口中，单击右键，选择【Clear logcat】命令清空窗口。然后单击“打开 Activity”按钮启动 SubActivity，可以查看已经执行的生命周期方法名称(如图 8-8 所示)。

图 8-8　Logcat 窗口输出结果(2)

在 Logcat 窗口中，单击右键，选择【Clear logcat】命令清空窗口。然后单击“关闭 Activity”按钮或者返回按钮(底部三角形的虚拟按钮)关闭 SubActivity，可以查看已经执行的生命周期方法名称(如图 8-9 所示)。

图 8-9　Logcat 窗口输出结果(3)

图 8-10　单击多任务按钮后的效果

在 Logcat 窗口中，单击右键，选择【Clear logcat】命令清空窗口。然后单击多任务按钮(底部正方形的虚拟按钮)后(如图 8-10 所示)，快速单击 C0803 返回到 MainActivity，可以查看已经执行的生命周期方法名称(如图 8-11 所示)。如果单击 C0803 的速度足够快，则不会调用 onRestart()方法。

在 Logcat 窗口中，单击右键，选择【Clear logcat】命令清空窗口。然后单击 Home 按钮(底部圆形的虚拟按钮)显示桌面，可以查看已经执行的生命周期方法名称(如图 8-12 所示)。

图 8-11　Logcat 窗口输出结果(4)

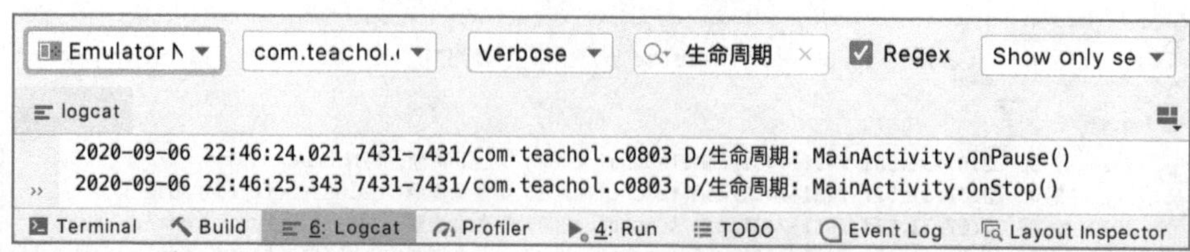

图 8-12　Logcat 窗口输出结果(5)

8.1.5　Activity 的启动模式

Activity 的启动模式本质区别在于任务栈的分配，共有 4 种启动模式：standard、singleTop、singleTask 和 singleInstance。启动模式需要在 AndroidManifest.xml 文件中设置 android:launchMode 属性进行设置，默认值是 standard。

提示：任务栈

任务栈是一种放置 Activity 实例的容器，使用先进后出的栈进行存储。因此，Activity 不支持重新排序，只能根据压栈和出栈操作更改 Activity 的顺序。

启动 App 时，系统会创建一个新的任务栈存储默认启动的 Activity，然后启动的其他 Activity 会被压入将其启动的 Activity 任务栈中并在前台显示出来。单击返回按钮，前台显示的 Activity 就会出栈。单击 Home 按钮回到桌面，再启动另一个 App，此时前一个 App 就被移到后台，其任务栈成为后台任务栈。而刚启动的 App 创建的任务栈被调到前台，成为前台任务栈，显示的 Activity 就是前台任务栈中的栈顶元素。

1．standard：标准模式

启动 standard 标准模式的 Activity 时，创建一个新的实例，放置在启动该 Activity 的任务栈顶部。

在 C0804 工程中，包含 MainActivity 和 StandardActivity（如图 8-13 所示）。在 AndroidManifest.xml 中，StandardActivity 设置的启动模式是 standard。运行 C0804 工程，启动默认项 MainActivity 后，单击相应的按钮依次启动 StandardActivity、StandardActivity 和 StandardActivity（如图 8-14 所示）。

```
/manifests/AndroidManifest.xml
<activity android:name = ".StandardActivity" android:launchMode = "standard" />
```

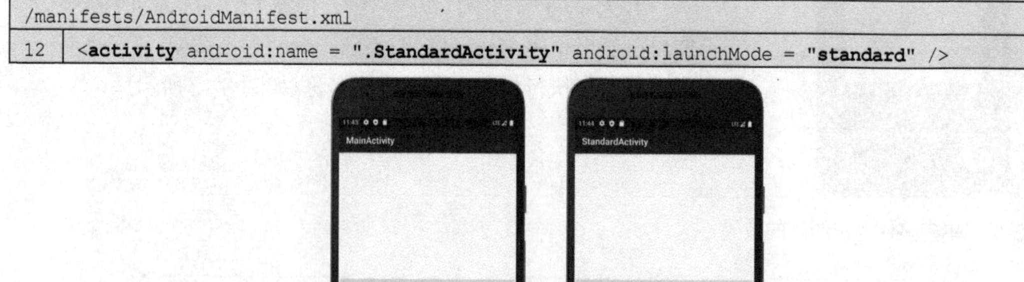

图 8-13　MainActivity 和 StandardActivity 的运行效果

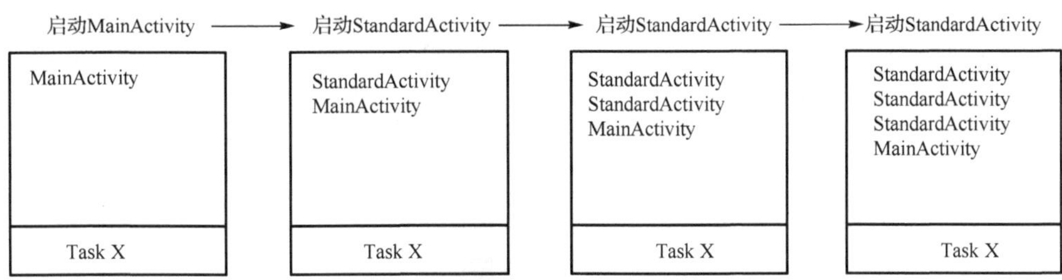

图 8-14　运行过程中任务栈的变化

在 Logcat 窗口中可以看到，每次启动的 StandardActivity 的任务栈 id 相同，但是 hashcode 都不相同(如图 8-15 所示)。此时单击 3 次返回按钮，才会返回到 MainActivity，说明启动了 3 个 StandardActivity。

图 8-15　Logcat 窗口输出结果

提示：standard 模式的使用场景

没有特殊需求的情况下，多使用标准模式启动 Activity。

2. singleTop：栈顶复用模式

当启动 singleTop 栈顶复用模式的 Activity 处于当前栈的顶部时，不会创建新的实例，而是直接启动该 Activity。onCreate()和 onStart()方法不会被调用，而是调用 onNewIntent()方法。当启动的 Activity 不在当前栈的顶部时，创建一个新的实例。

在 C0805 工程中，包含 MainActivity 和 SingleTopActivity(如图 8-16 所示)。在 AndroidManifest.xml 中，SingleTopActivity 设置的启动模式是 singleTop。运行 C0805 工程，启动默认项 MainActivity 后，单击相应的按钮依次启动 SingleTopActivity、SingleTopActivity、MainActivity、SingleTopActivity 和 SingleTopActivity(如图 8-17 所示)。

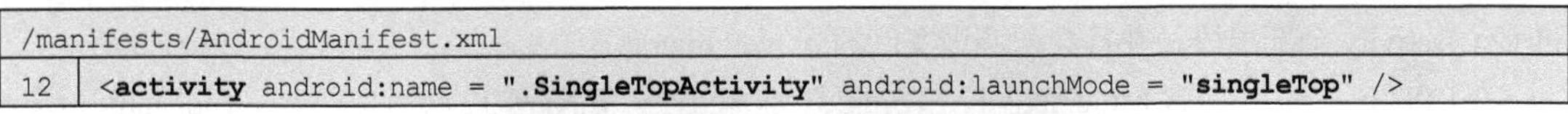

```
/manifests/AndroidManifest.xml
<activity android:name = ".SingleTopActivity" android:launchMode = "singleTop" />
```

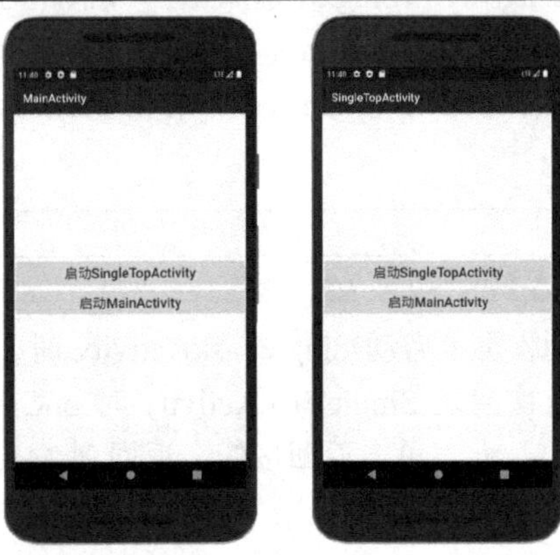

图 8-16　MainActivity 和 SingleTopActivity 的运行效果

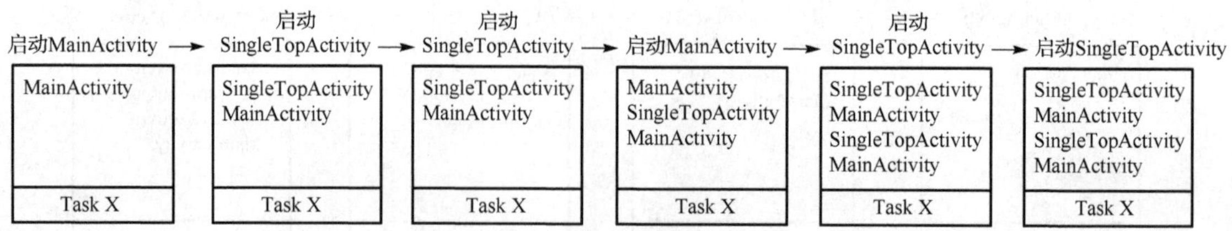

图 8-17　运行过程中任务栈的变化

在 Logcat 窗口中可以看到，第二次启动 SingleTopActivity 后，没有调用 onCreate()方法，而是调用 onNewIntent()方法，输出的任务栈 id 和 hashcode 没有变化。第三次启动 SingleTopActivity 后，hashcode 发生了变化，说明创建了一个新实例。第四次启动 SingleTopActivity 后，没有调用 onCreate()方法，而是调用 onNewIntent()方法，输出的任务栈 id 和 hashcode 与第三次启动的 SingleTopActivity 相同(如图 8-18 所示)。

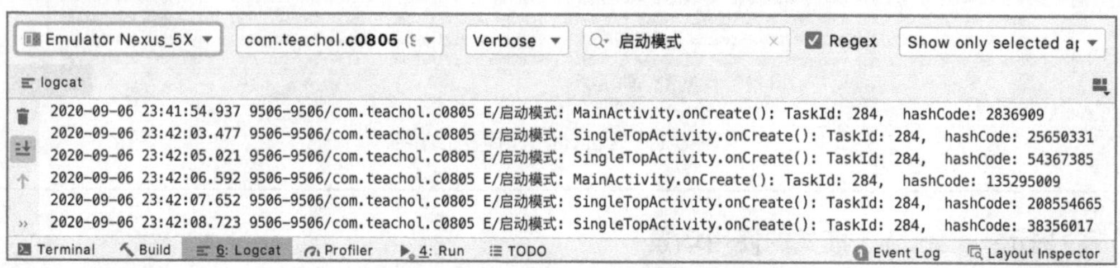

图 8-18　Logcat 窗口输出结果

提示：singleTop 模式的使用场景

如果 Activity 在栈顶运行时，需要启动同类型的 Activity，使用该模式能够减少 Activity 实例的创建并节省内存。例如，在通知栏收到了三条新闻的推送信息，单击推送信息会启动显示新闻详情的 Activity。单击第一条推送信息后，显示新闻详情的 Activity 已经处于栈顶。单击第二条和第三条推送信息时，只需通过 Intent 传入相应的数据即可，可以避免重复新建实例。

3. singleTask：栈内复用模式

任务栈中存在 singleTask 栈内复用模式的 Activity 时，再次启动该 Activity，栈内该 Activity 上面的所有 Activity 全部出栈，并且会回调该实例的 onNewIntent()方法。

在 C0806 工程中，包含 MainActivity 和 SingleTaskActivity(如图 8-19 所示)。在 AndroidManifest.xml 中，SingleTaskActivity 设置的启动模式是 singleTask。运行 C0806 工程，启动默认项 MainActivity 后，单击相应的按钮依次启动 SingleTaskActivity、MainActivity、MainActivity 和 SingleTaskActivity(如图 8-20 所示)。

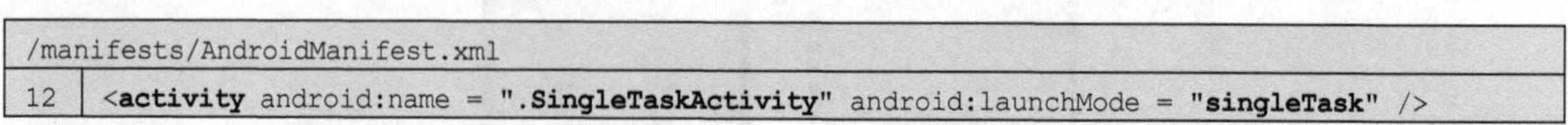

```
/manifests/AndroidManifest.xml
<activity android:name = ".SingleTaskActivity" android:launchMode = "singleTask" />
```

在 Logcat 窗口中可以看到，再次启动 SingleTaskActivity 时，调用 onNewIntent()方法输出的任务栈 id 和 hashcode，与首次启动 SingleTaskActivity 时 onCreate()方法输出的任务栈 id 和 hashcode 相同(如图 8-21 所示)。此时单击返回按钮会返回到 MainActivity，再单击返回按钮会返回到桌面。

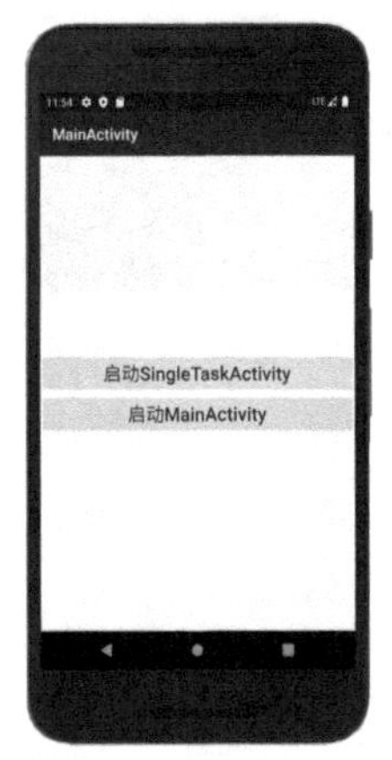

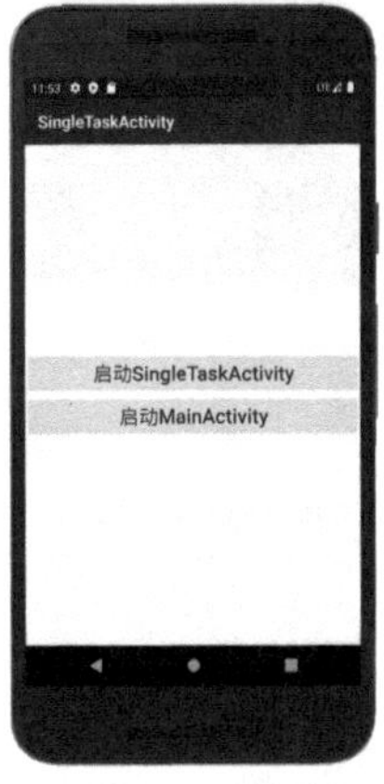

图 8-19　MainActivity 和 SingleTaskActivity 的运行效果

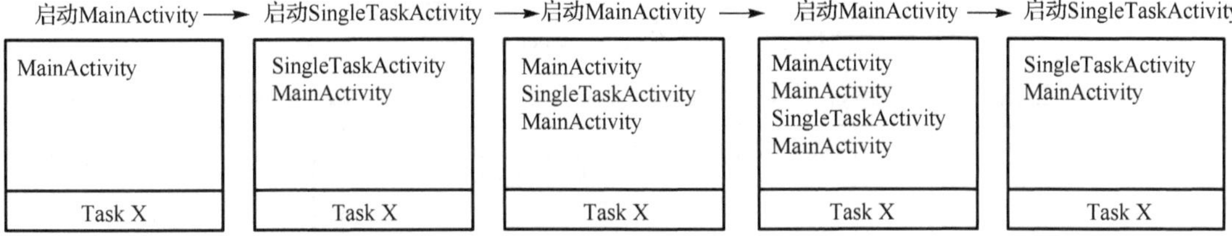

图 8-20　运行过程中任务栈的变化

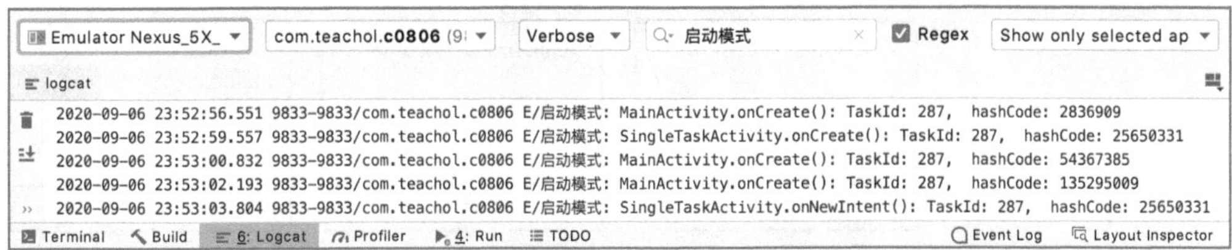

图 8-21　Logcat 窗口输出结果

提示：singleTask 模式的使用场景

此启动模式通常用于 App 首页的 Activity，且长时间保留在工作栈中，以保证 Activity 的唯一性。

4. singleInstance：单例模式

singleInstance 单例模式的 Activity 除具有 singleTask 模式特性外，还具有全局唯一性。系统中只能存在一个实例且单独占用一个任务栈，被该 Activity 开启的其他 Activity 会分配到其他任务栈内。首次启动该 Activity 时，会新建任务栈存储该 Activity 实例。再次启动该 Activity 时，除非该 Activity 实例已经销毁，否则不会在新栈内创建新的实例；重用该 Activity 实例，不会调用 onCreate()方法，会调用 onNewIntent()方法。

在 C0807 工程中，包含 MainActivity 和 SingleInstanceActivity（如图 8-22 所示）。在 AndroidManifest.xml 中，SingleInstanceActivity 设置的启动模式是 singleInstance。运行 C0807 工程，启动默认项 MainActivity 后，单击相应的按钮依次启动 SingleInstanceActivity、MainActivity 和 SingleInstanceActivity（如图 8-23 所示）。

/manifests/AndroidManifest.xml

```
<activity android:name = ".SingleInstanceActivity" android:launchMode = "singleInstance" />
```

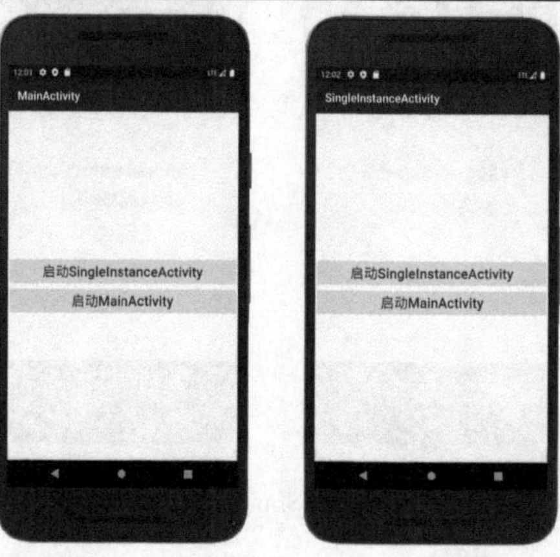

图 8-22　MainActivity 和 SingleInstanceActivity 的运行效果

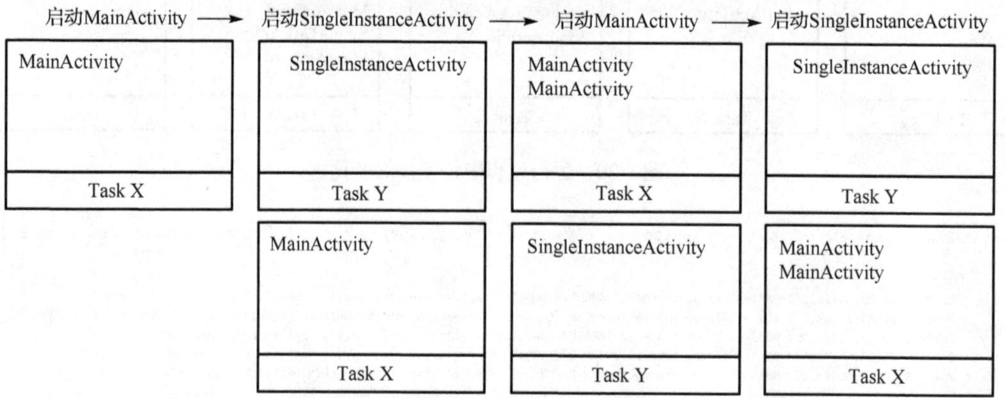

图 8-23　运行过程中任务栈的变化

在Logcat窗口中可以看到，启动SingleInstanceActivity时，为其创建了新的任务栈。然后启动MainActivity，将被放置在原有的任务栈内，而不是 SingleInstanceActivity 所在的任务栈。最后启动 SingleInstanceActivity，没有调用 onCreate() 方法，调用 onNewIntent() 方法输出的任务栈 id 和 hashcode，与第一次启动SingleInstanceActivity 时调用 onCreate()方法输出的任务栈 id 和 hashcode 相同（如图 8-24 所示）。

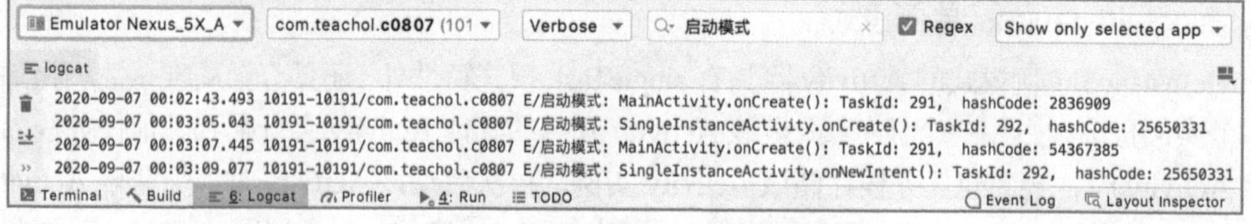

图 8-24　Logcat 窗口输出结果

提示：singleInstance 模式的使用场景

此启动模式通常用于工具类的 App，被其他 App 调用时能够保证全局唯一性，如拨号、短信、相机、地图等。

8.1.6　实例工程：Activity 的数据传递

本实例演示了简单数据类型的数据通过 Intent 对象在各 Activity 之间进行传递（如图 8-25 所示），Intent 类的 putExtra()方法提供了丰富的重载方法，可以传递不同类型的数据。关闭 Activity 时，还可以回传数据。

图 8-25　运行效果

1. 打开基础工程

打开“基础工程”文件夹中的“C0808”工程，该工程已经包含 MainActivity 和 ReceiveActivity 及资源文件。

2. 主界面的 Activity

```
/java/com/teachol/c0808/MainActivity.kt
class MainActivity : AppCompatActivity() {
    companion object{
        val REQUEST_CODE = 101
    }
    private lateinit var mResultEditText: EditText
    override fun onCreate(savedInstanceState: Bundle?) {
        mResultEditText = findViewById(R.id.edit_text_result)
        // 发送数据(没有回传数据)
        findViewById<View>(R.id.button_send).setOnClickListener(object : View.OnClickListener {
            override fun onClick(v: View?) {
                val intent = Intent(this@MainActivity, ReceiveActivity::class.java)
                intent.putExtra("id", 10407)
                intent.putExtra("msg", "MainActivity发送的信息1")
                startActivity(intent)
                mResultEditText.setText("")
            }
        })
        // 发送数据(接收回传数据)
        findViewById<View>(R.id.button_send_for_result).setOnClickListener(object : View.OnClickListener {
```

```
31            override fun onClick(view: View?) {
32                val intent = Intent(this@MainActivity, ReceiveActivity::class.java)
33                intent.putExtra("id", 20408)
34                intent.putExtra("msg", "MainActivity发送的信息2")
35                startActivityForResult(intent, REQUEST_CODE)
36                mResultEditText.setText("")
37            }
38        })
39    }
41    override fun onActivityResult(requestCode: Int, resultCode: Int, data: Intent?) { //处理回传数据
42        super.onActivityResult(requestCode, resultCode, data)
43        mResultEditText.setText("正在处理回传数据")
45        if (requestCode == REQUEST_CODE) { //判断请求码
47            if (resultCode == ReceiveActivity.RESULT_CODE) { //判断结果码
48                val result = data?.getStringExtra("data")
49                mResultEditText.setText(result)
50            } else {
51                mResultEditText.setText("没有回传数据")
52            }
53        }
54    }
55 }
```

第 10 ~ 12 行使用 companion object{}方法定义静态整型的 REQUEST_CODE 常量，用于回传数据的请求码。第 20-28 行“发送数据(没有回传数据)”按钮的单击事件使用 putExtra()方法将传递的数据存储在 intent 实例中，使用 startActivity(intent)启动 ReceiveActivity。第 30 ~ 38 行“发送数据(接收回传数据)”按钮的单击事件同样使用 putExtra()方法将传递的数据存储在 intent 实例中，使用 startActivityForResult(intent,REQUEST_CODE) 启动 ReceiveActivity。第 41 ~ 54 行重写 onActivityResult(Int,Int,Intent?)方法，根据 requestCode 和 resultCode 判断如何处理回传的 data 参数数据。第 47 行调用 ReceiveActivity 的静态整型常量 RESULT_CODE 判断是否 ReceiveActivity 使用该常量作为返回码回传数据。

3. 接收界面的 Activity

/java/com/teachol/c0808/ReceiveActivity.kt

```
class ReceiveActivity : AppCompatActivity() {
    companion object{
        val RESULT_CODE = 201
    }
    private lateinit var mIdEditText: EditText
    private lateinit var mMsgEditText: EditText
    override fun onCreate(savedInstanceState: Bundle?) {
        //获取传递数据
        val intent = intent
        val id = intent.getIntExtra("id", 0)
        val msg = intent.getStringExtra("msg")
        //显示传递数据
        mIdEditText = findViewById(R.id.edit_text_id)
```

```
24            mIdEditText.setText(id.toString())
25            mMsgEditText = findViewById(R.id.edit_text_name)
26            mMsgEditText.setText(msg)
28            findViewById<View>(R.id.button_finish).setOnClickListener { //关闭(没有回传数据)
29                finish() //关闭当前 Activity
30            }
32            findViewById<View>(R.id.button_finish_result).setOnClickListener { //关闭(发送回传数据)
33                val intent = Intent()
34                intent.putExtra("data", "已经查阅信息! ")
35                setResult(ReceiveActivity.RESULT_CODE, intent) //设置返回的数据
37                finish() //关闭当前 Activity
38            }
39        }
40    }
```

第 10 ~ 12 行使用 companion object{}方法定义静态整型的 RESULT_CODE 常量，用于回传数据的返回码。第 28 ~ 30 行“关闭(没有回传数据)”按钮的单击事件直接使用 finish()方法关闭当前 Activity，不会向启动当前 Activity 的 Activity 回传数据。第 32 ~ 38 行“关闭(发送回传数据)”按钮的单击事件通过 intent 实例存储回传的数据，然后使用 setResult(RESULT_CODE, intent)设置返回码和回传数据，最后使用 finish()方法关闭当前 Activity。

4. 测试运行

运行 C0808 工程。在 MainActivity 中，单击“发送数据(没有回传数据)”按钮启动 ReceiveActivity，ReceiveActivity 显示接收的数据，再单击“关闭(没有回传数据)”或“关闭(发送回传数据)”按钮关闭 ReceiveActivity，此时返回到 MainActivity 且没有获取回传数据(如图 8-26 所示)。

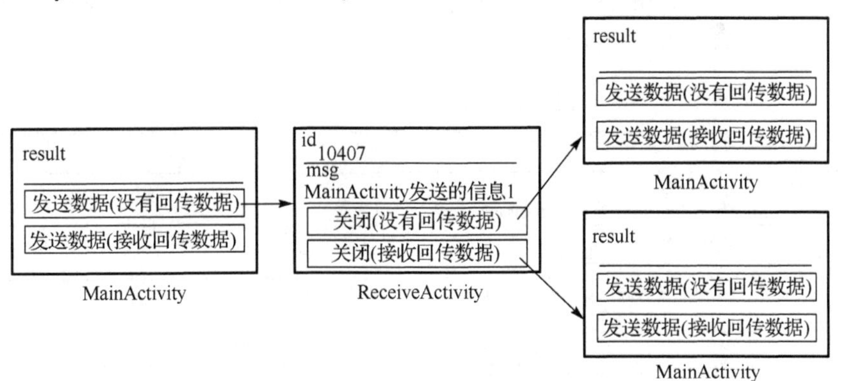

图 8-26 没有回传数据的效果

在 MainActivity 中，单击“发送数据(接收回传数据)”按钮启动 ReceiveActivity，ReceiveActivity 显示接收的数据。当单击“关闭(没有回传数据)”按钮时，关闭 ReceiveActivity 返回到 MainActivity，此时虽然调用了 onActivityResult(Int, Int, Intent?)方法，但是 ReceiveActivity 没有返回发送 resultCode 参数和 data 参数的数据，无法获取有效的回传数据。当单击“关闭(发送回传数据)”按钮时，关闭 ReceiveActivity 并返回到 MainActivity，调用了 onActivityResult(Int, Int, Intent?)方法且获取回传数据(如图 8-27 所示)。

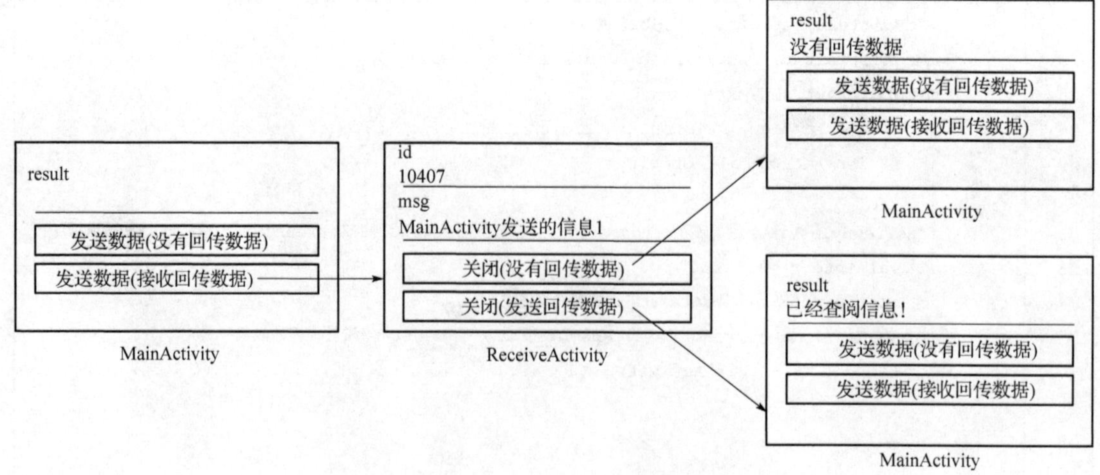

图 8-27　发送回传数据的效果

8.2　碎　片

8.2.1　Fragment 组件

Fragment（androidx.fragment.app.Fragment）是嵌入 Activity 的程序单元。Fragment 类的常用方法如表 8-3 所示，其必须依赖于 Activity，不能独立存在。但是因其有独立的生命周期，能接收输入事件，Activity 可以动态添加或删除 Fragment。一个 Activity 中可以包含多个 Fragment，一个 Fragment 可以被多个 Activity 重用。与 FragmentManager 类（常用方法如表 8-4 所示）和 FragmentTransaction 类（常用方法如表 8-5 所示）配合使用可以实现丰富的应用效果。

表 8-3　Fragment 类的常用方法

类型和修饰符	方　法
FragmentActivity?	getActivity() 获取与 Fragment 关联的 FragmentActivity。如果不关联，则返回 null
open Context?	getContext() 获取与 Fragment 关联的 Context
open Unit	onActivityCreated(@Nullable savedInstanceState: Bundle?) 当 Fragment 关联的 Activity 被创建且 Fragment 已经被实例化时，调用该方法
open Unit	onActivityResult(requestCode: Int, resultCode: Int, @Nullable data: Intent?) 接收前一次从 startActivityForResult(Intent!, Int) 调用的结果
open Unit	onAttach(@NonNull context: Context) 当 Fragment 首次附加到它的 context 时，调用该方法
open Unit	onAttachFragment(@NonNull childFragment: Fragment) 当一个 Fragment 附加到当前 Fragment 作为子对象时，调该方法
open Unit	onCreate(@Nullable savedInstanceState: Bundle?) 当 Fragment 初始创建时，调用该方法
open View?	onCreateView(@NonNull inflater: LayoutInflater, @Nullable container: ViewGroup?, @Nullable savedInstanceState: Bundle?) 当 Fragment 实例化它的用户界面视图时，调用该方法
open Unit	onDestroy() 当 Fragment 销毁时，调用该方法
open Unit	onDetach() 当 Fragment 不再附属于它的 Activity 时，调用该方法

续表

类型和修饰符	方　法
open Unit	onHiddenChanged (hidden: Boolean) 当 Fragment 的隐藏状态改变时，调用该方法
open Unit	onInflate (@NonNull context: Context, @NonNull attrs: AttributeSet, @Nullable savedInstanceState: Bundle?) 当 Fragment 被创建作为布局视图的一部分时，调用该方法
open Unit	onPause () 当 Fragment 不再是 resumed 状态时，调用该方法
open Unit	onPictureInPictureModeChanged (isInPictureInPictureMode: Boolean) 当 Fragment 关联的 Activity 改变画中画模式时，调用该方法
open Unit	onResume () 当 Fragment 运行呈现给用户时，调用该方法
open Unit	onSaveInstanceState (@NonNull outState: Bundle) 当要求 Fragment 保存当前动态状态时，调用该方法
open Unit	onStart () 当 Fragment 呈现给用户时，调用该方法
open Unit	onStop () 当 Fragment 不再是 started 状态时，调用该方法
open Unit	onViewCreated (@NonNull view: View, @Nullable savedInstanceState: Bundle?) 在 onCreateView (LayoutInflater, ViewGroup?, Bundle?) 返回后，且在恢复任何已保存的状态到视图前，调用该方法
open Unit	onViewStateRestored (@Nullable savedInstanceState: Bundle?) 当所有已保存状态恢复到 Fragment 的视图层级体系后，调用该方法
open Unit	startActivity (intent: Intent!) 从 Fragment 所在的 Activity 调用 Activity.startActivity (Intent!) 方法
open Unit	startActivity (intent: Intent!, @Nullable options: Bundle?) 从 Fragment 所在的 Activity 调用 Activity.startActivity (Intent!, Bundle?) 方法
open Unit	startActivityForResult (intent: Intent!, requestCode: Int) 从 Fragment 所在的 Activity 调用 Activity.startActivityForResult (Intent!,Int) 方法

表 8-4　FragmentManager 类的常用方法

类型和修饰符	方　法
open FragmentTransaction	beginTransaction () 对与 FragmentManager 关联的所有 Fragment 开启一系列操作
open Int	getBackStackEntryCount () 返回当前在后堆栈中加入的数量
open MutableList<Fragment!>	getFragments () 获取添加到 FragmentManager 中的 Fragment 列表

表 8-5　FragmentTransaction 类的常用方法

类型和修饰符	方　法
open FragmentTransaction	add (@IdRes containerViewId: Int, @NonNull fragment: Fragment, @Nullable tag: String?) 添加 Fragment
open FragmentTransaction	attach (@NonNull fragment: Fragment) 在先前使用 detach (Fragment) 从 UI 分离 Fragment 后，重新附加该 Fragment
abstract Int	commit () 提交事务
abstract Int	commitNow () 同步提交事务
open FragmentTransaction	detach (@NonNull fragment: Fragment) 从 UI 中分离 Fragment
open FragmentTransaction	hide (@NonNull fragment: Fragment) 隐藏指定的 Fragment
open Boolean	isEmpty () 判断是否有需要提交的事务
open FragmentTransaction	remove (@NonNull fragment: Fragment) 移除指定的 Fragment

续表

类型和修饰符	方　法
open FragmentTransaction	replace (@IdRes containerViewId: Int, @NonNull fragment: Fragment, @Nullable tag: String?) 替换已有的 Fragment 到指定容器
open FragmentTransaction	setCustomAnimations (@AnimatorRes @AnimRes enter: Int, @AnimatorRes @AnimRes exit: Int) 设置特定的动画资源作为进入和退出动画
open FragmentTransaction	show (@NonNull fragment: Fragment) 显示之前隐藏的 Fragment

8.2.2 Fragment 的生命周期

Fragment 是依赖 Activity 使用的，Fragment 的生命周期和 Activity 的生命周期是有对应关系的（如图 8-28 所示）。

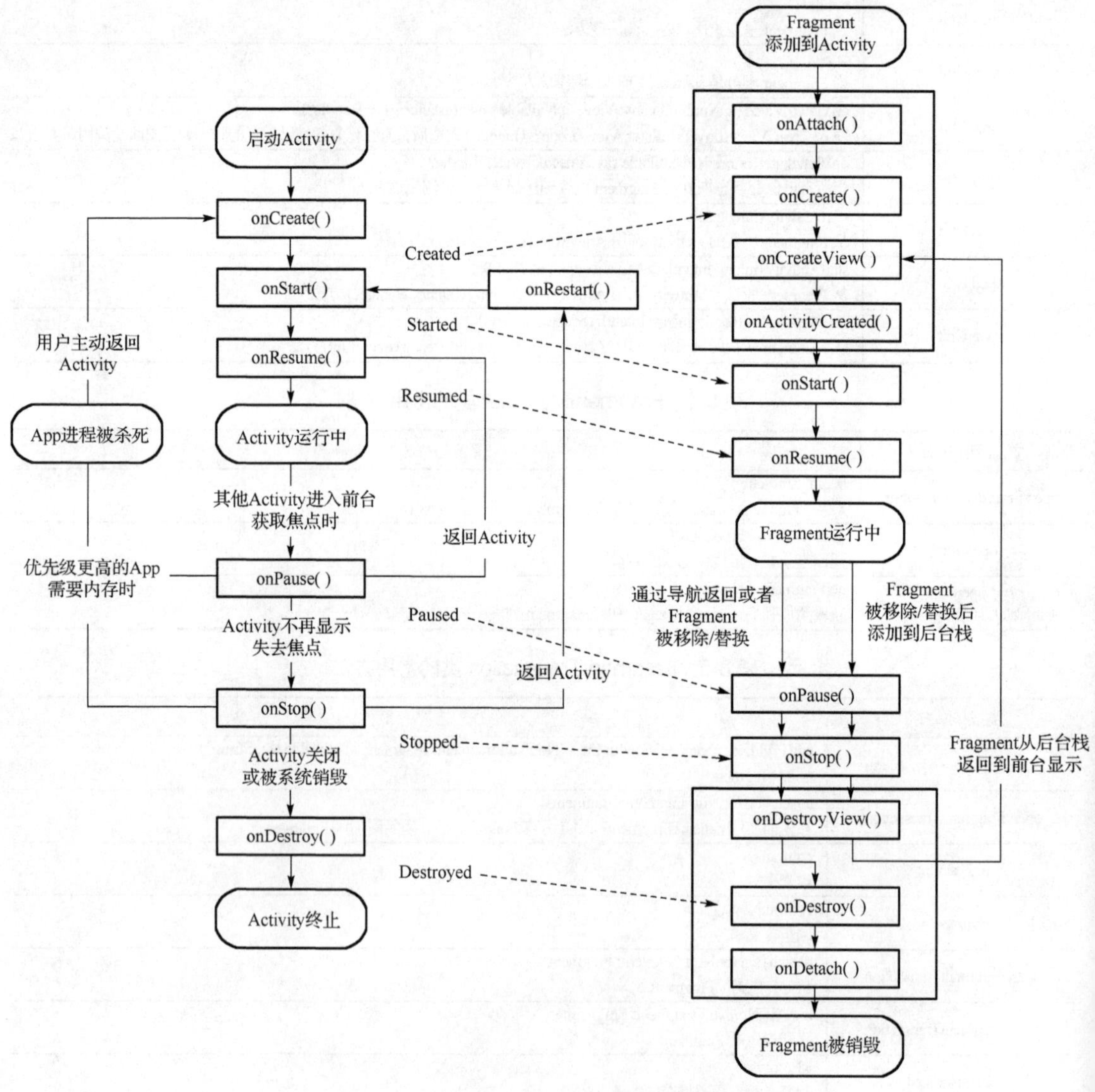

图 8-28　Fragment 和 Activit 的生命周期关系

8.2.3　实例工程：导航分页的主界面

本实例演示了常用的带导航功能的主界面，底部包含 5 个导航按钮，第 5 个按钮上默认显示数字圆点提示（如图 8-29 所示）。单击导航按钮后，显示数字圆点提示会消失，顶部的标题会相应改变，并且中间的区域会显示相应的 Fragment。每个 Fragment 都包含一个“显示 TAB 圆点”按钮，单击后相应的导航按钮会显示数字圆点提示。

图 8-29　运行效果

1. 打开基础工程

打开“基础工程”文件夹中的“C0809”工程，该工程已经包含 MainActivity 及资源文件，为了使用更加灵活，导航没有使用 FragmentTabHost。

2. OnShowTabNumListener 接口

右键单击“/java/com/teachol/c0509”文件夹，选择【New】→【Package】命令。在“New Package”对话框中，输入包名为“com.teachol. c0809.fragment”，回车完成创建（如图 8-30 所示）。

New Package
com.teachol.c0809.fragment

图 8-30　“New Package”对话框

右键单击“/java/com/teachol/c0809/fragment”文件夹，选择【New】→【New Kotlin File/Class】命令。在“New Kotlin File/Class”对话框中，输入名称为“OnShowTabListener”，双击“Interface”类型完成接口的创建（如图 8-31 所示）。

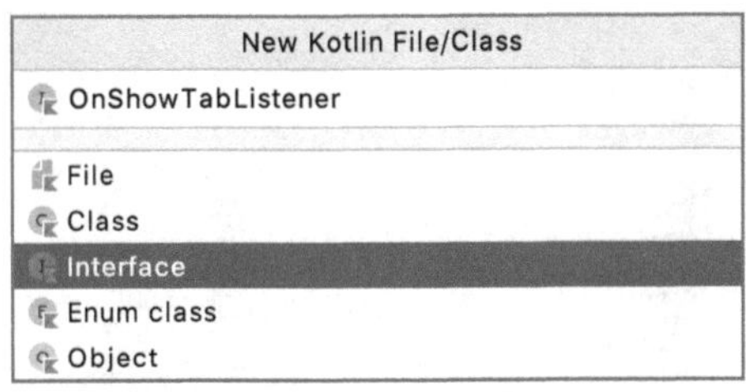

图 8-31　“New Kotlin File/Class”对话框

/java/com/teachol/c0809/fragment/OnShowTabListener.kt

```
03  interface OnShowTabNumListener {
04      fun onShowTabNum(index: Int, num: Int)
05  }
```

第 03 ~ 05 行定义 OnShowTabNumListener 接口及 onShowTabNum(index: Int, num: Int) 接口方法，index 参数表示 Fragment 的序号，num 参数表示图标上显示的提示数。

3. 分页的 Fragment 及其布局

右键单击“/java/com/teachol/c0509/fragment”文件夹，选择【New】→【Fragment】-【Fragment(Blank)】命令。在“New Android Component”对话框中，设置“Fragment Name”为“VicinityFragment”，单击“Finish”按钮（如图 8-32 所示）。

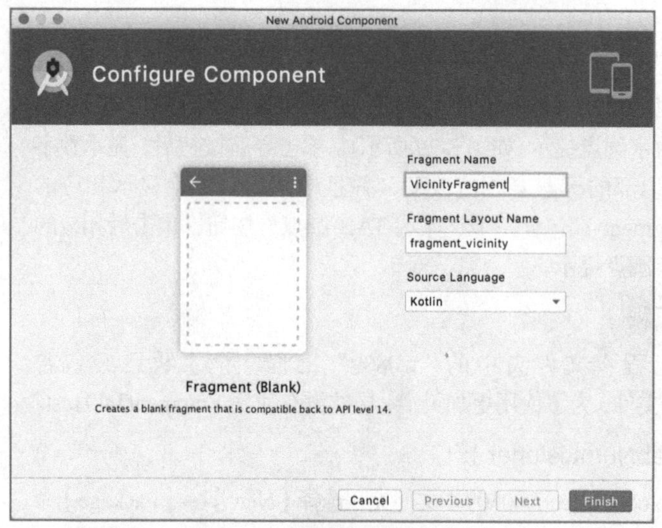

图 8-32 “New Android Component”对话框

按照相同的方法新建 FollowFragment、MeetFragment、RemindFragment 和 MineFragment，将这些 Fragment 的布局设置成相同的效果。下面的布局代码仅展示 fragment_vicinity.xml，其他 Fragment 布局文件的代码与其基本相同。

/res/layout/fragment_ vicinity.xml

```
<LinearLayout xmlns:android = "http://schemas.android.com/apk/res/android"
    android:gravity = "center"
    android:orientation = "vertical"
    tools:context = ".fragment.VicinityFragment">
    <TextView
        android:text = "VicinityFragment"
        android:textSize = "28sp" />
    <Button
        android:id = "@+id/button"
        android:text = "显示 tab 圆点"
        android:textSize = "28sp" />
</LinearLayout>
```

新建的 4 个 Fragment 中，MeetFragment 使用内部定义接口实现与 MainActivity 传递数据，其他 Fragment 使用外部接口与 MainActivity 传递数据。

/java/com/teachol/c0809/fragment/VicinityFragment.kt

```
import com.teachol.c0809.R
class VicinityFragment : Fragment() {
    override fun onCreateView(inflater: LayoutInflater, container: ViewGroup?, savedInstanceState: Bundle?): View? {
        val view: View = inflater.inflate(R.layout.fragment_vicinity, container, false)
        val button: Button = view.findViewById(R.id.button)
        button.setOnClickListener(View.OnClickListener {
            if (context is OnShowTabNumListener) {
                val onShowTabNumListener: OnShowTabNumListener?
                onShowTabNumListener = context as OnShowTabNumListener?
```

```
19                 onShowTabNumListener!!.onShowTabNum(0, 12)
20             } else {
21                 throw RuntimeException(context.toString() + " must implement OnShowTabNumListener")
22             }
23         })
24         return view
25     }
26 }
```

第 09 行导入“com.teachol.c0809.R”命名空间，这样才能调用相应的资源。第 16 行判断 Fragment 关联的 Activity 是不是 OnShowTabNumListener 接口的实例，如果没有关联的 Activity 则无法调用关联的 Activity 内的接口方法。第 17 ~ 19 行新建 onShowTabNumListener 实例，调用 OnShowTabNumListener 接口的 onShowTabNum (Int, Int) 方法显示圆点提示，需要将第一个参数设置成对应的 Fragment 的序号。第 21 行是当无法获取到关联的 Activity 时，抛出自定义异常。

```
/java/com/teachol/c0809/fragment/MeetFragment.kt
28 interface OnShowTabNumListener { //内部接口
29     fun onShowTabNum(index: Int, num: Int) //定义接口方法
30 }
```

第 28 ~ 30 行定义一个内部接口 OnShowTabNumListener，该接口名称和定义的接口方法与 Fragment 包内的 OnShowTabNumListener 接口名称和定义的接口方法相同。

4. 主界面的 Activity

在 MainActivity 中，实现对主界面导航的控制，并且实现 OnShowTabNumListener 和 MeetFragment.OnShowTabNumListener 的接口方法。

```
/java/com/teachol/c0809/MainActivity.kt
import com.teachol.c0809.fragment.*
class MainActivity : AppCompatActivity(), OnShowTabNumListener, MeetFragment.OnShowTabNumListener {
    private lateinit var mTitleTextView: TextView
    private lateinit var mFragmentManager: FragmentManager
    private val mTitles = arrayOf("附近", "关注", "偶遇", "提醒", "自己")
    private val mTab = arrayOfNulls<TextView>(5)
    private val mTabId = intArrayOf(R.id.tab_vicinity, R.id.tab_follow, R.id.tab_meet, R.id.tab_remind, R.id.tab_mine)
    private val mTabNumId = intArrayOf(R.id.tab_vicinity_num, R.id.tab_follow_num, R.id.tab_meet_num, R.id.tab_remind_num, R.id.tab_mine_num)
    private val mTabNum = arrayOfNulls<TextView>(5)
    private val mFragment: Array<Fragment> = arrayOf(VicinityFragment(), FollowFragment(), MeetFragment(), RemindFragment(), MineFragment())
    override fun onCreate(savedInstanceState: Bundle?) {
        mTitleTextView = findViewById(R.id.text_view_title)
        for (i in mTab.indices) {
            mTab[i] = findViewById(mTabId[i])
            mTab[i]!!.setTag(i)
            mTab[i]!!.setOnClickListener(object : View.OnClickListener {
                override fun onClick(v: View) {
                    showTabFragment(v.getTag() as Int)
                }
```

```
33              })
34              mTabNum[i] = findViewById(mTabNumId[i])
35          }
36          // 必须使用 FragmentActivity.getSupportFragmentManager()获取 FragmentManager 实例
37          mFragmentManager = supportFragmentManager
38          showTabFragment(2) //设置默认显示的 Fragment
39          showTabNum(4, 5) //设置 Tab 显示圆点数字
40      }
42      private fun hideTabNum(index: Int) { //隐藏 Tab 的圆点数字
43          mTabNum[index]!!.visibility = View.GONE
44      }
46      fun showTabNum(index: Int, num: Int) { //显示 Tab 的圆点数字
47          mTabNum[index]!!.text = num.toString()
48          mTabNum[index]!!.visibility = View.VISIBLE
49      }
51      private fun showTabFragment(index: Int) { //显示 Fragment
52          // 创建事务
53          val fragmentTransaction: FragmentTransaction = mFragmentManager.beginTransaction()
54          // 判断是否已经添加 Fragment
55          if (mFragmentManager.getFragments().isEmpty()) {
56              for (i in mTab.indices) {
57                  fragmentTransaction.add(R.id.fragment_content, mFragment[i]) //添加 Fragment
58              }
59          }
60          // 还原状态
61          for (i in mTab.indices) {
62              fragmentTransaction.hide(mFragment[i]) //隐藏 Fragment
63              mTab[i]!!.isSelected = false //取消 Tab 选中状态
64          }
65          mTitleTextView!!.text = mTitles[index] //设置标题
66          mTab[index]!!.isSelected = True //设置 Tab 选中状态
67          hideTabNum(index) //隐藏 tab 圆点
68          fragmentTransaction.show(mFragment[index]) //显示 Fragment
69          fragmentTransaction.commit() //提交事务
70      }
72      override fun onShowTabNum(index: Int, num: Int) { //实现接口方法
73          showTabNum(index, num)
74      }
75  }
```

第 10 行导入“com.teachol.c0809.fragment.*”命名空间。第 15 ~ 20 行使用数组创建实例并初始化导航功能所需的数据，这里使用数组的优势在于可以很容易地增减导航菜单并减少代码量。第 26 ~ 35 行使用 for 循环遍历数组的所有元素，初始化每个导航按钮。第 38 行设置了默认选中的导航按钮并显示相应的 Fragment。第 39 行设置第 5 个导航按钮显示圆点提示，圆点内显示的数字是 5，这行代码通常应该在从服务器端获取未读的信息数量后使用，这里只是一个模拟效果。第 55 ~ 59 行判断是否已经向 mFragmentManager 添加过 Fragment，因为重复通过 fragmentTransaction.add()方法向 mFragment Manager 添加 Fragment 会报错。第 69 行 fragmentTransaction.commit()提交事务才会执行 add()、hide()和 show()方法的事务。第 72 ~ 74 行实现接口的 onShowTabNum 方法。

第 9 章 Android 的后台服务与广播

Activity 和 Fragment 不在前台显示时是暂停运行的，此时如果有些功能需要能够继续运行，就需要通过后台服务来实现，后台服务甚至关闭 App 时还可以在后台继续运行。而广播为 App 之间提供了一种单向数据传递的方式，能够实现一对多的实时数据传递，其中，系统广播是系统根据设备的各种状态发送的广播。

9.1 服　　务

9.1.1 Service 组件

Service(android.app.Service)是在后台可以长时间运行且没有可视化视图的组件，Service 类的常用方法如表 9-1 所示。Service 不是线程，也不是在主线程外的方法。常用 Context.startService(Intent!)方法启动服务，常用 Context.stopService(Intent!)方法停止服务。

表 9-1 Service 类的常用方法

类型和修饰符	方　法
Application!	getApplication() 获取 Service 所属的 Application
abstract IBinder?	onBind(intent: Intent!) 绑定后调用该方法。返回 IBinder 实例，与客户端进行数据通信。没有绑定到服务时，可以返回 null
open Unit	onCreate() 创建时调用该方法
open Unit	onDestroy() 被销毁前调用该方法
open Unit	onRebind(intent: Intent!) 新客户端绑定时调用该方法，在此之前已通知其 onUnbind(Intent!)中的所有绑定已解除
open Int	onStartCommand(intent: Intent!, flags: Int, startId: Int) 调用 Context.startService(Intent!)方法启动服务时调用该方法
open Unit	onTaskRemoved(rootIntent: Intent!) Service 运行时，用户移除 Service 所属的 App 任务时，调用该方法
open Boolean	onUnbind(intent: Intent!) 当所有客户端都解除绑定时，调用该方法
Unit	startForeground(id: Int, notification: Notification!) 如果 Service 已通过 Context.startService(Intent!)启动，则持续显示 Notification
Unit	stopForeground(removeNotification: Boolean) 如果 removeNotification 为 true，则删除 Notification
Unit	stopSelf() 如果 Service 已经开始运行，则停止 Service
Boolean	stopSelfResult(startId: Int) 停止使用 startId 启动的 Service

Service 需要在 AndroidManifest.xml 中添加<service>标签后才能被调用，<service>标签属性(如表 9-2 所示)大多无法通过 Service 类的方法进行设置和获取。

表 9-2 <service>标签属性

属　性	说　明
android:description	设置描述 Service 的字符串
android:directBootAware	设置是否可以在用户解锁前直接启动
android:enabled	设置是否可实例化 Service
android:exported	设置是否允许其他 App 调用该服务
android:name	设置实现 Service 的类名
android:process	设置运行 Service 的进程名称。如果名称以“:”开头，在需要时创建专用新进程运行；如果名称以小写字母开头，在使用该名称的全局进程中运行

使用 Context.bindService(service: Intent!, conn: ServiceConnection, flags: Int)方法可以绑定服务，flags 参数使用 Service 类中与绑定相关的常量(如表 9-3 所示)。绑定的 Service 会随着 App 的退出而终止，未绑定的 Service 不会随着 App 的退出而终止。使用 Context.unbindService(conn: ServiceConnection)方法可以解除绑定 。

表 9-3 Service 类中与绑定相关的常量

Service 常量	说　明
BIND_AUTO_CREATE	绑定后自动创建服务，无须使用 startService(Intent!)方法后启动
BIND_ADJUST_WITH_ACTIVITY	如果从某个 Activity 绑定，则允许根据该 Activity 是否对用户可见来提高目标 Service 进程的重要性，不论是否使用另一个标志来减少使用客户端流程的总体重要性从而影响它的数量
BIND_ABOVE_CLIENT	指示绑定 Service 的客户端比 App 本身更重要。内存不足时，系统在结束 Service 之前先终止应用程序
BIND_ALLOW_OOM_MANAGEMENT	允许对绑定 Service 进行内存溢出管理，需要内存时允许系统结束绑定 Service 的进程
BIND_DEBUG_UNBIND	设置此标志，将保留 unbindService(ServiceConnection)调用的堆栈，如果稍后发生了不正确的解除绑定调用，则会打印该堆栈，建议只用于调试
BIND_EXTERNAL_SERVICE	绑定的 Service 是一个独立的外部服务
BIND_IMPORTANT	此服务对客户端非常重要，在客户端处于前台进程级别时应将其带到前台进程级别
BIND_INCLUDE_CAPABILITIES	如果绑定的 App 由于其前台状态(如 Activity 或前台 Service)而具有特定功能，只要 App 具有所需的权限，就允许绑定的 App 获取相同的功能
BIND_NOT_FOREGROUND	不允许此绑定提升 Service 进程到前台优先级，但仍会被提升到不高于客户端的优先级
BIND_NOT_PERCEPTIBLE	如果绑定来自可见或用户可感知的 App，则降低 Service 的重要性到可感知级别以下
BIND_WAIVE_PRIORITY	不要影响 Service 主进程的调度或内存管理优先级

9.1.2 Service 的生命周期

Service 分为非绑定 Service 和绑定 Service，二者生命周期有所不同(如图 9-1 所示)。非绑定 Service 包含 3 个生命周期方法，绑定 Service 包含 4 个生命周期方法。

两种 Service 都会调用 onCreate()方法来创建 Service，但绑定 Service 不会调用 onStartCommand()方法，而是调用 onBind()方法返回客户端一个 IBinder 接口。绑定 Service 通过 Context.unbindService()方法解除绑定时，回调 onUnbind()方法。如果所有绑定 Service 都调用了 unbindService()方法，绑定 Service 会被停止，然后回调 onDestroy()方法。

9.1.3　实例工程：Service 的开启和停止

本实例演示了启动和停止 Service 的方法(如图 9-2 所示)，单击“启动 SERVICE”按钮启动 MyService 类实例在后台运行，单击“停止 SERVICE”按钮停止 MyService 类实例的运行。

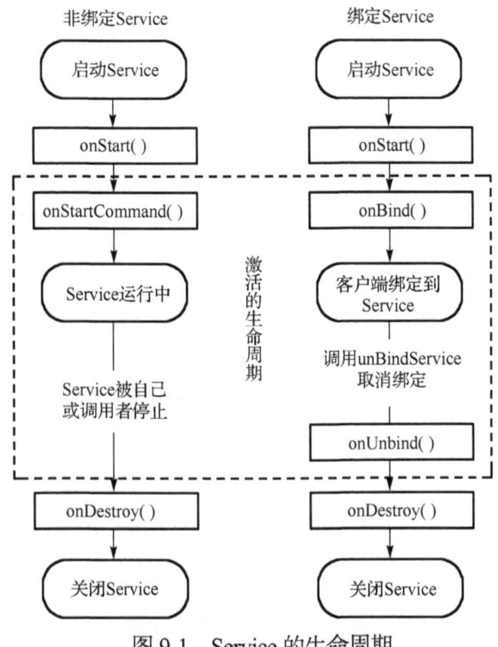

图 9-1　Service 的生命周期

图 9-2　运行效果

1．打开基础工程

打开“基础工程”文件夹中的“C0901”工程，该工程已经包含 MainActivity 及资源文件。

2．新建 Service

选择“/java/com/teachol/c0901”文件夹，单击右键，选择【New】→【Service】-【Service】命令。在“New Android Component”对话框中，设置“Class Name”为“MyService”，单击“Finish”按钮(如图 9-3 所示)。

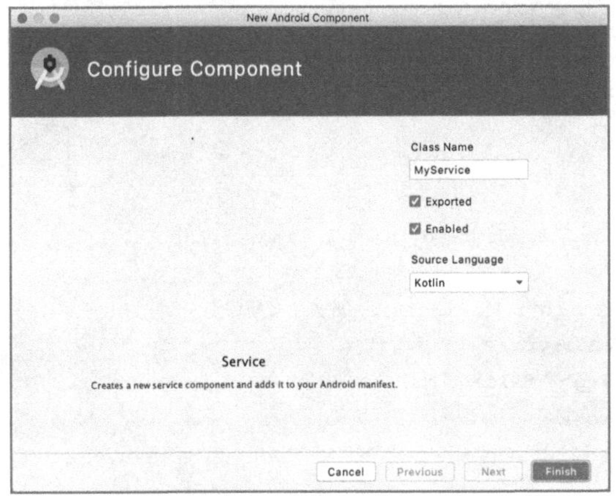

图 9-3　“New Android Component”对话框

此时，AndroidManifest.xml 中的<application>标签中添加<service>标签声明服务，否则无法使用。

```
/manifests/AndroidManifest.xml
12  <service
13      android:name = ".MyService"
14      android:enabled = "true"
15      android:exported = "true"/>
```

第 13 行 android:name 属性设置服务的类名。第 14 行 android:enabled 属性表示是否可以使用，默认值为 true；当父标签的 enable 属性也为 true 时，服务才会被激活，否则不会激活。第 15 行 android:exported 属性表示其他 App 组件是否可以唤醒 Service 或者和 Service 进行交互，默认值为 true。

```
/java/com/teachol/c0901/MyService.kt
class MyService : Service() {
    private val TAG = "MyService"
    private var mServiceRunning = false
    private var mSecond = 0
    override fun onBind(intent: Intent): IBinder? { //Service 绑定时调用该方法
        Log.i(TAG, "onBind 方法被调用!")
        return null
    }
    override fun onCreate() { //Service 被创建时调用
        Log.i(TAG, "onCreate 方法被调用!")
        super.onCreate()
    }
    override fun onStartCommand(intent: Intent, flags: Int, startId: Int): Int {//Service 被启动时
        Log.i(TAG, "onStartCommand 方法被调用!")
        object : Thread() {
            override fun run() {
                mServiceRunning = true
                while (mServiceRunning) {
                    Log.i(TAG, "新线程已经运行了" + mSecond + "秒")
                    try {
                        sleep(1000)
                        mSecond ++
                    } catch (e: InterruptedException) {
                        e.printStackTrace()
                    }
                }
            }
        }.start()
        return super.onStartCommand(intent, flags, startId)
    }
    override fun onDestroy() { //Service 被关闭之前回调
        Log.i(TAG, "onDestroy 方法被调用!")
        mServiceRunning = false
        super.onDestroy()
    }
}
```

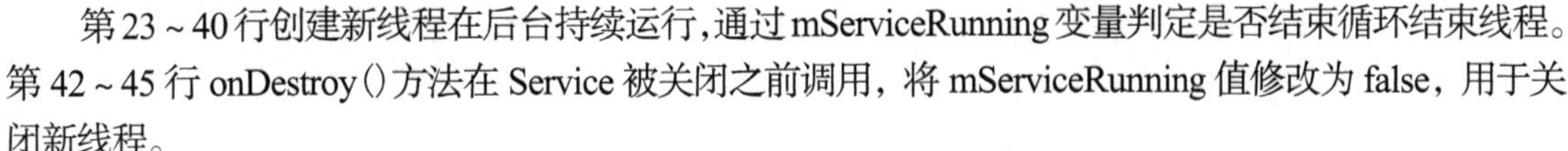

第 23 ~ 40 行创建新线程在后台持续运行，通过 mServiceRunning 变量判定是否结束循环结束线程。第 42 ~ 45 行 onDestroy() 方法在 Service 被关闭之前调用，将 mServiceRunning 值修改为 false，用于关闭新线程。

3. 主界面的 Activity

```
/java/com/teachol/c0901/MainActivity.kt
14  //创建启动 Service 的 Intent 及 Intent 属性
15  val intent = Intent(this, MyService::class.java)
16  findViewById<View>(R.id.button_start).setOnClickListener(object : View.OnClickListener {
17      override fun onClick(v: View?) {
18          startService(intent) //启动 Service
19          Log.i(TAG, "startService 方法被调用")
20      }
21  })
22  findViewById<View>(R.id.button_stop).setOnClickListener(object : View.OnClickListener {
23      override fun onClick(v: View?) {
24          stopService(intent) //停止 Service
25          Log.i(TAG, "stopService 方法被调用!")
26      }
27  })
```

第 18 行使用 startService(intent) 启动服务。第 24 行使用 stopService(intent) 停止服务。

4. 测试运行

运行工程，单击“启动 SERVICE”按钮后等待几秒，再单击“停止 SERVICE”按钮，会在 Logcat 窗口中观察到输出结果（如图 9-4 所示）。

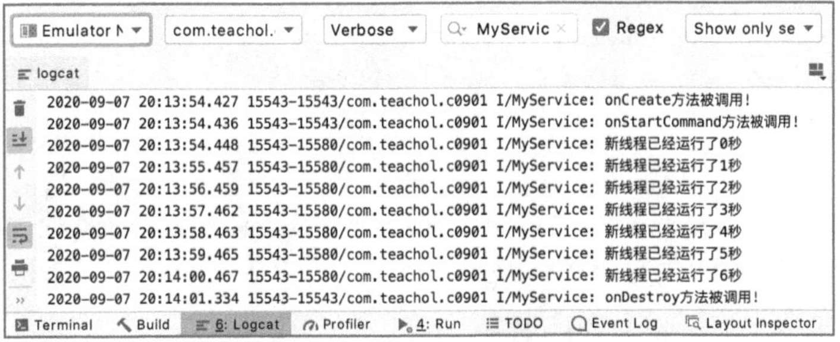

图 9-4　Logcat 窗口输出结果

9.1.4　实例工程：Service 的绑定和数据传递

本实例演示了 Service 的绑定和数据传递，以及关闭启动 Service 的 Activity 时如何停止运行 Service（如图 9-5 所示）。单击“启动 SubActivity”按钮启动 SubActivity。在 SubActivity 中，依次单击“启动 Service”和“绑定 Service”按钮后，再单击“获取 Service 运行时间”按钮会通过提示信息显示 Service 已经运行的时间。未停止 Service 时，关闭 SubActivity 会停止运行 Service。

1. 打开基础工程

打开“基础工程”文件夹中的“C0902”工程，该工程已经包含 MainActivity、SubActivity 类及资源文件。

图 9-5　运行效果

2. 新建 Service

在工程中，选择“/java/com/teachol/c0902”文件夹，单击右键，选择【New】→【Service】→【Service】命令。在“New Android Component”对话框中，设置“Class Name”为“MyService”，单击“Finish”按钮。

```
/java/com/teachol/c0902/MyService.kt
class MyService : Service() {
    private val TAG = "MyService"
    private val mBinder: MyBinder = MyBinder()
    private var mServiceRunning = false
    private var mSecond = 0
    inner class MyBinder : Binder(){ //创建MyBinder类，继承自Binder，用于与Activity传递数据
        fun getSecond(): Int {
            return mSecond
        }
    }
    override fun onBind(intent: Intent): IBinder? { //Service绑定时回调该方法
        Log.i(TAG, "onBind方法被调用!")
        return mBinder
    }
    override fun onCreate() { //Service被创建时回调
        super.onCreate()
        Log.i(TAG, "onCreate方法被调用!")
    }
    override fun onStartCommand(intent: Intent, flag: Int, startId: Int): Int {//Service被启动时调用
        object : Thread() {
            override fun run() {
                mServiceRunning = true
                while (mServiceRunning) {
                    Log.i(TAG, "新线程已经运行了" + mSecond + "秒")
                    try {
                        sleep(1000)
                        mSecond ++
```

```
40                } catch (e: InterruptedException) {
41                    e.printStackTrace()
42                }
43            }
44        }
45    }.start()
46    return super.onStartCommand(intent, flag, startId)
47 }
49 override fun onRebind(intent: Intent) { //Service 重新绑定时回调
50     super.onRebind(intent)
51     Log.i(TAG, "onRebind 方法被调用！")
52 }
54 override fun onUnbind(intent: Intent): Boolean { //Service 解除绑定时回调
55     Log.i(TAG, "onUnbind 方法被调用！")
56     mServiceRunning = false
57     return true
58 }
60 override fun onDestroy() { //Service 被关闭前回调
61     super.onDestroy()
62     Log.i(TAG, "onDestroyed 方法被调用！")
63     mServiceRunning = false
64 }
66 override fun onTaskRemoved(rootIntent: Intent) { //Service 所属的 App 被移除时回调该方法
67     Log.i(TAG, "onTaskRemoved 方法被调用！")
68 }
69 }
```

第 15 ~ 19 行创建 MyBinder 类，内部的 getSecond() 方法用于获取运行时间。第 23 行 mBinder 实例作为绑定的返回值，作为传递数据的对象。第 31 ~ 47 行 Service 被启动后开启新线程，实现不精确计时的功能，用于统计 Service 已经运行的时间。

3．控制 Service 的 Activity

```
/java/com/teachol/c0902/SubActivity.kt
import com.teachol.c0902.MyService.MyBinder
class SubActivity : AppCompatActivity() {
    private val TAG = "MainActivity"
    private var mIsBind = false
    //保持所启动的 Service 的 IBinder 对象，同时定义一个 ServiceConnection 对象
    private var mIBinder: MyBinder? = null
    private val mConn: ServiceConnection = object : ServiceConnection {
        //Activity 与 Service 连接成功时回调该方法
        override fun onServiceConnected(name: ComponentName, iBinder: IBinder) {
            Log.i(TAG, "onServiceConnected 方法被调用！")
            mIBinder = iBinder as MyBinder
        }
        //Activity 与 Service 断开连接时回调该方法
        override fun onServiceDisconnected(name: ComponentName) {
            Log.i(TAG, "onServiceDisconnected 方法被调用！")
            mIBinder = null
        }
```

```
31     }
32     override fun onCreate(savedInstanceState: Bundle?) {
36         //创建启动 Service 的 Intent
37         val intent = Intent(this, MyService::class.java)
38         findViewById<View>(R.id.button_start).setOnClickListener(object : View.OnClickListener {
39             override fun onClick(v: View?) {
40                 startService(intent) //启动 Service
41             }
42         })
43         findViewById<View>(R.id.button_stop).setOnClickListener(object : View.OnClickListener {
44             override fun onClick(v: View?) {
45                 stopService(intent) //停止 Service
46             }
47         })
48         findViewById<View>(R.id.button_bind).setOnClickListener(object : View.OnClickListener {
49             override fun onClick(v: View?) {
50                 bindService(intent, mConn, Service.BIND_IMPORTANT) //绑定 Service
51                 mIsBind = true
52             }
53         })
54         findViewById<View>(R.id.button_unbind).setOnClickListener(object : View.OnClickListener {
55             override fun onClick(v: View?) {
56                 unbindService(mConn) //解除绑定 Service
57                 mIsBind = false
58             }
59         })
60         findViewById<View>(R.id.button_get).setOnClickListener(object : View.OnClickListener {
61             override fun onClick(v: View?) {
62                 if (mIBinder != null) {
63                     Toast.makeText(applicationContext, "Service 已经运行了" + mIBinder!!.getSecond()
                           + "秒! ",Toast.LENGTH_SHORT).show()
64                 }
65             }
66         })
67     }
68     override fun onDestroy() {
69         super.onDestroy()
70         Log.i(TAG, "onDestroy 方法被调用!")
71         if (mIsBind) {
72             unbindService(mConn) //解除绑定 Service
73         }
74     }
75 }
```

第 20 ~ 31 行实例化 ServiceConnection 对象，重写 onServiceConnected()和 onServiceDisconnected()方法。第 24 行将 onBind()方法返回值强制类型转换为 MyService，然后赋给 mIBinder。第 68 ~ 74 行重写 onDestroy()方法，如果在 Activity 销毁前已经绑定了 Service，则解除绑定。在 Activity 销毁时，如果 Activity 绑定了 Service 会报错，所以要解除绑定。

4. 测试运行

运行后，单击“启动 SubActivity”按钮打开 SubActivity。依次单击“启动 Service”按钮、“绑定

Service”按钮、“解除绑定 Service”按钮和“停止 Service”按钮，单击返回按钮关闭 SubActivity，会在 Logcat 窗口中观察到输出结果(如图 9-6 所示)。

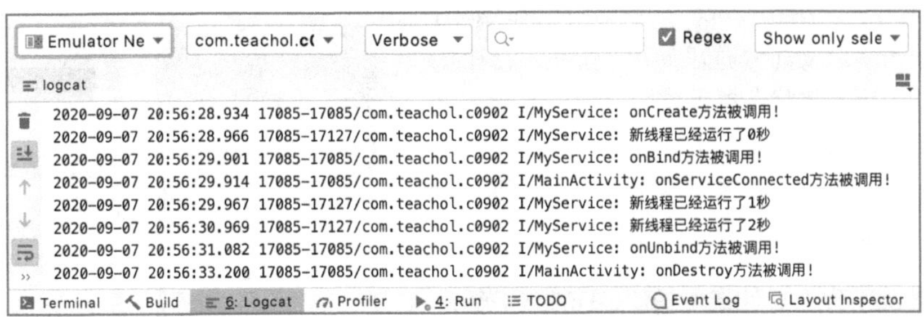

图 9-6　Logcat 窗口输出结果(1)

运行后，单击“启动 SubActivity”按钮打开 SubActivity。依次单击“启动 Service”按钮、“绑定 Service”按钮和“停止 Service”按钮，会在 Logcat 窗口中观察到输出结果(如图 9-7 所示)。

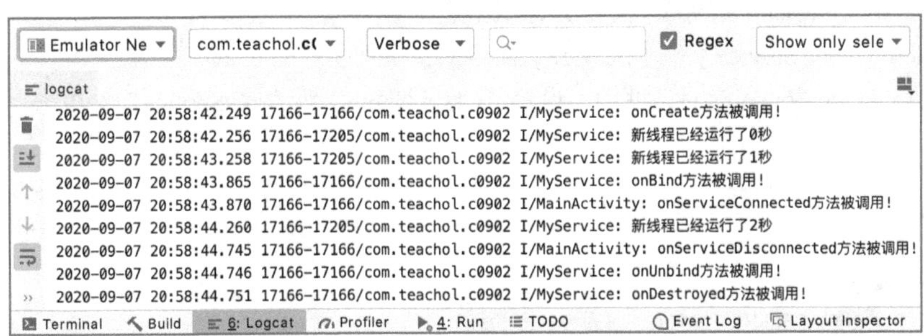

图 9-7　Logcat 窗口输出结果(2)

运行后，单击“启动 SubActivity”按钮打开 SubActivity。依次单击“启动 Service”按钮和“绑定 Service”按钮，单击返回按钮关闭 SubActivity。再单击“启动 SubActivity”按钮打开 SubActivity，单击“启动 Service”按钮，会在 Logcat 窗口中观察到输出结果(如图 9 ~ 8 所示)。再次打开 SubActivity 后单击“启动 Service”按钮，输出的运行时间并没有重置，而是连续的。这说明启动 Service 后没有使用 stopService(Intent!)方法停止该 Service，再次使用 startService(Intent!)方法启动该 Service 时，该 Service 并未销毁，因此不会执行 MyService 类的 onCreate()方法，但是会再次执行 MyService 类的 onStartCommand(Intent!,Int,Int)方法。

图 9-8　Logcat 窗口输出结果(3)

9.1.5 实例工程：Service 显示 Notification

本实例演示了使用 Service 显示前台通知(如图 9-9 所示)。单击“启动前台通知服务”按钮，会在状态栏显示通知图标，将状态栏下拉后会看到通知。只有单击“停止前台通知服务”按钮，才会关闭通知；否则，即使关闭该 App，通知也不会消失。

图 9-9 运行效果

1. 打开基础工程

打开“基础工程”文件夹中的“C0903”工程，该工程已经包含通知的布局文件及图标文件、MainActivity 及资源文件。

2. 新建 Service

选择“/java/com/teachol/c0903”文件夹，单击右键，选择【New】→【Service】→【Service】命令。在“New Android Component”对话框中，设置“Class Name”为“MyService”，单击“Finish”按钮完成创建。

/java/com/teachol/c0903/MyService.kt

```kotlin
class MyService : Service() {
    private val CHANNL_ID = "1231"
    private val CHANNL_NAME = "音乐"
    private lateinit var mNotificationManager: NotificationManager
    private lateinit var mNotificationChannel: NotificationChannel
    private lateinit var mNotificationBuilder: Notification.Builder
    private lateinit var mNotification: Notification
    override fun onBind(intent: Intent): IBinder? {
        throw UnsupportedOperationException("Not yet implemented")
    }
    override fun onCreate() {
        super.onCreate()
        //自定义远程视图
        val remoteView = RemoteViews(packageName, R.layout.notification_custom)
        remoteView.setImageViewResource(R.id.image_view_music, R.mipmap.ic_notification_music)
        remoteView.setImageViewResource(R.id.image_view_stop, R.mipmap.ic_notification_stop)
        //设置通知
        mNotificationChannel =
            NotificationChannel(CHANNL_ID, CHANNL_NAME, NotificationManager.IMPORTANCE_DEFAULT)
        mNotificationManager = getSystemService(NOTIFICATION_SERVICE) as NotificationManager
        mNotificationManager.createNotificationChannel(mNotificationChannel!!)
        mNotificationBuilder = Notification.Builder(this, CHANNL_ID)
        mNotificationBuilder.setSmallIcon(R.mipmap.ic_notification_small)
            .setCustomBigContentView(remoteView)
        mNotification = mNotificationBuilder.build()
        startForeground(1, mNotification) //启动前台通知
    }
}
```

第 25、26 行设置远程视图的图像，布局文件内的图像需要重新设置才会显示出来。第 34 行设置通知的自定义大视图。第 37 行使用 startForeground() 方法启动前台通知。

```
/manifests/AndroidManifest.xml
03 <uses-permission android:name = "android.permission.FOREGROUND_SERVICE" />
11 <service
12     android:name = ".MyService"
13     android:enabled = "true"
14     android:exported = "true"></service>
```

第 03 行添加前台服务的权限，否则 startForeground() 方法会报错。第 11～14 行是新建 MyService 后自动添加的标签。

3. 主界面的 Activity

```
/java/com/teachol/c0903/MainActivity.kt
12 val intent = Intent(this, MyService::class.java)
13 findViewById<View>(R.id.button_start_foreground).setOnClickListener(object :
   View.OnClickListener {
14     override fun onClick(v: View?) {
15         startService(intent)
16     }
17 })
18 findViewById<View>(R.id.button_stop_foreground).setOnClickListener(object : View.OnClickListener {
19     override fun onClick(v: View?) {
20         stopService(intent)
21     }
22 })
```

第 13～16 行设置“启动前台通知服务”按钮启动 MyService。第 18～22 行设置“停止前台通知服务”按钮停止 MyService。

9.2　广播接收器

9.2.1　BroadcastReceiver 组件

在 Android 中，可以通过发送广播通知其他 App 所发生的事件，以便其他 App 进行相应的处理。例如，电量低或者充足、刚启动完、插入耳机、输入法改变等发生时系统都会向所有 App 发送广播，这类广播称为系统广播（常量如表 9-4 所示）。用户也可以自定义广播，发送给指定的广播接收器。

表 9-4　常用的广播类型常量

广播类型	说明
ConnectivityManager.CONNECTIVITY_ACTION	网络连接状态改变的广播
Intent.ACTION_AIRPLANE_MODE_CHANGED	开关飞行模式时发送该广播
Intent.ACTION_BATTERY_CHANGED	电池状态变化时发送该广播，不能静态注册使用
Intent.ACTION_BOOT_COMPLETED	系统启动完成后发送该广播
Intent.ACTION_CAMERA_BUTTON	“拍照”按钮被按下时发送该广播
Intent.ACTION_CONFIGURATION_CHANGED	系统配置（方向、区域设置等）改变时发送该广播

续表

广播类型	说明
Intent.ACTION_DATE_CHANGED	修改日期后发送该广播
Intent.ACTION_DREAMING_STARTED	屏保开始后发送该广播
Intent.ACTION_DREAMING_STOPPED	屏保停止后发送该广播
Intent.ACTION_HEADSET_PLUG	耳机口上插拔耳机时发送该广播
Intent.ACTION_INPUT_METHOD_CHANGED	输入法改变时发送该广播
Intent.ACTION_LOCALE_CHANGED	系统设置的地区改变时发送该广播
Intent.ACTION_REBOOT	系统重新启动时发送该广播
Intent.ACTION_SCREEN_OFF	息屏时发送该广播
Intent.ACTION_SCREEN_ON	屏幕激活时发送该广播
Intent.ACTION_TIMEZONE_CHANGED	系统设置的时区改变时发送该广播
Intent.ACTION_TIME_CHANGED	系统设置的时间改变时发送该广播
Intent.ACTION_TIME_TICK	系统每分钟发送一次该广播，不能静态注册使用
Intent.ACTION_USER_PRESENT	屏幕唤醒且屏保消失后发送该广播
Intent.ACTION_USER_UNLOCKED	开机解锁后发送该广播，不能静态注册使用

BroadcastReceiver 类(android.content.BroadcastReceiver)是处理系统和应用程序之间通信的组件(常用方法如表 9-5 所示)。常用 ContextWrapper.registerReceiver(BroadcastReceiver?,IntentFilter!)方法注册广播接收器，常用 ContextWrapper.unregisterReceiver(BroadcastReceiver!)方法注销广播接收器。

表 9-5　BroadcastReceiver 类的常用方法

类型和修饰符	方法
	<init>() 主构造方法
Unit	abortBroadcast() 中止有序广播，仅适用于 sendOrderedBroadcast(Intent,String)发送的广播
Unit	clearAbortBroadcast() 恢复有序广播
Boolean	getAbortBroadcast() 获取是否中止有序广播
Int	getResultCode() 获取前一个接收器设置的返回码
String!	getResultData() 获取前一个接收器设置的 String 类型的返回结果
Bundle!	getResultExtras(makeMap: Boolean) 获取前一个接收器设置的 Bundle 类型的返回结果。makeMap 参数设置返回值是否可以为空
Boolean	isOrderedBroadcast() 判断接收到的广播是否是有序广播
abstract Unit	onReceive(context: Context!, intent: Intent!) 接收到广播后调用该方法
Unit	setResult(code: Int, data: String!, extras: Bundle!) 设置当前接收器广播的结果，仅适用于 sendOrderedBroadcast(Intent!,String)方法发送的广播
Unit	setResultCode(code: Int) 设置当前接收器广播的返回码，仅适用于 sendOrderedBroadcast(Intent!,String)方法发送的广播
Unit	setResultData(data: String!) 设置当前接收器广播的 String 类型的返回结果，仅适用于 sendOrderedBroadcast(Intent!,String)方法发送的广播
Unit	setResultExtras(extras: Bundle!) 设置当前接收器广播的 Bundle 类型的返回结果，仅适用于 sendOrderedBroadcast(Intent!,String)方法发送的广播

BroadcastReceiver 需要在 AndroidManifest.xml 中添加<receiver>标签后才能被调用，<receiver>标签属性（如表 9-6 所示）大多无法通过 BroadcastReceiver 类的方法进行设置和获取。

表 9-6　<receiver>标签的属性

属　性	说　明
android:directBootAware	设置是否可以在用户解锁前直接启动
android:enabled	设置是否可以被实例化
android:exported	设置是否可以接收来自其他 App 的消息
android:label	设置标签名称
android:name	设置实现 BroadcastReceivers 的类名
android:permission	设置向 BroadcastReceivers 发送消息须具备的权限名称
android:process	设置运行 BroadcastReceivers 的进程名称

9.2.2　接收广播

接收广播有两种方式：显式接收和隐式接收。显式接收使用 ContextWrapper.registerReceiver (BroadcastReceiver?, IntentFilter!) 方法注册广播接收器，隐式接收在 AndroidManifest.xml 内使用<intent-filter>标签设置接收的广播类型。

广播消息本身会被封装在一个 Intent 对象中，该对象的操作字符串会标识所发生的事件（如 android.intent.action.AIRPLANE_MODE）。该 Intent 可能还包含绑定到其 extra 字段中的附加信息。例如，飞行模式 intent 包含布尔值 extra 来指示是否已开启飞行模式。

提示：广播接收的限制

从 Android 8.0（API Level 26）开始，系统对 AndroidManifest.xml 声明的广播接收器进行了额外的限制，对大多数系统广播只能使用显式接收。

9.2.3　实例工程：显式和隐式接收广播

本实例演示了广播接收器使用显式和隐式接收广播（如图 9-10 所示）。单击“注册广播接收器”按钮，不但能够对网络、屏幕和电量等隐式广播进行接收，还能对自定义的显式广播进行接收。

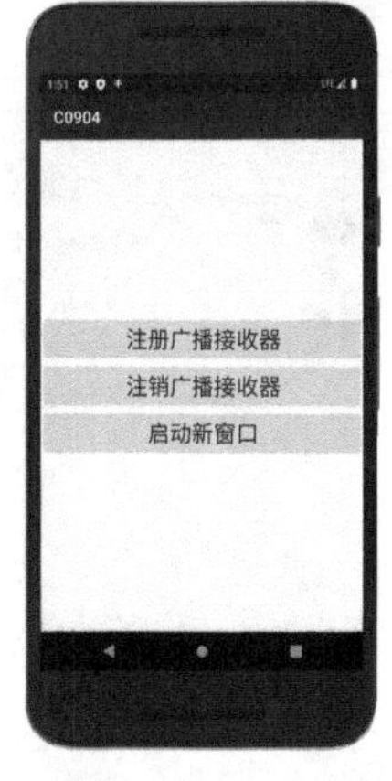

图 9-10　运行效果

1. 打开基础工程

打开“基础工程”文件夹中的“C0904”工程，该工程已经包含 MainActivity 类及资源文件。

2. 新建 BroadcastReceiver

选择“/java/com/teachol/c0904”文件夹，单击右键，选择【New】→【Other】→【Broadcast Receiver】命令。在“New Android Component”对话框中，设置“Class Name”为“MyBroadcastReceiver”，单击“Finish”按钮完成创建。

/java/com/teachol/c0904/MyBroadcastReceiver.kt

```
class MyBroadcastReceiver : BroadcastReceiver() {
    private val TAG = "MyBroadcastReceiver"
```

```
13    private var mBatteryLevel = 0 //电量值
14    override fun onReceive(context: Context, intent: Intent) {
15        Log.d(TAG, "-----------Receiver-----------")
16        Log.d(TAG, "Action: " + intent.action)
17        Log.d(TAG, "URI: " + intent.toUri(Intent.URI_INTENT_SCHEME))
19        if(intent.action == ConnectivityManager.CONNECTIVITY_ACTION) { //网络广播
20            val connectivity = context.getApplicationContext()
21                .getSystemService(Context.CONNECTIVITY_SERVICE) as ConnectivityManager
22            if (connectivity != null) {
23                val networks = connectivity.activeNetwork
24                val networkCapabilities = connectivity.getNetworkCapabilities(networks)
25                if (networkCapabilities != null) {
26                    if (networkCapabilities.hasTransport(NetworkCapabilities.TRANSPORT_WIFI)) {
27                        Log.d(TAG, "wifi")
28                    } else if (networkCapabilities.hasTransport(NetworkCapabilities.TRANSPORT_
CELLULAR)) {
29                        Log.d(TAG, "正在使用移动网络")
30                    }
31                } else {
32                    Log.d(TAG, "没有网络")
33                }
34            } {Log.d(TAG, "无法获取网络状态")}
35        }
37        if (intent.action == Intent.ACTION_SCREEN_ON) { //屏幕广播
38            Log.i(TAG, "屏幕打开")
39        } else if (intent.action == Intent.ACTION_SCREEN_OFF) {
40            Log.i(TAG, "屏幕关闭")
41        } else if (intent.action == Intent.ACTION_USER_PRESENT) {
42            Log.i(TAG, "屏幕解锁")
43        }
45        if (intent.action == Intent.ACTION_BATTERY_OKAY) { //电量
46            Log.i(TAG, "电量已满")
47        } else if (intent.action == Intent.ACTION_BATTERY_LOW) {
48            Log.i(TAG, "电量不足")
49        } else if (intent.action == Intent.ACTION_BATTERY_CHANGED) {
50            val level = intent.getIntExtra(BatteryManager.EXTRA_LEVEL, 0)
51            if (mBatteryLevel != level) {
52                Log.i(TAG, "电量: $level%")
53                mBatteryLevel = level
54            }
55        }
57        if (intent.action == "MyBroadcastReceiver.Custom") { //自定义广播
58            Log.d(TAG, "自定义广播的 info: " + intent.getStringExtra("info"))
59        }
60    }
61 }
```

第 14 ~ 55 行重写 onReceive(Context, Intent) 方法，用于接收并处理广播。第 16 行 intent.action 用于获取广播的类型。第 17 行 intent.toUri(Intent.URI_INTENT_SCHEME) 查看广播的完整信息。第 19 行判断是否接收到网络连接状态改变的广播。第 20 ~ 21 行 connectivity 变量用于获取网络连接的管理

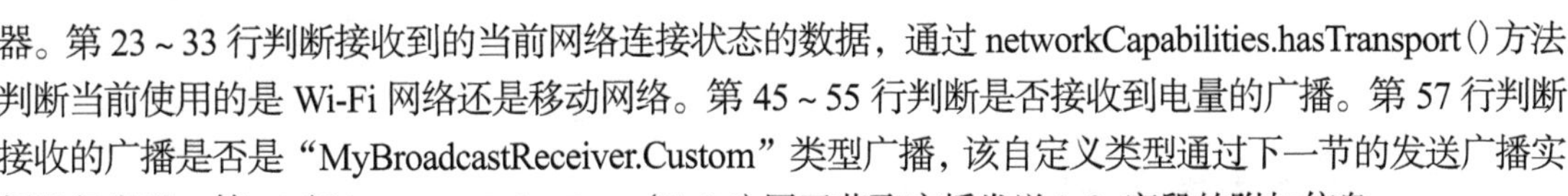

器。第 23 ~ 33 行判断接收到的当前网络连接状态的数据，通过 networkCapabilities.hasTransport()方法判断当前使用的是 Wi-Fi 网络还是移动网络。第 45 ~ 55 行判断是否接收到电量的广播。第 57 行判断接收的广播是否是“MyBroadcastReceiver.Custom”类型广播，该自定义类型通过下一节的发送广播实例进行发送。第 58 行 intent.getStringExtra("info")用于获取广播发送 info 字段的附加信息。

/manifests/AndroidManifest.xml

```
04 <uses-permission android:name = "android.permission.ACCESS_NETWORK_STATE" />
12 <receiver
13     android:name = ".MyBroadcastReceiver"
14     android:enabled = "true"
15     android:exported = "true">
16     <intent-filter>
17         <action android:name = "MyBroadcastReceiver.Custom" />
18     </intent-filter>
19 </receiver>
```

第 04 行添加访问网络状态的权限，否则接收到网络状态改变的通知后无法获取当前网络状态。第 16 ~ 18 行<intent-filter>标签用于隐式接收广播，设置接收广播的自定义类型为 MyBroadcastReceiver.Custom。

3. 主界面的 Activity

/java/com/teachol/c0904/MainActivity.kt

```
class MainActivity : AppCompatActivity() {
    private val myReceiver = MyBroadcastReceiver()
    private var mRegistered = false
    override fun onCreate(savedInstanceState: Bundle?) {
        //注册广播接收器
        findViewById<View>(R.id.button_register).setOnClickListener(object : View.OnClickListener {
            override fun onClick(v: View?) {
                val intentFilter = IntentFilter()
                intentFilter.addAction(ConnectivityManager.CONNECTIVITY_ACTION)
                intentFilter.addAction(Intent.ACTION_SCREEN_ON)
                intentFilter.addAction(Intent.ACTION_SCREEN_OFF)
                intentFilter.addAction(Intent.ACTION_USER_PRESENT)
                intentFilter.addAction(Intent.ACTION_BATTERY_OKAY)
                intentFilter.addAction(Intent.ACTION_BATTERY_LOW)
                intentFilter.addAction(Intent.ACTION_BATTERY_CHANGED)
                //intentFilter.addAction("com.teachol.c0901.MyBroadcastReceiver.ss");
                registerReceiver(myReceiver, intentFilter)
                mRegistered = true
            }
        })
        //注销广播接收器
        findViewById<View>(R.id.button_unregister).setOnClickListener(object : View.OnClickListener {
            override fun onClick(v: View?) {
                if (mRegistered) {
                    unregisterReceiver(myReceiver)
                    mRegistered = false
                }
            }
```

```
40          })
41          //启动新窗口
42          findViewById<View>(R.id.button_start_activity).setOnClickListener(object : View.OnClickListener {
43              override fun onClick(v: View?) {
44                  val i = Intent(this@MainActivity, MainActivity::class.java)
45                  startActivity(i)
46              }
47          })
48      }
49      override fun onDestroy() {
50          super.onDestroy()
51          if (mRegistered) {
52              unregisterReceiver(myReceiver)
53          }
54      }
55  }
```

第 11 行声明并实例化一个 MyBroadcastReceiver 类的 myReceiver 变量。第 12 行声明并初始化 mRegistered 变量，记录是否已经注册 myReceiver。第 20 ~ 26 行添加了 7 种类型的系统广播。第 27 行使用“//”注释了自定义类型的广播，当 AndroidManifest.xml 文件中没有隐式设置接收广播的自定义类型时，将“//”删除后就可以设置显式接收的自定义类型。第 28 行 registerReceiver(myReceiver, intentFilter) 注册 myReceiver 及过滤器 intentFilter。第 33 ~ 40 行注销广播接收器，先判断是否已经注册了 myReceiver，如果注册过了则使用 unregisterReceiver(myReceiver) 注销。这里添加判断的原因是如果未注册就注销 myReceiver，App 会报错并直接退出。第 42 ~ 47 行启动一个新的 MainActivity 实例，验证每个 Activity 都可以注册一个 myReceiver 实例，多个 MainActivity 实例的 myReceiver 实例可以同时注册运行。第 49 ~ 54 行重写 onDestroy() 方法，如果已经注册 myReceiver，则进行注销，因为未注销时关闭 Activity 会报错并直接退出。

4. 测试运行

运行后，单击“注册广播接收器”按钮会在 Logcat 窗口中输出网络连接和电量的相关信息；再单击电源按钮后屏幕会熄灭，在 Logcat 窗口中输出屏幕关闭的相关信息（如图 9-11 所示）。再次单击电源按钮后屏幕会亮起，然后单击“启动新窗口”按钮，启动新窗口，再单击新窗口中的“注册广播接收器”按钮，会在 Logcat 窗口中输出两条广播处理信息。后退关闭新窗口，会注销新窗口的广播接收器，此时 Logcat 窗口中只会输出一条广播处理信息。

图 9-11　Logcat 窗口输出结果

9.2.4　发送广播

系统广播只能由 Android 系统发出，用户只能发送自定义通知。自定义广播主要分为标准广播和有序广播。标准广播是指广播会被所有的接收者接收到，不可以被拦截和被修改，常用 Context.sendBroadcast (Intent!)方法发送标准广播。有序广播是指广播按照优先级逐级向下传递，接收者可以修改广播数据，也可以终止广播，常用 Context.sendOrderedBroadcast (Intent!,String)方法发送标准广播。

9.2.5　实例工程：发送标准广播和有序广播

本实例演示了发送标准广播和有序广播(如图 9-12 所示)。单击相应的按钮会在 Logcat 窗口中显示发送的相应信息。

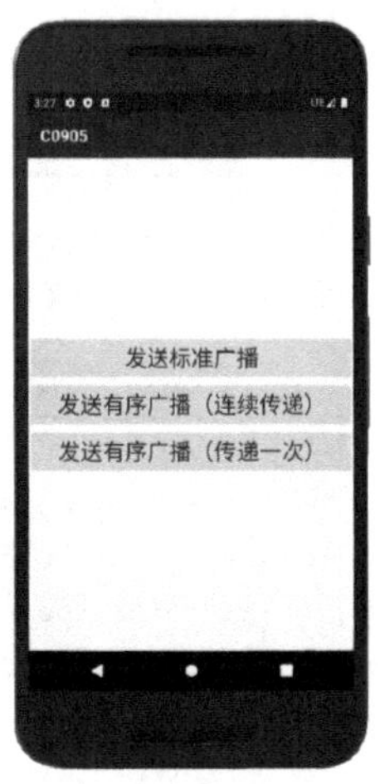

图 9-12　运行效果

1．打开基础工程

打开“基础工程”文件夹中的“C0905”工程，该工程已经包含 MainActivity 及资源文件。

2．新建 BroadcastReceiver

在“/java/com/teachol/c0905”文件夹中，新建两个自定义广播接收器类：MyBroadcastReceiver1 和 MyBroadcastReceiver2，用于接收有序广播。

/java/com/teachol/c0905/MyBroadcastReceiver1.kt

```
08 class MyBroadcastReceivers1 : BroadcastReceiver() {
09     companion object {
10         private const val TAG = "MyBroadcastReceiver"
11     }
12     override fun onReceive(context: Context, intent: Intent) {
13         Log.d(TAG, "-----------Receivers1-----------")
14         Log.d(TAG, "Action: " + intent.action)
15         Log.d(TAG, "URI: " + intent.toUri(Intent.URI_INTENT_SCHEME))
16         Log.d(TAG, "自定义广播的 info: " + intent.getStringExtra("info"))
17     }
18 }
```

第 12～17 行重写了 onReceive(Context context, Intent intent)方法，接收并处理广播。第 15 行 intent.toUri(Intent.URI_INTENT_SCHEME)用于接收广播的完整数据。第 16 行 intent.getStringExtra ("info")用于获取 info 字段的额外数据。

/java/com/teachol/c0905/MyBroadcastReceiver2.kt

```
class MyBroadcastReceivers2 : BroadcastReceiver() {
    companion object {
        private const val TAG = "MyBroadcastReceiver"
    }
    override fun onReceive(context: Context, intent: Intent) {
        Log.d(TAG, "-----------Receivers2-----------")
        Log.d(TAG, "Action: " + intent.action)
        Log.d(TAG, "URI: " + intent.toUri(Intent.URI_INTENT_SCHEME))
```

```
16            Log.d(TAG, "自定义广播的 info: " + intent.getStringExtra("info"))
17            if (intent.getBooleanExtra("stop", false)) {
18                abortBroadcast()
19            }
20        }
21    }
```

第 17 行 intent.getBooleanExtra("stop",false) 用于获取 Boolean 型 stop 字段的数据，如果无法获取到则返回 false。第 18 行 abortBroadcast() 终止有序广播继续传递。

/manifests/AndroidManifest.xml

```
12  <receiver
13      android:name = ".MyBroadcastReceivers1"
14      android:enabled = "true"
15      android:exported = "true">
16      <intent-filter android:priority = "1">
17          <action android:name = "OrderedBroadcast.Custom" />
18      </intent-filter>
19  </receiver>
20  <receiver
21      android:name = ".MyBroadcastReceivers2"
22      android:enabled = "true"
23      android:exported = "true">
24      <intent-filter android:priority = "200">
25          <action android:name = "OrderedBroadcast.Custom" />
26      </intent-filter>
27  </receiver>
```

第 16 行和第 24 行的 android:priority 属性用于设置有序广播的优先级，属性值范围是[−1000,1000]。第 17 行和第 25 行设置接收广播的类型为 OrderedBroadcast.Custom，该类型属于自定义广播类型，合法的字符串通常都可以作为自定义广播类型的名称。

3. 主界面的 Activity

/java/com/teachol/c0905/MainActivity.kt

```
//发送标准广播
findViewById<View>(R.id.button_send_broadcast).setOnClickListener(object :
View.OnClickListener {
    override fun onClick(v: View?) {
        val intent = Intent()
        intent.action = "MyBroadcastReceiver.Custom"
        intent.putExtra("info", "悄悄地告诉你一个秘密^6^")
        intent.addFlags(Intent.FLAG_ACTIVITY_PREVIOUS_IS_TOP)
        sendBroadcast(intent)
    }
})
//发送有序广播(连续传递)
findViewById<View>(R.id.button_send_ordered_broadcast).setOnClickListener(object : View.OnClickListener {
    override fun onClick(v: View?) {
        val intent = Intent()
        intent.action = "OrderedBroadcast.Custom"
```

```
27            intent.putExtra("info", "悄悄地告诉你一个秘密^6^")
28            intent.addFlags(Intent.FLAG_ACTIVITY_PREVIOUS_IS_TOP)
29            sendOrderedBroadcast(intent, null)
30        }
31    })
32    //发送有序广播(传递一次)
33    findViewById<View>(R.id.button_send_ordered_broadcast_stop).setOnClickListener(object :
      View.OnClickListener{
34        override fun onClick(v: View?) {
35            val intent = Intent()
36            intent.action = "OrderedBroadcast.Custom"
37            intent.putExtra("stop", true)
38            intent.putExtra("info", "悄悄地告诉你一个秘密 不要告诉别人哦^6^")
39            intent.addFlags(Intent.FLAG_ACTIVITY_PREVIOUS_IS_TOP)
40            sendOrderedBroadcast(intent, null)
41        }
42    })
```

第 13 ~ 21 行发送标准广播，intent.action = "MyBroadcastReceiver.Custom"设置发送广播的类型，intent.putExtra("info","悄悄地告诉你一个秘密^6^")设置 info 字段附加的字符串数据，intent.addFlags(Intent.FLAG_ACTIVITY_PREVIOUS_IS_TOP)设置 flag 标记，该标记是为了突破 Android 8.0 版本对隐式广播的限制。第 23 ~ 31 行和第 33 ~ 42 行发送的都是有序广播，接收到有序广播的对象能决定是否继续将广播进行传递。

4. 测试运行

使用“发送标准广播”按钮发送的标准广播，需要安装前一个实例才能正确接收到。运行后，自上而下依次单击三个按钮，将过滤器设置为“No Filters”，可以同时在 Logcat 窗口中观察到前一个实例输出结果(如图 9 ~ 13 所示)。单击 Logcat 窗口左侧工具栏的“Soft-Wrap”图标，可以取消输出信息的自动换行。

图 9-13　Logcat 窗口输出结果

第 10 章　Android 的数据存储与共享

App 通常需要保存用户的数据，有些保存在本地，有些则保存在服务器端。少量数据的本地存储适合使用共享偏好设置，大量数据的本地存储适合使用 SQLite。如果将数据保存到服务器端，通常会使用 JSON 格式发送数据，然后通过服务器端的程序进行保存。App 之间单向数据传递适合使用广播，而双向数据传递适合使用内容提供者。

10.1　共享偏好设置

共享偏好设置用于本地存储共享数据，使用 XML 文件保存简单数据存储在缓存文件夹中，特别适合本地存储用户的个人数据，敏感的个人数据最好进行加密后存储。

物理设备上可以直接通过文件浏览 App 查看该文件。Android Studio 中使用虚拟设备运行 App 时，需要通过“Device File Explorer”浏览虚拟设备的文件，在“/data/data”文件夹中可以找到对应 App 的文件夹，其内部的“shared_prefs”文件夹用于存储 SharedPreferences 创建的 XML 文件（如图 10-1 所示）。打开该 XML 文件后，可以看到存储数据的标签。由于 SharedPreferences 存储的数据是未经过加密的，而且用户是可以更改的，所以对重要数据应该进行加密处理或通过网络校验。

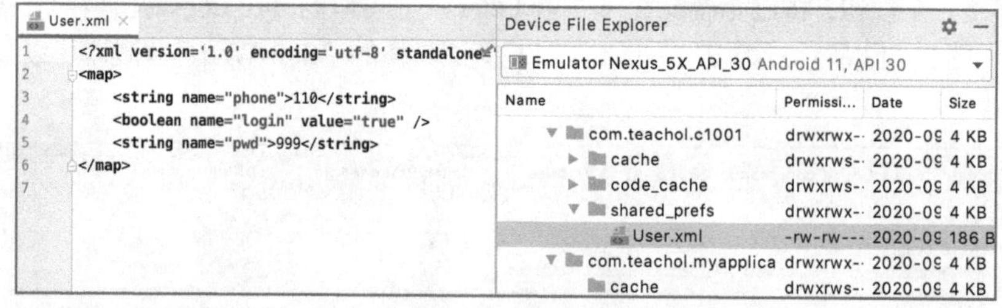

图 10-1 “Device File Explorer”窗口

10.1.1　SharedPreferences 组件

1. SharedPreferences 接口

SharedPreferences 接口（android.content.SharedPreferences）主要用于获取共享偏好设置数据（常用方法如表 10-1 所示）。使用 Context.getSharedPreferences（name: String!, mode: Int）方法获取本地存储的共享偏好设置数据，name 参数是存储共享偏好设置数据的文件名称，mode 参数多使用 Context.MODE_PRIVATE，其余 mode 参数已在 API Level 23 中过期。

2. SharedPreferences.Editor 接口

SharedPreferences.Editor 接口（android.content.SharedPreferences.Editor）主要用于写入、修改和清除共享偏好设置数据，常用方法如表 10-2 所示。

表 10-1　SharedPreferences 接口的常用方法

类型和修饰符	方　法
abstract Boolean	contains (key: String!) 检查是否包含指定名称的共享偏好设置
abstract SharedPreferences.Editor!	edit () 为共享偏好设置创建一个编辑器，通过该编辑器可以对偏好设置数据进行修改，并自动提交返回给 SharedPreferences 对象
abstract MutableMap<String!, *>!	getAll () 获取所有的偏好设置数据
abstract Boolean	getBoolean (key: String!, defValue: Boolean) 获取 Boolean 类型的键值数据，如果为空则返回 defValue 作为默认值
abstract Float	getFloat (key: String!, defValue: Float) 获取 Float 类型的键值数据，如果为空则返回 defValue 作为默认值
abstract Int	getInt (key: String!, defValue: Int) 获取 Int 类型的键值数据，如果为空则返回 defValue 作为默认值
abstract Long	getLong (key: String!, defValue: Long) 获取 Long 类型的键值数据，如果为空则返回 defValue 作为默认值
abstract String?	getString (key: String!, defValue: String?) 获取 String 类型的键值数据，如果为空则返回 defValue 作为默认值
abstract MutableSet<String!>?	getStringSet (key: String!, defValues: MutableSet<String!>?) 获取 Set<String>类型的键值数据，如果为空则返回 defValue 作为默认值
abstract Unit	registerOnSharedPreferenceChangeListener (listener: SharedPreferences.OnSharedPreferenceChangeListener!) 注册监听共享偏好设置数据改变的监听器
abstract Unit	unregisterOnSharedPreferenceChangeListener (listener: SharedPreferences.OnSharedPreferenceChangeListener!) 注销监听共享偏好设置数据改变的监听器

表 10-2　SharedPreferences.Editor 接口的常用方法

类型和修饰符	方　法
abstract Unit	apply () 异步提交共享偏好设置数据，数据会返回给 SharedPreferences 对象
abstract SharedPreferences.Editor!	clear () 清除共享偏好设置数据
abstract Boolean	commit () 提交共享偏好设置的操作，返回值为是否写入 XML 文件成功
abstract SharedPreferences.Editor!	putBoolean (key: String!, value: Boolean) 将 Boolean 类型值存入键值，通过 commit () 或 apply () 方法写入 XML 文件
abstract SharedPreferences.Editor!	putFloat (key: String!, value: Float) 将 Float 类型值存入键值，通过 commit () 或 apply () 方法写入 XML 文件
abstract SharedPreferences.Editor!	putInt (key: String!, value: Int) 将 Int 类型值存入键值，通过 commit () 或 apply () 方法写入 XML 文件
abstract SharedPreferences.Editor!	putLong (key: String!, value: Long) 将 Long 类型值存入键值，通过 commit () 或 apply () 方法写入 XML 文件
abstract SharedPreferences.Editor!	putString (key: String!, value: String?) 将 String 类型值存入键值，通过 commit () 或 apply () 方法写入 XML 文件
abstract SharedPreferences.Editor!	putStringSet (key: String!, values: MutableSet<String!>?) 将 Set<String>类型值存入键值，通过 commit () 或 apply () 方法写入 XML 文件
abstract SharedPreferences.Editor!	remove (key: String!) 在编辑器中标记应删除的键值，commit () 方法执行后移除该键值

10.1.2 实例工程：用户登录

本实例演示了 App 常用的登录功能(如图 10-2 所示)，登录后保存用户的登录数据，再次打开 App 时会自动登录。登录后单击“退出”按钮，可以清除用户登录数据，再次打开 App 时需要登录。

图 10-2　运行效果

1. 打开基础工程

打开“基础工程”文件夹中的“C1001”工程，该工程已经包含 MainActivity、SubActivity 及资源文件，添加了 drawable 文件、Strings 和 Styles 文件配置的内容，修改了 Colors 文件的颜色设置。

2. 主界面的 Activity

/java/com/teachol/c1001/MainActivity.kt

```
12  class MainActivity : AppCompatActivity() {
13      private lateinit var mContext: Context
14      override fun onCreate(savedInstanceState: Bundle?) {
17          mContext = this
18          checkLogin()
20          findViewById<View>(R.id.btn_login).setOnClickListener { //登录按钮单击事件
21              val phone = (findViewById<View>(R.id.edit_text_phone) as EditText).text.toString().trim { it <= ' ' }
22              val pwd = (findViewById<View>(R.id.edit_text_pwd) as EditText).text.toString().trim { it <= ' ' }
23              if (phone == "110" && pwd == "999") {
24                  val sp: SharedPreferences = mContext.getSharedPreferences("User", Context.MODE_PRIVATE)
25                  val editorSP = sp.edit()
26                  editorSP.putBoolean("login", true) //保存登录状态
27                  editorSP.putString("phone", phone)
28                  editorSP.putString("pwd", pwd)
29                  editorSP.commit() //提交保存
30                  val intent = Intent(mContext, SubActivity::class.java)
31                  startActivity(intent)
32              } else {
33                  (findViewById<View>(R.id.text_view_error) as TextView).text = "密码错误！"
34              }
35          }
36      }
38      private fun checkLogin() { //自动登录
39          val sp: SharedPreferences = mContext.getSharedPreferences("User", Context.MODE_PRIVATE)
40          if (sp.getBoolean("login", false)) { //获取登录状态
41              val intent = Intent(mContext, SubActivity::class.java)
42              startActivity(intent)
43          }
44      }
45  }
```

第 21、22 行 trim()方法删除从文本框获取的字符串两侧的空字符。第 23 行使用固定的字符串判断登录是否成功，真实项目中会将 phone 变量值和 pwd 变量值加密后发送到服务器端进行判断并返回结果。第 24 行 mContext.getSharedPreferences("User", Context.MODE_PRIVATE)获取 User.xml 文件中的共享偏好设置数据，如果 User.xml 文件不存在，则在 commit()方法执行后创建该文件。第 25 行

sp.edit()方法为共享偏好设置创建一个编辑器。第 26～28 行放入准备向共享偏好设置写入的键值数据。第 29 行 editorSP.commit()方法提交对共享偏好设置的操作。第 38～44 行是自动登录的方法，通过 SharedPreferences 存储的 login 键值判断是否已经登录，如果已经登录则直接启动 SubActivity。

3. 登录后界面的 Activity

```
/java/com/teachol/c1001/SubActivity.kt
10 class SubActivity : AppCompatActivity() {
11     private var mContext: Context = this
12     override fun onCreate(savedInstanceState: Bundle?) {
15         //获取登录用户数据
16         val sp: SharedPreferences = mContext.getSharedPreferences("User", Context.MODE_PRIVATE)
17         (findViewById<View>(R.id.text_view) as TextView).text = "欢迎${sp.getString("phone","")}访问! "
19         findViewById<View>(R.id.btn_quit).setOnClickListener { //退出登录
20             val sp: SharedPreferences = mContext.getSharedPreferences("User", Context.MODE_PRIVATE)
21             val editorSP = sp.edit()
22             editorSP.clear()
23             editorSP.commit()
24             finish()
25         }
26     }
27 }
```

第16行mContext.getSharedPreferences("User",Context.MODE_PRIVATE)用于获取User.xml文件中的共享偏好设置数据。第 17 行通过 sp.getString("phone","")用于获取存储的登录用户手机号码。第 22 行 editorSP.clear()用于清除共享偏好设置的所有数据。第23行editorSP.commit()用于提交对共享偏好设置的操作。

10.2　轻量级数据库

SQLite 是一个轻量级的关系型数据库，无服务器端，零配置，运算速度快，占用资源少，支持大部分标准 SQL 语法。

10.2.1　SQLite 的字段类型

每个字段所使用的存储类型都是动态的，共有 5 种存储类型：NULL、INTEGER、REAL、TEXT 和 BLOB，根据数据类型使用相应的存储类型进行存储(规则如表 10-3 所示)。SQLite 主要通过 SQLiteOpenHelper 类、SQLiteDatabase 类和 Cursor 类对其进行操作。

表 10-3　数据类型使用的存储类型规则

数 据 类 型	SQLite 使用的存储类型
null	NULL：存储 NULL 值
INT INTEGER TINYINT SMALLINT MEDIUMINT BIGINT UNSIGNED BIG INT INT2 INT8	INTEGER：存储有符号整数

续表

数据类型	SQLite 使用的存储类型
CHARACTER(20) VARCHAR(255) VARYING CHARACTER(255) NCHAR(55) NATIVE CHARACTER(70) NVARCHAR(100) TEXT CLOB	TEXT：存储字符串类型数据
REAL DOUBLE DOUBLE PRECISION FLOAT	REAL：存储浮点数
BLOB	BLOB：存储二进制大数据
NUMERIC DECIMAL(10,5) BOOLEAN DATE DATETIME	如果转换后完全可逆，则以 INTEGER 或 REAL 类型存储；如果转换后不可逆，则以 TEXT 类型存储

10.2.2 SQLite 组件

1. SQLiteOpenHelper 类

SQLiteOpenHelper 类(android.database.sqlite.SQLiteOpenHelper)主要用于 SQLite 数据库的创建和版本管理(常用方法如表 10-4 所示)，其子类需要实现 onCreate(SQLiteDatabase!)方法和 onUpgrade(SQLiteDatabase!, Int, Int)方法。如果数据库不存在，则自动创建扩展名为 db 的数据库文件。在“/data/data”文件夹中可以找到对应 App 的文件夹，其内部的“databases”文件夹用于存储 SQLite 数据库文件(如图 10-3 所示)。

表 10-4　SQLiteOpenHelper 类的常用方法

类型和修饰符	方　法
	<init>(context: Context?, name: String?, factory: SQLiteDatabase.CursorFactory?, version: Int) 主构造方法
open Unit	close() 关闭数据库
open String!	getDatabaseName() 获取数据库名称
open SQLiteDatabase!	getReadableDatabase() 以只读方式获取数据库对象
open SQLiteDatabase!	getWritableDatabase() 获取数据库对象，具有读写权限
abstract Unit	onCreate(db: SQLiteDatabase!) 创建数据库后调用该方法
open Unit	onDowngrade(db: SQLiteDatabase!, oldVersion: Int, newVersion: Int) 数据库版本降级时调用该方法
open Unit	onOpen(db: SQLiteDatabase!) 数据库打开后调用该方法
abstract Unit	onUpgrade(db: SQLiteDatabase!, oldVersion: Int, newVersion: Int) 数据库版本升级时调用该方法

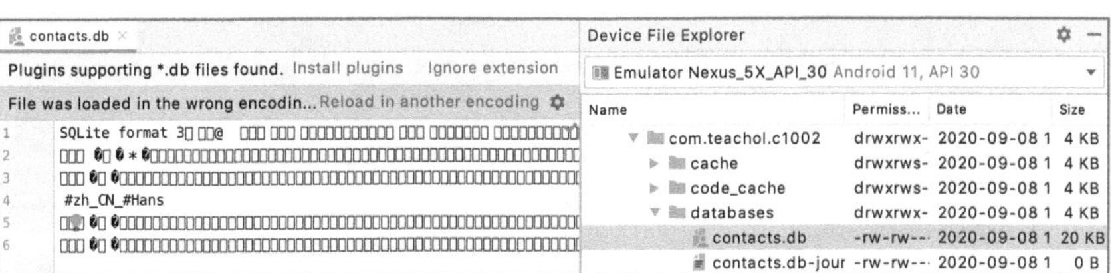

图 10-3 “Device File Explorer”窗口

2. SQLiteDatabase 类

SQLiteDatabase 类(android.database.sqlite.SQLiteDatabase)用于对 SQLite 数据库进行操作(常用方法如表 10-5 所示)，可以使用 SQL 语句，支持事务管理。

表 10-5　SQLiteDatabase 类的常用方法

类型和修饰符	方　法
Long	insert(table: String!, nullColumnHack: String!, values: ContentValues!) 插入数据的便捷方法，返回值为插入数据所在行的索引号，插入失败时返回-1
Int	delete(table: String!, whereClause: String!, whereArgs: Array<String!>!) 删除数据的便捷方法，返回值为删除的行数
Cursor!	query(table: String!, columns: Array<String!>!, selection: String!, selectionArgs: Array<String!>!, groupBy: String!, having: String!, orderBy: String!) 查询数据的便捷方法，返回值为数据集的指针
Int	update(table: String!, values: ContentValues!, whereClause: String!, whereArgs: Array<String!>!) 更新数据的便捷方法，返回值为更新的行数
Unit	beginTransaction() 以独占模式开始事务。事务可以嵌套，当外部事务结束时，该事务中完成的所有工作和所有嵌套事务都将提交或回滚。如果在未调用 setTransactionSuccessful()方法的情况下结束事务，则进行回滚
Unit	beginTransactionNonExclusive() 以立即模式开始事务
Unit	endTransaction() 结束事务
Unit	execSQL(sql: String!) 执行一个非 SELECT 或没有返回值的 SQL 语句
Unit	execSQL(sql: String!, bindArgs: Array<Any!>!) 执行一个不是 SELECT、INSERT、UPDATE 或 DELETE 的 SQL 语句
String!	getPath() 获取数据库文件的路径
Int	getVersion() 获取数据库的版本
Boolean	inTransaction() 当前线程是否有挂起的事务
Boolean	isOpen() 判断数据库是否已经打开
Boolean	isReadOnly() 判断数据库是否是只读的
Boolean	needUpgrade(newVersion: Int) 判断 newVersion 是否大于当前版本
Cursor!	rawQuery(sql: String!, selectionArgs: Array<String!>!) 执行 SQL 语句并返回结果数据集

续表

类型和修饰符	方　法
static Int	releaseMemory() 尝试释放缓存
Unit	setTransactionSuccessful() 将当前事务标记为成功。在调用该方法和调用 endTransaction()方法之间不要对数据库进行任何操作，遇到任何错误则提交事务
Unit	setVersion(version: Int) 设置数据库版本

3．Cursor 接口

Cursor 接口(android.database.Cursor)用于对数据库查询的数据集进行读取(常用方法如表 10-6 所示)，通过移动指针指向不同的数据。

表 10-6　Cursor 接口的常用方法

类型和修饰符	方　法
abstract Unit	close() 关闭指针，并释放资源
abstract ByteArray!	getBlob(columnIndex: Int) 以字节数组的形式返回请求列的值
abstract Int	getColumnCount() 获取列数和
abstract Int	getColumnIndex(columnName: String!) 通过列的字段名获取列的索引号
abstract String!	getColumnName(columnIndex: Int) 通过列的索引号获取列的字段名
abstract Array<String!>!	getColumnNames() 获取结果集中列的字段名称
abstract Int	getCount() 获取结果集中的行数
abstract Double	getDouble(columnIndex: Int) 根据列索引号获取 Double 类型的列值
abstract Float	getFloat(columnIndex: Int) 根据列索引号获取 Float 类型的列值
abstract Int	getInt(columnIndex: Int) 根据列索引号获取 Int 类型的列值
abstract Long	getLong(columnIndex: Int) 根据列索引号获取 Long 类型的列值
abstract Int	getPosition() 获取指针的位置
abstract Short	getShort(columnIndex: Int) 根据列索引号获取 Short 类型的列值
abstract String	getString(columnIndex: Int) 根据列索引号获取 String 类型的列值
abstract Boolean	isAfterLast() 判断指针是否指向最后一行之后的位置
abstract Boolean	isBeforeFirst() 判断指针是否指向第一行之前的位置

续表

类型和修饰符	方　法
abstract Boolean	isClosed() 判断指针是否关闭
abstract Boolean	isFirst() 判断指针是否指向第一行
abstract Boolean	isLast() 判断指针是否指向最后一行
abstract Boolean	move(int offset) 指针从当前位置以指定的偏移量进行移动
abstract Boolean	moveToFirst() 将指针移动到第一行
abstract Boolean	moveToLast() 将指针移动到最后一行
abstract Boolean	moveToNext() 将指针向下移动一行
abstract Boolean	moveToPosition(position: Int) 将指针移动到指定的位置
abstract Boolean	moveToPrevious() 将指针向上移动一行

10.2.3　实例工程：自定义通讯录

本实例演示了使用 SQLite 存储数据的通讯录，顶部使用 ListView 控件显示通讯录中的联系人，单击后在底部的 EditText 控件中显示该联系人的信息，可以通过底部的按钮对该联系人进行相应的操作或显示所有联系人（如图 10-4 所示）。

图 10-4　运行效果

1. 打开基础工程

打开“基础工程”文件夹中的“C1002”工程，该工程已经包含 MainActivity、ContactListAdapter、ContactModel 及资源文件。ContactListAdapter 类是显示联系人的 ListView 控件的适配器类，ContactModel 类是联系人的数据类。

2. ContactSQLiteOpenHelper 类

新建 ContactSQLiteOpenHelper 类，继承 SQLiteOpenHelper 类，用于创建和更新 SQLite 数据库。

```
/java/com/teachol/c1002/ContactSQLiteOpenHelper.kt
class ContactSQLiteOpenHelper(context: Context?, name: String?, factory: CursorFactory?, version: Int) :SQLiteOpenHelper(context, name, factory, version) {
    override fun onCreate(db: SQLiteDatabase) {
        db.execSQL("CREATE TABLE contacts(id INTEGER PRIMARY KEY AUTOINCREMENT,name VARCHAR(16),phone VARCHAR(11))") //创建 contacts 表
    }
    override fun onUpgrade(db: SQLiteDatabase, oldVersion: Int, newVersion: Int) {
```

```
14         if (newVersion > oldVersion) {
16             db.execSQL("ALTER TABLE contacts ADD notes VARCHAR(20)") //contacts表添加notes字段
17         }
18     }
19 }
```

第 09 ~ 12 行重写 onCreate(SQLiteDatabase)方法。第 11 行创建包含 id、name 和 phone 字段的数据表，通过 PRIMARY KEY AUTOINCREMENT 指定 id 字段为自动增长的主键。第 13 ~ 18 行重写 onUpgrade(SQLiteDatabase, Int, Int)方法，当数据库版本大于旧版本时，第 16 行为 contacts 表添加 notes 字段用于备注联系人的相关信息。

3. 主界面的 Activity

```
/java/com/teachol/c1002/MainActivity.kt
//SQLite相关变量
private lateinit var mSQLiteDataBase: SQLiteDatabase
private lateinit var mContactSQLiteOpenHelper: ContactSQLiteOpenHelper
private val mDataBaseName = "contacts.db"//数据库名称
private val mTableName = "contacts"//数据表名称
override fun onCreate(savedInstanceState: Bundle?) {
    mContext = this
    mContactModels = ArrayList<ContactModel>()
    mContactListView = findViewById(R.id.list_view_contacts)
    mIdEditText = findViewById(R.id.edit_text_id)
    mNameEditText = findViewById(R.id.edit_text_name)
    mPhoneEditText = findViewById(R.id.edit_text_phone)
    findViewById<View>(R.id.btn_list).setOnClickListener(this)
    findViewById<View>(R.id.btn_insert).setOnClickListener(this)
    findViewById<View>(R.id.btn_query).setOnClickListener(this)
    findViewById<View>(R.id.btn_update).setOnClickListener(this)
    findViewById<View>(R.id.btn_delete).setOnClickListener(this)
    init()
}
private fun init() { //初始化
    mContactSQLiteOpenHelper = ContactSQLiteOpenHelper(mContext, mDataBaseName, null, 1)
    mSQLiteDataBase = mContactSQLiteOpenHelper.getWritableDatabase()
    mContactListAdapter = ContactListAdapter(this, mContactModels)
    mContactListView.adapter = mContactListAdapter
    listAllContacts()
}
private fun listAllContacts() {//显示所有联系人
    mContactModels.clear()
    val cursor: Cursor = mSQLiteDataBase.rawQuery("SELECT * FROM $mTableName order by id desc",
null)
    while (cursor.moveToNext()) {
        val id: String = cursor.getString(cursor.getColumnIndex("id"))
        val name: String = cursor.getString(cursor.getColumnIndex("name"))
        val phone: String = cursor.getString(cursor.getColumnIndex("phone"))
        mContactModels.add(ContactModel(id, name, phone))
    }
    cursor.close()
```

```
63     mContactListAdapter.notifyDataSetChanged()//更新 ListView
64 }
66 override fun onClick(v: View) { //单击事件
67     val id = mIdEditText.text.toString()
68     val name = mNameEditText.text.toString()
69     val phone = mPhoneEditText.text.toString()
70     when (v.id) {
71         R.id.btn_list -> listAllContacts()//显示所有数据
72         R.id.btn_insert -> { //插入数据
73             mSQLiteDataBase.execSQL("INSERT   INTO   $mTableName   (name,phone)   values(?,?)", arrayOf(name, phone))
74             listAllContacts()
75         }
76         R.id.btn_query -> { //查询数据
77             mContactModels.clear()
78             val cursor: Cursor = mSQLiteDataBase.rawQuery("SELECT * FROM $mTableName WHERE id = ? or name = ? or phone = ?", arrayOf(id, name, phone))
79             while (cursor.moveToNext()) {
80                 val idQuery: String = cursor.getString(cursor.getColumnIndex("id"))
81                 val nameQuery: String = cursor.getString(cursor.getColumnIndex("name"))
82                 val phoneQuery: String = cursor.getString(cursor.getColumnIndex("phone"))
83                 mContactModels.add(ContactModel(idQuery, nameQuery, phoneQuery))
84             }
85             cursor.close()
86             mContactListAdapter.notifyDataSetChanged()
87         }
88         R.id.btn_update -> { //更新数据
89             mSQLiteDataBase.execSQL("UPDATE $mTableName SET name = ?,phone = ? WHERE id = ?", arrayOf(name, phone, id))
90             listAllContacts()
91         }
92         R.id.btn_delete -> { //删除数据
93             mSQLiteDataBase.execSQL("DELETE FROM $mTableName WHERE id = ?", arrayOf(id))
94             listAllContacts()
95         }
96     }
97 }
```

第 24 行声明 mDataBaseName 变量保存数据库文件的名称，数据库文件的扩展名必须为 db。第 25 行声明 mTableName 变量保存数据库的表名。第 44～50 行初始化 SQLite 相关的变量及 ListView 控件的适配器，然后调用 listAllContacts()方法查询并显示所有的联系人。第 54 行创建以 id 字段降序排列查询 contacts 表内所有联系人的 cursor 对象。第 55 行 cursor.moveToNext()向下移动指针，当指针指向的位置存在数据时返回 true，否则返回 false。第 56～58 行 cursor.getColumnIndex()方法根据字段名称获取字段的索引号，cursor.getString()方法根据字段的索引号获取字符型数据。第 61 行 cursor.close()关闭指针，释放缓存数据。第 73 行"INSERT INTO　$mTableName (name,phone) values(?,?)"合成插入数据的 SQL 语句，两个?使用 name 和 phone 的变量值替换，表名称不能使用?方式进行替换。第 74 行在插入数据后调用 listAllContacts()方法将新添加的联系人显示出来。第 78 行根据 id、name 或 phone 查询联系人，只要有一个字段相等即可。第 89 行根据 id 更新 name 和 phone 字段的数据。第 92 行删除指定 id 的联系人数据。

4．测试运行

运行工程后，添加 4 个联系人。在左下角的 Database Inspector 窗口中，显示当前工程创建的 SQLite 数据库，双击数据表后可以查看数据，如图 10-5 所示。

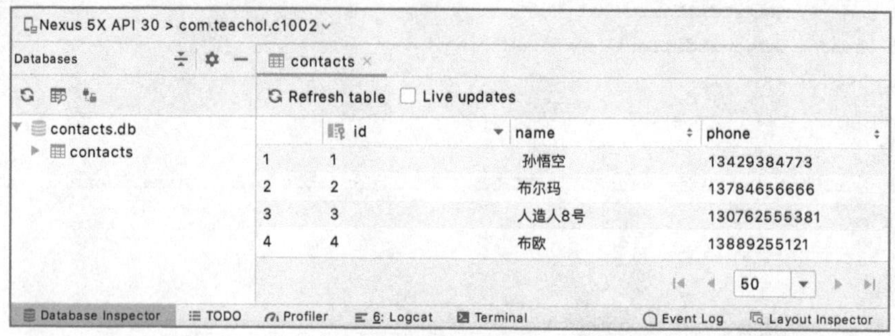

图 10-5　Database Inspector 窗口

10.3　内容提供者

App 可以通过 ContentProvider 向其他 App 提供内容服务，如系统通讯录经常被社交类 App 申请访问权限，读取联系人手机号码后提示用户保存的联系人也在使用该 App，询问用户是否在该 App 中添加好友。ContentProvider 提供的服务需要主动访问，所以需要设置访问的 URI。

10.3.1　URI

URI 是 Universal Resource Identifier（通用资源标志符）的缩写，Android 中的每种资源都可以用 URI 来表示，URI 有三种格式形式（如表 10-7 所示）。

表 10-7　URI 的格式形式

格 式 形 式	格 式 样 例
scheme://authority[/path]	content://com.teachol.contacts/people/47 content://media/external
scheme://host[:port][/path]	http://www.weiju2014.com http://www.meachol.com:80/weijian
scheme:scheme-specific-part	tel:13588888888 mailto:baizhe_22@qq.com

URI 作为 ContentProvider 授权的依据，还具有派生关系。例如，下面的三个 URI 中，第一个 URI 的 contact 表示 contact 数据表；第二个 URI 与第一个 URI 属于同一级别，其中#是一个通配符，表示数据表中的行号；第三个 URI 是前两个 URI 的派生 URI，类似于子类。

- content://com.teachol.contacts/contact
- content://com. teachol.contacts/contact/#
- content://com. teachol.contacts/contact/friend

10.3.2　数据交换原理

ContentProvider 不但可以向所在的 App 提供数据，还可以向其他 App 提供数据（如图 10-6 所示）。

请求数据的 App 使用 URI 向提供数据的 App 请求数据，提供数据的 App 通过 ContentProvider 对数据进行操作。ContentProvider 完成相应数据操作后，还可以选择是否向 ContentObserver 发送数据改变的通知，通过 ContentObserver 进行相应的响应处理。

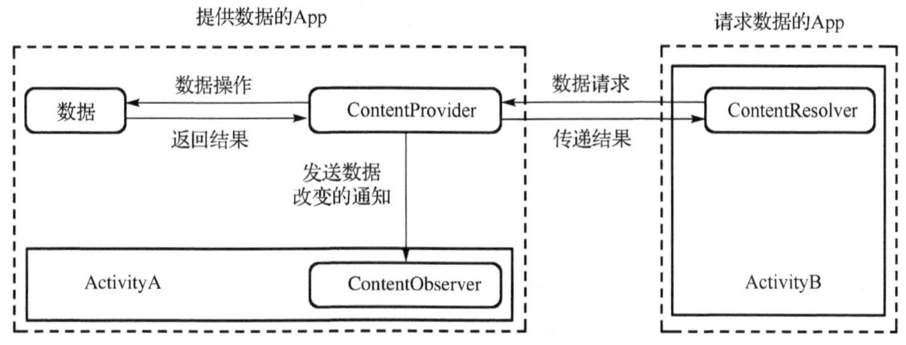

图 10-6　数据交换原理

10.3.3　ContentProvider 组件

1．ContentProvider 类

ContentProvider 类（android.content.ContentProvider）用于进程之间安全地访问和修改数据（方法如表 10-8 所示），适合为其他 App 提供复杂的数据或文件、向 Widget 公开应用数据或使用搜索框架提供自定义搜索建议。虽然 App 内部也可以使用 ContentProvider 访问或修改数据，但是并不推荐这样使用，因为内部直接访问或修改数据效率更高。

ContentProvider 可以精细控制数据访问权限，如选择仅在 App 内限制对 ContentProvider 的访问、授予访问其他应用数据的权限或配置读取和写入数据的不同权限。

表 10-8　ContentProvider 类的方法

类型和修饰符	方　法
	<init>() 主构造方法
abstract Boolean	onCreate() 创建时调用该方法
abstract Uri?	insert(uri: Uri, values: ContentValues?) 处理插入数据请求的方法，能够防止 SQL 注入式攻击
abstract Int	delete(uri: Uri, selection: String?, selectionArgs: Array<String!>?) 处理删除数据请求的方法，能够防止 SQL 注入式攻击
abstract Int	update(uri: Uri, values: ContentValues?, selection: String?, selectionArgs: Array<String!>?) 处理更新数据请求的方法，能够防止 SQL 注入式攻击
abstract Cursor?	query(uri: Uri, projection: Array<String!>?, selection: String?, selectionArgs: Array<String!>?, sortOrder: String?) 处理查询数据请求的方法，能够防止 SQL 注入式攻击
abstract String?	getType(uri: Uri) 处理通过 URI 获取 MIME 类型的方法
String?	getCallingPackage() 获取发送访问请求的包名称
Context?	getContext() 获取运行 ContentProvider 的 Context 对象

ContentProvider 需要在 AndroidManifest.xml 文件中添加<provider>标签后才能被调用，<provider>标签属性（如表 10-9 所示）大多无法通过 ContentProvider 类的方法进行设置和获取。

表 10-9　<provider>标签属性

属　性	说　明
android:authorities	设置用于标识 ContentProvider 提供数据的 URI，多个 URI 用“;”分隔
android:directBootAware	设置是否可以在用户解锁之前直接启动
android:name	设置实现 ContentProvider 的类名
android:enabled	设置是否可以实例化 ContentProvider
android:exported	设置是否可供其他 App 使用
android:initOrder	设置启动顺序优先级，数值越高，优先级越高
android:multiProcess	设置是否为每个进程提供一个 ContentProvider 对象
android:process	设置运行 ContentProvider 的进程名称
android:syncable	设置是否与服务器上的数据同步

2．ContentResolver 类

访问 ContentProvider 实现数据访问和修改，需要通过 ContentResolver 或 Intent 对象。ContentResolver 类（android.content.ContentResolver）用于对 URI 进行解析再将请求发送到对应的 ContentProvider，方法如表 10-10 所示。

表 10-10　ContentResolver 类的方法

类型和修饰符	方　法
open Array<ContentProviderResult!>	applyBatch（authority: String, operations: ArrayList<ContentProviderOperation!>） 批量执行 ContentProviderOperation 对象，并返回其结果的数组
Int	delete（url: Uri, where: String?, selectionArgs: Array<String!>?） 发送删除数据请求的方法
Uri	insert（url: Uri, values: ContentValues?） 发送插入数据请求的方法
Int	update（uri: Uri, values: ContentValues?, where: String?, selectionArgs: Array<String!>?） 发送更新数据请求的方法
Cursor?	query（uri: Uri, projection: Array<String!>?, selection: String?, selectionArgs: Array<String!>?, sortOrder: String?） 发送查询数据请求的方法
open Unit	notifyChange（uri: Uri, observer: ContentObserver?） 通知观察者数据更新的方法
Unit	registerContentObserver（uri: Uri, notifyForDescendants: Boolean, observer: ContentObserver） 注册观察者类对象的方法，该方法在调用 notifyChange（）方法后回调。如果 notifyForDescendants 参数指是否观察 uri 参数派生的子 URI
Unit	unregisterContentObserver（observer: ContentObserver） 注销观察者类对象的方法

3．ContentObserver 类

ContentObserver 类（android.database.ContentObserver）用于接收 ContentResolver 对象和自身发送的 URI 更新通知，作为 ContentProvider 对 URI 处理后是否响应相应处理的观察者，方法如表 10-11 所示。适合在 ContentProvider 所在的 App 内使用，观察通过 ContentProvider 插入、更新或删除的数据改变，以便 App 及时进行响应处理。

表 10-11　ContentObserver 类的方法

类型和修饰符	方　法
	<init>(handler: Handler!) 主构造方法
Unit	dispatchChange(selfChange: Boolean, uri: Uri?) 向观察者发送更改通知。如果构造方法提供了处理程序，则对 onChange(Boolean)方法的调用将被传递到处理程序的消息队列中；否则，调用 onChange(Boolean)方法
open Unit	onChange(selfChange: Boolean) 当内容发生更改时调用此方法。子类应重写此方法以处理内容更改
open Unit	onChange(selfChange: Boolean, uri: Uri?) 当内容发生更改时调用此方法。子类应重写此方法以处理内容更改，为了确保在未提供 uri 参数的旧版本框架上正确操作，应实现 onChange(Boolean)和 onChange(Boolean, Uri?)的重写

10.3.4　实例工程：自定义内容提供者

本实例演示了两个 App 之间的数据访问和修改，模仿了系统通讯录提供内容服务的方式(如图 10-7 所示)。C1003 工程使用 SQLite 存储联系人的数据，使用自定义的 ContentProvider 提供联系人的查询、插入、更新和删除功能。C1004 工程并没有存储联系人的数据，而是通过 ContentResolver 访问 C1003 工程提供的联系人数据的访问和修改功能。C1003 可以被看成服务器端，C1004 可以被看成客户端。

图 10-7　运行效果

1. 打开内容提供者的基础工程

打开“基础工程”文件夹中的“C1003”工程，该工程已经包含 MainActivity、ContactListAdapter、ContactModel 及资源文件。ContactListAdapter 是显示联系人的 ListView 控件的适配器类，ContactModel 是联系人的数据类。

2. 内容提供者的 ContactContentProvider 类

在“/java/com/teachol/c1003”文件夹上单击右键，选择【new】→【Other】→【Content Provider】命令。在打开的“New Android Component”对话框中(如图 10-8 所示)，设置“Class Name”为“ContactContentProvider”，“URI Authorities”为“com.teachol.contact”，单击“Finish”按钮完成 ContactContentProvider 类的创建。

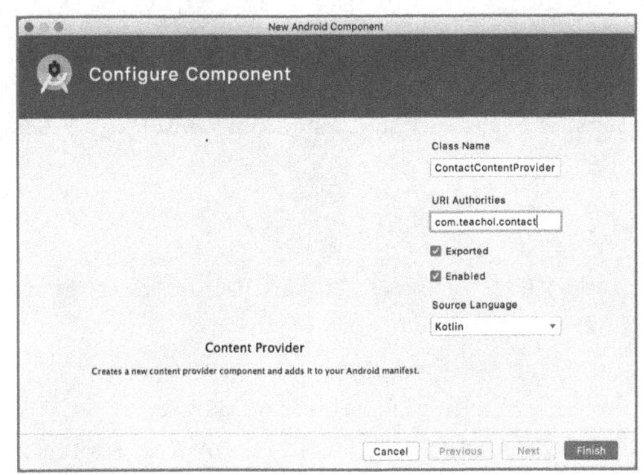

图 10-8　“New Android Component”对话框

```
/manifests/AndroidManifest.xml
12 <provider
13     android:name = ".ContactContentProvider"
14     android:authorities = "com.teachol.contact"
15     android:enabled = "true"
16     android:exported = "true"></provider>
```

第 12 ~ 16 行是新建 ContactContentProvider 类时自动生成的标签代码。第 14 行 android:authorities = "com.teachol.contact"表示能够接收包含"com.vt.contact"授权的 URI。

```
/java/com/teachol/c1003/ContactContentProvider.kt
class ContactContentProvider : ContentProvider() {
    companion object {
        private const val AUTHORITIES = "com.teachol.contacts"
        private lateinit var URI_MATCHER: UriMatcher //用于匹配 URI
        private const val CONTACTS_TABLE = 1 //URI 传递的数据包含数据表名
        private const val CONTACTS_TABLE_AND_ID = 2 //URI 传递的数据包含数据表名和 id
        private lateinit var mContactDBOpenHelper: ContactDBOpenHelper
        private lateinit var mSQLiteDatabase: SQLiteDatabase
        private const val mDataBaseName = "contacts.db"
    }
    init {
        URI_MATCHER = UriMatcher(UriMatcher.NO_MATCH)
        URI_MATCHER.addURI(AUTHORITIES, "contacts", CONTACTS_TABLE)
        URI_MATCHER.addURI(AUTHORITIES, "contacts/#", CONTACTS_TABLE_AND_ID)
    }
    override fun onCreate(): Boolean { //创建时调用
        //实例化 ContactDBOpenHelper
        mContactDBOpenHelper = ContactDBOpenHelper(this.context, mDataBaseName, null, 1)
        //获取具有写入权限的 SQLite 数据库
        mSQLiteDatabase = mContactDBOpenHelper.writableDatabase
        return true
    }
    override fun insert(uri: Uri, values: ContentValues?): Uri? { //插入
        var uri = uri
        val tableName = uri.pathSegments[0]
        //插入联系人成功后将联系人的 id 保存在 rowID 中
        val rowID = mSQLiteDatabase.insert(tableName, "", values)
        if (rowID > 0) { //rowID 值大于 0 表示插入成功
            uri = ContentUris.withAppendedId(uri, rowID) //将联系人 id 附加到 uri 中
            context!!.contentResolver.notifyChange(uri, null) //向观察者发送改变的通知
        }
        return uri
    }
    override fun update(uri: Uri, values: ContentValues?, selection: String?, selectionArgs: Array<String>?): Int { //更新
        var count = 0
        val id = uri.pathSegments[1] //获取 URI 中附加的联系人 id
        var where = "id = " + id + if (!TextUtils.isEmpty(selection)) " AND ($selection)" else ""
        //更新数据后将更新的行数赋给 count
        count = mSQLiteDatabase.update(uri.pathSegments[0], values, where, selectionArgs)
```

```
59         context!!.contentResolver.notifyChange(uri, null) //向观察者发送改变的通知
60         return count
61     }
63     override fun delete(uri: Uri, selection: String?, selectionArgs: Array<String>?): Int {//删除
64         var count = 0
65         val tableName = uri.pathSegments[0]
66         count = when (URI_MATCHER.match(uri)) {
67             //根据 name 和 phone 判断删除联系人并将删除的联系人数据行数赋给 count
68             CONTACTS_TABLE -> mSQLiteDatabase.delete(tableName, selection, selectionArgs)
69             CONTACTS_TABLE_AND_ID -> {
71                 val id = uri.pathSegments[1] //获取要删除的联系人 id
73                 mSQLiteDatabase.delete(tableName, "id = ?", arrayOf(id)) //根据 id 删除联系人
74             }
75             else -> throw IllegalArgumentException("Unknown URI $uri")
76         }
77         context!!.contentResolver.notifyChange(uri, null) //向观察者发送改变的通知
78         return count
79     }
81     override fun query(uri: Uri, columns: Array<String>?, selection: String?, selectionArgs:
   Array<String>?, sortOrder: String?): Cursor? { //查询
83         val tableName = uri.pathSegments[0] //获取联系人的 id
85         return mSQLiteDatabase.query(tableName, columns, selection, selectionArgs, null, null,
   sortOrder) //将查询结果的指针赋给 cursor
86     }
88     override fun getType(uri: Uri): String? { //通过 URI 获取 MIME 类型
89         val match = URI_MATCHER.match(uri)
90         return when (match) {
91             CONTACTS_TABLE -> "vnd.android.cursor.dir/contacts"//vnd.android.cursor.dir 表示数据集合
92             CONTACTS_TABLE_AND_ID -> "vnd.android.cursor.item/contacts"//vnd.android.cursor.item
                       表示一组数据
93             else -> null
94         }
95     }
96 }
```

第 23 ~ 25 行 URI_MATCHER 变量用于匹配 URI，content://com.teachol.contact/contacts 与 CONTACTS_TABLE 相匹配，content://com.teachol.contact/contacts/#与 CONTACTS_TABLE_AND_ID 相匹配。第 28 ~ 34 行只有在 ContentResolver 对象尝试访问 ContactContentProvider 时，系统才会创建它。此时初始化 SQLite 数据库的相关对象，返回值为 true 时表示 ContentResolver 对象初始化成功。第 36 ~ 49 行重写 insert(Uri, ContentValues?)方法，使用 mSQLiteDatabase.insert(tableName, "", values)插入数据，如果插入数据成功会将数据的行号保存在 rowID 中；rowID 大于 0 表示插入数据成功，context!!.ContentResolver().notifyChange(uri, null)向 MainActivity 内的观察者发送更新的通知，会调用 MainActivity 中的 ContactContentObserver 类的 onChange(Boolean)方法。第 51 ~ 61 行重写 update(Uri, ContentValues?, String?, String[]?)方法，更新联系人的数据。第 63 ~ 79 行重写 delete(Uri, String?, String[]?)方法，删除联系人的数据。第 81 ~ 86 行重写 query(Uri, String[]?, String?, String[]?, String?)方法，查找联系人。第 88 ~ 95 行对接收到的 URI 进行类型匹配。

3．内容提供者主界面的 Activity

```
/java/com/teachol/c1003/MainActivity.kt
52  //自定义 ContentObserver 类
53  internal inner class ContactContentObserver(handler: Handler?) : ContentObserver(handler) {
54      //ContentProvider.notifyChange()方法后回调该方法
55      override fun onChange(selfChange: Boolean) {
56          super.onChange(selfChange)
57          listAllContacts()
58      }
59  }
61  private fun init() { //初始化
63      listAllContacts() //显示所有联系人
65      val contactContentObserver = ContactContentObserver(null)
66      val contentResolver:ContentResolver = contentResolver
67      contentResolver.registerContentObserver(URI, true, contactContentObserver) //注册观察者
68  }
70  private fun listAllContacts() { //显示所有联系人
71      mContactModels.clear()
72      //查询所有的联系人并按照 id 降序排列
73      val cursor = mSQLiteDataBase.rawQuery("SELECT * FROM $TABLE_NAME order by id desc", null)
75      while (cursor.moveToNext()) { //遍历查询结果指针
76          val id = cursor.getString(cursor.getColumnIndex("id"))
77          val name = cursor.getString(cursor.getColumnIndex("name"))
78          val phone = cursor.getString(cursor.getColumnIndex("phone"))
79          mContactModels.add(ContactModel(id, name, phone))
80      }
81      cursor.close()
83      mContactListAdapter.notifyDataSetChanged() //更新 ListView
84  }
```

第 53 ~ 59 行 ContactContentObserver 继承自 ContentObserver，观察 ContentProvider 提供的数据变化，如果 ContactContentProvider 调用 Context.contentResolver.notifyChange(Uri, ContentObserver) 方法，则通知 ContactContentObserver 对象调用 onChange(Boolean) 方法重新查询并显示所有联系人。第 65 ~ 67 行通过 ContentResolver 类对象注册 ContactContentObserver 类对象，观察指定 URI 提供的数据变化。第 70 ~ 84 行 listAllContacts() 方法首先清空 mContactModels 对象，然后查询到联系人的数据指针，使用遍历的方式获取每个联系人的数据并存储到 mContactModels 对象，最后通过 mContactListAdapter.notifyDataSetChanged() 将 mContactModels 对象存储的联系人数据更新显示到 ListView 控件。

4．打开访问内容提供者的基础工程

打开“基础工程”文件夹中的“C1004”工程，该工程已经包含 MainActivity、ContactListAdapter、ContactModel 及布局文件。ContactListAdapter 类是显示联系人的 ListView 控件的适配器类，ContactModel 类是联系人的数据类。

5．访问内容提供者主界面的 Activity

```
/java/com/teachol/c1004/MainActivity.kt
private fun listAllContacts() { //显示所有联系人
    mContactModels.clear()
    val uri = Uri.parse("content://com.teachol.contacts/contacts")
```

```
    contentResolver.query(uri, null, null, null, "name desc")?.apply { //查询contacts表的所有数据
        Log.e("aa","query")
        while (moveToNext()) { //遍历查询到的数据
            val idQuery = getString(getColumnIndex("id"))
            val nameQuery = getString(getColumnIndex("name"))
            val phoneQuery = getString(getColumnIndex("phone"))
            mContactModels.add(ContactModel(idQuery!!, nameQuery!!, phoneQuery!!))
        }
        close()
    }
    mContactListAdapter.notifyDataSetChanged() //更新ListView
    Toast.makeText(this, "已经显示所有联系人", Toast.LENGTH_SHORT).show()
}
override fun onClick(v: View) { //单击事件
    val id = mIdEditText.text.toString()//获取id输入文本框的字符串
    val name = mNameEditText.text.toString()//获取姓名输入文本框的字符串
    val phone = mPhoneEditText.text.toString()//获取手机号输入文本框的字符串
    val uri: Uri
    var selection: String
    var selectionArgs: Array<String?>
    var selectionList: ArrayList<String>
    val resolver:ContentResolver = contentResolver
    when (v.id) {
        R.id.btn_list -> listAllContacts()
        R.id.btn_insert -> {
            uri = Uri.parse("content://com.teachol.contacts/contacts")
            val values = ContentValues() //将要插入的数据
            values.put("name", name)
            values.put("phone", phone)
            val resUri = resolver.insert(uri, values) //插入数据
            if (resUri === uri) {
                Toast.makeText(this, "插入失败", Toast.LENGTH_SHORT).show()
            } else {
                listAllContacts() //重新显示所有联系人
                Toast.makeText(this, "插入成功", Toast.LENGTH_SHORT).show()
            }
        }
        R.id.btn_query -> {
            mContactModels.clear()
            uri = Uri.parse("content://com.teachol.contacts/contacts")
            //合成查询数据条件的变量值
            selection = (if (!TextUtils.isEmpty(id)) "id = ?" else "") + (if (!TextUtils.isEmpty(id) && !TextUtils.isEmpty(name)) " and " else "") + (if (!TextUtils.isEmpty(name)) "name = ?" else "") + (if ((!TextUtils.isEmpty(id) || !TextUtils.isEmpty(name)) && !TextUtils.isEmpty(phone)) " and " else "") + if (!TextUtils.isEmpty(phone)) "phone = ?" else ""
            selectionList = ArrayList()
            if (id != "") selectionList.add(id)
            if (name != "") selectionList.add(name)
            if (phone != "") selectionList.add(phone)
            selectionArgs = arrayOfNulls(selectionList.size)
```

```
            selectionList.toArray(selectionArgs)
            resolver.query(uri, null, selection, selectionArgs, "name desc")?.apply { //查询数据
                while (moveToNext()) { //遍历查询到的数据
                    val idQuery = getString(getColumnIndex("id"))
                    val nameQuery = getString(getColumnIndex("name"))
                    val phoneQuery = getString(getColumnIndex("phone"))
                    mContactModels.add(ContactModel(idQuery!!, nameQuery!!, phoneQuery!!))
                }
                close()
            }
            if (mContactModels.size>0) {
                mContactListAdapter.notifyDataSetChanged() //刷新显示的数据
                Toast.makeText(this, "查询完成", Toast.LENGTH_SHORT).show()
            } else {
                Toast.makeText(this, "查询失败", Toast.LENGTH_SHORT).show()
            }
        }
        R.id.btn_update -> if (id != "") {
            uri = Uri.parse("content://com.teachol.contacts/contacts/$id")
            val values1 = ContentValues() //将要更新的数据
            values1.put("name", name)
            values1.put("phone", phone)
            val resUpdate = resolver.update(uri, values1, null, null) //更新数据
            if (resUpdate > 0) {
                listAllContacts() //重新显示所有联系人
                Toast.makeText(this, "更新成功", Toast.LENGTH_SHORT).show()
            } else {
                Log.d("aa", "更新失败")
                Toast.makeText(this, "更新失败", Toast.LENGTH_SHORT).show()
            }
        }
        R.id.btn_delete -> {
            var resDelete = 0
            if (id != "") {
                //根据 id 删除联系人
                uri = Uri.parse("content://com.teachol.contacts/contacts/$id")
                resDelete = resolver.delete(uri, null, null)
            } else if (name != "") {
                //根据姓名或者同时和电话号码删除联系人
                uri = Uri.parse("content://com.teachol.contacts/contacts")
                var where = "name = ?" + if (!TextUtils.isEmpty(phone)) " AND phone = ?" else
""
                resDelete = resolver.delete(uri, where, arrayOf(name, phone))
            } else if (phone != "") {
                //根据电话号码删除联系人
                uri = Uri.parse("content://com.teachol.contacts/contacts")
                resDelete = resolver.delete(uri, "phone = ?", arrayOf(phone))
            }
            if (resDelete > 0) {
                listAllContacts()
```

```
153                 Toast.makeText(this, "删除成功", Toast.LENGTH_SHORT).show()
154             } else {
155                 Toast.makeText(this, "删除失败", Toast.LENGTH_SHORT).show()
156             }
157         }
158     }
    }
```

第 49 行通过 contentResolver.query(uri, null, null, null, "name desc") 获取联系人数据的指针。第 51 ~ 56 行遍历指针存储的联系人数据并存储在 mContactModels 变量中。第 76 ~ 88 行通过 URI 向 C1003 发送添加联系人的数据请求，并处理返回结果。第 91 ~ 116 行清空 mContactModels 对象存储的联系人数据后，通过 URI 向 C1003 发送查询联系人的数据请求，并将查询到的联系人数据显示出来。第 119 ~ 132 行根据联系人的 id 通过 URI 向 C1003 发送更新联系人的数据请求，如果更新成功则更新显示的联系人数据。第 135 ~ 155 行根据联系人的 id 通过 URI 向 C1003 发送删除联系人的数据请求，如果删除成功则更新显示的联系人数据。

10.3.5　实例工程：访问和修改系统通讯录数据

本实例演示了使用 ContentResolver 访问和修改系统通讯录的联系人，首次运行时会弹出获取访问系统通讯录权限的对话框。添加联系人后，可以在系统通讯录中查看到添加的联系人（如图 10-9 所示）。

图 10-9　运行效果

提示：系统通讯录

系统通讯录是系统自带的 App，使用 raw_contacts 表和 data 表保存联系人的数据。

- raw_contacts 表：存储联系人。_id 字段为主键，声明为 autoincrement，存储联系人的 ID；display_name 字段存储联系人的姓名。对应的 URI 是 content://com.android.contacts/raw_contacts。
- data 表：存储联系人具体的数据。raw_contact_id 字段存储联系人的 ID，与 raw_contacts 表内联系人的 ID 相对应；mimetype 字段存储记录数据的类型；data1-data9 字段存储具体的数据。对应 URI 是 content://com.android.contacts/data。

每个联系人在 raw_contacts 表内只有一条记录；在 data 表会有多条记录，每条记录保存一种形式的数据，通过 mimetype 字段进行区分。常用的 mimetype 字段值如下。

- vnd.android.cursor.item/phone_v2：电话。

- vnd.android.cursor.item/name：姓名。
- vnd.android.cursor.item/email_v2：邮件。
- vnd.android.cursor.item/postal-address_v2：地址。
- vnd.android.cursor.item/organization：组织。
- vnd.android.cursor.item/photo：照片。

1．打开基础工程

打开“基础工程”文件夹中的“C1005”工程，该工程已经包含 MainActivity、ContactListAdapter、ContactModel 及资源文件。ContactListAdapter 类是显示联系人的 ListView 控件的适配器类，ContactModel 类是联系人的数据类。

提示：ContentProviderOperation 类

ContentProvider 类封装了数据的访问接口，其底层数据一般都是保存在本地或者服务器端。需要操作多行数据时，可以选择多次调用 ContentResolver 类的相关方法。为了使批量更新、插入、删除数据更加方便，Android 提供了 ContentProviderOperation 类，优势如下：

- 在一个事务中执行所有操作，只需打开和关闭一个事务，能够保证数据完整性。
- 减少占用 CPU 的时间，提升性能，同时减少电量的消耗。

2．设置权限

/manifests/AndroidManifest.xml

```
03  <uses-permission android:name = "android.permission.READ_CONTACTS"/>
04  <uses-permission android:name = "android.permission.WRITE_CONTACTS"/>
```

第 03 行读取系统通讯录联系人的权限。第 04 行添加、删除和更新系统通讯录联系人的权限。

提示：动态权限

为了防止 Android 权限滥用危害用户信息安全，从 Android 6.0（API Level 23）开始，在使用权限时，不但需要在 AndroidManifest.xml 文件中设置权限，还要在使用权限时进行动态授权。动态权限牺牲了使用的便捷性，提升了系统的安全性。首次动态授权后，在应用设置中对授权进行修改。目前很多 App 强制用户开启权限，否则无法使用部分功能甚至全部功能。动态授权主要涉及以下三个方法。

- ActivityCompat.checkSelfPermission（permission: String）：检测是否已经获得该权限。
- ActivityCompat.requestPermissions（permissions: String[],requestCode: Int）：请求单个或多个权限，此时会弹出系统动态授权的对话框。
- onRequestPermissionsResult（requestCode:Int, permissions:String[], grantResults:Int[]）：Activity 对请求权限结果的回调方法。

3．主界面的 Activity

/java/com/teachol/c1005/MainActivity.kt

```
companion object {
    private const val OPERATE_LIST = 0
    private const val OPERATE_QUERY = 1
    private const val OPERATE_INSERT = 2
    private const val OPERATE_UPDATE = 3
    private const val OPERATE_DELETE = 4
```

```
    private val rawContactsUri: Uri = Uri.parse("content://com.android.contacts/raw_
contacts")
    private val dataUri: Uri = Uri.parse("content://com.android.contacts/data")
}
private lateinit var mContactModels: MutableList<ContactModel>
private lateinit var mContactListAdapter: ContactListAdapter
private lateinit var mContactListView: ListView
lateinit var mIdEditText: EditText
lateinit var mNameEditText: EditText
lateinit var mPhoneEditText: EditText
//查询所有联系人
private fun listContacts() {
    mContactModels.clear()
    val resolver = contentResolver
    //查询raw_contacts表中存储的联系人
    val rawContactsCursor: Cursor? = resolver.query(rawContactsUri, null, null, null, null)
    while (rawContactsCursor!!.moveToNext()) { //遍历查询结果
        //获取raw_contacts表中存储的联系人id和显示名字
        val rawContactId: String = rawContactsCursor.getString(rawContactsCursor.
getColumnIndex(ContactsContract.CommonDataKinds.Phone._ID))
        val displayname: String = rawContactsCursor.getString(rawContactsCursor
.getColumnIndex(ContactsContract.CommonDataKinds.Phone.DISPLAY_NAME))
        //根据联系人id查询data表中存储的联系人电话号码
        val dataCursor: Cursor? = resolver.query(dataUri, null, "raw_contact_id = ? AND mimetype
= 'vnd.android.cursor.item/phone_v2'", arrayOf(rawContactId), null)
        while (dataCursor!!.moveToNext()) {
            val phone: String = dataCursor.getString(dataCursor.getColumnIndex("data1"))
            mContactModels.add(ContactModel(rawContactId, displayname, phone))
        }
    }
    rawContactsCursor.close()
    mContactListAdapter.notifyDataSetChanged()
}
//根据显示名字查询联系人
private fun queryContact() {
    val name = mNameEditText.text.toString().trim { it <= ' ' }
    val resolver = contentResolver
    //根据联系人的显示名字搜索
    val rawContactsCursor: Cursor? = resolver.query(rawContactsUri, null, "display_name = ?",
arrayOf(name), null)
    if (rawContactsCursor!!.getCount() > 0) { //判断是否搜索到联系人
        mContactModels.clear()
        while (rawContactsCursor.moveToNext()) {
            //获取raw_contacts表中存储的联系人id和显示名字
            val rawContactId: String = rawContactsCursor.getString(rawContactsCursor.
getColumnIndex(ContactsContract.CommonDataKinds.Phone._ID))
            val displayName: String = rawContactsCursor.getString(rawContactsCursor.
getColumnIndex(ContactsContract.CommonDataKinds.Phone.DISPLAY_NAME))
            //根据联系人id查询data表中存储的联系人电话号码
            val dataCursor: Cursor? = resolver.query(dataUri, null, "raw_contact_id = ? AND
```

```
mimetype = 'vnd.android.cursor.item/phone_v2'", arrayOf(rawContactId), null)
            while (dataCursor!!.moveToNext()) {
                val phone: String = dataCursor.getString(dataCursor.getColumnIndex("data1"))
                mContactModels.add(ContactModel(rawContactId, displayName, phone))
            }
        }
        Toast.makeText(this, "已经查询到该联系人", Toast.LENGTH_SHORT).show()
        mContactListAdapter.notifyDataSetChanged()
    } else {
        Toast.makeText(this, "没有查询到该联系人", Toast.LENGTH_SHORT).show()
    }
    rawContactsCursor.close()
}
//添加联系人
private fun insertContact(useContentProviderOperation: Boolean) {
    val name = mNameEditText.text.toString()
    val phone = mPhoneEditText.text.toString()
    //是否使用 ContentProviderOperation
    if (useContentProviderOperation) {
        //向 raw_contacts 表中插入联系人的操作
        val operations: ArrayList<ContentProviderOperation>= ArrayList()
        val op1 = ContentProviderOperation.newInsert(rawContactsUri)
                .withValue("account_name", null)
                .build()
        operations.add(op1)
        //向 data 表中插入联系人的显示名字的操作
        val op2 = ContentProviderOperation.newInsert(dataUri)
                .withValueBackReference("raw_contact_id", 0)
                .withValue("mimetype", "vnd.android.cursor.item/name")
                .withValue("data2", name)
                .build()
        operations.add(op2)
        //向 data 表中插入联系人的电话号码的操作
        val op3 = ContentProviderOperation.newInsert(dataUri)
                .withValueBackReference("raw_contact_id", 0)
                .withValue("mimetype", "vnd.android.cursor.item/phone_v2")
                .withValue("data1", phone)
                .build()
        operations.add(op3)
        try {
            val resolver = contentResolver
            val contentProviderResult = resolver.applyBatch("com.android.contacts", operations)
            if (contentProviderResult.size == operations.size) { //判断是否添加成功
                Toast.makeText(this, "添加联系人成功", Toast.LENGTH_SHORT).show()
                listContacts()
            } else {
                Toast.makeText(this, "添加联系人失败", Toast.LENGTH_SHORT).show()
            }
        } catch (e: OperationApplicationException) {
            e.printStackTrace()
```

```
            Toast.makeText(this, "添加联系人异常: $e", Toast.LENGTH_SHORT).show()
        } catch (e: RemoteException) {
            e.printStackTrace()
            Toast.makeText(this, "添加联系人异常: $e", Toast.LENGTH_SHORT).show()
        }
    } else {
        val resolver = contentResolver
        //向 raw_contacts 表中插入联系人
        val value1 = ContentValues()
        value1.putNull("account_name")
        val resultUri: Uri? = resolver.insert(rawContactsUri, value1)
        val rawContactsId: Int = Integer.valueOf(resultUri!!.getPathSegments().get(1))
        //向 data 表中插入联系人的显示名字
        val value2 = ContentValues()
        value2.put("raw_contact_id", rawContactsId)
        value2.put("mimetype", "vnd.android.cursor.item/name")
        value2.put("data2", name)
        val dataNameUri: Uri? = resolver.insert(dataUri, value2)
        val dataNameId: Int = Integer.valueOf(dataNameUri!!.getPathSegments().get(1))
        //向 data 表中插入联系人的电话号码的操作
        val value3 = ContentValues()
        value3.put("raw_contact_id", rawContactsId)
        value3.put("mimetype", "vnd.android.cursor.item/phone_v2")
        value3.put("data1", phone)
        val dataPhoneUri: Uri? = resolver.insert(dataUri, value3)
        val dataPhoneId: Int = Integer.valueOf(dataPhoneUri!!.getPathSegments().get(1))
        if (rawContactsId > 0 && dataNameId > 0 && dataPhoneId > 0) { //判断是否添加成功
            Toast.makeText(this, "添加联系人成功", Toast.LENGTH_SHORT).show()
            listContacts()
        } else {
            Toast.makeText(this, "添加联系人失败或部分失败", Toast.LENGTH_SHORT).show()
        }
    }
}
//修改联系人
private fun updateContact(useContentProviderOperation: Boolean) {
    val id = mIdEditText.text.toString().trim { it <= ' ' }
    val name = mNameEditText.text.toString().trim { it <= ' ' }
    val phone = mPhoneEditText.text.toString().trim { it <= ' ' }
    //是否使用 ContentProviderOperation
    if (useContentProviderOperation) {
        val operations: ArrayList<ContentProviderOperation>= ArrayList()
        //根据联系人的 id 更新 raw_contacts 表的操作
        val op1 = ContentProviderOperation.newUpdate(rawContactsUri)
                .withValue("display_name", name)
                .withSelection("_id = ?", arrayOf(id))
                .build()
        operations.add(op1)
        //根据联系人的 id 更新 data 表
        val op2 = ContentProviderOperation.newUpdate(dataUri)
```

```
                .withValue("data1", phone)
                .withSelection("raw_contact_id = ? AND mimetype = 'vnd.android.cursor.item/phone_v2'", arrayOf(id))
                .build()
        operations.add(op2)
        try {
            val resolver = contentResolver
            //批量执行操作，并将结果返回
            val contentProviderResult = resolver.applyBatch("com.android.contacts", operations)
            //判断是否添加成功
            if (contentProviderResult.size == operations.size) {
                Toast.makeText(this, "修改联系人成功", Toast.LENGTH_SHORT).show()
                listContacts()
            } else {
                Toast.makeText(this, "修改联系人失败", Toast.LENGTH_SHORT).show()
            }
        } catch (e: OperationApplicationException) {
            e.printStackTrace()
            Toast.makeText(this, "修改联系人异常: $e", Toast.LENGTH_SHORT).show()
        } catch (e: RemoteException) {
            e.printStackTrace()
            Toast.makeText(this, "修改联系人异常: $e", Toast.LENGTH_SHORT).show()
        }
    } else {
        val resolver = contentResolver
        val values1 = ContentValues() //raw_contacts 表将要更新的数据
        values1.put("display_name", name)
        //根据联系人的 id 更新 raw_contacts 表
        val resUpdateName = resolver.update(rawContactsUri, values1, "_id = ?", arrayOf(id))
        val values2 = ContentValues() //data 表将要更新的数据
        values2.put("data1", phone)
        val resUpdatePhone = resolver.update(dataUri, values2, "raw_contact_id = ? AND mimetype = 'vnd.android.cursor.item/phone_v2'", arrayOf(id)) //根据联系人的 id 更新 data 表
        if (resUpdateName > 0 && resUpdatePhone > 0) {
            Toast.makeText(this, "修改联系人成功", Toast.LENGTH_SHORT).show()
            listContacts()
        } else {
            Toast.makeText(this, "修改联系人失败或部分失败", Toast.LENGTH_SHORT).show()
        }
    }
}
//删除联系人
private fun deleteContact(useContentProviderOperation: Boolean) {
    val idDelete = mIdEditText.text.toString()
    //是否使用 ContentProviderOperation
    if (useContentProviderOperation) {
        val operations: ArrayList<ContentProviderOperation>= ArrayList()
        //删除 raw_contacts 表中联系人数据的操作
        val op1 = ContentProviderOperation.newDelete(rawContactsUri)
                .withSelection("_id = ?", arrayOf(idDelete))
```

```
319                 .build()
320         operations.add(op1)
321         //删除 data 表中联系人数据的操作
322         val op2 = ContentProviderOperation.newDelete(dataUri)
323                 .withSelection("raw_contact_id = ?", arrayOf(idDelete))
324                 .build()
325         operations.add(op2)
326         try {
327             val resolver = contentResolver
328             //批量执行操作，并将结果返回
329             val contentProviderResult = resolver.applyBatch("com.android.contacts", operations)
331             if (contentProviderResult.size == operations.size) { //判断是否添加成功
332                 Toast.makeText(this, "删除联系人成功", Toast.LENGTH_SHORT).show()
333                 listContacts()
334             } else {
335                 Toast.makeText(this, "删除联系人失败", Toast.LENGTH_SHORT).show()
336             }
337         } catch (e: OperationApplicationException) {
338             e.printStackTrace()
339             Toast.makeText(this, "删除联系人异常: $e", Toast.LENGTH_SHORT).show()
340         } catch (e: RemoteException) {
341             e.printStackTrace()
342             Toast.makeText(this, "删除联系人异常: $e", Toast.LENGTH_SHORT).show()
343         }
344     } else {
345         val resolver = contentResolver
346         //删除 raw_contacts 表中联系人数据，返回值大于 0 说明有被删除的数据
347         val rawContactsDeleteResult = resolver.delete(rawContactsUri, "_id = ?", arrayOf(idDelete))
348         //删除 data 表中联系人数据，返回值大于 0 说明有被删除的数据
349         val dataDeleteResult = resolver.delete(dataUri, "raw_contact_id = ?", arrayOf(idDelete))
351         if (rawContactsDeleteResult > 0 && dataDeleteResult > 0) { //判断是否删除成功
352             Toast.makeText(this, "删除联系人成功", Toast.LENGTH_SHORT).show()
353             listContacts()
354         } else {
355             Toast.makeText(this, "删除联系人失败或部分失败", Toast.LENGTH_SHORT).show()
356         }
357     }
358 }
```

第 132 ~ 151 行查询系统通讯录中所有的联系人并显示出来。第 153 ~ 178 行根据联系人的名字查询系统通讯录中是否存在此人，如果存在则显示查询结果。第 186 ~ 221 行使用 ContentProviderOperation 方式向系统通讯录中添加联系人，这种方式便于理解系统联系人的数据表结构。第 223 ~ 249 行使用 ContentValues 的方式向系统通讯录中添加联系人，这种方式比较简洁。第 253 ~ 309 行 updateContact(Boolean) 方法使用 ContentProviderOperation 和 ContentValues 方式修改系统通讯录中的联系人。第 311 ~ 358 行 deleteContact(Boolean) 方法使用 ContentProviderOperation 和 ContentValues 方式删除系统通讯录中的联系人。

10.4　JavaScript 对象表示法

JSON(JavaScript Object Notation) 是一种轻量级的数据交换方法，是基于 JavaScript 的一个子集，

主要用于网络或程序之间传递数据。JSON 使用字符串形式保存数据，结构层次简洁、清晰，易于解析和生成，占用空间小，有效地提升网络传输效率。通常情况下，JSON 数据并不会直接呈现给用户，而是在后台与服务器端进行数据通信。

10.4.1 JSON 的数据结构

JSON 对象是 JSON 的基本构成单位，JSON 数组用于存储一类 JSON 对象集合（如表 10-12 所示）。JSON 对象表示为键值对，数据由逗号分隔，花括号保存对象，使用双引号保存键名和键值，键名和键值之间由冒号分隔。如果值是 String 类型且含有双引号或冒号，则需要使用“\”转义。

表 10-12 JSON 对象和 JSON 数组的实例对比

JSON 对象	JSON 数组
{"name": "小白魔", "content": "今天你吃了吗？"}	[{"name": "小白魔", "content": "今天你吃了吗？"}, {"name": "天使", "content": "关你什么事！"}]

XML（Extensible Markup Language）是可扩展标记语言，是标准通用标记语言的子集。XML 诞生于 1998 年，早于 2001 年诞生的 JSON。JSON 与 XML 相比（如表 10-13 所示），数据的描述性较差，数据可读性基本相同，数据的体积更小，传输与解析速度更快。

表 10-13 XML 与 JSON 数据实例对比

XML 数据	JSON 数据

多行形式：
<?xml version = "1.0" encoding = "utf-8"?>
<country>
<name>中国</name>
<province>
<name>辽宁</name>
<cities>
<city>沈阳</city>
<city>本溪</city>
</cities>
</province>
<province>
<name>新疆</name>
<cities>
<city>乌鲁木齐</city>
<city>喀什</city>
</cities>
</province>
</country>
	多行形式：
	{
	"name": "中国",
	"province": [
	{
	"name": "辽宁",
	"cities": {"city": ["沈阳", "本溪"]}
	},
	{
	"name": "新疆",
	"cities": {"city": ["乌鲁木齐", "喀什"]}
	}
	]
	}

单行形式： <?xml version = "1.0" encoding = "utf-8"?><country><name>中国</name><province><name>辽宁</name><cities><city>沈阳</city><city>本溪</city></cities></province><province><name>新疆</name><cities><city>乌鲁木齐</city><city>喀什</city></cities></province></country>	单行形式： {"name": "中国","province": [{"name": "辽宁","cities": {"city": ["沈阳", "本溪"]}}, {"name": "新疆","cities": {"city": ["乌鲁木齐", "喀什"]}}]}

JSON 有很多支持库，常见的有 GSON、FastJSON 和 Jackson。Android 自带的 JSON 库是 org.json，

可以直接使用，能够满足大多数的使用需求，比第三方库更加方便，提供了 JSONObject、JSONArray、JSONStringer 和 JSONTokener 四个类。实际工程中推荐直接使用 JSONObject、JSONArray 合成和解析 JSON，解析后赋给 Array 或 ArrayList 对象再进行其他操作。

10.4.2　JSONObject 类

JSONObject 类（org.json.JSONObject）用于新建、读取、解析和操作 JSON 对象，常用方法如表 10-14 所示。键名是非空字符串，键值可以是 JSONObject、JSONArray、String、Boolean、Integer、Long、Double 或 NULL 的任意组合。调用 putXXX（name,null）将从对象中移除键名，但是调用 putXXX（name, JSONObject.null）会存储 JSONObject.null 值，getXXX（）方法失败时返回 JSONObject.null。JSONArray（org.json.JSONArray）与 JSONObject 所提供的方法类似，没有提供 names（）方法，putXXX（）方法的返回值类型为 JSONArray。

表 10-14　JSONObject 类的常用方法

类型和修饰符	方　法
	<init>（） 主构造方法
	<init>（json: String） 主构造方法，通过 JSON 字符串实例化
open Any	get（name: String） 根据键名获取 Object 类型键值，如果不存在则抛出异常
open Boolean	getBoolean（name: String） 根据键名获取 Boolean 类型键值，如果不存在则抛出异常
open Double	getDouble（name: String） 根据键名获取 Double 类型键值，如果不存在则抛出异常
open Int	getInt（name: String） 根据键名获取 Int 类型键值，如果不存在则抛出异常
open JSONArray	getJSONArray（name: String） 根据键名获取 JSONArray 类型键值，如果不存在则抛出异常
open JSONObject	getJSONObject（name: String） 根据键名获取 JSONObject 类型键值，如果不存在则抛出异常
open Long	getLong（name: String） 根据键名获取 Long 类型键值，如果不存在则抛出异常
open String	getString（name: String） 根据键名获取 String 类型键值，如果不存在则抛出异常
open Boolean	has（name: String?） 判断是否包含该键名的数据
open Boolean	isNull（name: String?） 判断键值是否为 NULL
open Int	length（） 返回包含对象数量
open JSONArray?	names（） 返回包含键名的 JSONArray 对象
open Any?	opt（name: String?） 根据键名获取 String 类型键值，如果不存在则返回 null
open Boolean	optBoolean（name: String?） 根据键名获取 Boolean 类型或转换为 Boolean 类型后的键值，否则返回 fallback

续表

类型和修饰符	方 法
open Double	optDouble (name: String?) 根据键名获取 Double 类型或转换为 Double 类型后的键值，否则返回 fallback
open Int	optInt (name: String?) 根据键名获取 Int 类型或转换为 Int 类型后的键值，否则返回 fallback
open JSONArray?	optJSONArray (name: String?) 根据键名获取 JSONArray 类型的键值，否则返回 null
open JSONObject?	optJSONObject (name: String?) 根据键名获取 JSONObject 类型的键值，否则返回 null
open Long	optLong (name: String?, fallback: Long) 根据键名获取 Long 类型或转换为 Long 类型后的键值，否则返回 fallback
open String	optString (name: String?, fallback: String) 根据键名获取 String 类型或转换为 String 类型后的键值，否则返回 fallback
open JSONObject	put (name: String, value: Double) 将 Double 型键值存入 name 键名内，能够覆盖该键名之前存储的键值
open JSONObject	put (name: String, value: Boolean) 将 Boolean 型键值存入 name 键名内，能够覆盖该键名之前存储的键值
open JSONObject	put (name: String, value: Int) 将 Int 型键值存入 name 键名内，能够覆盖该键名之前存储的键值
open JSONObject	put (name: String, value: Long) 将 Long 型键值存入 name 键名内，能够覆盖该键名之前存储的键值
open JSONObject	put (name: String, value: Any?) 将 Object 型键值存入 name 键名内，能够覆盖该键名之前存储的键值
open JSONObject	putOpt (name: String?, value: Any?) 当两个参数都非空时，与 put (name，value) 等价；否则不执行任何操作
open Any?	remove (name: String?) 移除键值，否则不执行任何操作
open JSONArray?	toJSONArray (names: JSONArray?) 返回值与名称对应的数组
open String	toString () 转换为单行的 JSON 字符串
open String	toString (indentSpaces: Int) 转换为多行的 JSON 字符串，indentSpaces 表示每层嵌套缩进的空格数

10.4.3 实例工程：合成和解析 JSON 数据

本实例模拟了发布朋友圈或微博动态前合成 JSON（如图 10-10 所示），实例中没有演示合成后数据发往服务器端的过程。从服务器端获取到朋友圈或微博动态的 JSON 数据后，对 JSON 数据进行解析后显示，实例中没有演示从服务器端获取 JSON 数据的过程，而是使用合成后的数据直接进行解析。单击“合成单行”或“合成多行”按钮，会在下方显示合成后的单行或多行 JSON 字符串。单击“清空”按钮，清空顶部 EditText 控件的内容。单击“解析”按钮，将合成的 JSON 数据解析后显示在顶部相应的 EditText 控件中。

1. 打开基础工程

打开“基础工程”文件夹中的“C1006”工程，该工程已经包含 MainActivity 及资源文件。

图 10-10　运行效果

2. 主界面的 Activity

/java/com/teachol/c1006/MainActivity.kt

```
//合成 JSON
private fun encodeJSON(isSingle: Boolean) {
    val jsonObject = JSONObject()
    try {
        //将昵称和内容存入 jsonObject 对象中
        jsonObject.put("name", mNameEditText.text)
        jsonObject.put("content", mContentEditText.text)
        //将图像分别存入新建的 JSONObject 类对象中
        val pic0JsonObject = JSONObject()
        pic0JsonObject.put("type", "pic")
        pic0JsonObject.put("url", mExtraEditText[0]!!.text)
        val pic1JsonObject = JSONObject()
        pic1JsonObject.put("type", "pic")
        pic1JsonObject.put("url", mExtraEditText[1]!!.text)
        val pic2JsonObject = JSONObject()
        pic2JsonObject.put("type", "pic")
        pic2JsonObject.put("url", mExtraEditText[2]!!.text)
        //将存入图像的三个 JSONObject 类对象存入 jsonArray 对象中
        val jsonArray = JSONArray()
        jsonArray.put(pic0JsonObject)
        jsonArray.put(pic1JsonObject)
        jsonArray.put(pic2JsonObject)
        //将保存图像存入 jsonObject 对象
        jsonObject.put("extra", jsonArray)
        //判断是单行还是多行显示
        if (isSingle) {
            mResultTextView.text = jsonObject.toString()
        } else {
            mResultTextView.text = jsonObject.toString(4)
        }
    } catch (e: JSONException) {
```

```
78            e.printStackTrace()
79        }
80    }
82    private fun decodeJSON() { //解析 JSON
83        try {
84            val jsonObject = JSONObject(mResultTextView.text.toString().trim { it <= ' ' })
86            val name = jsonObject.getString("name") //获取 name 键值的数据
87            mNameEditText.setText(name)
89            val content = jsonObject.getString("content") //获取 content 键值的数据
90            mContentEditText.setText(content)
92            val jsonArray = jsonObject.getJSONArray("extra") //获取 extra 键值的数据
94            for (i in 0 until jsonArray.length()) { //遍历 extra 存储的 JSONArray 数据
95                val extraJSONObject = jsonArray.getJSONObject(i)
96                mExtraEditText[i]!!.setText(extraJSONObject.optString("url", "images/default.jpg"))
97            }
98        } catch (e: JSONException) {
99            e.printStackTrace()
100       }
101   }
103   override fun onClick(v: View) { //单击事件
104       when (v.id) {
105           R.id.button_encode_json_single -> encodeJSON(true)
106           R.id.button_encode_json_multi -> encodeJSON(false)
107           R.id.button_decode_json -> decodeJSON()
108           R.id.button_empty -> {
109               mNameEditText.setText("")
110               mContentEditText.setText("")
111               mExtraEditText[0]!!.setText("")
112               mExtraEditText[1]!!.setText("")
113               mExtraEditText[2]!!.setText("")
114           }
115       }
116   }
```

第 52、53 行将 name 和 content 元素存储在 jsonObject 对象中。第 55 ~ 63 行新建 pic0JsonObject、pic1JsonObject 和 pic2JsonObject 对象，分别存储三组 type 和 url 元素。第 65 ~ 68 行新建 jsonArray 对象，然后将 pic0JsonObject、pic1JsonObject 和 pic2JsonObject 存入。第 70 行将 jsonArray 对象存入 jsonObject 对象的 extra 元素中。第 86 行 jsonObject.getString("name") 解析出 String 类型的 name 元素。第 92 ~ 97 行通过 jsonObject.getJSONArray("extra") 解析出 JSONArray 类型的 extra 元素，并遍历 extra 元素中的所有子元素。

第 11 章　Android 的多媒体与传感器

除通过触摸屏、网络可以与外部进行交互外，还可以使用摄像头、麦克风和传感器获取外部的信息。从本质上来看，摄像头和麦克风也是一种传感器。但是 Android 将两者独立出来，不仅因为它们功能丰富，还因为它们是用户最能直接感知到的传感器。因此，Android 系统还提供了独立的两个 App——相机和录音机。

11.1　系统相机和相册

Android 系统中预置了相机和相册两个 App，可以通过 Intent 对象调用这两个 App，进行拍照、录制视频、选取图片和选取视频。微信、微博、小红书、抖音、大众点评等并没有使用内置的功能选取照片或视频，而是先通过 ContentProvider 获取相册中的照片数据并重新显示出来，再实现多选的功能，甚至可以实现添加标签和编辑功能。

11.1.1　实例工程：拍照、选取和显示图片

本实例演示了使用系统相机 App 拍摄照片及使用系统相册选取图片。首次运行时会弹出权限请求的系统对话框，单击“拒绝”按钮会弹出自定义的对话框，提示需要开启的权限，再单击“打开应用设置”按钮(如图 11-1 所示)。单击“拍照”按钮会调用系统相机，拍照后返回的拍摄照片画质较低，不适合实际使用；单击“拍照(指定位置保存)”按钮会调用系统相机，拍照后返回的拍摄照片画质正常，保存在指定文件夹中，并且能够显示在系统相册中；单击“选取图片”按钮会调用系统相册选取照片，选取照片后调用剪裁的功能，进行剪裁后显示方形图像(如图 11-2 所示)。调用系统相机拍照不会自动将照片保存显示在系统相册中，而是需要创建保存照片的文件，扫描媒体文件并发送广播才会显示在系统相册中。

图 11-1　权限请求和手动设置权限的运行效果

图 11-2　拍照和选取图片的运行效果

1. 打开基础工程

打开“基础工程”文件夹中的“C1101”工程，该工程已经包含 MainActivity 及资源文件。

2. Util 类

在“/java/com/teachol/c1101”文件夹中，新建 Util 类，用于存储几种能够重复使用的公共方法。

/java/com/teachol/c1101/Util.kt

```
10  object Util {
12      fun createFile(parentPath: File?, childPath: String, extension: String): File { //创建文件
14          val format = SimpleDateFormat("yyyyMMdd_hhmmss") //使用时间生成文件名
15          val fileName: String = format.format(Date())
17          val storageDir = File(parentPath, childPath) //存储路径
18          if (!storageDir.exists()) { //判断文件夹是否存在
19              storageDir.mkdirs() //创建文件夹
20          }
22          val saveRecorderFile = File(storageDir, "IMG_$fileName.$extension") //创建文件
23          if (saveRecorderFile.exists()) { //判断文件是否存在
24              saveRecorderFile.delete() //创建文件
25          }
26          return saveRecorderFile
27      }
29      fun showInAlbum(context: Context, path: String) { //将图片或视频显示在系统相册中
30          //Android 中分割字符串需要在分隔符左右两侧加上方括号
31          val str = path.split("[.]".toRegex()).toTypedArray()
32          //获取扩展名对应的文件类型值
33          val mimeType = MimeTypeMap.getSingleton().getMimeTypeFromExtension(str[str.size - 1])
34          //根据路径和扩展名的类型扫描媒体文件
35          MediaScannerConnection.scanFile(context, arrayOf(path), arrayOf(mimeType), null)
36      }
37  }
```

第 12 行 createFile()方法用于创建存储照片的文件，设置了三个参数，parentPath 参数表示存储新建文件的父路径，childPath 参数表示存储新建文件的子路径，extension 参数表示新建文件的扩展名。第 17 行为了增加控制的灵活性，使用父路径和子路径合成存储照片的文件夹路径，即 parentPath 和

childPath。第 22 行创建保存图片的文件对象，文件名的前缀为“IMG_”。第 29 行 showInAlbum 方法用于将图片显示在系统相册中，设置了两个参数，context 参数表示上下文，path 参数表示图片的路径。第 33 行 str[str.size - 1]获取文件的扩展名，MimeTypeMap.getSingleton().getMimeTypeFromExtension()方法获取扩展名对应的 MIME 类型。

提示：MIME

MIME(Multipurpose Internet Mail Extensions，多用途互联网邮件扩展)类型用来表示图片、视频、音频、文件或字节流的性质和格式，最初用于电子邮件。标准格式为 type/subtype，不区分英文字母大小写，type 表示类别，subtype 表示类别的细分类型。如果 subtype 使用*，则表示可以是 type 的所有细分类型。常用的 MIME 类型包括 text/html、image/jpeg、image/png、video/mp4、audio/mp3、application/json 等。

3. Permissions 类

在“/java/com/teachol/c1101”文件夹中，新建 Permissions 类，用于封装权限申请，便于重复使用。

/java/com/teachol/c1101/Permissions.kt

```
object Permissions {
    const val REQUEST_PERMISSIONS = 1 //权限请求码
    //是否已经获取权限
    fun hasPermissionsGranted(context: Context?, permissions: Array<String?>): Boolean {
        for (permission in permissions) { //只要有一个未允许申请权限就返回 false
            if (ActivityCompat.checkSelfPermission(context!!, permission!!) != PackageManager.PERMISSION_GRANTED) {
                return false
            }
        }
        return true
    }
    //请求权限
    fun requestPermissions(context: Context?, permissions: Array<String?>?) {
        //如果已经拒绝了权限申请，则弹出对话框提示手动开启权限；否则显示权限请求的系统对话框
        if (shouldShowRequestPermissionRationale(context, permissions!!)) {
            var msg = ""
            for (i in permissions.indices) {
                msg = if (i == 0) {permissions[i]!!} else {"$msg\n${permissions[i]}".trimIndent()}
            }
            //自定义对话框
            val adBuilder: AlertDialog.Builder = AlertDialog.Builder(context)
            adBuilder.setIcon(R.mipmap.ic_launcher)
            adBuilder.setTitle("需要手动开启以下权限")
            adBuilder.setMessage(msg)
            //单击确认按钮事件
            adBuilder.setPositiveButton("打开应用设置",
                DialogInterface.OnClickListener { dialog, which -> //打开应用设置
                    val intent = Intent()
                    intent.action = "android.settings.APPLICATION_DETAILS_SETTINGS"
                    intent.data = Uri.fromParts("package", context!!.getPackageName(), null)
                    context.startActivity(intent)
                    dialog.dismiss()
```

```
46                })
47            adBuilder.show()
48        } else {
50            ActivityCompat.requestPermissions((context as Activity), permissions, REQUEST_PERMISSIONS) //显示权限请求的系统对话框
51        }
52    }
53    //获取是否拒绝过权限请求
54    fun shouldShowRequestPermissionRationale(context: Context?, permissions: Array<String?>): Boolean {
55        for (permission in permissions) { //如果有拒绝的申请权限返回 true
57            if (ActivityCompat.shouldShowRequestPermissionRationale((context as Activity?)!!, permission!!)) {
58                return true
59            }
60        }
61        return false
62    }
63 }
```

第 16 ~ 24 行判断 permissions 参数中是否包含未允许的权限，只要包含任意一个未允许的权限就会返回 false。第 19 行 PackageManager.PERMISSION_GRANTED 表示允许的权限。第 26 ~ 52 行请求权限，如果 permissions 参数中有被用户拒绝过的权限，会弹出对话框提示手动开启权限；否则，显示所有权限请求的系统对话框。第 54 ~ 62 行判断 permissions 参数中是否包含被用户拒绝的权限请求。

4. FileProvider 的权限路径

在“/res”文件夹中，新建“xml”文件夹。然后在“/res/xml”文件夹中，新建“file_paths.xml”文件，用于设置 FileProvider 的权限路径。

```
/res/xml/file_paths.xml
02 <paths>
03     <external-media-path name = "media_images" path = "C1101"/>
04 </paths>
```

第 03 行所表示的路径是 Context.externalMediaDirs[0].absolutePath+"images"，该路径下的文件夹可以添加到系统相册中进行显示。external-media-path 表示 Context.externalMediaDirs[0].absolutePath，是扩展媒体目录的绝对路径。name 属性表示引用名称。path 属性表示子文件夹的名称。如果要在系统相册中显示出媒体文件，需要将路径设置为 external-media-path，path 属性，作为相册内文件夹的名称。

提示：FileProvider

Android 7.0（API Level 24）禁止对外部 App 公开 file://路径格式的内部文件，否则会抛出 FileUriExposedException 异常。需要使用 FileProvider 类生成 content://类型的 URI 供外部 App 访问，并且需要为其提供临时的文件访问权限。

在 AndroidManifest.xml 文件中使用<provider>标签声明供外部 App 访问路径，具体的路径保存在 XML 文件中。在 XML 文件中使用<paths>作为顶层标签，<paths>子标签指定文件夹。<paths>子标签的 name 属性是 path 属性指定路径的替代名称，用于隐藏真实的目录；<paths>子标签的 path 属性表示指定路径下的共享目录。常用的<paths>子标签如下。

- <files-path>: Context.filesDir 所指向的目录。
- <cache-path>: Context.cacheDir 所指向的目录。
- <external-path>: Environment.externalStorageDirectory 所指向的目录。
- <external-files-path>: Context.externalFilesDir 所指向的目录。
- <external-cache-path>: Context.externalCacheDir 所指向的目录。
- <external-media-path>: Context.externalMediaDirs 所指向的目录(以 API Level 21 开始支持)。

5. 设置权限和 FileProvider 的权限路径

```
/manifests/AndroidManifest.xml
03 <uses-permission android:name = "android.permission.CAMERA" />
04 <uses-permission android:name = "android.permission.READ_EXTERNAL_STORAGE" />
05 <uses-permission android:name = "android.permission.WRITE_EXTERNAL_STORAGE" />
19    <provider
20       android:name = "androidx.core.content.FileProvider"
21       android:authorities = "com.teachol.c1101.fileprovider"
22       android:exported = "false"
23       android:grantUriPermissions = "true">
24       <meta-data
25          android:name = "android.support.FILE_PROVIDER_PATHS"
26          android:resource = "@xml/file_paths"></meta-data>
27    </provider>
```

第 03 ~ 05 行设置摄像头和外部存储读写权限。第 19 ~ 27 行设置 FileProvider 的相关标签属性，android:authorities = "com.teachol.c1101.fileprovider" 设置授权字符串，android:resource = "@xml/file_paths"指定设置权限文件夹的文件，这样可以将调用系统相机拍摄的照片保存到当前 App 的缓存文件夹中。

6. 主界面的 Activity

```
/java/com/teachol/c1101/MainActivity.kt
class MainActivity : AppCompatActivity(), View.OnClickListener {
   companion object {
      private const val REQUEST_CODE_TAKE_PHOTO_DEFAULT = 0
      private const val REQUEST_CODE_TAKE_PHOTO_CUSTOM = 1
      private const val REQUEST_CODE_SELECT = 2
      private const val REQUEST_CODE_CROP = 3
      private val PERMISSIONS = arrayOf<String?>( Manifest.permission.CAMERA, Manifest.permission.READ_EXTERNAL_STORAGE, Manifest.permission.WRITE_EXTERNAL_STORAGE) //需要的权限
   }
   private val mContext: Context = this
   private lateinit var mPhotoFile: File
   override fun onCreate(savedInstanceState: Bundle?) {
      findViewById<View>(R.id.button_take_photo_default).setOnClickListener(this)
      findViewById<View>(R.id.button_take_photo_custom).setOnClickListener(this)
      findViewById<View>(R.id.button_select_photo).setOnClickListener(this)
      if (!hasPermissionsGranted(mContext, PERMISSIONS)) { //判断是否已经取得权限
         Permissions.requestPermissions(mContext, PERMISSIONS) //请求权限
      }
   }
   override fun onRequestPermissionsResult(requestCode: Int, permissions: Array<String>, grantResults: IntArray) { //请求权限的回调
```

```
        when (requestCode) {
            Permissions.REQUEST_PERMISSIONS -> if (!hasPermissionsGranted(mContext, PERMISSIONS)) {
                Permissions.requestPermissions(mContext, PERMISSIONS)
            }
        }
    }
    override fun onClick(v: View) { //单击事件
        when (v.id) {
            R.id.button_take_photo_default -> takePhotoDefault()
            R.id.button_take_photo_custom -> takePhotoCustom()
            R.id.button_select_photo -> selectPhoto()
        }
    }
    private fun takePhotoDefault() { //系统相机拍照，使用默认保存路径
        val intent = Intent(MediaStore.ACTION_IMAGE_CAPTURE)
        startActivityForResult(intent, REQUEST_CODE_TAKE_PHOTO_DEFAULT) //调用系统相机拍照
    }
    private fun takePhotoCustom() { //系统相机拍照，照片保存到指定路径
        mContext.filesDir
        mPhotoFile = createFile(mContext.externalCacheDir, mContext.resources.getString(R.string.app_name), "jpg") //创建文件
        val photoURI: Uri = FileProvider.getUriForFile(this, "com.teachol.c1101.fileprovider", mPhotoFile) //获取"com.teachol.c1101.fileprovider"授权路径下的mPhotoFile文件的Uri
        val intent = Intent(MediaStore.ACTION_IMAGE_CAPTURE)
        intent.putExtra(MediaStore.EXTRA_OUTPUT, photoURI) //设置保存文件的Uri地址
        startActivityForResult(intent, REQUEST_CODE_TAKE_PHOTO_CUSTOM) //调用系统相机拍照
    }
    private fun selectPhoto() { //调用系统相册选择图片
        val intent = Intent(Intent.ACTION_PICK, MediaStore.Images.Media.EXTERNAL_CONTENT_URI)
        intent.type = "image/*"
        startActivityForResult(intent, REQUEST_CODE_SELECT)
    }
    override fun onActivityResult(requestCode: Int, resultCode: Int, intent: Intent?) { //回调
        super.onActivityResult(requestCode, resultCode, intent)
        if (requestCode == REQUEST_CODE_TAKE_PHOTO_DEFAULT && resultCode == RESULT_OK) {
            //从data中取出传递回来缩略图的信息，图片质量差，适合传递小图片
            val bundle = intent!!.extras
            val bitmap = bundle!!.getParcelable<Bitmap>("data")
            (findViewById<View>(R.id.image_view) as ImageView).setImageBitmap(bitmap) //显示照片
        } else if (requestCode == REQUEST_CODE_TAKE_PHOTO_CUSTOM && resultCode == RESULT_OK) {
            (findViewById<View>(R.id.image_view) as ImageView).setImageURI(Uri.fromFile(mPhotoFile)) //显示照片
            showInAlbum(mContext, mPhotoFile.absolutePath) //扫描更新相册
        } else if (requestCode == REQUEST_CODE_SELECT && resultCode == RESULT_OK) {
            mPhotoFile = createFile(mContext.externalCacheDir, mContext.resources.getString(R.string.app_name), "jpg") //新建保存裁剪照片的文件
            val cropIntent = Intent("com.android.camera.action.CROP") //裁剪照片
            cropIntent.setDataAndType(intent!!.data, "image/*")
            cropIntent.putExtra("crop", "true")
            cropIntent.putExtra("aspectX", 1) //设置裁剪的比例
```

```
107             cropIntent.putExtra("aspectY", 1)
109             cropIntent.putExtra("outputX", 500) //设置裁剪后保存图片的尺寸
110             cropIntent.putExtra("outputY", 500)
111             //设置保存图片的文件格式
112             cropIntent.putExtra("outputFormat", Bitmap.CompressFormat.JPEG.toString())
114             cropIntent.putExtra(MediaStore.EXTRA_OUTPUT, Uri.parse("file://${mPhotoFile.absolutePath}"))
    //设置保存图片的文件路径
115             cropIntent.putExtra("return-data", false)
116             startActivityForResult(cropIntent, REQUEST_CODE_CROP)
118             showInAlbum(mContext, mPhotoFile.absolutePath) //扫描更新相册
119         } else if (requestCode == REQUEST_CODE_CROP && resultCode == RESULT_OK) {
121             (findViewById<View>(R.id.image_view) as ImageView).setImageURI(Uri.fromFile
    (mPhotoFile)) //显示照片
122         }
123     }
124 }
```

第 39～42 行判断是否已经获取授权，如果首次运行时没有获取所有授权，或请求的权限在应用设置中设置为“询问”，则弹出系统授权对话框。第 45～51 行重写处理系统授权对话框对权限选择结果的回调，如果拒绝了权限申请，则出自定义对话框提示用户手动开启权限。第 63 行 MediaStore.ACTION_IMAGE_CAPTURE 表示拍照的行为，拍照时可以选择系统相机或第三方相机 App。第 72 行生成的 Uri 格式为“content://com.teachol.c1101.fileprovider/media_images/”+文件名称，如 content://com.teachol.c1101.fileprovider/media_images/IMG_20200213_010345.jpg。第 80 行 Intent.ACTION_PICK 表示选取的行为，MediaStore.Images.Media.EXTERNAL_CONTENT_URI 表示提供图片的 Uri。第 100 行 mContext.externalCacheDir 用于获取扩展缓存文件夹的路径，mContext.resources.getString(R.string.app_name)用于获取 App 的名称。第 102～118 行在选择图片后调用系统的图片剪裁功能，并将剪裁后图片保存。第 114 行设置剪裁文件保存的文件路径，使用 Uri.parse()方法将"file://"形式的路径转换为 Uri。

7. 测试运行

运行工程后，单击“拍照(指定位置保存)”按钮调用系统相机，拍摄照片后会在系统相册中显示出来，如图 11-3 所示。照片保存的位置可以通过 Device File Explorer 窗口查看，如图 11-4 所示。

图 11-3　拍照后在指定位置保存并显示在系统相册中

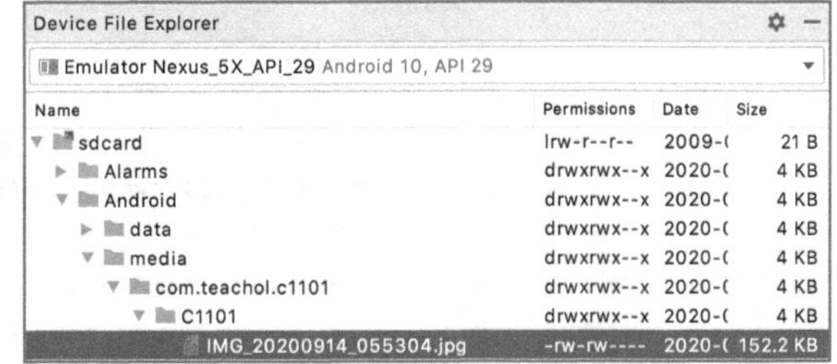

图 11-4　照片保存的位置

11.1.2 实例工程：录制、选取和播放视频

本实例演示了使用系统相机 App 录制视频及使用系统相册选取视频，使用 VideoView 控件播放视频，录制的视频会自动保存到系统相册中。使用系统相机录制视频同样需要获取相应的权限，单击“录制视频”按钮调用系统相机可以录制 15 秒内的视频，单击“选取视频”按钮调用系统相册选取视频后在 VideoView 控件中播放，单击 VideoView 控件可以显示出播放控制器(如图 11-5 所示)。

图 11-5　运行效果

1. 打开基础工程

打开“基础工程”文件夹中的“C1102”工程，该工程已经包含 MainActivity、Permissions 及资源文件。Permissions 类与 C1101 工程的 Permissions 类相同，是可以直接复制过来进行重复使用的。activity_layout.xml 布局文件中包含一个 VideoView 控件用于播放视频。

2. 主界面的 Activity

```
/java/com/teachol/c1102/MainActivity.kt
fun recordVideo() { //调用系统相机录制视频
    val intent = Intent(MediaStore.ACTION_VIDEO_CAPTURE)
    intent.putExtra(MediaStore.EXTRA_VIDEO_QUALITY, 1) //设置视频质量(1 为高画质)
    intent.putExtra(MediaStore.EXTRA_DURATION_LIMIT, 15) //设置最大时长为 15 秒
    startActivityForResult(intent, REQUEST_CODE_RECORD)
}
fun selectVideo() { //调用系统相册选择视频
    val intent: Intent? = Intent(Intent.ACTION_PICK, MediaStore.Video.Media.EXTERNAL_CONTENT_URI)
    startActivityForResult(intent, REQUEST_CODE_SELECT)
}
override fun onActivityResult(requestCode: Int, resultCode: Int, intent: Intent?) { //回调
    super.onActivityResult(requestCode, resultCode, intent)
    if ((requestCode == REQUEST_CODE_RECORD || requestCode == REQUEST_CODE_SELECT) &&
resultCode == Activity.RESULT_OK) {
        val videoUri: Uri? = intent?.getData()
        val videoView: VideoView = findViewById(R.id.video_view)
        videoView.setVideoURI(videoUri)
        videoView.setMediaController(android.widget.MediaController(this))
        videoView.setOnPreparedListener(object : OnPreparedListener {
```

```
68                override fun onPrepared(mp: MediaPlayer?) {
69                    videoView.start() //播放录制的视频
70                }
71            })
72        }
73    }
```

第 49 行 MediaStore.ACTION_VIDEO_CAPTURE 是录制视频的行为。第 50 行设置录制视频的画质，1 代表高画质，0 代表低画质。第 51 行设置录制视频的最长时间，时间单位是秒，录制界面会出现倒计时的数字提示。第 56 行 Intent.ACTION_PICK 表示选取的行为，android.provider.MediaStore.Video.Media.EXTERNAL_CONTENT_URI 表示提供视频的 URI。第 63 行 intent?.getData()用于获取选取或录制的视频 URI。第 65 行设置 VideoView 控件播放的视频 URI。第 66 行为 VideoView 控件设置媒体控制器，当 VideoView 控件播放视频时，单击视频会出现媒体控制器，再次单击媒体控制器隐藏。第 67～71 行设置视频准备完毕事件的监听器，重写 onPrepared(MediaPlayer?)方法，当视频准备完毕后播放。

11.2　拍摄照片和录制视频

调用系统相机虽然能够拍摄照片和录制视频，但是所提供的功能有限且灵活性差。使用 API 提供的类直接控制摄像头更加灵活，可以编写出自定义功能的相机。目前 Android 提供的摄像头 API 共有三个版本，分别是 Camera 类、Camera2 类和 CameraX 类，使用这三个类都需要在 AndroidManifest.xml 文件中设置 Manifest.permission.camera 权限，官方推荐使用 Camera2 类。因为 Camera 类在 API Level 21 时已过时，CameraX 类目前还是 alpha 版本，官方暂时不推荐在实际工程中使用。CameraX 类是为了解决 Camera2 类的复杂性和硬件设备的兼容性，很容易实现人像、HDR、夜间模式和美颜等效果，是未来使用摄像头的 API。

11.2.1　Camera2 组件

Camera2 组件(android.hardware.camera2)包含涉及摄像头管理和使用的类。Camera2 包所提供的功能比 Camera 包要丰富得多，增加了使用的复杂度。Camera2 包将摄像头作为管道，该管道接收输入请求以捕获单个帧，每个请求捕获单个图像，然后输出一个捕获结果元数据包及该请求的一组输出图像缓冲区。请求需要按顺序处理，多个请求可以一次运行。

1. CameraManager 类

CameraManager 类(android.hardware.camera2.CameraManager)用于检测、描述、查询和连接可用的摄像头，常用方法如表 11-1 所示。需要使用 Context.getSystemService(String)或者 Context.getSystemService(Class<T>)方法获取该实例。

表 11-1　CameraManager 类的常用方法

类型和修饰符	方　法
CameraCharacteristics	getCameraCharacteristics(cameraId: String) 查询摄像头的功能
Array<String!>	getCameraIdList() 获取可用的摄像头 id 列表

续表

类型和修饰符	方　法
Unit	openCamera (cameraId: String, callback: CameraDevice.StateCallback, handler: Handler?) 打开摄像头
Unit	registerAvailabilityCallback (callback: CameraManager.AvailabilityCallback, handler: Handler?) 注册摄像头优先级和可用性的回调
Unit	registerTorchCallback (callback: CameraManager.TorchCallback, handler: Handler?) 注册闪光灯的回调
Unit	setTorchMode (cameraId: String, enabled: Boolean) 设置闪光灯的闪光灯模式
Unit	unregisterAvailabilityCallback (callback: CameraManager.AvailabilityCallback) 注销摄像头优先级和可用性的回调
Unit	unregisterTorchCallback (callback: CameraManager.TorchCallback) 注销闪光灯的回调

2. CameraManager.AvailabilityCallback 类

CameraManager.AvailabilityCallback 类（android.hardware.camera2.CameraManager.AvailabilityCallback）用于处理摄像头的优先级改变或可用性改变的回调，常用方法如表 11-2 所示。需要使用 CameraManager 类的 registerAvailabilityCallback (CameraManager.AvailabilityCallback, Handler?) 方法和 unregister AvailabilityCallback (CameraManager.AvailabilityCallback) 方法进行注册和注销。

表 11-2　CameraManager.AvailabilityCallback 类的常用方法

类型和修饰符	方　法
	<init>() 主构造方法
open Unit	onCameraAccessPrioritiesChanged () 摄像头访问优先级改变时的回调方法
open Unit	onCameraAvailable (cameraId: String) 摄像头可用时的回调方法
open Unit	onCameraUnavailable (cameraId: String) 当前可用的摄像头不可用时的回调方法
open Unit	onPhysicalCameraAvailable (cameraId: String, physicalCameraId: String) 物理摄像头可用时的回调方法
open Unit	onPhysicalCameraUnavailable (cameraId: String, physicalCameraId: String) 当前可用的物理摄像头不可用时的回调方法

3. CameraManager.TorchCallback 类

CameraManager.TorchCallback 类（android.hardware.camera2.CameraManager.TorchCallback）用于闪光灯模式可用性改变时或不可用时的回调，常用方法如表 11-3 所示，需要使用 CameraManager 类的 registerTorchCallback (CameraManager.TorchCallback, Handler?) 方法和 unregisterTorchCallback (CameraManager.TorchCallback) 方法进行注册和注销。

表 11-3　CameraManager.TorchCallback 类的常用方法

类型和修饰符	方　法
	<init>() 主构造方法

续表

类型和修饰符	方　法
open Unit	onTorchModeChanged(cameraId: String, enabled: Boolean) 摄像头闪光灯模式改变时的回调
open Unit	onTorchModeUnavailable(cameraId: String) 摄像头闪光灯模式不可用时的回调

4. CameraDevice 类

CameraDevice 类(android.hardware.camera2.CameraDevice)用于摄像头对图像捕捉的请求和会话进行处理，常用方法如表 11-4 所示。

表 11-4　CameraDevice 类的常用方法

类型和修饰符	方　法
abstract Unit	close() 关闭与摄像头的连接
open CaptureRequest.Builder	createCaptureRequest(templateType: Int, physicalCameraIdSet: MutableSet<String!>!) 创建一个捕捉请求。templateType 参数表示模版类型，使用以下常量来表示。 ● CameraDevice.TEMPLATE_MANUAL：手动模式 ● CameraDevice.TEMPLATE_PREVIEW：摄像头预览模式 ● CameraDevice.TEMPLATE_RECORD：视频录制模式 ● CameraDevice.TEMPLATE_STILL_CAPTURE：拍照模式 ● CameraDevice.TEMPLATE_VIDEO_SNAPSHOT：录制视频时的拍照模式 ● CameraDevice.TEMPLATE_ZERO_SHUTTER_LAG：零快门滞后仍然捕获模式
open CaptureRequest.Builder	createCaptureRequest(templateType: Int) 创建捕捉请求。templateType 参数与 createCaptureRequest(Int, MutableSet<String!>!)方法的 templateType 参数相同
open Unit	createCaptureSession(config: SessionConfiguration!) 创建捕捉会话
abstract Unit	createCaptureSession(outputs: MutableList<Surface!>, callback: CameraCaptureSession.StateCallback, handler: Handler?) 创建捕捉会话
open Int	getCameraAudioRestriction() 获取摄像机音频限制模式
abstract String	getId() 获取摄像头的 id
open Boolean	isSessionConfigurationSupported(sessionConfig: SessionConfiguration) 检查相机设备是否支持特定会话配置
open Unit	setCameraAudioRestriction(mode: Int) 设置摄像机音频限制模式

5. CameraDevice.StateCallback 类

CameraDevice.StateCallback 类(android.hardware.camera2.CameraDevice.StateCallback)用于对摄像头的开启、关闭、断开连接和报错时的回调，常用方法如表 11-5 所示。使用 CameraManager.openCamera(String, CameraDevice.StateCallback, Handler?)方法打开摄像头后，才能处理回调。

表 11-5　CameraDevice.StateCallback 类的常用方法

类型和修饰符	方　法
open Unit	onClosed(camera: CameraDevice) 关闭摄像头时调用该方法

续表

类型和修饰符	方　法
abstract Unit	onDisconnected (camera: CameraDevice) 摄像头断开连接不再可用时调用该方法
abstract Unit	onError (camera: CameraDevice, error: Int) 摄像头遇到错误时调用该方法。error 参数表示错误类型，使用以下常量来表示。 ● CameraDevice.StateCallback.ERROR_CAMERA_DEVICE：摄像头硬件错误 ● CameraDevice.StateCallback.ERROR_CAMERA_DISABLED：摄像头不可用 ● CameraDevice.StateCallback.ERROR_CAMERA_IN_USE：摄像头正在使用 ● CameraDevice.StateCallback.ERROR_CAMERA_SERVICE：摄像头服务错误 ● CameraDevice.StateCallback.ERROR_MAX_CAMERAS_IN_USE：到达最大使用数量
abstract Unit	onOpened (camera: CameraDevice) 摄像头被打开后调用该方法

6. CameraCaptureSession 类

CameraCaptureSession 类(android.hardware.camera2.CameraCaptureSession)用于设置和管理捕捉图像会话，常用方法如表 11-6 所示。使用 CameraDevice 的 createCaptureSession()方法创建的捕捉图像会话属于异步操作，因为需要配置摄像头的内部管道，并分配内存缓冲区，以便将捕捉的图像发送到所需的目标。捕捉图像会话创建后就会处于活动状态，并激活 CameraCaptureSession.StateCallback 的回调方法。

表 11-6　CameraCaptureSession 类的常用方法

类型和修饰符	方　法
	<init>() 主构造方法
abstract Unit	abortCaptures() 丢弃当前挂起和正在进行的所有捕获图像会话
abstract Int	capture(request: CaptureRequest, listener: CameraCaptureSession.CaptureCallback?, handler: Handler?) 提交捕捉图像请求的会话
abstract Int	captureBurst(List<CaptureRequest> requests, CameraCaptureSession.CaptureCallback listener, Handler handler) 提交捕捉图像序列请求的会话，序列按照顺序执行
open Int	captureBurstRequests(requests: MutableList<CaptureRequest!>, executor: Executor, listener: CameraCaptureSession.CaptureCallback) 提交捕捉图像序列请求的会话
open Int	captureSingleRequest(request: CaptureRequest, executor: Executor, listener: CameraCaptureSession.CaptureCallback) 提交单个捕捉图像请求的会话
abstract Unit	close() 异步关闭此捕捉图像会话
abstract CameraDevice	getDevice() 获取为此会话创建的摄像头设备
abstract Surface?	getInputSurface() 获取输入捕捉图像会话关联的 Surface
abstract Boolean	isReprocessable() 判断是否可以重新提交捕捉请求的会话
abstract Unit	prepare(surface: Surface) 准备捕捉，为用于捕捉图像输出的 Surface 预先分配缓冲区
abstract int	setRepeatingBurst(requests: MutableList<CaptureRequest!>, listener: CameraCaptureSession.CaptureCallback?, handler: Handler?) 设置重复捕捉图像的序列

续表

类型和修饰符	方　法
Int	setRepeatingBurstRequests (List<CaptureRequest> requests, Executor executor, CameraCaptureSession.CaptureCallback listener) 设置重复捕捉图像的序列请求
abstract Int	setRepeatingRequest (request: CaptureRequest, listener: CameraCaptureSession.CaptureCallback?, handler: Handler?) 设置重复捕捉图像的请求
open Int	setSingleRepeatingRequest (request: CaptureRequest, executor: Executor, listener: CameraCaptureSession.CaptureCallback) 设置单个重复捕捉图像的请求
abstract Unit	stopRepeating () 取消使用 setRepeatingRequest () 方法或 setRepeatingBurst () 方法设置的重复捕捉图像请求
open Unit	updateOutputConfiguration (config: OutputConfiguration!) 更新最终确定的输出配置

7. CameraCaptureSession.CaptureCallback 类

CameraCaptureSession.CaptureCallback 类（android.hardware.camera2.CameraCaptureSession.CaptureCallback）用于跟踪提交的 CaptureRequest 对象，常用方法如表 11-7 所示。

表 11-7　CameraCaptureSession.CaptureCallback 类的常用方法

类型和修饰符	方　法
	<init> () 主构造方法
open Unit	onCaptureBufferLost (session: CameraCaptureSession, request: CaptureRequest, target: Surface, frameNumber: Long) 当无法将单个捕捉图像缓冲发送到 Surface 时，调用此方法
open Unit	onCaptureCompleted (session: CameraCaptureSession, request: CaptureRequest, result: TotalCaptureResult) 当捕捉图像完成后且所有结果元数据都可用时，调用此方法
open Unit	onCaptureFailed (session: CameraCaptureSession, request: CaptureRequest, failure: CaptureFailure) 当捕捉图像失败无法生成 CaptureResult 时，调用此方法
open Unit	onCaptureProgressed (session: CameraCaptureSession, request: CaptureRequest, partialResult: CaptureResult) 当捕捉图像部分进度完成后，调用此方法，可以使用已经捕捉的图像结果
open Unit	onCaptureSequenceAborted (session: CameraCaptureSession, sequenceId: Int) 当捕捉图像序列被中止后，调用此方法
open Unit	onCaptureSequenceCompleted (session: CameraCaptureSession, sequenceId: Int, frameNumber: Long) 当捕捉图像序列完成后，调用此方法
open Unit	onCaptureStarted (session: CameraCaptureSession, request: CaptureRequest, timestamp: Long, frameNumber: Long) 当捕捉图像请求开始曝光照片或开始处理输入的图像后，调用此方法

8. CameraCaptureSession.StateCallback 类

CameraCaptureSession.StateCallback 类（android.hardware.camera2.CameraCaptureSession.StateCallback）用于接收捕捉图像会话状态的改变，常用方法如表 11-8 所示。

表 11-8　CameraCaptureSession.StateCallback 类的常用方法

类型和修饰符	方　法
	<init> () 主构造方法
open Unit	onActive (session: CameraCaptureSession) 当会话开始主动处理捕捉图像请求时，调用此方法
open Unit	onCaptureQueueEmpty (session: CameraCaptureSession) 当捕捉图像队列为空并准备接收下一个请求时，调用此方法

续表

类型和修饰符	方　法
open Unit	onClosed(session: CameraCaptureSession) 当捕捉图像会话关闭后，调用此方法
abstract Unit	onConfigureFailed(session: CameraCaptureSession) 当配置失败后，调用此方法
abstract Unit	onConfigured(session: CameraCaptureSession) 当配置成功后，调用此方法
open Unit	onReady(session: CameraCaptureSession) 当会话不再有任何请求时，调用此方法
open Unit	onSurfacePrepared(session: CameraCaptureSession, surface: Surface) 当输出 Surface 的缓冲区预分配已完成后，调用此方法

9．CaptureRequest 类

CaptureRequest 类(android.hardware.camera2.CaptureRequest)用来配置捕捉单个图像，常用方法如表 11-9 所示。CaptureRequest 对象调用 CameraCaptureSession 的 capture(CaptureRequest, CameraCaptureSession. CaptureCallback, Handler?)或者 setRepeatingRequest(CaptureRequest, CameraCaptureSession.CaptureCallback, Handler?) 方法从摄像头捕捉拍摄的照片数据。

每个请求可以为摄像头指定一个 Surface，将捕捉的图像数据发送到该 Surface。 当请求提交到会话时，请求中使用的所有 Surface 都必须包含在最后一次调用 CameraDevice.createCaptureSession (MutableList<Surface!>, CameraCaptureSession.StateCallback, Handler?)方法的 Surface 列表参数中。例如，一个请求用于低分辨率图像的预览，另一个用于高分辨率图像的捕捉。

重新处理捕获请求允许将先前从相机设备捕获的图像发送回设备做进一步处理，实现双缓冲的功能。除通过 CameraDevice.createCaptureSession()方法创建的捕捉图像会话外，还可以通过 CameraDevice. createReprocessableCaptureSession()方法创建可重新处理的捕捉图像会话，能够在提交常规捕捉图像请求外的重新处理捕捉图像请求。重新处理捕捉图像请求从会话的输入 Surface 中获取下一个可用缓冲区，然后通过摄像头的管道将其发送作为 Surface 的缓冲区，此时没有为重新处理请求捕捉新的图像数据。

表 11-9　CaptureRequest 类的常用方法

类型和修饰符	方　法
T?	get(key: CaptureRequest.Key<T>!) 获取键值
MutableList<CaptureRequest.Key<*>!>	getKeys() 获取此映射中包含的键列表
Any?	getTag() 获取标签
Boolean	isReprocess() 判断是否重新处理捕捉图像请求

10．CaptureRequest.Builder 类

CaptureRequest.Builder 类(android.hardware.camera2.CaptureRequest.Builder)用于构建捕捉图像请求，常用方法如表 11-10 所示。可以使用 CameraDeviced.createCaptureRequest(int)方法获得构建器实例，该方法将请求字段初始化为在 CameraDevice 中定义的一个模板。

表 11-10　CaptureRequest.Builder 类的常用方法

类型和修饰符	方　法
Unit	addTarget(outputTarget: Surface) 添加捕捉图像输出的 Surface，添加的 Surface 必须包含在 CameraDevice.createCaptureSession(MutableList<Surface!>, CameraCaptureSession.StateCallback, Handler?)方法的参数中
CaptureRequest	build() 构建捕捉图像请求
T?	get(key: CaptureRequest.Key<T>!) 获取键值
T?	getPhysicalCameraKey(key: CaptureRequest.Key<T>!, physicalCameraId: String) 获取特定物理摄像机 Id 的捕获请求字段值
Unit	removeTarget(outputTarget: Surface) 移除捕捉图像输出的 Surface
Unit	set(key: CaptureRequest.Key<T>, value: T) 设置键值
CaptureRequest.Builder!	setPhysicalCameraKey(key: CaptureRequest.Key<T>, value: T, physicalCameraId: String) 将捕获请求字段设置为值
Unit	setTag(tag: Any?) 获取标签

11. CameraCharacteristics 类

CameraCharacteristics 类(android.hardware.camera2.CameraCharacteristics)用于描述摄像头设备的属性，常用方法如表 11-11 所示。使用 CameraManager.getCameraCharacteristics(String)方法获取单个摄像头的 CameraCharacteristics 对象。

表 11-11　CameraCharacteristics 类的常用方法

类型和修饰符	方　法
T?	get(key: CameraCharacteristics.Key<T>!) 获取键值
MutableList<CaptureRequest.Key<*>!>	getAvailableCaptureRequestKeys() 获取有效的捕捉图像请求的键列表
MutableList<CaptureResult.Key<*>!>	getAvailableCaptureResultKeys() 返回此 CameraDevice 支持的键列表，以便使用 CaptureResult 进行查询
MutableList<CaptureRequest.Key<*>!>	getAvailablePhysicalCameraRequestKeys() 获取有效的物理摄像头请求的键列表
MutableList<CaptureRequest.Key<*>!>!	getAvailableSessionKeys() 获取有效的会话键列表
MutableList<CameraCharacteristics.Key<*>!>	getKeys() 获取键列表
MutableList<CameraCharacteristics.Key<*>!>	getKeysNeedingPermission() 获取需要 Manifest.permission.CAMERA 权限支持的所有键
MutableSet<String!>	getPhysicalCameraIds() 获取物理摄像头 id
RecommendedStreamConfigurationMap?	getRecommendedStreamConfigurationMap(usecase: Int) 检索给定用例的摄影机设备推荐流配置映射 RecommendedStreamConfigurationMap

11.2.2　ImageReader 类

Camera 类通过 Camera.takePicture()方法的参数——Camera.PictureCallback 实例回调获取拍摄的照片数据。而 Camera2 类则通过 ImageReader 类读取拍摄时的照片数据。

ImageReader 类（android.media.ImageReader）用于读取 Surface 对象存储的图像数据，常用方法如表 11-12 所示。调用 ImageReader.acquireLatestImage() 方法或者 ImageReader.acquireNextImage() 方法读取 Surface 对象提供的图像队列。由于内存限制，如果不能以相同的速率获取和释放图像，那么 Surface 对象就停止提供或删除图像。

表 11-12　ImageReader 类的常用方法

类型和修饰符	方　法
open Image!	acquireLatestImage() 从 ImageReader 的队列中获取最新的图像，删除较旧的图像
open Image!	acquireNextImage() 从 ImageReader 的队列中获取下一个图像
open Unit	close() 关闭并释放资源
open Unit	discardFreeBuffers() 丢弃此 ImageReader 拥有的所有可用缓冲区
open Int	getHeight() 获取图像高度
open Int	getImageFormat() 获取图像格式
open Int	getMaxImages() 获取可获取的最大图像数量
open Surface!	getSurface() 获取读取图像的 Surface
open Int	getWidth() 获取图像宽度
open static ImageReader	newInstance(width: Int, height: Int, format: Int, maxImages: Int) 创建 ImageReader 的新实例
open static ImageReader	newInstance(width: Int, height: Int, format: Int, maxImages: Int, usage: Long) 创建 ImageReader 的新实例
open Unit	setOnImageAvailableListener(listener: ImageReader.OnImageAvailableListener!, handler: Handler!) 设置从 ImageReader 获得新图像时调用的监听器

11.2.3　MediaRecorder 类

MediaRecorder 类（android.media.MediaRecorder）用于设置录制视频的参数及录制、暂停等操作（常用方法如表 11-13 所示），录制的视频中可以包含音频，同时适用于 Camera 类和 Camera2 类录制视频。可以录制摄像头捕捉的视频，也可以录制其他视频源捕获的视频。

表 11-13　MediaRecorder 类的常用方法

类型和修饰符	方　法
	<init>() 主构造方法
open MutableList<MicrophoneInfo!>!	getActiveMicrophones() 返回当前活动麦克风的 microponeinfo 列表
open AudioRecordingConfiguration?	getActiveRecordingConfiguration() 返回当前活动音频录的配置
open Surface!	getSurface() 获取录制视频源的 Surface

续表

类型和修饰符	方　法
open Unit	pause() 暂停录制
open Unit	prepare() 准备开始捕捉和编码数据
open Unit	registerAudioRecordingCallback(executor: Executor, cb: AudioManager.AudioRecordingCallback) 注册音频录制的回调
open Unit	release() 释放资源
open Unit	reset() 重置
open Unit	resume() 恢复录制
open Unit	setAudioEncoder(audio_encoder: Int) 设置音频编码器。audio_encoder 的常量如下。 ● MediaRecorder.AudioEncoder.AAC：高级音频编码(Advanced Audio Coding)，与 mp3 相比，AAC 格式的音质更佳，文件更小 ● MediaRecorder.AudioEncoder.AAC_ELD：低延迟的 AAC ● MediaRecorder.AudioEncoder.AMR_NB：窄频 AMR ● MediaRecorder.AudioEncoder.AMR_WB：宽频 AMR ● MediaRecorder.AudioEncoder.DEFAULT：默认编码 ● MediaRecorder.AudioEncoder.HE_AAC：高保真 AAC ● MediaRecorder.AudioEncoder.OPUS：Opus 编码器。低码率下 Opus 完胜 HE AAC，中码率下与码率高出 30%左右的 AAC 格式相当，高码率下更接近原始音频 ● MediaRecorder.AudioEncoder.VORBIS：Ogg Vorbis 编码器，不用重新编码便可调节文件的位速率
open Unit	setAudioEncodingBitRate(bitRate: Int) 设置音频编码比特率
open Unit	setAudioSamplingRate(samplingRate: Int) 设置音频采样率
open Unit	setAudioSource(audioSource: Int) 设置音源
open Unit	setCaptureRate(fps: Double) 设置捕捉的帧速率
open Unit	setInputSurface(surface: Surface) 设置录制视频源的 Surface
open Unit	setLocation(latitude: Float, longitude: Float) 设置 GPS 坐标
open Unit	setMaxDuration(max_duration_ms: Int) 设置录制的最长时间(毫秒)
open Unit	setMaxFileSize(max_filesize_bytes: Long) 设置最大文件尺寸(字节)
open Unit	setNextOutputFile(fd: FileDescriptor!) 设置到达最大文件尺寸后的下一个输出文件
open Unit	setNextOutputFile(file: File!) 设置到达最大文件尺寸后的下一个输出文件
open Unit	ssetOnErrorListener(l: MediaRecorder.OnErrorListener!) 设置捕捉错误的监听器

续表

类型和修饰符	方　法
open Unit	setOnInfoListener (listener: MediaRecorder.OnInfoListener!) 设置信息事件的监听器。监听的信息事件如下。 ● MediaRecorder.MEDIA_RECORDER_INFO_UNKNOWN：未知信息 ● MediaRecorder.MEDIA_RECORDER_INFO_MAX_DURATION_REACHED：到达最长录制时间 ● MediaRecorder.MEDIA_RECORDER_INFO_MAX_FILESIZE_REACHED：录制的输出文件已经到达最大文件尺寸
open Unit	setOrientationHint (degrees: Int) 设置输出视频回放的方向提示
open Unit	setOutputFile (fd: FileDescriptor!) 设置录制的输出文件
open Unit	setOutputFile (path: String!) 设置录制的输出文件的路径
open Unit	setOutputFile (file: File!) 设置录制的输出文件
open Unit	setOutputFormat (output_format: Int) 设置录制的输出格式
open Boolean	setPreferredDevice (deviceInfo: AudioDeviceInfo!) 设置首选麦克风的音频录制设备信息
open Boolean	setPreferredMicrophoneDirection (direction: Int) 设置首选麦克风
open Boolean	setPreferredMicrophoneFieldDimension (zoom: Float) 设置首选麦克风声音大小的缩放
open Unit	setPreviewDisplay (sv: Surface!) 设置预览视频显示的 Surface
open Unit	setVideoEncoder (video_encoder: Int) 设置录制视频的编码。video_encoder 的常量如下。 ● MediaRecorder.VideoEncoder.DEFAULT ● MediaRecorder.VideoEncoder.H263 ● MediaRecorder.VideoEncoder.H264 ● MediaRecorder.VideoEncoder.HEVC ● MediaRecorder.VideoEncoder.MPEG_4_SP ● MediaRecorder.VideoEncoder.VP8
open Unit	setVideoEncodingBitRate (bitRate: Int) 设置录制视频的编码比特率
open Unit	setVideoFrameRate (rate: Int) 设置录制视频的帧速率
open Unit	setVideoSize (width: Int, height: Int) 设置录制视频的尺寸
open Unit	setVideoSource (video_source: Int) 设置录制视频源
open Unit	start () 开始录制视频，捕捉视频进行编码并写入输出文件
open Unit	stop () 停止录制
open Unit	unregisterAudioRecordingCallback (cb: AudioManager.AudioRecordingCallback) 注销声音录制的回调

11.2.4　实例工程：使用 Camera2 类拍摄照片

本实例演示了使用 Camera2 类开发第三方拍照 App，单击“拍照”按钮后拍摄照片并显示照片。在 C1101 调用系统相机时，该 App 可以作为备选项供用户选择(如图 11-6 所示)。

图 11-6　运行效果

1．打开基础工程

打开“基础工程”文件夹中的“C1103”工程，该工程已经包含 MainActivity、PreviewActivity、Permissions、PreviewTextureView、Util 类及资源文件。

2．AndroidManifest 文件

```
/manifests/AndroidManifest.xml
03  <uses-permission android:name = "android.permission.CAMERA" />
04  <uses-permission android:name = "android.permission.READ_EXTERNAL_STORAGE" />
05  <uses-permission android:name = "android.permission.WRITE_EXTERNAL_STORAGE" />
06  <application android:theme = "@style/AppTheme">
14      <activity android:name = ".MainActivity">
15          <intent-filter>
16              <action android:name = "android.intent.action.MAIN" />
17              <action android:name = "android.media.action.IMAGE_CAPTURE" />
18              <action android:name = "android.media.action.STILL_IMAGE_CAMERA" />
19              <category android:name = "android.intent.category.DEFAULT" />
20              <category android:name = "android.intent.category.LAUNCHER" />
21          </intent-filter>
22      </activity>
23  </application>
```

第 03 ~ 05 行设置拍摄和存储照片所需的权限。第 17、18 行设置隐性启动的行为，作为可以被系统调用的拍照 App。第 19 行设置作为隐性启动的 Activity。

3．Util 类

打开 Util 类，添加两个可重用的静态方法：imageToBytes() 和 chooseOptimalSize()，以及一个可重用的内部类 CompareSizesByArea。

```
/java/com/teachol/c1103/Util.kt
//bitmap转byte[]
fun imageToBytes(image: Image?): ByteArray? {
    val buffer: ByteBuffer = image!!.getPlanes().get(0).getBuffer()
    val bytes = ByteArray(buffer.remaining())
    buffer.get(bytes)
    return bytes
}
//获取最佳尺寸
fun chooseOptimalSize(context: Context?, choices: Array<Size>?, previewWidth: Int?, previewHeight: Int?, maxPreviewWidth: Int?, maxPreviewHeight: Int?, aspectRatio: Size?): Size {
    var maxWidth:Int
    var maxHeight:Int
    if (Build.VERSION.SDK_INT >= Build.VERSION_CODES.R) {
        val metrics = DisplayMetrics()
        val display: Display = (context as Activity).display!!
        display.getRealMetrics(metrics)
        maxWidth = metrics.widthPixels
        maxHeight = metrics.heightPixels
    } else {
        val point = Point()
        val manager = context!!.getSystemService(Context.WINDOW_SERVICE) as WindowManager
        val display = manager.getDefaultDisplay()
        display.getSize(point)
        maxWidth = point.x
        maxHeight = point.y
    }
    if (maxWidth > maxPreviewWidth!!) maxWidth = maxPreviewWidth
    if (maxHeight > maxPreviewHeight!!) maxHeight = maxPreviewHeight
    val bigEnough: MutableList<Size>= ArrayList() //保存支持的不小于预览图像的尺寸
    val notBigEnough: MutableList<Size>= ArrayList() //保存支持的小于预览图像的尺寸
    val w: Int = aspectRatio!!.width
    val h: Int = aspectRatio.height
    for (option in choices!!) { //将设备支持的尺寸分类保存
        if (option.width <= maxWidth && option.height <= maxHeight && option.height == option.width * h / w) {
            if (option.width >= previewWidth!! && option.height >= previewHeight!!) {
                bigEnough.add(option)
            } else {
                notBigEnough.add(option)
            }
        }
    }
    return when { //选择足够大的最小尺寸。如果没有足够大的尺寸，就从不够大的尺寸中挑选最大的尺寸
        bigEnough.size > 0 -> Collections.min(bigEnough, CompareSizesByArea())
        notBigEnough.size > 0 -> Collections.max(notBigEnough, CompareSizesByArea())
        else -> choices[0]
    }
}
class CompareSizesByArea : Comparator<Size> { //比较两个Size大小的规则
```

```
152     override fun compare(lhs: Size?, rhs: Size?): Int {
153         return java.lang.Long.signum(lhs!!.width.toLong() * lhs.height - rhs!!.width.toLong() * rhs.height)
154     }
155 }
```

第 103 ~ 108 行是 Bitmap 实例转 byte[]实例的静态方法，用于隐式启动的回调数据传递。第 110 ~ 149 行是根据摄像头支持的尺寸、预览控件尺寸和最大预览尺寸限制，获取预览控件最佳尺寸的方法。第 151 ~ 155 行是用于比较 Size 实例大小运算规则的内置类，重写 compare()方法计算出最大面积。

4．主界面的 Activity

/java/com/teachol/c1103/MainActivity.kt

```
private val TAG = "MainActivity"
private val mContext: Context = this
private val MAX_PREVIEW_WIDTH = 1080 //预览的最大宽度限制
private val MAX_PREVIEW_HEIGHT = 1920 //预览的最大高度限制
private lateinit var mPreviewSize: Size //预览尺寸
private lateinit var mPreviewView: PreviewTextureView//预览摄像头画面的控件
private lateinit var mButton: Button //拍照按钮
private val mSemaphore = Semaphore(1) //设置信号许可数量
private var mBackgroundThread: HandlerThread? = null //后台线程
private var mBackgroundHandler: Handler? = null //后台线程的句柄
private val mCameraId = "0" //摄像头 id
private lateinit var mCameraManager: CameraManager //摄像头管理器
private var mCaptureSession: CameraCaptureSession? = null //摄像头捕捉会话
private var mCameraDevice: CameraDevice? = null //摄像头设备
private lateinit var mCaptureRequestBuilder: CaptureRequest.Builder //捕捉图像请求构建器
private lateinit var mCaptureRequest: CaptureRequest //捕捉图像请求
private var mImageReader: ImageReader? = null //捕获图像的读取器
private lateinit var mPhotoFile: File //保存图像的文件
private val PERMISSIONS:Array<String?>  = arrayOf(
    Manifest.permission.CAMERA,
    Manifest.permission.READ_EXTERNAL_STORAGE,
    Manifest.permission.WRITE_EXTERNAL_STORAGE
)
//TextureView 的生命周期事件
private val mSurfaceTextureListener: SurfaceTextureListener = object : SurfaceTextureListener {
    override fun onSurfaceTextureAvailable(texture: SurfaceTexture, width: Int, height: Int) {
        Log.d(TAG, "TextureView.onSurfaceTextureAvailable()")
        openCamera(width, height)
    }
    override fun onSurfaceTextureDestroyed(texture: SurfaceTexture): Boolean { return true }
}
//摄像头设备状态的回调
private val mCameraDeviceStateCallback: CameraDevice.StateCallback =
    object : CameraDevice.StateCallback() {
        override fun onOpened(cameraDevice: CameraDevice) {
            Log.d(TAG, "CameraDevice.onOpened()")
            mCameraDevice = cameraDevice
            startPreview()
```

```
            mSemaphore.release() //释放 1 个信号许可
        }
        override fun onDisconnected(cameraDevice: CameraDevice) {
            Log.d(TAG, "CameraDevice.onDisconnected()")
            mSemaphore.release() //释放 1 个信号许可
            cameraDevice.close()
            mCameraDevice!!.close()
        }
        override fun onError(cameraDevice: CameraDevice, error: Int) {
            Log.d(TAG, "CameraDevice.onError()")
            mSemaphore.release() //释放 1 个信号许可
            cameraDevice.close()
            mCameraDevice!!.close()
        }
    }
//ImageReader 的监听器
private val mOnImageAvailableListener: ImageReader.OnImageAvailableListener =
    ImageReader.OnImageAvailableListener { reader ->
        Log.d(TAG, "ImageReader.onImageAvailable()")
        mImageReader!!.setOnImageAvailableListener(null, null) //解除 mImageReader 的监听器
        (mContext as Activity).runOnUiThread { handlePhoto(reader) }//读取到图像后使用新线程处理图像数据
    }
override fun onCreate(savedInstanceState: Bundle?) {
    super.onCreate(savedInstanceState)
    setContentView(R.layout.activity_main)
    mPreviewView = findViewById(R.id.texture_view)
    mButton = findViewById(R.id.button)
    mButton.setOnClickListener(this)
}
override fun onResume() {
    super.onResume()
    startBackgroundThread()
    if (mPreviewView.isAvailable) { //预览显示时打开摄像头，否则通过监听器监测打开摄像头
        openCamera(mPreviewView.width, mPreviewView.height)
    } else {
        mPreviewView.surfaceTextureListener = mSurfaceTextureListener
    }
}
override fun onPause() {
    closeCamera()
    stopBackgroundThread()
    super.onPause()
}
private fun startBackgroundThread() { //开启后台线程
    Log.d(TAG, "startBackgroundThread()")
    mBackgroundThread = HandlerThread("CameraBackground")
    mBackgroundThread!!.start()
    mBackgroundHandler = Handler(mBackgroundThread!!.looper)
}
```

```
private fun stopBackgroundThread() { //停止后台线程
    Log.d(TAG, "stopBackgroundThread()")
    mBackgroundThread!!.quitSafely()
    try {
        mBackgroundThread!!.join()
        mBackgroundThread = null
        mBackgroundHandler = null
    } catch (e: InterruptedException) {
        e.printStackTrace()
    }
}
override fun onClick(view: View?) { //单击事件
    takePhoto()
}
private fun openCamera(width: Int, height: Int) { //打开摄像头
    Log.d(TAG, "openCamera()")
    if (!hasPermissionsGranted(mContext, PERMISSIONS)) {
        Permissions.requestPermissions(mContext, PERMISSIONS)
        return
    }
    try { //2500 毫秒内请求获取 1 个许可，否则抛出异常
        if (!mSemaphore.tryAcquire(2500, TimeUnit.MILLISECONDS)) {
            throw RuntimeException("打开摄像头超时")
        }
        mCameraManager = getSystemService(Context.CAMERA_SERVICE) as CameraManager
        val characteristics: CameraCharacteristics = mCameraManager.getCameraCharacteristics(mCameraId) //获取摄像头支持的属性特征
        //获取摄像头支持的可用流配置，包括每种格式、大小组合的最小帧持续时间和停顿持续时间
        val map = characteristics.get(CameraCharacteristics.SCALER_STREAM_CONFIGURATION_MAP)
        val outPutSizeList:MutableList<Size>= mutableListOf()
        for (value in map!!.getOutputSizes(ImageFormat.JPEG)) { //获取摄像头捕捉的最大尺寸读取图像
            outPutSizeList.add(value)
        }
        val largest: Size = Collections.max(outPutSizeList, CompareSizesByArea())
        mImageReader = ImageReader.newInstance(largest.width, largest.height, ImageFormat.JPEG, 2)
        //选择最佳预览尺寸
        mPreviewSize = chooseOptimalSize(mContext, map.getOutputSizes(SurfaceTexture::class.java), width, height, MAX_PREVIEW_WIDTH, MAX_PREVIEW_HEIGHT, largest)
        mPreviewView.setAspectRatio(mPreviewSize.height, mPreviewSize.width) //设置预览尺寸
        //打开摄像头
        mCameraManager.openCamera(mCameraId, mCameraDeviceStateCallback, mBackgroundHandler)
    } catch (e: CameraAccessException) {
        e.printStackTrace()
    } catch (e: InterruptedException) {
        throw RuntimeException("打开摄像头被中断", e)
    }
}
private fun startPreview() { //创建摄像头预览会话
    Log.d(TAG, "startPreview()")
    if (mCameraDevice == null || !mPreviewView.isAvailable) { return }
```

```
    try {
        closePreview()
        val texture = mPreviewView.surfaceTexture //设置预览的缓冲区大小
        texture!!.setDefaultBufferSize(mPreviewSize.width, mPreviewSize.height)
        val surface = Surface(texture) //实例化预览输出的 Surface
        mCaptureRequestBuilder = mCameraDevice!!.createCaptureRequest(CameraDevice.
TEMPLATE_PREVIEW)
        mCaptureRequestBuilder.addTarget(surface)
        mCameraDevice!!.createCaptureSession( //创建摄像头的捕获会话
            Arrays.asList(surface, mImageReader!!.getSurface()),
            object : CameraCaptureSession.StateCallback() {
                override fun onConfigured(cameraCaptureSession: CameraCaptureSession) {
                    mCaptureSession = cameraCaptureSession
                    updatePreview()
                }
                override fun onConfigureFailed(cameraCaptureSession: CameraCaptureSession) {
                    Toast.makeText(mContext, "摄像头配置失败", Toast.LENGTH_LONG).show()
                }
            },
            mBackgroundHandler
        )
    } catch (e: CameraAccessException) { e.printStackTrace() }
}
//拍照
private fun takePhoto() {
    Log.d(TAG, "takePhoto()")
    if (null == mCameraDevice || !mPreviewView.isAvailable) {
        return
    }
    try {
        closePreview()
        mCaptureRequestBuilder = mCameraDevice!!.createCaptureRequest(CameraDevice.
TEMPLATE_STILL_CAPTURE)
        val texture = mPreviewView.surfaceTexture
        texture!!.setDefaultBufferSize(mPreviewSize.width, mPreviewSize.height) //设置预览的
缓冲区大小
        val surfaces: MutableList<Surface>= ArrayList()
        val previewSurface = Surface(texture) //设置预览的 Surface
        surfaces.add(previewSurface)
        mCaptureRequestBuilder.addTarget(previewSurface)
        val imageReaderSurface: Surface = mImageReader!!.surface //设置拍照的 Surface
        surfaces.add(imageReaderSurface)
        mCaptureRequestBuilder.addTarget(imageReaderSurface)
        mCameraDevice!!.createCaptureSession( //创建捕捉会话
            surfaces,
            object : CameraCaptureSession.StateCallback() {
                override fun onConfigured(cameraCaptureSession: CameraCaptureSession) {
                    Log.d(TAG, "CameraCaptureSession.onConfigured()")
                    mCaptureSession = cameraCaptureSession
                    mImageReader!!.setOnImageAvailableListener( //添加 ImageReader 监听器处理拍摄照片
```

```
                    mOnImageAvailableListener,
                    mBackgroundHandler
                )
                mCaptureRequest = mCaptureRequestBuilder.build()
                try { //拍摄照片
                    mCaptureSession!!.capture(mCaptureRequest, null, null)
                } catch (e: CameraAccessException) {
                    e.printStackTrace()
                }
            }
            override fun onConfigureFailed(cameraCaptureSession: CameraCaptureSession) {
                Toast.makeText(mContext, "摄像头配置失败", Toast.LENGTH_LONG).show()
            }
        },
        mBackgroundHandler
    )
  } catch (e: CameraAccessException) { e.printStackTrace() }
}
private fun handlePhoto(reader: ImageReader) { //处理拍摄的照片
    Log.d(TAG, "handlePhoto()")
    var photoByte = imageToBytes(reader.acquireNextImage())
    mPhotoFile = creatFile(externalMediaDirs[0].absoluteFile
, mContext.resources.getString(R.string.app_name), "jpg")
    photoByte = saveImage(photoByte!!, mPhotoFile, 90) //保存拍摄的照片
    showInAlbum(mContext, mPhotoFile.absolutePath) //将拍摄的照片显示在系统相册中
    startPreview() //恢复预览
    var intent = intent //获取启动该 Activity 的 Intent
    if (intent.action == "android.media.action.IMAGE_CAPTURE" || intent.action ==
"android.media.action.STILL_IMAGE_CAMERA") { //判断是否通过外部 App 隐式启动
        val uri: Uri? = intent.getParcelableExtra(MediaStore.EXTRA_OUTPUT)
        if (uri != null) { //判断是否指定了照片保存路径的 Uri
            val resolver: ContentResolver = mContext.contentResolver
            try { //向 Uri 指定路径的文件写入照片数据
                val descriptor = resolver.openFileDescriptor(uri, "rw")
                val output = FileOutputStream(descriptor!!.fileDescriptor)
                output.write(photoByte)
                descriptor.close()
                output.close()
            } catch (e: FileNotFoundException) {
                e.printStackTrace()
            } catch (e: IOException) {
                e.printStackTrace()
            }
            setResult(RESULT_OK)
            finish()
        } else {
            val options = BitmapFactory.Options() //设置照片压缩的参数
            options.inPreferredConfig = Bitmap.Config.RGB_565
            options.inSampleSize = 16
            val bitmap = BitmapFactory.decodeFile(mPhotoFile.absolutePath, options) //压缩照片
```

```
316             //返回RESULT_OK，并包含一个Intent对象，其中Extra中key为data，value是保存照片的bitmap对象
317             setResult(RESULT_OK, Intent().putExtra("data", bitmap))
318             finish()
319         }
320     } else {
322         intent = Intent(this@MainActivity, PreviewActivity::class.java)
323         intent.putExtra("path", mPhotoFile.absolutePath)
324         startActivity(intent) //启动预览照片的Activity
325     }
326 }
328 private fun updatePreview() { //更新预览
329     try {
331         mCaptureRequestBuilder.set(CaptureRequest.CONTROL_AF_MODE, CaptureRequest.CONTROL_
    AF_MODE_CONTINUOUS_PICTURE) //设置自动对焦
332         mCaptureRequestBuilder.set(CaptureRequest.CONTROL_MODE, CameraMetadata.CONTROL_MODE_AUTO)
334         mCaptureRequest = mCaptureRequestBuilder.build() //显示预览
335         //捕捉图像会话设置重复请求
336         mCaptureSession!!.setRepeatingRequest(mCaptureRequest, null, mBackgroundHandler)
337     } catch (e: CameraAccessException) {
338         e.printStackTrace()
339     }
340 }
342 private fun closePreview() { //关闭预览会话
343     if (mCaptureSession != null) {
344         mCaptureSession!!.close()
345         mCaptureSession = null
346     }
347 }
349 private fun closeCamera() { //关闭摄像头
350     Log.d(TAG, "closeCamera()")
351     if (null != mCaptureSession) {
352         mCaptureSession!!.close()
353         mCaptureSession = null
354     }
355     if (null != mCameraDevice) {
356         mCameraDevice!!.close()
357         mCameraDevice = null
358     }
359     if (null != mImageReader) {
360         mImageReader!!.close()
361         mImageReader = null
362     }
363     mSemaphore.release() //释放1个信号许可
364 }
```

第 69 ~ 77 行是用于 PreviewTextureView 控件的监听器，当 PreviewTextureView 控件的 SurfaceTexture 可用时调用 openCamera(int, int)方法。第 79 ~ 99 行是摄像头状态的回调，当摄像头打开后开始预览摄像头的画面，当摄像头断开连接或出现错误时关闭设备。第 101 ~ 108 行是 ImageReader 的图像可用监听器，当 ImageReader 的图像可用时，开启新线程调用 handlePhoto(ImageReader)方法处

理获取的图像数据。第 134 ~ 139 行创建后台线程，mBackgroundHandler 作为打开摄像头和图像可用监听器的线程，避免主线程的阻塞。第 158 ~ 192 行检查摄像头权限，设置最佳画面尺寸，开启摄像头。第 209 ~ 221 行创建预览时摄像头的捕获会话，用于更新预览画面。第 248 ~ 271 行创建拍照时摄像头的捕捉会话，当配置完成后捕捉图像，通过 mCaptureRequest 对象将捕捉的图像数据传递给 mImageReader。第 277 ~ 326 行处理拍摄的照片。第 328 ~ 340 行更新预览视图的图像。第 349 ~ 364 行关闭摄像头，并释放资源。

11.2.5　实例工程：使用 Camera2 类录制视频

本实例演示了使用 Camera2 类开发第三方视频录制 App。与前一个实例相比，本实例将录制的功能封装到了单独的类中，便于重复使用。单击“录制”按钮后开始录制视频，此时按钮文字显示为“停止”，再次单击停止录制(如图 11-7 所示)。在其他 App 调用系统相机录制视频时，该 App 可以作为备选项供用户选择。

图 11-7　运行效果

1. 打开基础工程

打开“基础工程”文件夹中的 C1104 工程，该工程已经包含 MainActivity、PreviewActivity、Permissions、PreviewTextureView、Util 类及布局和权限路径(/res/xml/file_paths.xml)的文件，AndroidManifest.xml 文件中还添加了录制和存储视频所需的权限及 FileProvider 权限。

2. Util 类

打开 Util 类，添加 chooseVideoSize()方法，用于选取录制视频的尺寸。

```
/java/com/teachol/c1104/Util.kt
151 fun chooseVideoSize(choices: Array<Size>?): Size? { //选择视频尺寸
152     for (size in choices!!) {
153         println(size.width.toString() + "*" + size.height)
154         if (size.width == size.height * 4 / 3 && size.width <= 1080) return size
155     }
156     return choices[choices.size - 1]
157 }
```

第 152 ~ 155 行遍历所有支持的视频尺寸。第 154 行判断尺寸是不是 4∶3 比例的画幅，并且宽度

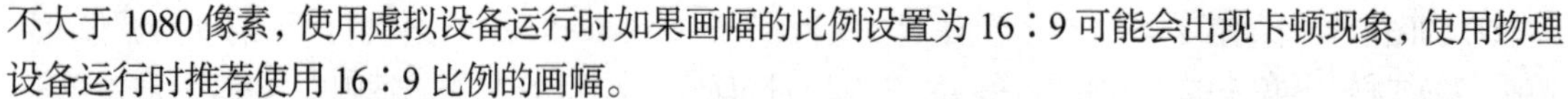

不大于 1080 像素，使用虚拟设备运行时如果画幅的比例设置为 16∶9 可能会出现卡顿现象，使用物理设备运行时推荐使用 16∶9 比例的画幅。

3．Camera 类

在“/java/com/teachol/c1104”文件夹中，新建 Camera 类，用于封装录制的权限请求、录制的开始和停止、录制后的关闭和回放等功能。

/java/com/teachol/c1104/Camera.kt

```
class Camera(val mContext: AppCompatActivity, val mPreviewView: PreviewTextureView, val mButton: Button, val mAuthority:String, var mPreviewActivity:Class<*>?) {
    companion object {
        val STOP_PAUSE = 0 //切换到后台停止
        val STOP_ERROR = 1 //异常停止
        val STOP_NORMAL = 2 //正常停止
    }
    private val TAG = "Camera"
    private val MSG_START = 0 //开始录制
    private val MSG_STOP = 1 //停止录制
    private var mBackgroundThread: HandlerThread? = null //后台线程
    private var mBackgroundHandler: Handler? = null //后台线程的句柄
    private val MAX_PREVIEW_WIDTH = 1080 //预览的最大宽度限制
    private val MAX_PREVIEW_HEIGHT = 1920 //预览的最大高度限制
    private var mCameraManager: CameraManager? = null //摄像头管理器
    private var mCameraDevice: CameraDevice? = null //摄像头设备
    private var mCaptureSession: CameraCaptureSession? = null //摄像头捕捉会话
    private var mCaptureRequestBuilder: CaptureRequest.Builder? = null
    private var mCaptureRequest: CaptureRequest? = null //捕捉图像请求
    private var mMediaRecorder: MediaRecorder? = null //媒体录制器
    private var mPreviewSize: Size = Size(1080, 1920) //预览画面尺寸
    private lateinit var mVideoSize: Size //录制画面尺寸
    var mIsRecordingVideo = false //是否在录制视频
    private lateinit var mVideoFile: File //保存视频的文件
    private val mSemaphore = Semaphore(1) //设置信号许可数量
    private val PERMISSIONS = arrayOf<String?>( //所需权限
        Manifest.permission.CAMERA,
        Manifest.permission.RECORD_AUDIO,
        Manifest.permission.READ_EXTERNAL_STORAGE,
        Manifest.permission.WRITE_EXTERNAL_STORAGE
    )
    private val mSurfaceTextureListener: TextureView.SurfaceTextureListener = object: TextureView. SurfaceTextureListener { //TextureView 的生命周期事件
        override fun onSurfaceTextureAvailable(surfaceTexture: SurfaceTexture, width: Int, height: Int) {
            Log.d(TAG, "TextureView.onSurfaceTextureAvailable()")
            openCamera(width, height)
        }
        override fun onSurfaceTextureDestroyed(surfaceTexture: SurfaceTexture): Boolean { return true }
    }
    private val mCameraDeviceStateCallback: CameraDevice.StateCallback = object : CameraDevice.StateCallback() { //摄像头状态的回调
```

```
        override fun onOpened(cameraDevice: CameraDevice) {
            Log.d(TAG, "CameraDevice.onOpened()")
            mCameraDevice = cameraDevice
            startPreview()
            mSemaphore.release() //释放 1 个信号许可
        }
        override fun onDisconnected(cameraDevice: CameraDevice) {
            Log.d(TAG, "CameraDevice.onDisconnected()")
            mSemaphore.release() //释放 1 个信号许可
            cameraDevice.close()
            mCameraDevice = null
        }
        override fun onError(cameraDevice: CameraDevice, error: Int) {
            Log.d(TAG, "CameraDevice.onError()")
            mSemaphore.release() //释放 1 个信号许可
            cameraDevice.close()
            mCameraDevice = null
        }
    }
    fun onResume() { //Activity 恢复
        startBackgroundThread()
        //预览显示时打开摄像头，否则通过监听器监测打开摄像头
        if (mPreviewView.isAvailable) {
            openCamera(mPreviewView.width, mPreviewView.height)
        } else {
            mPreviewView.surfaceTextureListener = mSurfaceTextureListener
        }
    }
    fun onPause() { //Activity 后台运行
        if(mIsRecordingVideo) stopRecordingVideo(STOP_PAUSE)
        closeCamera()
        stopBackgroundThread()
    }
    private fun startBackgroundThread() { //开启后台线程
        Log.d(TAG, "startBackgroundThread()")
        mBackgroundThread = HandlerThread("CameraBackground")
        mBackgroundThread!!.start()
        mBackgroundHandler = Handler(mBackgroundThread!!.looper)
    }
    private fun stopBackgroundThread() { //停止后台线程
        Log.d(TAG, "stopBackgroundThread()")
        mBackgroundThread!!.quitSafely()
        try {
            mBackgroundThread!!.join()
            mBackgroundThread = null
            mBackgroundHandler = null
        } catch (e: InterruptedException) {
            e.printStackTrace()
        }
    }
```

```
    private fun openCamera(width: Int, height: Int) { //打开摄像头
        Log.d(TAG, "openCamera()")
        if (!Permissions.hasPermissionsGranted(mContext, PERMISSIONS)) {
            Permissions.requestPermissions(mContext, PERMISSIONS)
            return
        }
        try {
            if (!mSemaphore.tryAcquire(2500, TimeUnit.MILLISECONDS)) throw RuntimeException
("打开摄像头超时")//2500 毫秒内请求获取 1 个许可，否则抛出异常
            mCameraManager = mContext.getSystemService(CAMERA_SERVICE) as CameraManager
            val cameraId = mCameraManager!!.cameraIdList[0] //获取后置摄像头
            //获取预览和录制视频的尺寸
            val characteristics = mCameraManager!!.getCameraCharacteristics(cameraId)
            val map = characteristics.get(CameraCharacteristics.SCALER_STREAM_CONFIGURATION_MAP)
            mVideoSize  =  Util.chooseVideoSize(map!!.getOutputSizes(MediaRecorder::class.
java))!!
            mPreviewSize = Util.chooseOptimalSize(mContext, map.getOutputSizes(SurfaceTexture::class.java),
width, height, MAX_PREVIEW_WIDTH, MAX_PREVIEW_HEIGHT, mVideoSize)
            mPreviewView.setAspectRatio(mPreviewSize.height, mPreviewSize.width)
            mMediaRecorder = MediaRecorder()
            mCameraManager!!.openCamera(cameraId, mCameraDeviceStateCallback, mBackgroundHandler)
        } catch (e: CameraAccessException) {
            Toast.makeText(mContext, "摄像头不可用", Toast.LENGTH_LONG).show()
        } catch (e: InterruptedException) {
            throw RuntimeException("打开摄像头被中断", e)
        }
    }
    private fun setUpMediaRecorder() { //设置录制参数
        Log.d(TAG, "setUpMediaRecorder()")
        mVideoFile = Util.creatFile(mContext.externalMediaDirs[0].absolutePath, mContext.resources.
getString(R.string.app_name), "mp4")
        mMediaRecorder!!.setVideoSource(MediaRecorder.VideoSource.SURFACE)
        mMediaRecorder!!.setAudioSource(MediaRecorder.AudioSource.MIC)
        mMediaRecorder!!.setOutputFormat(MediaRecorder.OutputFormat.MPEG_4)
        mMediaRecorder!!.setOutputFile(mVideoFile.absolutePath)
        mMediaRecorder!!.setVideoEncoder(MediaRecorder.VideoEncoder.H264)
        mMediaRecorder!!.setAudioEncoder(MediaRecorder.AudioEncoder.AAC)
        mMediaRecorder!!.setVideoEncodingBitRate(8 * mVideoSize.width * mVideoSize.height)
        mMediaRecorder!!.setVideoFrameRate(30)
        mMediaRecorder!!.setVideoSize(mVideoSize.width, mVideoSize.height)
        mMediaRecorder!!.setOrientationHint(90) //设置视频的角度
        mMediaRecorder!!.setOnErrorListener { mr, what, extra -> stopRecordingVideo(STOP_ERROR) }
        mMediaRecorder!!.prepare()
    }
    private fun startPreview() { //开始预览
        Log.d(TAG, "startPreview()")
        if (null == mCameraDevice || !mPreviewView.isAvailable) return
        try {
            closePreview()
            val texture = mPreviewView.surfaceTexture //设置预览的缓冲区大小
```

```
            texture!!.setDefaultBufferSize(mPreviewSize.width, mPreviewSize.height)
            val previewSurface = Surface(texture) //设置预览输出的 Surface
            mCaptureRequestBuilder = mCameraDevice!!.createCaptureRequest(CameraDevice.TEMPLATE_PREVIEW)
            mCaptureRequestBuilder!!.addTarget(previewSurface)
            //创建摄像头预览视频的捕获会话
            mCameraDevice!!.createCaptureSession(listOf(previewSurface), object : CameraCaptureSession.StateCallback() {
                override fun onConfigured(session: CameraCaptureSession) {
                    mCaptureSession = session
                    updatePreview()
                }
                override fun onConfigureFailed(session: CameraCaptureSession) {
                    Toast.makeText(mContext, "摄像头配置失败", Toast.LENGTH_LONG).show()
                }
            }, mBackgroundHandler)
        } catch (e: CameraAccessException) {
            e.printStackTrace()
        }
    }
    fun startRecordingVideo() { //开始录制视频
        Log.d(TAG, "startRecordingVideo()")
        mButton.isEnabled = false
        mIsRecordingVideo = true
        if (null == mCameraDevice || !mPreviewView.isAvailable) return
        try {
            closePreview()
            setUpMediaRecorder()
            mCaptureRequestBuilder = mCameraDevice!!.createCaptureRequest(CameraDevice.TEMPLATE_RECORD)
            val texture = mPreviewView.surfaceTexture //设置预览的缓冲区大小
            texture!!.setDefaultBufferSize(mPreviewSize.width, mPreviewSize.height)
            val surfaces: MutableList<Surface>= ArrayList() //设置预览的 Surface
            val previewSurface = Surface(texture)
            surfaces.add(previewSurface)
            mCaptureRequestBuilder!!.addTarget(previewSurface)
            val recorderSurface = mMediaRecorder!!.surface //设置录制视频的 Surface
            surfaces.add(recorderSurface)
            mCaptureRequestBuilder!!.addTarget(recorderSurface)
            //创建摄像头录制视频的捕捉会话
            mCameraDevice!!.createCaptureSession(surfaces, object : CameraCaptureSession.
StateCallback() {
                override fun onConfigured(session: CameraCaptureSession) {
                    mCaptureSession = session
                    updatePreview()
                    mContext.runOnUiThread {
                        mMediaRecorder!!.start()
                        mHandler.obtainMessage(MSG_START).sendToTarget()
                    }
                }
                override fun onConfigureFailed(cameraCaptureSession: CameraCaptureSession) {
```

```
                    Toast.makeText(mContext, "摄像头配置失败", Toast.LENGTH_LONG).show()
                }
            }, mBackgroundHandler)
        } catch (e: CameraAccessException) {
            e.printStackTrace()
        } catch (e: IOException) {
            e.printStackTrace()
        }
    }
    fun stopRecordingVideo(state: Int) { //停止录制视频
        Log.d(TAG, "stopRecordingVideo()")
        mButton.isEnabled = false
        if (mMediaRecorder != null) {
            try {
                mMediaRecorder!!.stop()
                mMediaRecorder!!.reset()
                mIsRecordingVideo = false
                if (state == STOP_ERROR) return //如果因为错误停止，则直接返回不进行后续处理
                Util.showInAlbum(mContext, mVideoFile.absolutePath)
                val videoUri = FileProvider.getUriForFile(mContext, mAuthority, mVideoFile)
                var intent = mContext.intent
                //判断是否被其他App隐式启动
                if (intent.action == "android.media.action.VIDEO_CAPTURE") {
                    intent = Intent()
                    intent.setDataAndType(videoUri, "video/mp4") //设置返回的数据及其类型
                    intent.flags = Intent.FLAG_GRANT_READ_URI_PERMISSION //授予临时读取权限
                    mContext.setResult(RESULT_OK, intent)
                    mContext.finish()//关闭返回到其他App
                } else {
                    if(mPreviewActivity != null) {
                        intent = Intent(mContext, mPreviewActivity)
                        intent.setDataAndType(videoUri, "video/mp4")
                        mContext.startActivity(intent) //启动播放录制视频的Activity
                        mHandler.obtainMessage(MSG_STOP).sendToTarget()
                    }
                }
                if (state == STOP_PAUSE) return
                startPreview()
            } catch (e: Exception) {
                mMediaRecorder = null
                mMediaRecorder = MediaRecorder()
                setUpMediaRecorder()
            }
            mMediaRecorder!!.release()
            mMediaRecorder = null
        }
    }
    private val mHandler: Handler = object : Handler(Looper.myLooper()!!) { //处理UI
        override fun handleMessage(msg: Message) {
            when (msg.what) {
```

```
286                 MSG_START -> mButton.text = "停止"
287                 MSG_STOP -> mButton.text = "录制"
288             }
289             mButton.isEnabled = true
290         }
291     }
293     private fun updatePreview() { //更新预览
294         try {
296             mCaptureRequestBuilder!!.set(CaptureRequest.CONTROL_MODE, CameraMetadata.
    CONTROL_MODE_AUTO) //设置自动对焦
298             mCaptureRequest = mCaptureRequestBuilder!!.build() //显示预览
299             val thread = HandlerThread("CameraPreview")
300             thread.start()
301             //捕捉图像会话设置重复请求
302             mCaptureSession!!.setRepeatingRequest(mCaptureRequest!!, null, mBackgroundHandler)
303         } catch (e: CameraAccessException) {
304             e.printStackTrace()
305         }
306     }
332     fun onRequestPermissionsResult(requestCode: Int) { //请求权限
333         Log.d(TAG, "onRequestPermissionsResult()")
334         if(requestCode == Permissions.REQUEST_PERMISSIONS && !Permissions.hasPermissionsGranted
    (mContext, PERMISSIONS))
335             Permissions.requestPermissions(mContext, PERMISSIONS)
336     }
337 }
```

第 29 行声明 Camera 类，包含 5 个参数。mContext 参数是包调用摄像头的 AppCompatActivity，nPreviewView 参数是预览摄像头画面的视图，mButton 参数是控制录制的按钮，mAuthority 参数是保字视频路径的授权，mPreviewActivity 参数是录制视频的预览 AppCompatActivity 的类。第 61～69 行是 PreviewTextureView 控件的监听器，当 PreviewTextureView 控件的 SurfaceTexture 可用时调用 ›penCamera(Int, Int)方法。第 71～90 行是摄像头状态的回调，当摄像头打开后开始预览摄像头的画面，当摄像头断开连接或出现错误时关闭设备。第 108～113 行开启后台线程，mBackgroundHandler 作为 J开摄像头、预览捕捉会话和录制捕捉会话的线程，以避免主线程的阻塞。第 185～193 行创建摄像头预览视频的捕捉会话，当配置完成后，使用 updatePreview()方法更新预览图像，实现视频的预览。第 221～233 行创建摄像头录制视频的捕捉会话，当配置完成后，调用 mMediaRecorder.start()方法录制视频。第 261～267 行判断 mPreviewActivity 不为 null 时，将视频地址传递给 Intent 对象，打开预览的 Activity 进行预览。由于 Intent 是 Java 类，构造方法的第二个参数类型是 Class<?>，所以声明该类时 nPreviewActivity 参数需要将类型转换为 Class<*>，Class<*>表示 Java 类。

4. 主界面的 Activity

/java/com/teachol/c1104/MainActivity.kt

```
class MainActivity : AppCompatActivity(), View.OnClickListener {
    private lateinit var mPreviewView: PreviewTextureView //预览拍摄画面
    private lateinit var mButton: Button //录制按钮
    private lateinit var mCamera:Camera
    override fun onCreate(savedInstanceState: Bundle?) {
        super.onCreate(savedInstanceState)
```

```
14          setContentView(R.layout.activity_main)
17          mPreviewView = findViewById(R.id.texture_view)
18          mButton = findViewById(R.id.button)
19          mButton.setOnClickListener(this)
20          mCamera = Camera(this,mPreviewView,mButton,"com.teachol.c1104.fileprovider",
    PreviewActivity::class.java)
21      }
23      override fun onResume() { //Activity恢复到前台事件
24          super.onResume()
25          mCamera.onResume()
26      }
28      override fun onPause() { //Activity后台运行事件
29          super.onPause()
30          mCamera.onPause()
31      }
33      override fun onClick(v: View?) { //单击事件
34          if ( mCamera.mIsRecordingVideo) mCamera.stopRecordingVideo(Camera.STOP_NORMAL)
35          else mCamera.startRecordingVideo()
36      }
38      override fun onRequestPermissionsResult(requestCode: Int, permissions: Array<String?>,
    grantResults: IntArray) { //请求权限回调
39          mCamera.onRequestPermissionsResult(requestCode)
40      }
41  }
```

第 20 行 PreviewActivity::class.java 表示 PreviewActivity 是 Java 类参数。第 25 行 Activity 恢复到前台时调用 Camera 类的 onResume()方法。第 30 行 Activity 后台运行时调用 Camera 类的 onResume(方法。第 39 行权限请求回调事件时调用 Camera 类的 onRequestPermissionsResult (Int)方法。

11.3 录制音频

麦克风采集到的音频可以通过 MediaRecorder 类或 AudioRecord 类进行录制。使用 MediaRecorde类录制音频的方法和录制视频的方法是类似的，只需在录制参数中删除视频相关的内容即可。而AudioRecord 类提供的方法更加丰富，能够直接获取采集到的音频流数据。音频流数据是原始的 PCM格式音频数据，保存到文件后，需要添加头信息才能转为可以播放的 wav 格式文件。MP3 等压缩格式的音频文件需要进行转码，转为 MP3 格式通常使用第三方类库——Lame 库。

播放音频可以使用 MediaPlayer 类、AudioTrack 类、SoundPool 类或 Ringtone 类，它们的使用场景有所不同。MediaPlayer 类播放音频和播放视频的方法是一样的，支持添加 MediaController 类的控制器。AudioTrack 类用于播放 PCM 音乐流，支持流模式，播放其他压缩格式需要先进行解码。SoundPool类用于多个短音频文件交错密集播放，适合游戏、乐器类的 App。Ringtone 类用于铃声、通知声的播放，可以通过 RingtoneManager 类获取系统铃声列表。

11.3.1 AudioRecord 类

AudioRecord 类(android.media.AudioRecord)用于读取麦克风采集的音频数据流(常用方法如表11-14 所示)。创建后，AudioRecord 对象将初始化其关联的音频缓冲区，它将用新的音频数据填充

在构造期间指定的此缓冲区大小确定 AudioRecord 在尚未读取的“超速运行”数据之前可以记录多长时间。注意，应从音频硬件读取数据，其大小应小于总记录缓冲区的大小。

表 11-14　AudioRecord 类的常用方法

类型和修饰符	方　法
	<init>(audioSource: Int, sampleRateInHz: Int, channelConfig: Int, audioFormat: Int, bufferSizeInBytes: Int) 主构造方法。audioSource 参数使用以下 MediaRecorder.AudioSource 类定义的常量表示不同的音源模式。 ● DEFAULT：默认模式 ● MIC：麦克风模式 ● VOICE_UPLINK：电话上行音频模式 ● VOICE_DOWNLINK：电话下行音频模式 ● VOICE_CALL：电话模式 ● CAMCORDER：麦克风方向模式，根据开启的摄像头方向选择 ● VOICE_RECOGNITION：识别模式，先进行声音识别，然后进行录制 ● VOICE_COMMUNICATION：交流模式，开启回声消除和自动增益 ● REMOTE_SUBMIX：远程混合模式，录制系统内置声音
open Int	getAudioSource() 获取音源
open AudioFormat	getFormat() 获取音频录制的格式
open static Int	getMinBufferSize(sampleRateInHz: Int, channelConfig: Int, audioFormat: Int) 获取最小缓冲区大小
open Int	getSampleRate() 获取音频采样率
open Int	read(byte[] audioData, int offsetInBytes, int sizeInBytes) 从音频输入设备读取音频数据流并保存到 byte 数组中
open Unit	release() 释放 AudioRecord 实例的占用资源
open Unit	startRecording() 开始录制音频
open Unit	stop() 停止录制音频

11.3.2　AudioTrack 类

AudioTrack 类(android.media.AudioTrack)用于管理和播放单个音频资源(常用方法如表 11-15 所示)，支持静态和流式两种方式播放音频。

表 11-15　AudioTrack 类的常用方法

类型和修饰符	方　法
	<init>(attributes: AudioAttributes!, format: AudioFormat!, bufferSizeInBytes: Int, mode: Int, sessionId: Int) 主构造方法
open static Float	getMaxVolume() 获取最大音量
open static Int	getMinBufferSize(sampleRateInHz: Int, channelConfig: Int, audioFormat: Int) 获取最小缓冲区大小
open static Float	getMinVolume() 获取最小音量

续表

类型和修饰符	方　法
open Int	getPlaybackRate() 获取回放采样频率
open Int	getSampleRate() 获取音频采样频率
open Unit	pause() 暂停播放音频
open Unit	play() 播放音频
open Unit	release() 释放资源
open Int	setPlaybackRate(sampleRateInHz: Int) 设置回放的采样速率
open Int	setVolume(gain: Float) 设置音量
open Unit	stop() 停止播放音频
open Int	write(byte[] audioData, int offsetInBytes, int sizeInBytes) 写入播放的音频流数

11.3.3　实例工程：使用 AudioRecord 类录音

本实例演示了使用不同方式通过 AudioRecord 录制音频，然后使用 AudioTrack 播放录音（如图 11-8 所示）。单击“开始录音”按钮开始录音后会显示已经录制的时间，底部显示录制音频文件保存的文件路径。该按钮文字变成“暂停”，单击后暂停录音，按钮文字变成“继续”，再次单击后继续录音。单击“结束录音”按钮停止录音，单击“播放声音”按钮播放录制的音频。由于虚拟设备无法使用麦克风，所以需要使用物理设备才能录制和播放音频。

图 11-8　运行效果

1. 打开基础工程

打开“基础工程”文件夹中的“C1105”工程，该工程已经包含 MainActivity、Permissions、Util 类及资源文件，AndroidManifest.xml 文件中还添加了录音和存储文件所需的权限。Util 类中已经添加了两个可重用的静态方法——pcmToWave (String?,String?,long?,int?) 方法和 writeWaveFileHeader(FileOutputStream?, long?, long?, long?, int?, long?)方法，用于转为 WAV 文件。

2. 主界面的 Activity

/java/com/teachol/c1105/MainActivity.kt

```
override fun onCheckedChanged(group: RadioGroup?, checkedId: Int) { //单选按钮选项改变监听事件
    when (checkedId) {
        R.id.default_radio_button -> mAudioSource = MediaRecorder.AudioSource.DEFAULT
        R.id.mic_radio_button -> mAudioSource = MediaRecorder.AudioSource.MIC
        R.id.call_radio_button -> mAudioSource = MediaRecorder.AudioSource.VOICE_CALL
```

```
        R.id.communication_radio_button -> mAudioSource = MediaRecorder.AudioSource.VOICE_COMMUNICATION
    }
}
private fun startRecording() { //开始录音
    mIsStartRecording = true
    mIsPauseRecording = false
    mPlayButton.isEnabled = false
    mStartButton.text = "暂停"
    mPCMFile = Util.creatFile(getExternalFilesDir("audio")!!.absolutePath, "", "pcm")
    mWAVFile = Util.creatFile(getExternalFilesDir("audio")!!.absolutePath, "", "wav")
    mLogTextView.text = mPCMFile!!.absolutePath
    mBufferSize = AudioRecord.getMinBufferSize(mInSampleRate, mInChannelConfig, mInAudioFormat)
    mAudioData = ByteArray(mBufferSize)
    mAudioRecord = AudioRecord(mAudioSource, mInSampleRate, mInChannelConfig, mInAudioFormat, mBufferSize)
    mAudioRecord!!.startRecording() //开始录音(此后能够读取音频数据)
    mTimerThread = Thread(this) //实例化计时线程
    mTimerThread!!.start()//开启计时线程
    mExecutor.execute { //录制音频的线程
        try {
            val outputStream = FileOutputStream(mPCMFile!!.absoluteFile)
            while (mIsStartRecording) { //开始录制时循环
                while (mIsStartRecording && mIsPauseRecording) { //暂停时循环
                    try {
                        Thread.sleep(100) //休眠 100 毫秒，避免暂停循环次数过多
                    } catch (e: InterruptedException) {
                        e.printStackTrace()
                    }
                }
                mAudioRecord!!.read(mAudioData, 0, mAudioData.size)
                outputStream.write(mAudioData)
            }
            outputStream.close()
            Util.pcmToWave(mPCMFile!!.absolutePath, mWAVFile!!.absolutePath, mInSampleRate.toLong(), mBufferSize) //将 PCM 文件转为 WAV 文件
        } catch (e: FileNotFoundException) {
            e.printStackTrace()
        } catch (e: IOException) {
            e.printStackTrace()
        }
    }
}
private fun pauseRecord() { //暂停录音
    mAudioRecord!!.stop()
    mIsPauseRecording = true
    mStartButton.text = "继续"
}
private fun continueRecord() { //继续录音
    mAudioRecord!!.startRecording()
```

```
        mIsPauseRecording = false
        mStartButton.text = "暂停"
    }
    private fun stopRecord() { //停止录音
        mAudioRecord!!.stop()
        mIsStartRecording = false
        mIsPauseRecording = true
        mStartButton.text = "开始录音"
        Toast.makeText(this, "录音已结束", Toast.LENGTH_SHORT).show()
        mPlayButton.isEnabled = true
    }
    private fun createAudioTrack() { //创建音轨
        val mBufferSizeInBytes = AudioTrack.getMinBufferSize(mOutSampleRate, mOutChannelConfig,
mOutAudioFormat)
        check(mBufferSizeInBytes > 0) { "最小缓冲区尺寸: $mBufferSizeInBytes" }
        mAudioTrack = AudioTrack.Builder()
            .setAudioAttributes( //设置属性
                AudioAttributes.Builder()
                    .setUsage(AudioAttributes.USAGE_MEDIA)
                    .setContentType(AudioAttributes.CONTENT_TYPE_MUSIC)
                    .setLegacyStreamType(AudioManager.STREAM_MUSIC)
                    .build()
            )
            .setAudioFormat( //设置格式
                AudioFormat.Builder()
                    .setEncoding(mInAudioFormat)
                    .setSampleRate(mInSampleRate)
                    .setChannelMask(mInChannelConfig)
                    .build()
            )
            .setTransferMode(AudioTrack.MODE_STREAM) //设置转换模式
            .setBufferSizeInBytes(mBufferSizeInBytes) //设置缓冲区大小
            .build()
        mAudioTrack!!.setVolume(1.0f)
    }
    private fun play() { //播放录音
        mExecutor.execute {
            try {
                playAudioData(mWAVFile) //播放
            } catch (e: IOException) {
                mMainHandler.post { Toast.makeText(mContext, "录音播放出现错误",
Toast.LENGTH_SHORT).show() }
            }
        }
    }
    private fun playAudioData(audioFile: File?) { //播放音频数据
        val fis = FileInputStream(audioFile)
        val dis = DataInputStream(BufferedInputStream(fis))
        val bytes = ByteArray(mInSampleRate)
        var len: Int
```

```
226     mAudioTrack!!.play() //播放音轨(此时不会播放出声音)
227     while (dis.read(bytes).also { len = it } != -1) mAudioTrack!!.write(bytes, 0, len) //写入音轨播放音频
228     mMainHandler.post { Toast.makeText(mContext, "录音播放完毕", Toast.LENGTH_SHORT).show() }
229     dis.close()
230 }
```

第 113 ~ 120 行是单选按钮选项改变监听事件，选择相应的选项后设置不同的音源类型。第 122 ~ 164 行通过 mExecutor 对象开启新线程循环读取音频流，直到停止录制为止，当暂停时执行内部循环暂停 100 毫秒。第 187 ~ 209 行实例化 mAudioTrack 对象创建音轨，并设置最大音量。第 211 ~ 219 行开启线程池播放录制音频，以避免线程阻塞。第 221 ~ 230 行通过循环方式读取音频数据并进行播放。

11.3.4 实例工程：使用 MediaRecorder 类录音

本实例演示了使用不同方式通过 MediaRecorder 录制 AMR 格式音频，然后使用 MediaPlayer 播放录音（如图 11-9 所示）。按钮用法与上一个实例相同。

图 11-9　运行效果

提示：AMR 格式

AMR 格式(Adaptive Multi-Rate)是欧洲通信标准化委员会提出的音频标准格式，比 MP3 格式的压缩比更大。

1. 打开基础工程

打开“基础工程”文件夹中的“C1106”工程，该工程已经包含 MainActivity、Permissions、Util 类及资源文件，AndroidManifest.xml 中已经添加录音和存储文件所需的权限。

2. 主界面的 Activity

/java/com/teachol/c1106/MainActivity.kt

```
private fun startRecord() { //开始录音
    try {
        if (!mIsStartRecording) {
            mAudioFile = Util.creatFile(getExternalFilesDir("audio")!!.absolutePath, "", "amr")
            mLogTextView.text = mAudioFile.absolutePath
            mMediaRecorder = MediaRecorder()
```

```
99              mMediaRecorder.setAudioSource(mAudioSource) //音频输入源
100             mMediaRecorder.setOutputFormat(MediaRecorder.OutputFormat.AMR_WB) //设置输出格式
101             mMediaRecorder.setAudioEncoder(MediaRecorder.AudioEncoder.AMR_WB) //设置编码格式
102             mMediaRecorder.setOutputFile(mAudioFile.absolutePath) //设置输出文件
103             mMediaRecorder.prepare() //录音准备
104             mIsStartRecording = true
105             mIsPauseRecording = false
106         }
107         if (mIsPauseRecording) {
108             mMediaRecorder.resume() //继续录音
109             mIsPauseRecording = false
110         } else mMediaRecorder.start() //开始录音
111         mStartButton.text = "暂停"
113         mTimerThread = Thread(this) //实例化计时线程
114         mTimerThread.start() //开启计时线程
115     } catch (e: IOException) { e.printStackTrace() }
118 }
120 private fun pauseRecord() { //暂停
121     mMediaRecorder.pause() //暂停录音
122     mIsPauseRecording = true
123     mStartButton.text = "继续"
124 }
126 private fun stopRecord() { //停止
127     mMediaRecorder.stop() //停止录音
128     mMediaRecorder.release()
129     mIsStartRecording = false
130     mIsPauseRecording = true
131     mStartButton.text = "开始录音"
132     Toast.makeText(this, "录音结束", Toast.LENGTH_SHORT).show()
133 }
135 private fun play() { //播放录音
136     try {
137         val audioAttributes = AudioAttributes.Builder()
138             .setUsage(AudioAttributes.USAGE_MEDIA)
139             .setContentType(AudioAttributes.CONTENT_TYPE_SPEECH)
140             .build()
141         val uri = Uri.fromFile(mAudioFile)
142         val mediaPlayer = MediaPlayer()
143         mediaPlayer.setAudioAttributes(audioAttributes)
144         mediaPlayer.setDataSource(applicationContext, uri)
145         mediaPlayer.prepare()
146         mediaPlayer.start()
147     } catch (e: IOException) {
148         e.printStackTrace()
149     }
150 }
```

第 95 ~ 114 行开始录音前配置相关的参数，使用 mMediaRecorder.start()方法开始录音或暂停后使用 mMediaRecorder.resume()方法继续录音，然后通过 mTimerThread 对象开启新线程计时。第 121 行使用 mMediaRecorder.pause()方法暂停录音。第 127 行使用 mMediaRecorder.stop()方法停止录音，停

止录音后会自动保存设置的输出文件。第 137 ~ 140 行创建音频属性。第 142 ~ 146 行通过 mediaPlayer 对象播放指定的 URI 路径的录音文件。

11.4 传 感 器

传感器是用于感知外部环境数据的设备。在广义范畴上，摄像头、麦克风和 GPS 都属于传感器。Android 系统所指的传感器是狭义范畴的，是指除摄像头、麦克风和 GPS 外的能够感知外部环境数据的设备，这些传感器都能使用 SensorManager 类进行管理。Android 设备内置的传感器用于测量运动、屏幕方向和各种环境条件，能提供高度精确的原始数据用来监测 Android 设备的位置、运动或周围环境的变化。

11.4.1 Sensor 组件

1. SensorManager 类

SensorManager 类(android.hardware.SensorManager)用于获取和访问传感器、注册和注销传感器事件监听器，以及获取屏幕方向的数据(常用方法如表 11-16 所示)。它还提供了几个传感器常量，用于报告传感器精确度，设置数据采集频率和校准传感器。

表 11-16　SensorManager 类的常用方法

类型和修饰符	方　法
open static Float	getAltitude(p0: Float, p: Float) 获取海平面大气压力 p0 和指定大气压力 p 计算以 m 为单位的海拔高度
open static Unit	getAngleChange(angleChange: FloatArray!, R: FloatArray!, prevR: FloatArray!) 计算两个旋转矩阵之间的角度变化，使用 angleChange 参数进行存储
open Sensor!	getDefaultSensor(type: Int) 获取指定类型的默认传感器
open Sensor!	getDefaultSensor(type: Int, wakeUp: Boolean) 获取具有指定类型和唤醒属性的传感器
open static FloatArray!	getOrientation(R: FloatArray!, values: FloatArray!) 根据旋转矩阵计算设备的方向，使用 values 参数进行存储
open static Boolean	getRotationMatrix(R: FloatArray!, I: FloatArray!, gravity: FloatArray!, geomagnetic: FloatArray!) 通过重力矩阵 gravity 和地磁矩阵 geomagnetic 从设备坐标系转换为世界坐标系的倾角矩阵 I 和旋转矩阵 R
open static Unit	getRotationMatrixFromVector(R: FloatArray!, rotationVector: FloatArray!) 将旋转矢量转换为旋转矩阵，使用 R 参数进行存储
open Unit	registerDynamicSensorCallback(callback: SensorManager.DynamicSensorCallback!) 注册动态传感器的回调
open Boolean	registerListener(listener: SensorEventListener!, sensor: Sensor!, samplingPeriodUs: Int) 注册指定采样频率的传感器监听器
open Unit	unregisterDynamicSensorCallback(callback: SensorManager.DynamicSensorCallback!) 注销动态传感器的回调
open Unit	unregisterListener(listener: SensorListener!) 注销传感器的监听器

2. Sensor 类

Sensor 类(android.hardware.Sensor)用于获取和判断传感器的特性(常用方法如表 11-17 所示)，需要通过 SensorManager.getDefaultSensor(Int)方法或 SensorManager.getDefaultSensor(Int, Boolean)方法获

取指定类型的传感器实例。传感器的类型使用 Sensor 类的常量表示，有些是指硬件传感器，有些是指通过单个或多个硬件传感器虚拟的软件传感器。

表 11-17　Sensor 类的常用方法

类型和修饰符	方　法
Int	getMaxDelay() 获取两个传感器事件之间的最大延迟，仅针对连续变化传感器
Int	getMinDelay() 获取传感器测量数据更改时返回数值的最小延迟时间（μs）
String	getStringType() 获取传感器类型的字符串
Int	getType() 获取传感器的类型
Int	getVersion() 获取传感器的版本号

3．SensorEvent 类

SensorEvent 类（android.hardware.SensorEvent）提供有关传感器事件的信息，包括发生事件的传感器的数据、传感器类型、数据精度和时间戳（属性如表 11-18 所示）。

表 11-18　SensorEvent 类的属性

类型和修饰符	属　性
Int	accuracy 事件的数据精度。数据精度的数值可用以下常量表示。 ● SensorManager.SENSOR_STATUS_ACCURACY_HIGH：高精度值 ● SensorManager.SENSOR_STATUS_ACCURACY_LOW：低精度值 ● SensorManager.SENSOR_STATUS_ACCURACY_MEDIUM：中等精度值 ● SensorManager.SENSOR_STATUS_ACCURACY_UNRELIABLE：精度值不可靠
Sensor!	sensor 事件的传感器类型
Long	timestamp 事件的时间戳
FloatArray!	values 事件的数据

4．SensorEventListener 接口

SensorEventListener 接口（android.hardware.SensorEventListener）用于监听传感器获取的数值或传感器精确度的变化（方法如表 11-19 所示）。

表 11-19　SensorEventListener 接口的方法

类型和修饰符	方　法
abstract Unit	onAccuracyChanged(sensor: Sensor!, accuracy: Int) 当传感器精度改变时调用
abstract Unit	onSensorChanged(event: SensorEvent!) 当传感器值改变时调用

11.4.2　运动类传感器

运动类传感器用于测量设备的运动（如倾斜、晃动、旋转或摆动），包含加速度传感器、重力传感器、陀螺仪传感器和旋转矢量传感器，如表 11-20 所示。

表 11-20　Sensor 类的运动类传感器类型常量

传感器类型	传感器事件数据	说　明
TYPE_ACCELEROMETER 加速度传感器(单位：m/s^2)	SensorEvent.values[0]	沿 x 轴的加速力(包括重力)
	SensorEvent.values[1]	沿 y 轴的加速力(包括重力)
	SensorEvent.values[2]	沿 z 轴的加速力(包括重力)
TYPE_ACCELEROMETER_UNCALIBRATED 未经校准的加速度传感器(单位：m/s^2)	SensorEvent.values[0]	沿 x 轴的加速度，无偏差补偿
	SensorEvent.values[1]	沿 y 轴的加速度，无偏差补偿
	SensorEvent.values[2]	沿 z 轴的加速度，无偏差补偿
	SensorEvent.values[3]	沿 x 轴的加速度，有估算的偏差补偿
	SensorEvent.values[4]	沿 y 轴的加速度，有估算的偏差补偿
	SensorEvent.values[5]	沿 z 轴的加速度，有估算的偏差补偿
TYPE_GRAVITY 重力传感器(单位：m/s^2)	SensorEvent.values[0]	沿 x 轴的重力
	SensorEvent.values[1]	沿 y 轴的重力
	SensorEvent.values[2]	沿 z 轴的重力
TYPE_GYROSCOPE 陀螺仪传感器(单位：弧度/秒)	SensorEvent.values[0]	绕 x 轴的旋转速率
	SensorEvent.values[1]	绕 y 轴的旋转速率
	SensorEvent.values[2]	绕 z 轴的旋转速率
TYPE_GYROSCOPE_UNCALIBRATED 未经校准的陀螺仪传感器(单位：弧度/秒)	SensorEvent.values[0]	绕 x 轴的旋转速率，无漂移补偿
	SensorEvent.values[1]	绕 y 轴的旋转速率，无漂移补偿
	SensorEvent.values[2]	绕 z 轴的旋转速率，无漂移补偿
	SensorEvent.values[3]	绕 x 轴的估算漂移
	SensorEvent.values[4]	绕 y 轴的估算漂移
	SensorEvent.values[5]	绕 z 轴的估算漂移
TYPE_LINEAR_ACCELERATION 线性加速度传感器(单位：m/s^2)	SensorEvent.values[0]	沿 x 轴的加速力(不包括重力)
	SensorEvent.values[1]	沿 y 轴的加速力(不包括重力)
	SensorEvent.values[2]	沿 z 轴的加速力(不包括重力)
TYPE_ROTATION_VECTOR 旋转矢量传感器	SensorEvent.values[0]	沿 x 轴的旋转矢量分量(x*sin(θ/2))
	SensorEvent.values[1]	沿 y 轴的旋转矢量分量(y*sin(θ/2))
	SensorEvent.values[2]	沿 z 轴的旋转矢量分量(z*sin(θ/2))
	SensorEvent.values[3]	旋转矢量的标量分量((cos(θ/2))

11.4.3　实例工程：摇一摇比大小

本实例演示了摇动手机产生 0～9 随机数的聚会小游戏。单击“开始”按钮，摇动手机会自动生成数字。如果使用虚拟设备运行，需要打开虚拟传感器的控制界面，左右拖曳沿 y 轴移动的滑动条模拟上下摇晃效果(如图 11-10 所示)。

1. 打开基础工程

打开“基础工程”文件夹中的“C1107”工程，该工程已经包含 MainActivity 类及资源文件。

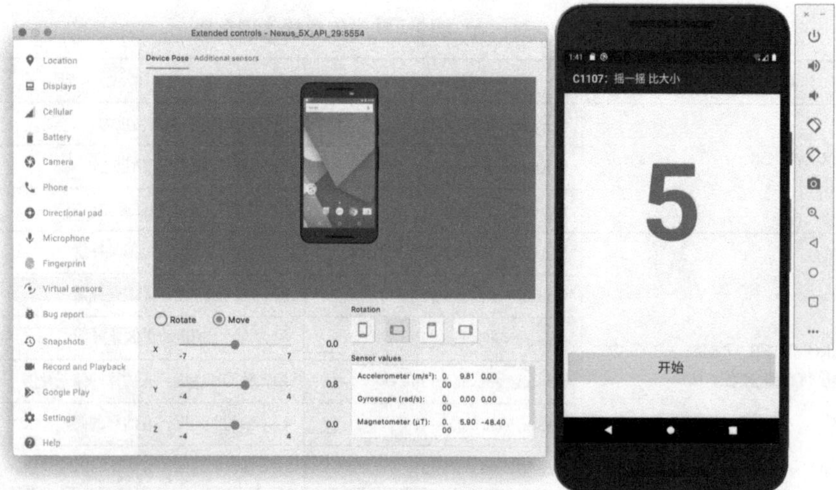

图 11-10　运行效果

2. 主界面的 Activity

/java/com/teachol/c1107/MainActivity.kt

```
private fun init() { //初始化传感器
    mSensorManager = getSystemService(SENSOR_SERVICE) as SensorManager //实例化传感器管理者
    mSensor = mSensorManager.getDefaultSensor(Sensor.TYPE_ACCELEROMETER) //初始化加速度传感器
    mSensorManager.registerListener(this, mSensor, SensorManager.SENSOR_DELAY_UI) //注册传感器监听器
}
override fun onClick(v: View) {
    mNumTextView.text = "0"
    if (mIsStart) {
        mStartButton.text = "开始"
        mIsStart = false
    } else {
        mStartButton.text = "停止"
        mIsStart = true
        mTimerThread = Thread(this) //实例化计时线程
        mTimerThread.start() //开启计时线程
    }
}
override fun onSensorChanged(event: SensorEvent) { //传感器数值改变事件
    if (!mIsStart) { return }
    val value = event.values
    mCurrentValue = getMod(value[0], value[1], value[2])
    if (!mMaxState) { //监测正向峰值
        if (mCurrentValue >= mLastValue) {
            mLastValue = mCurrentValue
        } else {
            if (Math.abs(mCurrentValue - mLastValue) > mMotionRange) {
                mMaxState = true
            }
        }
```

```
71          }
73          if (mMaxState) { //监测反向峰值
74              if (mCurrentValue <= mLastValue) {
75                  mLastValue = mCurrentValue
76              } else {
77                  if (Math.abs(mCurrentValue - mLastValue) > mMotionRange) {
78                      mCount++
79                      mMaxState = false
80                  }
81              }
82          }
83      }
87      private fun getMod(x: Float, y: Float, z: Float): Double { //向量求模
88          return Math.sqrt(x * x + y * y + (z * z).toDouble())
89      }
90      override fun run() {
92          while (mCount < mStopCount) { //达到摇晃次数结束循环
93              try {
94                  Thread.sleep(1000)
95              } catch (e: InterruptedException) {
96                  e.printStackTrace()
97              }
98              if (!mIsStart) { return }
99          }
101         mMainHandler.post { //摇晃结束
102             mCount = 0
103             mIsStart = false
104             mStartButton.text = "开始"
105             mNumTextView.setText((Math.random() * 10).toInt().toString())
106         }
107     }
108     override fun onDestroy() {
109         super.onDestroy()
110         mSensorManager.unregisterListener(this)
111     }
```

第 39 ~ 43 行初始化加速度传感器，并注册监听器。第 58 ~ 83 行重写 SensorEventListener 接口的 onSensorChanged(SensorEvent) 方法，当检测到 y 轴的正向峰值到负向峰值后累加摇动次数 mCount。第 90 ~ 107 行重写 Runnable 接口的 run() 方法，如果摇动次数 mCount 等于设定的停止摇动次数 mStopCount 则退出循环，并通过 mMainHandler.post(Runnable) 方法对 UI 进行操作显示生成的随机数。第 108 ~ 111 行重写 onDestroy() 方法，当 MainActivity 销毁时注销传感器的监听器，避免关闭 MainActivity 时报错。

11.4.4　位置类传感器

位置类传感器用于测量设备的物理位置或与附近物体的距离，包含旋转矢量传感器、地磁场传感器和近程传感器（如表 11-21 所示）。

表 11-21　Sensor 类的位置类传感器类型常量

传感器类型	传感器事件数据	说　明
TYPE_GAME_ROTATION_VECTOR 游戏旋转矢量传感器	SensorEvent.values[0]	沿 x 轴的旋转矢量分量($x * \sin(\theta/2)$)
	SensorEvent.values[1]	沿 y 轴的旋转矢量分量($y * \sin(\theta/2)$)
	SensorEvent.values[2]	沿 z 轴的旋转矢量分量($z * \sin(\theta/2)$)
TYPE_GEOMAGNETIC_ROTATION_VECTOR 地磁旋转矢量传感器	SensorEvent.values[0]	沿 x 轴的旋转矢量分量($x * \sin(\theta/2)$)
	SensorEvent.values[1]	沿 y 轴的旋转矢量分量($y * \sin(\theta/2)$)
	SensorEvent.values[2]	沿 z 轴的旋转矢量分量($z * \sin(\theta/2)$)
TYPE_MAGNETIC_FIELD 地磁场传感器(单位：μT)	SensorEvent.values[0]	沿 x 轴的地磁场强度
	SensorEvent.values[1]	沿 y 轴的地磁场强度
	SensorEvent.values[2]	沿 z 轴的地磁场强度
TYPE_MAGNETIC_FIELD_UNCALIBRATED 未经校准的磁力计(单位：μT)	SensorEvent.values[0]	沿 x 轴的地磁场强度(无硬铁校准)
	SensorEvent.values[1]	沿 y 轴的地磁场强度(无硬铁校准)
	SensorEvent.values[2]	沿 z 轴的地磁场强度(无硬铁校准)
	SensorEvent.values[3]	沿 x 轴的铁偏差估算
	SensorEvent.values[4]	沿 y 轴的铁偏差估算
	SensorEvent.values[5]	沿 z 轴的铁偏差估算
TYPE_PROXIMITY 近程传感器(单位：cm)	SensorEvent.values[0]	与物体的距离

11.4.5　实例工程：指南针

本实例演示了指南针及手机姿态的数据，单击虚拟设备右侧的旋转按钮可以改变手机的方向，观察到方向和数据的变化(如图 11-11 所示)。手机方位的数据还可以用于拍摄照片和录制视频时保存文件的方向，适用于解决之前实例工程中横向拍摄保存的照片方向错误的问题。

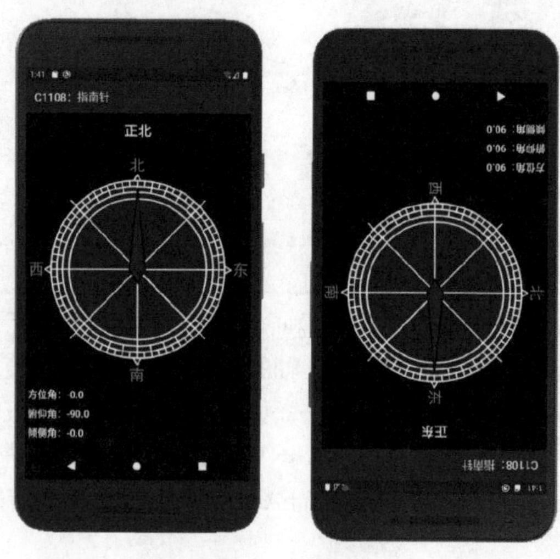

图 11-11　运行效果

1. 打开基础工程

打开“基础工程”文件夹中的“C1108”工程，该工程已经包含 MainActivity 类及资源文件。

2. 主界面的 Activity

/java/com/teachol/c1108/MainActivity.kt

```
private fun init() { //初始化
    mSensorManager = getSystemService(SENSOR_SERVICE) as SensorManager //实例化传感器管理者
    mAccelerometer = mSensorManager.getDefaultSensor(Sensor.TYPE_ACCELEROMETER) //初始化加速度传感器
    mMagnetic = mSensorManager.getDefaultSensor(Sensor.TYPE_MAGNETIC_FIELD) //初始化地磁场传感器
    calculateOrientation() //计算方向
}
override fun onResume() {
    mSensorManager.registerListener(OrientationSensorEventListener(), mAccelerometer, SensorManager.SENSOR_DELAY_NORMAL)
    mSensorManager.registerListener(OrientationSensorEventListener(), mMagnetic, SensorManager.SENSOR_DELAY_NORMAL)
    super.onResume()
}
override fun onPause() {
    mSensorManager.unregisterListener(OrientationSensorEventListener())
    super.onPause()
}
private fun calculateOrientation() { //计算方向
    val R = FloatArray(9) //旋转矩阵
    val values = FloatArray(3) //方位数组
    SensorManager.getRotationMatrix(R, null, mAccelerometerValues, mMagneticFieldValues) //获取旋转矩阵
    SensorManager.getOrientation(R, values) //获取方位数组
    values[0] = Math.toDegrees(values[0].toDouble()).toFloat()
    values[1] = Math.toDegrees(values[1].toDouble()).toFloat()
    values[2] = Math.toDegrees(values[2].toDouble()).toFloat()
    mTextView1.text = "方位角: ${values[0]}"
    mTextView2.text = "俯仰角: ${values[1]}"
    mTextView3.text = "倾侧角: ${values[2]}"
    if (values[0] >= -5 && values[0] < 5) { //判断手机正前方水平方向
        mOrientationTextView.text = "正北"
    } else if (values[0] >= 5 && values[0] < 85) {
        mOrientationTextView.text = "东北"
    } else if (values[0] >= 85 && values[0] <= 95) {
        mOrientationTextView.text = "正东"
    } else if (values[0] >= 95 && values[0] < 175) {
        mOrientationTextView.text = "东南"
    } else if (values[0] >= 175 && values[0] <= 180 || values[0] >= -180 && values[0] < -175) {
        mOrientationTextView.text = "正南"
    } else if (values[0] >= -175 && values[0] < -95) {
        mOrientationTextView.text = "西南"
    } else if (values[0] >= -95 && values[0] < -85) {
        mOrientationTextView.text = "正西"
    } else if (values[0] >= -85 && values[0] < -5) {
        mOrientationTextView.text = "西北"
    }
    //转换指南针方向
    var degree = (-values[0]).toInt()
```

```
86      if (degree - mCurrentDegree > 180) {
87          degree -= 360
88      } else if (degree - mCurrentDegree < -180) {
89          degree += 360
90      }
92      if (Math.abs(degree - mCurrentDegree) > 2) { //改变值大于2度改变指南针方向(避免频繁抖动)
93          val ra = RotateAnimation(mCurrentDegree.toFloat(), degree.toFloat(), Animation.RELATIVE_
    TO_SELF, 0.5f, Animation.RELATIVE_TO_SELF, 0.5f)
94          ra.duration = 200 //旋转持续时间
95          ra.fillAfter = true //旋转结束后停留
96          mCompassImageView.startAnimation(ra) //开始旋转动画
97          mCurrentDegree = degree //更新当前指针角度
98      }
99  }
101 internal inner class OrientationSensorEventListener : SensorEventListener { //传感器事件监听器
102     override fun onSensorChanged(event: SensorEvent) {
103         if (event.sensor.type == Sensor.TYPE_ACCELEROMETER) {
104             mAccelerometerValues = event.values
105         }
106         if (event.sensor.type == Sensor.TYPE_MAGNETIC_FIELD) {
107             mMagneticFieldValues = event.values
108         }
109         calculateOrientation()
110     }
112 }
```

第 38～43 行初始化加速度传感器和地磁场传感器，并调用 calculateOrientation()方法计算手机指向的方向。第 44～48 行重写 onResume()方法，当 MainActivity 显示到前台后，注册加速度传感器和地磁场传感器的监听器。第 49～52 行重写 onPause()方法，注销传感器的监听器。第 55～65 行根据从传感器获取的方位角、俯仰角、倾侧角的数据，并根据方位角旋转指南针。第 85～90 行由于方位角的数值范围是[−180,180]，手机水平旋转一周时方位角会从 180 直接变成−180，所以要对指南针方向进行转换。第 92～98 行由于方位角的精度很高，为了保证指南针的不抖动，且能够平滑旋转，所以设置方位角的数值变化大于 2 时才进行旋转。第 101～112 行自定义实现 SensorEventListener 接口的 OrientationSensorEventListener 类，重写 onSensorChanged(SensorEvent)方法获取加速度传感器和地磁场传感器的数据，最后调用 calculateOrientation()方法计算并显示方位。

11.4.6 环境类传感器

环境类传感器用于测量环境气温、气压、照度和湿度等参数，包含温度传感器、气压传感器、光度传感器和湿度传感器(如表 11-22 所示)。

表 11-22 Sensor 类的环境类传感器类型常量

传 感 器	传感器常量	传感器事件数据	数 据 说 明
温度传感器(单位：摄氏度)	TYPE_AMBIENT_TEMPERATURE	SensorEvent.values[0]	温度
光度传感器(单位：勒克斯)	TYPE_LIGHT	SensorEvent.values[0]	光度
气压传感器(单位：百帕斯卡)	TYPE_PRESSURE	SensorEvent.values[0]	空气压力
湿度传感器(单位：百分比)	TYPE_RELATIVE_HUMIDITY	SensorEvent.values[0]	相对湿度

11.4.7 实例工程：光照计和气压计

本实例演示了光照计和气压计，能够根据光照强度改变屏幕亮度。可以在虚拟设备中设置虚拟传感器的参数，观察模拟效果(如图 11-12 所示)。海拔高度是通过气压计算得到的，由于温度和风速会对气压产生影响，所以使用这种方式计算海拔高度的精度不高。

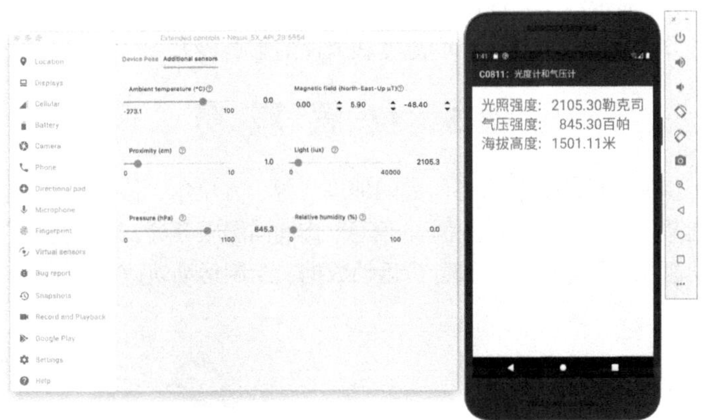

图 11-12 运行效果

1. 打开基础工程

打开“基础工程”文件夹中的“C1109”工程，该工程已经包含 MainActivity 类及资源文件。

2. 主界面的 Activity

/java/com/teachol/c1109/MainActivity.kt

```
private fun init() { //初始化
    mWindow = this.window
    mSensorManager = getSystemService(SENSOR_SERVICE) as SensorManager
    mLightSensor = mSensorManager.getDefaultSensor(Sensor.TYPE_LIGHT) //光度传感器
    mPressureSensor = mSensorManager.getDefaultSensor(Sensor.TYPE_PRESSURE) //气压传感器
    mSensorManager.registerListener(object : SensorEventListener { //注册光度传感器
        override fun onSensorChanged(event: SensorEvent) {
            val lux = event.values[0] //获取光照强度
            mLightTextView.text = String.format("%.2f",lux.toDouble())
            val lp = mWindow.getAttributes() //获取屏幕属性
            lp.screenBrightness = lux / 500 //重新计算亮度
            mWindow.setAttributes(lp) //设置屏幕亮度
        }
        override fun onAccuracyChanged(sensor: Sensor, accuracy: Int) {}
    }, mLightSensor, SensorManager.SENSOR_DELAY_NORMAL)
    mSensorManager.registerListener(object : SensorEventListener { //注册气压传感器
        override fun onSensorChanged(event: SensorEvent) {
            val pa = event.values[0].toDouble() //获取大气压力
            mPressureTextView.text = String.format("%.2f",pa)
            mAltitudeTextView.text = String.format("%.2f",altitude(pa))
        }
        override fun onAccuracyChanged(sensor: Sensor, accuracy: Int) {}
    }, mPressureSensor, SensorManager.SENSOR_DELAY_NORMAL)
}
private fun altitude(pa: Double): Double { //通过气压计算海拔高度(需要手机包含气压传感器)
```

```
59        val p0 = 1013.21f //海平面大气压
60        return 44300 * (1 - Math.pow(pa / p0, 1 / 5.256))
61    }
```

第 36 ~ 46 行注册光度传感器的监听器，并重写 onSensorChanged(SensorEvent)方法，用于格式化显示光照强度和根据光照强度改变屏幕亮度。第 48 ~ 55 行注册气压传感器，并重写 onSensorChanged(SensorEvent)方法，用于显示气压和通过气压计算的高度。第 58 ~ 61 行 altitude(double)通过气压计算海拔高度，使用这种方式计算的海拔高度受温度和风速的影响较大。

11.5 位 置 服 务

位置服务通过卫星或者网络定位设备所处的地理位置、方位角、移动速度等信息。社交类和本地服务类的 App 通过地理位置向用户推送相应的信息、查询周围服务等，运动类和户外类的 App 将地理位置、方位角、移动速度等信息记录下来分析运动数据、绘制运动轨迹等。

11.5.1 Location 组件

1. LocationManager 类

LocationManager 类(android.location.LocationManager)用于系统位置服务的访问管理(常用方法如表 11-23 所示)，系统位置服务定期更新位置或者进入指定地理位置发送通知。定位方式有 LocationManager.GPS_PROVIDER(GPS 获取定位信息)、LocationManager.NETWORK_PROVIDER(网络获取定位信息)和 LocationManager PASSIVE_PROVIDER(其他 App 提供定位信息)三种。

表 11-23 LocationManager 类的常用方法

类型和修饰符	方 法
open Boolean	addNmeaListener(listener: OnNmeaMessageListener, handler: Handler?) 添加一个 NMEA 监听器
open Unit	addProximityAlert(latitude: Double, longitude: Double, radius: Float, expiration: Long, intent: PendingIntent) 添加接近警报，根据位置(纬度、经度)和半径指定的位置设置接近警报
open MutableList<String!>	getAllProviders() 获取所有位置提供者的名称列表
open Unit	getCurrentLocation(provider: String, cancellationSignal: CancellationSignal?, executor: Executor, consumer: Consumer<Location!>) 异步获取当前位置
open String?	getGnssHardwareModelName() 获取 GNSS(全球导航卫星系统)硬件驱动程序的型号名称
open Location?	getLastKnownLocation(provider: String) 获取最后一个已知位置
open Boolean	isLocationEnabled() 判断位置服务是否启用
open Unit	removeNmeaListener(listener: OnNmeaMessageListener) 移除 NMEA 监听器
open Unit	removeProximityAlert(intent: PendingIntent) 移除接近警报
open Unit	removeUpdates(listener: LocationListener) 移除指定的 LocationListener 的位置更新监听器
open Unit	requestLocationUpdates(provider: String, minTimeMs: Long, minDistanceM: Float, listener: LocationListener) 注册位置提供者的位置更新监听器。minTimeMs 参数定位时间精度，到达该时间后无论是否达到定位距离精度都会更新定位数据。minDistanceM 参数定位距离精度，移动位置超过该距离后更新定位数据。这两个参数都设置为 0 时，只要位置发生改变就会更新定位数据

2. Location 类

Location 类（android.location.Location）用于存储、计算和转换地理位置（常用方法如表 11-24 所示），存储纬度、经度、高度、方位角和速度等数据。

表 11-24　Location 类的常用方法

类型和修饰符	方　法
	<init>(provider: String!) 主构造方法
open Float	bearingTo(dest: Location!) 沿此位置与给定位置之间的最短路径行驶时，返回近似初始方位角
open static String!	convert(coordinate: Double, outputType: Int) 将坐标转换为字符串形式
open static Double	convert(coordinate: String!) 将字符串形式的坐标转换为双精度型
open static Unit	distanceBetween(startLatitude: Double, startLongitude: Double, endLatitude: Double, endLongitude: Double, results: FloatArray!) 计算两个位置之间的近似距离（以 m 为单位），以及两个位置之间最短路径的初始和最终方位
open Float	distanceTo(dest: Location!) 返回此位置和给定位置之间的近似距离（以 m 为单位）
open Float	getAccuracy() 获取精度（以 m 为单位）
open Double	getAltitude() 获取海拔高度
open Float	getBearing() 获取方位角（以度为单位）
open Float	getBearingAccuracyDegrees() 获取方位角精度（以度为单位）
open Double	getLatitude() 获取纬度（以度为单位）
open Double	getLongitude() 获取经度（以度为单位）
open Float	getSpeed() 获取速度（以 m/s 为单位）
open Unit	reset() 清除位置的内容
open Unit	set(l: Location!) 将位置的内容设置为给定位置的值
open Unit	setAltitude(altitude: Double) 设置高度（以 m 为单位）
open Unit	setBearing(bearing: Float) 设置方位角（以度为单位）
open Unit	setLatitude(latitude: Double) 设置纬度（以度为单位）
open Unit	setLongitude(longitude: Double) 设置经度（以度为单位）
open Unit	setSpeed(speed: Float) 设置速度（以 m/s 为单位）

3. LocationListener 接口

LocationListener 接口（android.location.LocationListener）用于监听定位改变和位置服务改变的事件（方法如表 11-25 所示）。LocationManager.requestLocationUpdates（String,long,float,LocationListener）方法用于注册监听器，LocationManager.removeUpdates（LocationListener 方法）用于移除监听器。

表 11-25 LocationListener 接口的方法

类型和修饰符	方 法
abstract Unit	onLocationChanged（location: Location） 定位发生改变时调用该方法
open Unit	onProviderDisabled（provider: String） 关闭位置服务时调用该方法
open Unit	onProviderEnabled（provider: String） 开启位置服务时调用该方法

提示：反向地理编码

反向地理编码，又称为逆向地理编码，是根据 GPS 坐标反向查询所在的街道名称或建筑名称，在社交 App 中最为常见。该功能需要大量的数据采集和定期的数据更新，因此推荐使用第三方提供的服务实现该功能。国内推荐使用百度或高德开放平台，国外可以使用 Google Map。

11.5.2 实例工程：获取经纬度坐标

本实例演示了使用 GPS 卫星和网络进行定位，模拟器上切换定位方式时数据并没有发生变化（如图 11-13 所示）。因此，最好使用真机进行调试，将调试数据输出到 MainActivity 界面中，以便在室外环境下观察调试的数据。两种定位方式各有利弊：卫星定位精度较高，搜索到足够数量的卫星后获取定位，但是室内往往搜索不到卫星，无法使用卫星定位；网络定位精度较低，而且通常无法准确获取高度、速度和方位角，优势在于耗电量小。

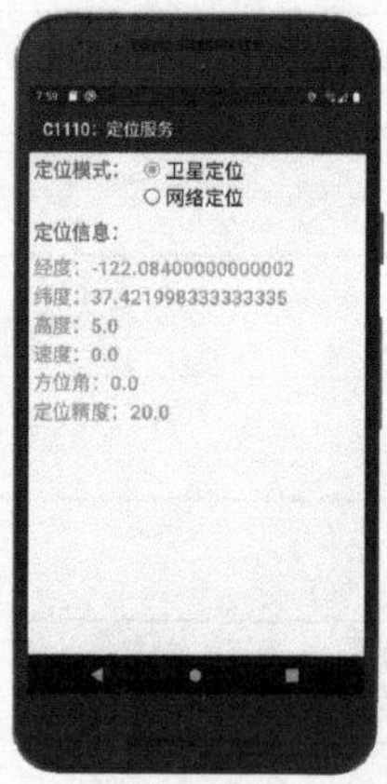

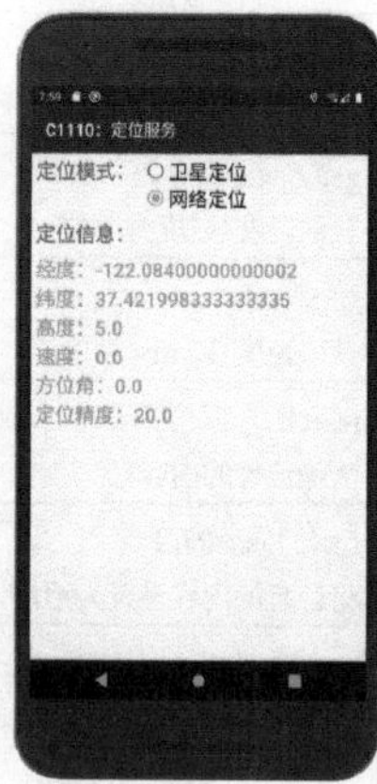

图 11-13 运行效果

1. 打开基础工程

打开“基础工程”文件夹中的“C1110”工程，该工程已经包含 Permissions 类、MainActivity 类及资源文件。在 AndroidManifest.xml 文件中添加了<uses-permission android:name = "android.permission ACCESS_FINE_LOCATION"/>权限，用于获取位置服务权限。

2. 主界面的 Activity

/java/com/teachol/c1110/MainActivity.kt

```
@SuppressLint ("MissingPermission")
private fun init() { //初始化
    mRadioGroup = findViewById(R.id.radio_group)
    mRadioGroup.setOnCheckedChangeListener(this)
    mLocationTextView = findViewById(R.id.location_text_view)
    //获取定位服务
    mLocationManager = getSystemService(LOCATION_SERVICE) as LocationManager
    //判断是否开启位置服务
    if (!mLocationManager!!.isProviderEnabled(LocationManager.GPS_PROVIDER)) {
        Toast.makeText(this@MainActivity, "请开启位置服务", Toast.LENGTH_SHORT).show()
        Thread.sleep(1000)
        val intent = Intent(Settings.ACTION_LOCATION_SOURCE_SETTINGS)
        startActivityForResult(intent, 0)
    } else { //默认使用 GPS 卫星进行定位
        mLocationManager!!.requestLocationUpdates(LocationManager.GPS_PROVIDER, 0, 0f, this)
    }
}
//OnCheckedChangeListener 接口方法，单选改变时调用
override fun onCheckedChanged(group: RadioGroup, checkedId: Int) {
    mLocationManager!!.removeUpdates(this) //移除更新监听器
    when (checkedId) {
        R.id.gnss_button -> mLocationManager!!.requestLocationUpdates(LocationManager.GPS_PROVIDER, 0, 0f, this)
        R.id.network_button -> mLocationManager!!.requestLocationUpdates(LocationManager.NETWORK_PROVIDER, 0, 0f, this)
    }
}
//LocationListener 接口方法，定位发生改变时调用
override fun onLocationChanged(location: Location) {
    Log.e("onLocationChanged", location.toString())
    displayLocationData(location)
}
//LocationListener 接口方法，开启位置服务后调用
override fun onProviderEnabled(provider: String) {
    Log.e("onProviderEnabled", mRadioGroup.checkedRadioButtonId.toString())
    mLocationManager!!.removeUpdates(this)
    if (mRadioGroup.checkedRadioButtonId == R.id.gnss_button) {
        mLocationManager!!.requestLocationUpdates(LocationManager.GPS_PROVIDER, 0, 0f, this)
    } else if (mRadioGroup.checkedRadioButtonId == R.id.network_button) {
        mLocationManager!!.requestLocationUpdates(LocationManager.NETWORK_PROVIDER, 0, 0f, this)
    }
    Toast.makeText(this@MainActivity, "LocationListener : $provider onProviderEnabled", Toast.LENGTH_SHORT).show()
}
//LocationListener 接口方法，关闭位置服务后调用
override fun onProviderDisabled(provider: String) {
    displayLocationData(null)
```

```
93        Toast.makeText(this@MainActivity, "LocationListener : $provider onProviderDisabled", Toast.LENGTH_SHORT).show()
94    }
95    //显示定位数据
96    private fun displayLocationData(location: Location?) {
97        if (location != null) {
98            val sb = StringBuilder()
99            sb.append("经度: ${location.longitude}\n")
100           sb.append("纬度: ${location.latitude}\n")
101           sb.append("高度: ${location.altitude}\n")
102           sb.append("速度: ${location.speed}\n")
103           sb.append("方位角: ${location.bearing}\n")
104           sb.append("定位精度: ${location.accuracy}")
105           mLocationTextView.text = sb.toString()
106       } else {
107           mLocationTextView.text = "无法获取定位数据"
108       }
109   }
```

第 46 行使用 SuppressLint 注解允许没有权限申请，否则第 60 行会报错。第 52 行获取定位服务后强制转换为 LocationManager 实例。第 54～61 行使用 isProviderEnabled()方法判断是否开启了 GPS 定位，如果没有开启则打开位置服务的设置，否则请求 GPS 定位更新并设置其监听器。第 65～71 行重写 onCheckedChanged(RadioGroup, Int)方法监听单选按钮的选项改变，用于切换 GPS 卫星定位和网络定位。第 73～76 行重写 onLocationChanged(Location)方法监听定位的改变，定位改变后调用 displayLocationData(Location)方法将定位信息显示出来。第 79～88 行重写 onProviderEnabled(String)方法根据定位模式的选项开启响应的位置服务，请求定位更新并设置其监听器。第 90～93 行重写 onProviderDisabled(String)方法关闭位置服务后显示信息。第95～108行 displayLocationData(Location?)方法用于显示定位数据，当数据为 null 时提示无法获取定位数据。

第 12 章 Android 的 HTTP 网络通信

HTTP 网络通信是实现起来较便捷的网络通信方式，特别适合对低延迟要求不高，且与服务器端非持续连接的数据通信。HTTP 网络通信需要客户端发送请求，服务器端才能发送数据。HTTP 除了不适合移动设备作为服务器端，还特别不适合应用于低数据延迟的网络游戏和实时音视频传输。

12.1 HttpURLConnection 类

HttpURLConnection 类（java.net.HttpURLConnection）用于使用 HTTP 进行数据的传输（常用方法如表 12-1 所示），GET 和 POST 方法是 HTTP 使用较频繁的方法。通过 HTTP 向服务器端发送请求后，服务器端会发送回状态码用于快速判断响应结果（如表 12-2 所示）。

表 12-1 HttpURLConnection 类的常用方法

类型和修饰符	方　法
	<init>(u: URL!) 主构造方法
Unit	addRequestProperty(key: String!, value: String!) 添加请求属性
Unit	connect() 打开连接
abstract Unit	disconnect() 断开连接
open String!	getHeaderField(n: Int) 获取消息头的字段数据
open Long	getHeaderFieldDate(name: String!, Default: Long) 获取消息头的日期字段数据
open String!	getHeaderFieldKey(n: Int) 获取消息头的字段
open String!	getRequestMethod() 获取请求方式
open Int	getResponseCode() 获取 HTTP 响应消息的状态代码
open String!	getResponseMessage() 获取 HTTP 的响应消息
Unit	setConnectTimeout(timeout: Int) 设置连接的超时时间（以 ms 为单位）
Unit	setReadTimeout(timeout: Int) 设置下载数据的超时时间（以 ms 为单位）
open Unit	setRequestMethod(method: String!) 设置请求方法（GET、POST、HEAD、OPTIONS、PUT、DELETE 或 TRACE）

表 12-2　HTTP 的常用状态码

状 态 码	HttpURLConnection 类的常量	含　义
200	HTTP_OK	请求已成功，请求的响应头或数据体将随之返回
400	HTTP_BAD_REQUEST	请求无效
404	HTTP_NOT_FOUND	请求失败
408	HTTP_CLIENT_TIMEOUT	请求超时
500	HTTP_INTERNAL_ERROR	服务器内部错误，多源于服务器端的源代码出现错误
505	HTTP_VERSION	服务器版本不支持

提示：HTTP 和 HTTPS

HTTP(Hyper Text Transfer Protocol)是从 Web 服务器传输超文本到本地浏览器的传送协议，基于 TCP/IP 来传递数据。浏览器作为客户端通过 URL 向服务器端的 Web 服务器发送所有请求，Web 服务器根据接收到的请求，向客户端发送响应信息。

HTTPS(Hyper Text Transfer Protocol over SecureSocket Layer)是由 HTTP 加上 TLS/SSL 协议构建的可进行加密传输、身份认证的网络协议，主要通过数字证书、加密算法、非对称密钥等技术完成互联网数据传输加密，实现互联网传输安全保护。相同网络环境下，HTTPS 会使页面的加载时间延长近 50%，且增加 10%到 20%的耗电。

12.2　实例工程：加载网络图片(带缓存)

本实例演示了使用 HttpURLConnection 类下载图片并进行本地缓存，以及清除本地缓存(如图 12-1 所示)。单击“加载图片”按钮，开始下载图片，图片下载完成后保存在本地缓存文件夹中，然后显示在 ImageView 控件中。由于没有下载提示，所以只有图片下载后才能看到下载是否成功。单击“清除缓存”按钮，清除本地缓存，ImageView 控件中显示 R.mipmap.img_error_m 资源图片。

图 12-1　运行效果

提示：模拟器无法联网

模拟器的控制器中，选择 Settings 菜单下的 Proxy 选项卡，取消勾选“Use Android Studio HTTP proxy settings”复选框，单击 APPLY 按钮取消代理，如图 12-2 所示。

图 12-2　模拟器的控制器

1. 打开基础工程

打开“基础工程”文件夹中的“C1201”工程，该工程已经包含 MainActivity、FileHelper 类及资源文件。

2. 设置权限

/manifests/AndroidManifest.xml

```
03 <uses-permission android:name = "android.permission.INTERNET"/>
04 <uses-permission android:name = "android.permission.ACCESS_NETWORK_STATE"/>
05 <application
12    android:usesCleartextTraffic = "true">
```

第 03 行 android.permission.INTERNET 是访问互联网权限。第 04 行 android.permission.ACCESS_NETWORK_STATE 是访问网络状态权限。第 12 行 android:usesCleartextTraffic = "true"表示允许使用明文访问网络，在 Android 9 或更高版本需要将该属性设置为 true，否则只能访问 HTTPS 地址，访问 HTTP 地址时会报错。

提示：添加网络权限调试无效

如果未添加网络权限时已经在模拟器中进行了调试，那么添加网络权限后可能会报错。此时在模拟器中将应用删除，重新调试即可。

3. Http 类

在“/java/com/teachol/c1201/”文件夹中，新建 Http.kt 文件，用于添加下载文件的静态方法。

/java/com/teachol/c1201/Http.kt

```
object Http {
   private const val EXCEPTION_FAILURE = "请求 URL 失败"
   //获取缓存文件
```

```
11      fun getCacheFile(path: String, cachePath: String): File{
12          val storageDir = File(cachePath) //本地缓存文件
13          if (!storageDir.exists()) storageDir.mkdirs()//如果缓存文件夹不存在则新建
14          val temp = path.split("/".toRegex()).toTypedArray()//拆分路径字符串
15          val fileName = temp[temp.size - 1] //获取文件名
16          return File(storageDir, fileName) //返回缓存文件
17      }
18      //下载文件
19      fun downloadFile(path: String, cachePath: String): File {
20          val cacheFile = getCacheFile(path, cachePath) //缓存文件
21          //下载请求
22          val url = URL(path) //将字符串格式的路径转为 URL
23          val conn = url.openConnection() as HttpURLConnection //创建连接
24          conn.connectTimeout = 5000 //设置连接超时时间
25          conn.readTimeout = 5000 //设置下载超时时间
26          conn.requestMethod = "GET" //设置请求类型为 Get 类型
27          //判断请求 Url 是否成功
28          if (conn.responseCode != 200) throw RuntimeException(EXCEPTION_FAILURE)
29          //如果有本地缓存文件则不进行下载而直接使用
30          if (cacheFile.exists()) { //判断是否缓存
31              //如果缓存文件与下载文件大小相等则直接返回缓存文件
32              if (cacheFile.length() == conn.contentLengthLong) return cacheFile
33          }
34          //下载文件
35          val inputStream = conn.inputStream //获取输入流
36          if (inputStream != null) {
37              val fileOutputStream = FileOutputStream(cacheFile) //文件输出流
38              val buf = ByteArray(1024)
39              var ch: Int
40              while (inputStream.read(buf).also { ch = it } != -1) {
41  //              Thread.sleep(10);//下载过快时使用进程休眠便于观察下载进度
42                  fileOutputStream.write(buf, 0, ch) //将获取到的流写入文件中
43              }
44              //释放资源
45              fileOutputStream.flush()
46              fileOutputStream.close()
47          }else throw RuntimeException(EXCEPTION_FAILURE)
48          conn.disconnect()
49          return cacheFile
50      }
51  }
```

第 11 ~ 17 行获取本地缓存文件，如果缓存目录不存在，则创建缓存目录。从路径中分离出文件名，在缓存文件夹中创建相同文件名的缓存文件。第 23 ~ 28 行通过 URL 发送请求，如果请求失败，则抛出异常。第 30 ~ 33 行如果缓存文件存在且文件大小相同，则直接加载缓存文件。第 35 ~ 47 行如果读取到了数据流，则将文件数据写入缓存文件，否则抛出异常。

4. ImageLoader 类

在“/java/com/teachol/c1201/”文件夹中，新建 ImageLoader.kt 文件，用于使用独立线程下载图片并显示在指定的 ImageView 控件中。

```
/java/com/teachol/c1201/ImageLoader.kt
10 class ImageLoader {
11     companion object {
12         const val MSG_OK = 0
13         const val MSG_ERROR = 1
14     }
15     private lateinit var mImageView: ImageView
16     private lateinit var mDownloadFile: File
17     //使用独立线程加载图片
18     fun displayImage(imageView: ImageView, imageUrl: String, cacheDir: String) {
19         mImageView = imageView
20         mImageView.setImageBitmap(null)
21         Thread {
22             try {
23                 mDownloadFile = Http.downloadFile(imageUrl, cacheDir)
24                 if (mDownloadFile.exists()) {
25                     mHandler.obtainMessage(MSG_OK).sendToTarget()
26                 } else {
27                     mHandler.obtainMessage(MSG_ERROR).sendToTarget()
28                 }
29             } catch (e: Exception) {
30                 e.printStackTrace()
31                 mDownloadFile = Http.getCacheFile(imageUrl, cacheDir) //缓存文件
32                 if (mDownloadFile.exists()) {
33                     mHandler.obtainMessage(MSG_OK).sendToTarget()
34                 } else {
35                     mHandler.obtainMessage(MSG_ERROR).sendToTarget()
36                 }
37             }
38         }.start()
39     }
40     //更新 UI
41     private val mHandler: Handler = object: Handler(Looper.getMainLooper()) {
42         override fun handleMessage(msg: Message) {
43             when(msg.what){
44                 MSG_OK -> mImageView.setImageURI(Uri.fromFile(mDownloadFile))
45                 MSG_ERROR -> //根据控件宽度显示不同的错误图片
46                     when {
47                         mImageView.width > 900 -> mImageView.setImageResource(R.mipmap.img_error_h)
48                         mImageView.width > 300 -> mImageView.setImageResource(R.mipmap.img_error_m)
49                         else -> mImageView.setImageResource(R.mipmap.img_error_s)
50                     }
51             }
52         }
53     }
54 }
```

第 21 ~ 38 行开启新线程加载图片，访问网络需要阻塞线程，所以不能在主线程执行。如果加载成功，使用 mHandler.obtainMessage(MSG_OK).sendToTarget() 发送成功的信息。第 41 ~ 53 行 mHandler 对象用于接收发送的信息，通过 handleMessage(Message) 方法根据发送的信息进行分类处理。

5. 主界面的 Activity

/java/com/teachol/c1201/MainActivity.kt

```
08 class MainActivity : AppCompatActivity() {
09     private lateinit var mUrl: String
10     private lateinit var mImageLoader: ImageLoader
11     private lateinit var mImageView: ImageView
12     override fun onCreate(savedInstanceState: Bundle?) {
16         mUrl = "http://www.weiju2014.com/teachol/android/IMG225.jpg"
17         //初始化加载图片
18         mImageView = findViewById(R.id.image_view)
19         mImageLoader = ImageLoader()
21         val displayButton = findViewById<Button>(R.id.display_button)
22         displayButton.setOnClickListener { //加载图片
23             mImageLoader.displayImage(mImageView, mUrl, cacheDir.absolutePath + "/image")
24         }
26         val cleanCacheButton = findViewById<Button>(R.id.clean_cache_button)
27         cleanCacheButton.setOnClickListener { //清除缓存
28             FileHelper.delete(cacheDir.absolutePath + "/image")
29             mImageView.setImageResource(R.mipmap.img_error_m)
30         }
31     }
32 }
```

第 23 行使用 mImageLoader 加载图片，如果加载成功，则将图片显示在 mImageView 控件中。第 28 行调用 FileHelper.delete(String)静态方法，用于删除缓存文件夹中的所有文件。

扩展实例：下载提示

在下载时还可以使用等待提示或者下载进度对用户进行提示(如图 12-3 所示)，C1202 工程和 C1203 工程分别使用这两种效果提示下载图片，由于篇幅所限，读者可打开工程文件自行查看，其中包含详细的注释。

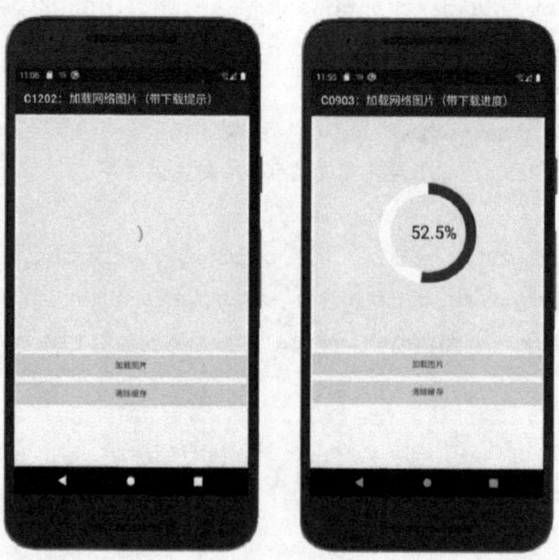

图 12-3　带下载提示和下载进度的效果

12.3 实例工程：发布动态(POST 方式)

本实例演示了使用 HttpURLConnection 类发送带附件的 POST 请求，服务器端检验后返回成功值并关闭 Activity(如图 12-4 所示)。单击“动态列表”按钮，启动 DailyListActivity。单击 + 按钮，使用系统提供的功能选取图片，选取图片后对图片进行压缩并保存在本地缓存文件夹中。这样处理不但可以避免 OOM 异常的出现，还能减小上传的数据量。单击“发布”按钮，将控件输入的昵称和内容及选择后缓存在本地的缓存图片合成为 POST 数据流后进行发送。

图 12-4 运行效果

1. 打开基础工程

打开“基础工程”文件夹中的“C1204”工程，该工程已经包含 MainActivity、DailyAddActivity、Http、CompressImage、Util 类及资源文件，并在 AndroidManifest.xml 文件中添加了访问网络的权限。CompressImage 类和 Util 类预先添加了一些功能性的方法，由于篇幅所限，详细的代码可以在工程文件中查看，其中有详细的注释帮助理解代码。

CompressImage 类除构造方法外，还包含三个方法：compress(ContentResolver, Uri) 方法用于压缩位图；getSampleSize(Bitmap) 方法用于获取缩放采样；compressByQuality(Bitmap) 方法用于压缩质量。

Util 类添加了三个方法：UriToString(ContentResolver, Uri) 方法用于将 URI 转为 String；getImageOrientation(Context, Uri) 方法用于获取图片旋转角度；saveCacheBitmap(String, String, Bitmap) 方法用于保存缓存图片。

2. Http 类

/java/com/teachol/c1204/Http.kt

```
//post 请求
fun post(requestURL: String, files: Map<String, File?>, params: Map<String, String>): Int {
    val url = URL(requestURL)
    val BOUNDARY = UUID.randomUUID().toString() //边界标识(随机生成)
    val PREFIX = "--" //前缀字符串
    val LINE_END = "\r\n" //换行字符串
    val CONTENT_TYPE = "multipart/form-data" //内容类型
    //创建连接
    val conn = url.openConnection() as HttpURLConnection
```

```
conn.connectTimeout = CONNECT_TIME_OUT
conn.readTimeout = READ_TIME_OUT
conn.doInput = true //允许输入流
conn.doOutput = true //允许输出流
conn.useCaches = false //不允许使用缓存
conn.requestMethod = "POST" //请求方式
conn.setRequestProperty("Charset", CHARSET) //设置编码
conn.setRequestProperty("connection", "keep-alive")
conn.setRequestProperty("Content-Type", "$CONTENT_TYPE;boundary = $BOUNDARY")
//设置数据流
val outputSteam = conn.outputStream
val dos = DataOutputStream(outputSteam)
val sb = StringBuffer()
//添加参数
for ((key, value) in params) {
    sb.append(PREFIX)
    sb.append(BOUNDARY)
    sb.append(LINE_END)
    sb.append("Content-Disposition: form-data; name = \"$key\"$LINE_END")
    sb.append("Content-Type: text/plain; charset = $CHARSET$LINE_END")
    sb.append("Content-Transfer-Encoding: 8bit$LINE_END")
    sb.append(LINE_END)
    sb.append(value)
    sb.append(LINE_END)
    dos.write(sb.toString().toByteArray())
    Log.e(TAG, "$key: $value")
}
//添加文件
for ((key, value) in files) {
    sb.append(PREFIX)
    sb.append(BOUNDARY)
    sb.append(LINE_END)
    sb.append("Content-Disposition: form-data; name = \"uploadinput[]\"; filename =
\"$key\ "$LINE_END") //uploadinput[]是服务器端用于接收图片文件的变量数组，filename是文件名
    sb.append("Content-Type: multipart/form-data; charset = $CHARSET$LINE_END")
    sb.append(LINE_END)
    dos.write(sb.toString().toByteArray())
    Log.e(TAG, "file:$key")
    //写入图片文件数据
    val inputStream: InputStream = FileInputStream(value)
    val buffer = ByteArray(1024)
    var len: Int
    while (inputStream.read(buffer).also { len = it } != -1) {
        dos.write(buffer, 0, len)
    }
    inputStream.close()
    dos.write(LINE_END.toByteArray())
}
val endData = (PREFIX + BOUNDARY + PREFIX + LINE_END).toByteArray()
dos.write(endData) //写入结束标志
```

```
116         dos.flush()//发送数据流
118         return if (conn.responseCode == 200) { //获取服务器端的响应码，200 表示发送成功
119             val br = BufferedReader(InputStreamReader(conn.inputStream))
120             val state = br.readLine() //读取服务器返回的数据
121             Log.e(TAG, "response getInputStream:$state")
122             Log.e(TAG, "response Message:" + conn.responseMessage)
123             conn.responseCode
124         } else throw RuntimeException(EXCEPTION_FAILURE)
125     }
```

第 65 ~ 73 行设置 POST 类型的 HTTP 请求。第 79 ~ 91 行遍历 params 将所有上传的参数写入请求的数据流中。第 93 ~ 111 行遍历 files 将所有上传的附件数据写入请求的数据流中。第 118 ~ 124 行根据返回的响应码进行相应的处理。如果响应码等于 200，则返回响应码，否则抛出异常。

3. 发布动态的 Activity

/java/com/teachol/c1204/DailyActivity.kt

```
//单击事件
override fun onClick(v: View) {
    when (v.id) {
        R.id.image_view -> {
            i = 0
            val intent = Intent(Intent.ACTION_PICK)
            intent.type = "image/*"
            startActivityForResult(intent, RESULT_CANCELED)
        }
        R.id.send_button -> if (mNameEditText.text.toString() != "" && mContentEditText.text.toString() != "") {
            //设置上传地址
            val url = "http://www.weiju2014.com/teachol/android/DailyAdd.php"
            //post 参数
            val params: MutableMap<String, String>= HashMap()
            params["name"] = mNameEditText.text.toString()
            params["content"] = mContentEditText.text.toString()
            //将选择的图片存储在数组中
            val selectFilePath = ArrayList<String?>()
            for (path in mImagePath) if (path != null) selectFilePath.add(path)
            val uploadFilePath = arrayOfNulls<String>(selectFilePath.size)
            selectFilePath.toArray(uploadFilePath)
            //post 上传
            pd = ProgressDialog.show(mContext, "", "正在发布中……", false, false)
            mHttpPostThread = HttpPostThread()
            mHttpPostThread.url = url
            mHttpPostThread.params = params
            mHttpPostThread.filePath = uploadFilePath
            mHttpPostThread.start()
            hideSoftInput() //关闭软键盘
        } else {
            Toast.makeText(this, "亲，写点什么吧!", Toast.LENGTH_SHORT).show()
        }
    }
```

```
99    }
100   //上传动态数据进程
101   inner class HttpPostThread : Thread() {
102       lateinit var url: String
103       lateinit var params: Map<String, String>
104       lateinit var filePath: Array<String?>
105       override fun run() {
107           var cacheFile: File //缓存文件
108           val postFiles: MutableMap<String, File>= HashMap() //上传图片文件
109           for (i in filePath.indices) {
110               //对大图片进行压缩和旋转
158               val cr = mContext.contentResolver
159               val ci = CompressImage( 500 * 1024)
160               lateinit var bitmap: Bitmap
161               try {
162                   bitmap = ci.compress(cr, mImageUri[i])!! //压缩图片
163               } catch (e: IOException) {
164                   e.printStackTrace()
165               }
166               bitmap = Util.rotateBitmapByDegree(bitmap, mImageDegrees[i])!!
167               cacheFile = Util.saveCacheBitmap(cacheDir, "IMG$i.jpg", bitmap) //缓存图片
169               postFiles[cacheFile.name] = cacheFile //添加上传图片文件
170           }
172           try {
173               Http.post(url, postFiles, params) //post 方式上传
174               mHandler.obtainMessage(MSG_HTTP_SUCCESS).sendToTarget()
175           } catch (e: Exception) {
176               e.printStackTrace()
177               mHandler.obtainMessage(MSG_HTTP_FAILURE).sendToTarget()
178           }
179       }
180   }
```

第71～73行调用系统功能选取图片。第77～94行开启独立线程发送POST请求发布动态。第101～108行 HttpPostThread 类继承 Thread 类，实现压缩附件文件并发送 POST 请求发布动态。

4．测试运行

运行工程，单击“发布动态”按钮启动 DailyActivity，输入昵称和内容，再选择图片，然后单击“发布”按钮。发送成功后关闭 DailyActivity，并在 Logcat 窗口中显示发送的内容及响应码和响应数据(如图 12-5 所示)。

图 12-5　Logcat 窗口输出结果

12.4　实例工程：动态列表(GET 方式)

本实例演示了使用 HttpURLConnection 类发送 GET 请求获取动态列表的数据,服务器端返回动态列表的 JSON 数据，对 JSON 数据进行解析后使用 ListView 控件显示动态的昵称和内容，下载图片并显示(如图 12-6 所示)。

GET 请求访问的地址为 http://www.weiju2014.com/teachol/android/DailyList.php，返回的 JSON 数据包含一个 JOSN 对象和一个 JOSN 数组,键名分别为 state 和 dailyList，如表 12-3 所示。dailyList 数组包含 10 个 JOSN 对象，使用索引和字段名作为 JOSN 对象的键名，这样就可以分别使用索引和字段名获取键值数据。

图 12-6　运行效果

表 12-3　获取的 JSON 数据样例

JSON 数据

```
{
    "state":"ok",
    "dailyList":[
        {
        "0":"63","Id":"63",
        "1":"小白魔","Name":"小白魔",
        "2":"今天你吃了么","Content":"今天你吃了么",
        "3":"upLoad\/20200918212922_381.jpg","Image":"upLoad\/20200918212922_381.jpg",
        "4":"2020-09-18 21:29:22","CreateTime":"2020-09-18 21:29:22",
        "5":"175.172.20.225","IP":"175.172.20.225"
        }
    ]
}
```

1．打开基础工程

打开“基础工程”文件夹中的“C1205”工程，该工程已经包含 MainActivity、DailyAddActivity、DailyListActivity、DailyModel、DailyListViewAdapter、Http、ImageLoader、CompressImage、Util 类及资源文件，并在 AndroidManifest.xml 文件中添加了访问网络的权限。

2．Http 类

/java/com/teachol/c1205/Http.kt

```
//get 请求
fun get(requestUrl: String, params: String): String {
    val url = URL(requestUrl + params)
    val conn = url.openConnection() as HttpURLConnection
    conn.connectTimeout = CONNECT_TIME_OUT
    conn.readTimeout = READ_TIME_OUT
    conn.requestMethod = "GET"
```

```
133         return if (conn.responseCode == 200) {
134             val br = BufferedReader(InputStreamReader(conn.inputStream))
135             val state = br.readLine() //读取服务器返回的数据
136             Log.e(TAG, "response getInputStream:$state")
137             Log.e(TAG, "response Message:" + conn.responseMessage)
138             state
139         } else throw RuntimeException(EXCEPTION_FAILURE)
140     }
```

第 130 ~ 132 行设置 GET 请求的 HTTP 连接。第 133 ~ 139 行发送请求并读取响应码，当响应码等于 200 时表示服务器响应成功，读取服务器返回的数据会作为方法的返回值，否则抛出异常。

3. 动态列表的 Activity

/java/com/teachol/c1205/DailyListActivity.kt

```
class DailyListActivity : AppCompatActivity() {
    companion object {
        const val DOMAIN = "http://www.weiju2014.com/teachol/android/"
        private const val MSG_NETERROR = 0//网络连接失败
        private const val MSG_SUCCESS = 1//获取数据成功
        private const val MSG_NONDATA = 2//无数据
    }
    private val url = DOMAIN + "DailyList.php"
    private var mContext: Context = this
    private lateinit var mListView: ListView
    private lateinit var mAdapter: DailyListViewAdapter
    private val mDailyModel = ArrayList<DailyModel>()
    var mCount = 20 //设置加载数量
    var mStartIndex = 0 //设置开始序号
    override fun onCreate(savedInstanceState: Bundle?) {
        mContext = this
        mAdapter = DailyListViewAdapter(DOMAIN, mContext, mDailyModel, R.layout.list_view_
item_daily_list)
        mListView = findViewById(R.id.list_view)
        mListView.adapter = mAdapter
        loader(mCount, mStartIndex)
    }
    private fun loader(num: Int, startIndex: Int) { //启动加载线程
        var loaderThread = LoaderThread()
        loaderThread.num = num
        loaderThread.startIndex = startIndex
        loaderThread.start()
    }
    inner class LoaderThread : Thread() { //加载线程
        var num = 0
        var startIndex = 0
        override fun run() {
            addLists(num, startIndex)
        }
    }
    private fun addLists(num: Int, startIndex: Int): Int { //获取加载数据
```

```
57        var length = 0
58        try {
59            val res: String? = Http.get(url, "?&num = $num&start = $startIndex") //获取动态列表数据
60            //解析 JSON 数据
61            val jsonParser = JSONTokener(res!!.trim { it <= ' ' })
62            val person = jsonParser.nextValue() as JSONObject
63            Log.e("json", person.getString("state"))
64            if (person.getString("state") == "ok") {
65                val jsonObject = JSONArray(person.getString("dailyList"))
66                length = jsonObject.length()
67                for (i in 0 until length) {
68                    val jo = jsonObject.opt(i) as JSONObject
69                    println("Name:" + jo.getString("Content"))
70                    mDailyModel.add(
71                        DailyModel(
72                            jo.getString("Id"),
73                            jo.getString("Name"),
74                            jo.getString("Content"),
75                            jo.getString("Image"),
76                            jo.getString("CreateTime")
77                        )
78                    )
79                }
80                mHandler.obtainMessage(MSG_SUCCESS).sendToTarget() //发送信息
81            } else {
82                mHandler.obtainMessage(MSG_NONDATA).sendToTarget() //发送信息
83            }
84        } catch (e: Exception) {
85            e.printStackTrace()
86            mHandler.obtainMessage(MSG_NETERROR).sendToTarget() //发送信息
87        }
88        return length
89    }
90    //更新 UI
91    private val mHandler: Handler = object : Handler(Looper.myLooper()!!) {
92        override fun handleMessage(msg: Message) { //处理信息
93            when (msg.what) { //判断信息类型
94                MSG_NETERROR -> Toast.makeText(mContext, "网络连接失败", Toast.LENGTH_LONG).show()
95                MSG_SUCCESS -> mAdapter!!.notifyDataSetChanged()
96                MSG_NONDATA -> Toast.makeText(mContext, "无数据", Toast.LENGTH_LONG).show()
97            }
98        }
99    }
100 }
```

第 41～46 行启动独立线程用于加载动态列表的数据。第 48～54 行 LoaderThread 类继承 Thread 类，用于调用 addLists(int, int)方法加载数据。第 56～79 行通过 Http.get(String,String)方法获取服务器返回的 JSON 数据。如果 JSON 数据中的 state 元素值为 ok，则对动态列表数据进行解析并保存到 mDailyModel 对象，然后通过 mHandler.obtainMessage(MSG_SUCCESS).sendToTarget()发送加载成功

的消息；否则发送加载失败的消息。第 92 ~ 98 行通过 handleMessage(Message) 方法处理发送过来的消息，当接收到 MSG_SUCCESS 消息时，调用 mAdapter.notifyDataSetChanged() 方法更新 ListView 控件显示的动态列表。

4. 测试运行

运行 C1205 工程，单击“动态列表”按钮启动 DailyListActivity。列表数据加载完成后，在 Logcat 窗口中可以查看输出的 JSON 数据，如图 12-7 所示。

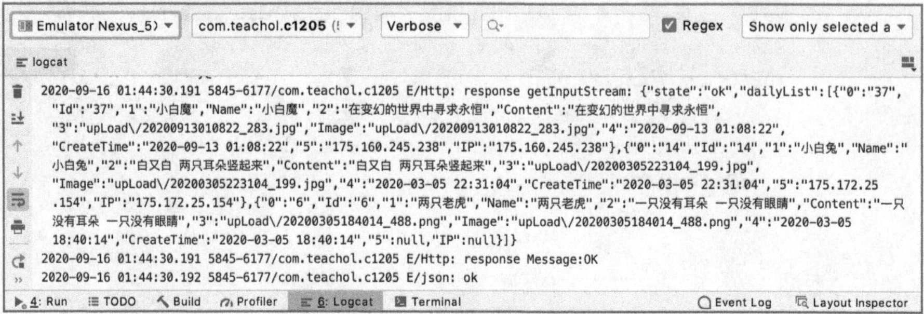

图 12-7　Logcat 窗口输出结果

第 13 章 Android 的快速开发套件

Jetpack 是由官方二次开发的多个控件和组件的套件，使用标准 API 进行了封装，提供了更方便快捷的开发方式；解决了以往 Android API 版本升级后需要添加或者修改不同版本的支持库以实现老版本 API 未提供功能的问题，可以在各种 Android 版本和设备中实现一致的运行效果。同样的功能使用 Jetpack 进行开发，可以减少约 50%的代码量和 40%的 bug 数量。

13.1 Jetpack 简介

由于 Android 开发中可能充斥了大量不规范的操作和重复代码，如生命周期的管理、开发过程的重复、项目架构的选择等，所以 Google 为了规范开发行为，在 2018 年 Google I/O 大会上正式发布了 Jetpack，使开发者能够更好、更快、更规范地进行开发。**Jetpack 并不是全新的功能，也不是 Android 开发过程中必须使用的。其中很多组件都在不断完善中，甚至有些组件官方都不推荐在实际项目中使用。**

Jetpack 包含了数十种控件和组件，以 **androidx 作为统一的一级包名**，按其功能可分为基础、架构、行为和界面四大类别库。基础库提供向后兼容、测试和 Kotlin 语言支持等横向功能；架构库提供设计稳健、可测试且易于维护的功能；行为库提供与标准 Android 服务（如通知、权限、分享和 Google 助理）相集成的功能；界面库提供生物识别、动画、表情符号、布局和调色板等功能。

Jet 的英文原意为喷气发动机，可以看出其最主要的目的是提高开发效率。其中的 RecyclerView、Lifecycle、ViewModel、LiveData、Room 等控件或组件的实用性较强，因而十分受到开发人员的喜爱。但其中也包含一些侵入性较高且实用性较弱的组件，如 databinding，navigation、paging 等。

提示：工程模板中包含的 Jetpack 包

目前使用工程模板创建的工程包含部分 JetPack 包，例如，Empty Activity 模板包含 androidx.core:core-ktx、androidx.appcompat:appcompat 和 androidx.constraintlayout:constraintlayout 包的依赖，可以在 build.gradle(:app) 文件中查看到这三个依赖库。在前文中已经使用了部分组件和控件，本章介绍的组件和控件都需要单独下载和添加依赖。

Jetpack 与 Android API 是分别发布的，其控件和组件的版本更新更加频繁且无规律。每个版本都要经历三个预发布阶段：Alpha 版、Beta 版和 RC 版（候选版），最终才能成为稳定版。目前 JetPack 远没有到达成熟的阶段，可以在官网查看当前的组件版本（如图 13-1 所示）。

提示：Alpha 版、Beta 版和 RC 版

Alpha 版是预览版，一般用于内部测试。API 的功能可能不完整，后续版本可能会添加、移除或更改 API。

Beta 版是测试版，一般用于公开测试。API 的功能完整，但可能包含错误，且无法使用实验性编译器功能（如@UseExperimental），不允许与 Alpha 版的 JetPack 库同时使用。

RC 版是未来的稳定版，可能会进行重要修复，但是 API 不会再更改，只允许与 Beta 版或稳定版的 JetPack 库同时使用。

图 13-1　当前的组件版本

13.2　回 收 视 图

13.2.1　RecyclerView 控件

RecyclerView 控件(androidx.recyclerview.widget.RecyclerView)是可以滚动显示列表信息并能够缓存子视图重复调用的控件，是 android.view.ViewGroup 的子类；比 ListView 控件相比，支持数据集、缓存子视图、横向显示及瀑布流布局的功能。RecyclerView 类的常用方法如表 13-1 所示。

表 13-1　RecyclerView 类的常用方法

类型和修饰符	方　法
	<init>(@NonNull context: Context) 主构造方法
open Unit	addOnChildAttachStateChangeListener(@NonNull listener: RecyclerView.OnChildAttachStateChangeListener) 添加子视图附加到或分离事件的监听器
open Unit	addOnItemTouchListener(@NonNull listener: RecyclerView.OnItemTouchListener) 添加子视图触摸事件的监听器
open Unit	addOnScrollListener(@NonNull listener: RecyclerView.OnScrollListener) 添加滚动事件的监听器
open Unit	clearOnChildAttachStateChangeListeners() 清除所有的子视图附加到或分离事件监听器
open Unit	clearOnScrollListeners() 清除所有的滚动事件监听器

续表

类型和修饰符	方　法
open Int	computeHorizontalScrollExtent() 计算水平滚动条滑块的显示内容范围，即显示子视图区域的宽度
open Int	computeHorizontalScrollOffset() 计算水平滚动条滑块的水平偏移值，即当前水平方向滚动的位置
open Int	computeHorizontalScrollRange() 计算水平滚动条的范围，即子视图的宽度和
open Int	computeVerticalScrollExtent() 计算垂直滚动条滑块的显示内容范围，即显示子视图区域的高度
open Int	computeVerticalScrollOffset() 计算垂直滚动条滑块的垂直偏移值，即当前垂直方向滚动的位置
open Int	computeVerticalScrollRange() 计算垂直滚动条的范围
open View?	findChildViewUnder(x: Float, y: Float) 根据坐标查找子视图
open Boolean	fling(velocityX: Int, velocityY: Int) 根据初始速度进行抛滑，然后逐渐减速直至停止滑动
open RecyclerView.Adapter <RecyclerView. ViewHolder!>?	getAdapter() 获取适配器
open Int	getScrollState() 获取滚动状态，包含以下三种常量。 ● SCROLL_STATE_IDLE：静止 ● SCROLL_STATE_DRAGGING：拖动 ● SCROLL_STATE_SETTLING：运动到最终位置
open Unit	removeOnChildAttachStateChangeListener(@NonNull listener: RecyclerView.OnChildAttachStateChangeListener) 添加所有的子视图附加到或分离事件监听器
open Unit	removeOnItemTouchListener(@NonNull listener: RecyclerView.OnItemTouchListener) 移除视图项的触摸事件监听器
open Unit	removeOnScrollListener(@NonNull listener: RecyclerView.OnScrollListener) 清除指定的滚动事件监听器
open Unit	onScrollStateChanged(state: Int) 滚动状态改变的事件发生后调用
open Unit	onScrolled(@Px dx: Int, @Px dy: Int) 滚动位置更改时调用
open Unit	scrollBy(x: Int, y: Int) 立即滚动到指定的相对坐标位置
open Unit	scrollTo(x: Int, y: Int) 立即滚动到指定的坐标位置
open Unit	scrollToPosition(position: Int) 立即滚动到指定的子视图位置
open Unit	setAdapter(@Nullable adapter: RecyclerView.Adapter<RecyclerView.ViewHolder!>?) 设置适配器
open Unit	setOnFlingListener(@Nullable onFlingListener: RecyclerView.OnFlingListener?) 设置抛滑监听器
open Unit	setRecyclerListener(@Nullable listener: RecyclerView.RecyclerListener?) 注册一个监听器，当子视图被回收时，它将被通知

续表

类型和修饰符	方　法
open Unit	smoothScrollBy (@Px dx: Int, @Px dy: Int) 平滑地滚动到指定的相对坐标位置
open Unit	smoothScrollToPosition (position: Int) 平滑地滚动到指定的坐标位置
open Unit	stopScroll () 停止任何当前正在进行的滚动，如 smoothScrollBy (Int，Int)、fling (Int，Int) 或触摸启动的滚动
open Unit	swapAdapter (@Nullable adapter: RecyclerView.Adapter<RecyclerView.ViewHolder!>?, removeAndRecycleExistingViews: Boolean) 将当前适配器与提供的适配器交换

RecyclerView.Adapter 类（常用方法如表 13-2 所示）为 RecyclerView 控件进行数据适配，需要创建一个子类重写 getItemCount ()、onBindViewHolder (VH,Int) 和 onCreateViewHolder (ViewGroup, Int) 抽象方法以实现对子视图的数据适配。

表 13-2　RecyclerView.Adapter 类的常用方法

类型和修饰符	方　法
	<init> () 主构造方法
abstract Int	getItemCount () 获取子视图的数量
abstract Unit	onBindViewHolder (@NonNull holder: ViewHolder, position: Int) 在指定位置绑定 ViewHolder 时调用
abstract VH	onCreateViewHolder (@NonNull parent: ViewGroup, viewType: Int) 创建一个指定类型的 ViewHolder 显示子视图时调用
Unit	bindViewHolder (@NonNull holder: ViewHolder, position: Int) 调用 onBindViewHolder (ViewGroup,Int) 方法为指定位置的子视图绑定 ViewHolder
VH	createViewHolder (@NonNull parent: ViewGroup, viewType: Int) 调用 onCreateViewHolder (ViewGroup, Int) 方法创建一个新的 ViewHolder，并初始化一些私有字段供 RecyclerView 使用
open Long	getItemId (position: Int) 根据位置获取子视图的 Id
Unit	notifyDataSetChanged () 通知数据集已经更新，并更新视图
Unit	notifyItemChanged (position: Int) 通知指定位置的子视图数据更新，并更新子视图
open Unit	onAttachedToRecyclerView (@NonNull recyclerView: RecyclerView) 当 RecyclerView 开始观察适配器时调用
open Unit	onDetachedFromRecyclerView (@NonNull recyclerView: RecyclerView) 当 RecyclerView 停止观察适配器时调用
open Unit	onFailedToRecycleView (@NonNull holder: ViewHolder) 当适配器创建的 ViewHolder 由于其瞬态状态而无法回收时调用
open Unit	onViewRecycled (@NonNull holder: ViewHolder) 当适配器创建的视图被回收时调用
open Unit	registerAdapterDataObserver (@NonNull observer: RecyclerView.AdapterDataObserver) 注册一个新的观察者监听数据的变化
open Unit	unregisterAdapterDataObserver (@NonNull observer: RecyclerView.AdapterDataObserver) 注销当前监听数据更改的观察者

ViewHolder 类用于缓存 RecyclerView 控件的子视图（常用方法如表 13-3 所示），需要创建一个子

类，并在子类内添加缓存的控件属性。

表 13-3　ViewHolder 类的常用方法

类型和修饰符	方　法
	<init>(@NonNull itemView: View) 主构造方法
RecyclerView.Adapter<out RecyclerView.ViewHolder!>?	getBindingAdapter() 获取绑定的适配器
Int	getLayoutPosition() 获取最近一级布局中的位置
Long	getItemId() 获取子视图的 Id
Boolean	isRecyclable() 判断是否被回收
Unit	setIsRecyclable(recyclable: Boolean) 设置是否被回收

13.2.2　实例工程：瀑布流动态列表

本实例演示了使用 RecyclerView 类和 RecyclerView.Adapter 类以瀑布流的形式显示动态列表，单击子视图中的图片会通过提示信息显示当前子视图的位置(如图 13-2 所示)。

图 13-2　运行效果

1. 打开基础工程

打开“基础工程”文件夹中的“C1301”工程，该工程已经包含 MainActivity、DailyModel、Http 和 ImageLoader 类及资源文件，并在 AndroidManifest.xml 文件中添加了访问网络的权限。

2. 下载依赖库

在布局文件的 Palette 面板中，RecyclerView 控件后方有一个下载的按钮，如图 13-3 所示，单击后进行下载并自动更新 Gradle。

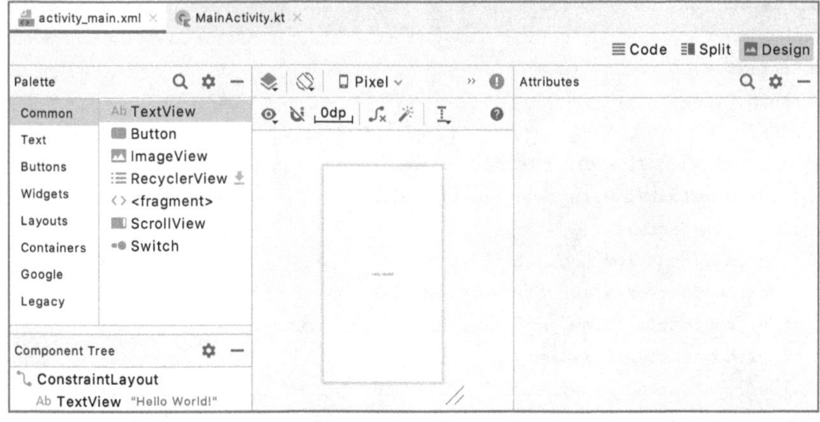

图 13-3　Palette 面板中 RecyclerView 控件

更新 Gradle 后，会自动在 build.gradle 文件中第 33 行使用 implementation 关键字添加该控件的依赖库，依赖库包含包名和版本号。

```
/app/build.gradle
```

```
dependencies {
    implementation fileTree(dir: "libs", include: ["*.jar"])
    implementation "org.jetbrains.kotlin:kotlin-stdlib:$kotlin_version"
    implementation 'androidx.core:core-ktx:1.3.1'
    implementation 'androidx.appcompat:appcompat:1.2.0'
    implementation 'androidx.constraintlayout:constraintlayout:2.0.1'
    implementation 'androidx.recyclerview:recyclerview:1.1.0'
    testImplementation 'junit:junit:4.12'
    androidTestImplementation 'androidx.test.ext:junit:1.1.2'
    androidTestImplementation 'androidx.test.espresso:espresso-core:3.3.0'
}
```

此后在新工程中使用拖曳方式添加该控件，会弹出自动添加工程依赖库的对话框，如图 13-4 所示，单击“OK”按钮后自动添加依赖库的代码。也可以手动添加依赖库的代码。

Add Project Dependency
This operation requires the library androidx.recyclerview:recyclerview:+.
Would you like to add this now?
Cancel OK

图 13-4　添加工程依赖库的对话框

3. DailyRecyclerViewAdapter 类

在“/java/com/teachol/c1301/”文件夹中，新建 DailyRecyclerViewAdapter.kt 文件，用于声明自定义的 RecyclerView.Adapter<ViewHolder>类。

```
/java/com/teachol/c1301/DailyRecyclerViewAdapter.kt
class DailyRecyclerViewAdapter(private val mDonmain: String, private val dailyModels: List<DailyModel>, private val layoutResource: Int) : RecyclerView.Adapter<DailyRecyclerViewAdapter.ViewHolder>() {
    private lateinit var mContext: Context
    override fun getItemCount(): Int { //获取子视图数量
        return dailyModels.size
    }
    //创建 ViewHolder 时调用
    override fun onCreateViewHolder(parent: ViewGroup, viewType: Int): ViewHolder {
        val view: View = LayoutInflater.from(parent.context).inflate(layoutResource, parent,false)
        mContext = parent.context
        return ViewHolder(view)
    }
    override fun onBindViewHolder(holder: ViewHolder, position: Int) { //绑定 ViewHolder 时调用
        holder.nameTextView.text = "${dailyModels[position].name}: " //姓名
        holder.nameTextView.id = position
        holder.nameTextView.setOnClickListener { v -> displayPosition(v.id) }
        holder.contentTextView.text = dailyModels[position].content //内容
        holder.contentTextView.setLineSpacing(1.0f, 1.0f)
        holder.contentTextView.id = position
        holder.contentTextView.setOnClickListener { v -> displayPosition(v.id) }
        if (dailyModels[position].image != "") { //图片
            holder.imageImageView.visibility = View.VISIBLE
            holder.imageImageView.id = position
            holder.imageImageView.setOnClickListener { v -> displayPosition(v.id) }
            val url = mDonmain + dailyModels[position].image.replace("d/", "d/_M_")
            val mImageLoader = ImageLoader()
            mImageLoader.displayImage(
```

```
42              holder.imageImageView, url, mContext.externalCacheDir!!.absolutePath + "/image" )
46          } else {
47              holder.imageImageView.visibility = View.GONE
48          }
50          holder.creatTimeTextView.text = dailyModels[position].createTime //发布时间
51      }
53      private fun displayPosition(position: Int?) { //显示位置
54          Toast.makeText(mContext, "第" + position + "个动态", Toast.LENGTH_LONG).show()
55      }
56      //缓存子视图控件的自定义 ViewHolder
57      open class ViewHolder(view: View) : RecyclerView.ViewHolder(view) {
58          var nameTextView: TextView = view.findViewById(R.id.name_text_view)
59          var contentTextView: TextView = view.findViewById(R.id.content_text_view)
60          var imageImageView: ImageView = view.findViewById(R.id.image_image_view)
61          var creatTimeTextView: TextView = view.findViewById(R.id.create_time_text_view)
62      }
63  }
```

第 12 行继承自 RecyclerView.Adapter<DailyRecyclerViewAdapter.ViewHolder>，通过主构造方法定义 4 个属性。第 15 ~ 18 行重写 getItemCount() 方法，通过 dailyModels 属性获取子视图的数量。第 19 ~ 23 行重写 onCreateViewHolder(ViewGroup, Int) 方法，通过 parent.context 获取到使用该适配器的 Activity，将子视图传递给 ViewHolder 对象作为返回值。第 25 ~ 51 行重写 onBindViewHolder(ViewHolder, Int) 方法，将数据适配具有缓存功能的 ViewHolder 对象。第 57 ~ 62 行自定义的 ViewHolder 类继承 RecyclerView.ViewHolder 类，定义 4 个属性用于缓存子视图的控件。

4. 主界面的 Activity

/java/com/teachol/c1301/MainActivity.kt

```
private val mDailyModel = ArrayList<DailyModel>()
private lateinit var mRecyclerView: RecyclerView
private lateinit var mAdapter: DailyRecyclerViewAdapter
override fun onCreate(savedInstanceState: Bundle?) {
    super.onCreate(savedInstanceState)
    setContentView(R.layout.activity_main)
    this.title = "C1301: 瀑布流动态列表"
    mContext = this
    //val layoutManager = LinearLayoutManager(this)
    //layoutManager.orientation = LinearLayoutManager.HORIZONTAL
    val layoutManager = StaggeredGridLayoutManager(2,StaggeredGridLayoutManager.VERTICAL)
    mAdapter = DailyRecyclerViewAdapter(DOMAIN, mDailyModel, R.layout.list_view_item_daily_list)
    mRecyclerView = findViewById(R.id.recycler_view)
    mRecyclerView.layoutManager = layoutManager
    mRecyclerView.adapter = mAdapter
    mRecyclerView.addOnScrollListener(object:RecyclerView.OnScrollListener(){
        override fun onScrolled(recyclerView: RecyclerView, dx: Int, dy: Int){
            Log.e("computeVerticalScrollExtent:",
                recyclerView.computeVerticalScrollExtent().toString())
            Log.e("computeVerticalScrollOffset:",
                recyclerView.computeVerticalScrollOffset().toString())
            Log.e("computeVerticalScrollRange:",
```

```
50                  recyclerView.computeVerticalScrollRange().toString()
51              }
52          })
53          loader(mCount, mStartIndex)
54      }
104     //更新 UI
105     private val mHandler: Handler = object : Handler(Looper.myLooper()!!) {
106         override fun handleMessage(msg: Message) {
107             when (msg.what) {
108                 MSG_NETERROR -> Toast.makeText(mContext, "网络连接失败", Toast.LENGTH_LONG).show()
109                 MSG_SUCCESS -> mAdapter.notifyDataSetChanged()
110                 MSG_NONDATA -> Toast.makeText(mContext, "无数据", Toast.LENGTH_LONG).show()
111             }
112         }
113     }
```

第 36、37 行注释掉了水平滚动的设置，如果解除注释且将第 38 行注释掉，可以实现水平方向的滑动效果。第 38 行声明 layoutManager 变量为两列纵向瀑布流效果的布局管理对象。第 41 行将 layoutManager 变量设置为 mRecyclerView 变量的布局管理。第 43 ~ 52 行 mRecyclerView 变量添加滚动监听器，重写 onScrolled(RecyclerView, Int, Int)方法。当滚动控件时会在 Logcat 窗口中输出垂直方向的滚动显示区域高度、滚动的位置和滚动区域高度信息。第 109 行从网络上获取数据后，使用 mAdapter.notifyDataSetChanged()更新 RecyclerView 控件中的子视图。

13.3 滑动刷新布局

滑动刷新布局可以让用户用垂直滑动手势刷新视图内容，每次再次完成手势时都会通知监听器刷新内容。该布局只能添加一个子标签，但是子标签内可以再添加其子标签。

13.3.1 SwipeRefreshLayout 组件

SwipeRefreshLayout 组件(androidx.swiperefreshlayout.widget.SwipeRefreshLayout)是进行滑动刷新的容器组件(常用方法如表 13-4 所示)，是 android.view.ViewGroup 的子类。

表 13-4 SwipeRefreshLayout 类的常用方法

类型和修饰符	方 法
	<init>(@NonNull context: Context) 主构造方法
open Boolean	isRefreshing() 判断是否刷新状态
open Unit	setOnRefreshListener(@Nullable listener: SwipeRefreshLayout.OnRefreshListener?) 设置刷新的监听器
open Unit	setRefreshing(refreshing: Boolean) 设置刷新状态

13.3.2 实例工程：下拉刷新和上拉加载的动态列表

本实例演示了使用 SwipeRefreshLayout 类实现上拉刷新和下拉加载功能的瀑布流动态列表(如图 13-5 所示)。在顶部下拉刷新动态列表；在底部上拉加载更多的动态到列表中；打开飞行模式，再

上拉加载更多时，页脚子视图显示“网络故障 稍后再试”。

图 13-5　运行效果

1. 打开基础工程

打开“基础工程”文件夹中的“C1302”工程，该工程已经包含 MainActivity、DailyModel、Http 和 ImageLoader 类及资源文件，并在 AndroidManifest.xml 文件中添加了访问网络的权限。

2. 添加依赖库

```
/app/build.gradle
33  implementation 'androidx.recyclerview:recyclerview:1.1.0'
34  implementation "androidx.swiperefreshlayout:swiperefreshlayout:1.1.0"
```

第 33 行使用 implementation 关键字添加 Recyclerview 组件的依赖。第 34 行使用 implementation 关键字添加 Swiperefreshlayout 组件的依赖，版本为 1.1.0，然后更新 Gradle。

3. 加载页脚的布局

在“/ res/layout /”文件夹中，新建 recycler_view_footer.xml 布局文件，实现加载页脚的布局。

```
/res/layout/recycler_view_footer.xml
11  <ProgressBar
12      android:id = "@+id/progress_bar"
13      style = "?android:attr/progressBarStyle"
14      android:layout_width = "wrap_content"
15      android:layout_height = "30dp" />
16  <TextView
17      android:id = "@+id/footer_text_view"
18      android:layout_width = "wrap_content"
19      android:layout_height = "wrap_content"
20      android:layout_margin = "5dp"
21      android:textColor = "#000000"
22      tools:text = "正在加载" />
```

第 11～15 行添加 ProgressBar 控件，提示用户正在加载数据，样式设置为 Android 自带的 ?android:attr/progressBarStyle"，即环形旋转的进度条。第 16～22 行添加 TextView 控件显示提示等待的文字。

4．DailyRecyclerViewAdapter 类

在“/java/com/teachol/c1302/”文件夹中，新建 DailyRecyclerViewAdapter.kt 文件，用于添加自定义的 RecyclerView.Adapter<ViewHolder>类。

```
/java/com/teachol/c1302/DailyRecyclerViewAdapter.kt
const val TYPE_ITEM = 0 //动态子视图
const val TYPE_FOOTER = 1 //页脚子视图
class DailyRecyclerViewAdapter(
    private val mDonmain: String, //域名
    private val mDailyModels: List<DailyModel>, //动态数据模型的列表
    private val mItemLayoutResource: Int, //动态子视图布局资源
    private val mFooterLayoutResource: Int //页脚子视图布局资源
) : RecyclerView.Adapter<RecyclerView.ViewHolder>() {
    private lateinit var mContext: Context //获取子视图数量
    var mCurrentState = 0 //当前状态
    companion object{
        const val STATUS_REFRESHING = 0 //上拉刷新状态
        const val STATUS_LOADING_MORE = 1 //正在加载状态
        const val STATUS_FINISH = 2 //加载完成状态
        const val STATUS_NET_ERROR = 3 //网络错误状态
        const val STATUS_NONDATA = 4 //已经加载全部数据状态
        const val STATUS_MAX_COUNT = 5 //达到加载上限状态
    }
    //获取子视图数量
    override fun getItemCount(): Int {
        return mDailyModels.size + 1//增加一个子视图用于显示加载提示子视图
    }
    //创建 ViewHolder 时调用
    override fun onCreateViewHolder(parent: ViewGroup, viewType: Int): RecyclerView.ViewHolder {
        mContext = parent.context
        val holder:RecyclerView.ViewHolder
        //返回缓存子视图的 ViewHolder
        if (viewType == TYPE_ITEM) {
            holder = ItemViewHolder(LayoutInflater.from(parent.context).inflate(mItemLayoutResource, parent, false))
        } else {
            holder = FooterViewHolder(LayoutInflater.from(parent.context).inflate(mFooterLayoutResource, parent, false)).also { (it.linearLayout.layoutParams as StaggeredGridLayoutManager.LayoutParams).isFullSpan = true}
        }
        return holder
    }
    //绑定 ViewHolder 时调用
    override fun onBindViewHolder(holder: RecyclerView.ViewHolder, position: Int) {
        //为缓存子视图的 ViewHolder 匹配数据
        if (holder is ItemViewHolder) { //日记子视图
            val itemViewHolder: ItemViewHolder = holder
            with(itemViewHolder) {
                nameTextView.text = String.format("%1s:",mDailyModels[position].name)
```

```
                nameTextView.id = position
                nameTextView.setOnClickListener { v -> displayPosition(v.id) }
                //内容
                contentTextView.text = mDailyModels[position].content
                contentTextView.setLineSpacing(1.0f, 1.0f)
                contentTextView.id = position
                contentTextView.setOnClickListener { v -> displayPosition(v.id) }
                //图片
                if (mDailyModels[position].image != "") {
                    imageImageView.visibility = View.VISIBLE
                    imageImageView.id = position
                    imageImageView.setOnClickListener { v -> displayPosition(v.id) }
                    val url = mDonmain + mDailyModels[position].image.replace("d/", "d/_M_")
                    val mImageLoader = ImageLoader()
                    mImageLoader.displayImage(imageImageView, url, mContext.cacheDir!!.absolutePath
+ "/image")
                } else { //加载提示子视图
                    imageImageView.visibility = View.GONE
                }
                //发布时间
                creatTimeTextView.text = mDailyModels[position].createTime
            }
        } else if (holder is FooterViewHolder) {
            val footerViewHolder: FooterViewHolder = holder
            when (mCurrentState) { //根据当前状态显示页脚子视图的内容
                STATUS_FINISH -> footerViewHolder.linearLayout.visibility = View.GONE
                STATUS_REFRESHING -> footerViewHolder.linearLayout.visibility = View.GONE
                STATUS_LOADING_MORE -> {
                    footerViewHolder.footerTextView.text = "正在加载"
                    footerViewHolder.progressBar.visibility = View.VISIBLE
                    footerViewHolder.linearLayout.visibility = View.VISIBLE
                }
                STATUS_NONDATA -> {
                    footerViewHolder.footerTextView.text = "已经全部显示"
                    footerViewHolder.progressBar.visibility = View.GONE
                    footerViewHolder.linearLayout.visibility = View.VISIBLE
                }
                STATUS_MAX_COUNT -> {
                    footerViewHolder.footerTextView.text = "已经到达显示上限"
                    footerViewHolder.progressBar.visibility = View.GONE
                    footerViewHolder.linearLayout.visibility = View.VISIBLE
                }
                STATUS_NET_ERROR -> {
                    footerViewHolder.footerTextView.text = "网络故障 稍后重试"
                    footerViewHolder.progressBar.visibility = View.GONE
                    footerViewHolder.linearLayout.visibility = View.VISIBLE
                }
            }
        }
    }
```

```
101         //获取子视图类型
102         override fun getItemViewType(position: Int): Int {
103             return if (position + 1 == itemCount) TYPE_FOOTER else TYPE_ITEM
104         }
105         //显示位置
106         private fun displayPosition(position: Int?) {
107             Toast.makeText(mContext, "第" + position + "个动态", Toast.LENGTH_LONG).show()
108         }
109     //改变当前状态
110         fun changeStatus(state: Int) {
111             mCurrentState = state
112             notifyDataSetChanged() //通知 RecyclerView 控件数据集改变，然后刷新视图
113         }
114         //缓存动态子视图的自定义 ViewHolder
115         open class ItemViewHolder(view: View) : RecyclerView.ViewHolder(view) {
116             var nameTextView: TextView = view.findViewById(R.id.name_text_view)
117             var contentTextView: TextView = view.findViewById(R.id.content_text_view)
118             var imageImageView: ImageView = view.findViewById(R.id.image_image_view)
119             var creatTimeTextView: TextView = view.findViewById(R.id.create_time_text_view)
120         }
121         //缓存页脚子视图的自定义 ViewHolder
122         open class FooterViewHolder(view: View) : RecyclerView.ViewHolder(view) {
123             var linearLayout: LinearLayout = view.findViewById(R.id.linear_layout)
124             var footerTextView: TextView = view.findViewById(R.id.footer_text_view)
125             var progressBar: ProgressBar = view.findViewById(R.id.progress_bar)
126         }
127     }
```

第 30 ~ 32 行重写 getItemCount() 方法，获取子视图的数量，在显示的动态数量基础上加 1，即增加一个子视图，用于显示页脚。第 38 ~ 42 行根据子视图类型，返回对应类型的缓存实例。第 41 行将页脚子视图缓存类的 linearLayout.layoutParams 转换为 StaggeredGridLayoutManager.LayoutParams 类型，将其 isFullSpan 属性设置为 true，将页脚子视图独占一行显示。第 103 行判断当前位置子视图类型，position 是从 0 开始的，因此 position+1 等于 itemCount 时代表最后一个子视图。第 110 ~ 113 行改变当前状态并通知 RecyclerView 控件数据集已经改变，然后刷新视图。

5. 主界面的布局

```
/res/layout/activity_main.xml
<androidx.swiperefreshlayout.widget.SwipeRefreshLayout
    android:id = "@+id/swipere_fresh_layout"
    android:layout_height = "match_parent"
    android:layout_width = "match_parent"
    xmlns:android = "http://schemas.android.com/apk/res/android">
    <androidx.recyclerview.widget.RecyclerView
        android:id = "@+id/recycler_view"
        android:layout_height = "match_parent"
        android:layout_margin = "2dp"
        android:layout_width = "match_parent" />
</androidx.swiperefreshlayout.widget.SwipeRefreshLayout>
```

第 07 ~ 11 行在 SwipeRefreshLayout 标签内添加唯一的子标签——RecyclerView，用于显示瀑布流的动态。

6．主界面的 Activity

```
/java/com/teachol/c1302/MainActivity.kt
companion object {
    const val DOMAIN = "http://www.weiju2014.com/teachol/android/"
    private const val MSG_NULL = 0//网络连接失败
    private const val MSG_NETERROR = 1//网络连接失败
    private const val MSG_FINISH = 2//获取数据成功
    private const val MSG_MAXCOUNT = 3//到达最大加载数量
    private const val MSG_NONDATA = 4//无数据
}
private val url = DOMAIN + "DailyList.php"
private var mContext: Context = this
private var mMaxCount = 20 //设置最大加载数量
private var mNum = 5 //设置加载数量
private var mStartIndex = 0 //设置开始序号
private val mDailyModel = ArrayList<DailyModel>() //动态的数据模型
private var mLastPositions = 1 //最后一个视图项的位置
private lateinit var swiperefreshLayout: SwipeRefreshLayout
private lateinit var mRecyclerView: RecyclerView
private lateinit var mAdapter: DailyRecyclerViewAdapter
override fun onCreate(savedInstanceState: Bundle?) {
    super.onCreate(savedInstanceState)
    setContentView(R.layout.activity_main)
    mContext = this
    //实例化适配器
    mAdapter = DailyRecyclerViewAdapter(DOMAIN, mDailyModel, R.layout.recycler_view_item_diary, R.layout.recycler_view_footer)
    //设置滑动刷新布局
    swiperefreshLayout = findViewById(R.id.swipere_fresh_layout)
    swiperefreshLayout.setOnRefreshListener { //刷新事件
        mAdapter.changeStatus(DailyRecyclerViewAdapter.STATUS_REFRESHING) //刷新状态
        mStartIndex = 0 //重置开始序号
        LoaderThread(mNum, mStartIndex, true).start() //开启新线程加载数据
    }
    //设置瀑布流布局为纵向两列
    val layoutManager = StaggeredGridLayoutManager(2, StaggeredGridLayoutManager.VERTICAL)
    //设置回收视图
    mRecyclerView = findViewById(R.id.recycler_view)
    mRecyclerView.layoutManager = layoutManager
    mRecyclerView.adapter = mAdapter
    mRecyclerView.addOnScrollListener(object : RecyclerView.OnScrollListener() { //添加滚动监听器
        //重写滚动完成事件
        override fun onScrolled(recyclerView: RecyclerView, dx: Int, dy: Int) {
            super.onScrolled(recyclerView, dx, dy)
            val positionArray = IntArray((recyclerView.layoutManager as StaggeredGridLayoutManager).spanCount) //存储最后一行子视图位置的数组
```

```
62            layoutManager.findLastVisibleItemPositions(positionArray) //获取显示的最后一行子视图位置的数组
63            mLastPositions = findMax(positionArray) //最后一个子视图的位置
64            if (mLastPositions + 1 == mAdapter.itemCount) { //显示的最后一个子视图是否是最后一个子视图
65                if (mAdapter.mCurrentState == DailyRecyclerViewAdapter.STATUS_LOADING_MORE) return
66                mAdapter.changeStatus(DailyRecyclerViewAdapter.STATUS_LOADING_MORE)//设置加载状态
67                LoaderThread(mNum, mStartIndex).start()//加载
68            }
69        }
70        //获取最后一个视图项的位置
71        fun findMax(positionArray: IntArray): Int {
72            var max = positionArray[0]
73            for (value: Int in positionArray) if (value > max) max = value
74            return max
75        }
76    })
77    swiperefreshLayout.isRefreshing = true //显示刷新状态
78 }
131 //更新 UI
132 private val mHandler: Handler = object : Handler(Looper.myLooper()!!) {
133     override fun handleMessage(msg: Message) {
134         swiperefreshLayout.isRefreshing = false //关闭刷新状态
135         when (msg.what) {
136             MSG_NETERROR -> mAdapter.changeStatus(DailyRecyclerViewAdapter.STATUS_NET_ERROR)
137             MSG_FINISH -> mAdapter.changeStatus(DailyRecyclerViewAdapter.STATUS_FINISH)
138             MSG_NONDATA -> mAdapter.changeStatus(DailyRecyclerViewAdapter.STATUS_NONDATA)
139             MSG_MAXCOUNT -> mAdapter.changeStatus(DailyRecyclerViewAdapter.STATUS_MAX_COUNT)
140         }
141     }
142 }
```

第 46 ~ 50 行设置下拉刷新的监听器，当刷新事件发生后，重置查询起始位置的序号为 0，开启新线程下载数据。第 59 ~ 69 行重写滚动完成事件，该事件在 mRecyclerView 对象加载后用户没有滚动视图的情况下也会调用一次，因此进入后的首次加载也是调用该事件实现的。第 61 行声明存储最后一行子视图位置的数组，数组的长度为 mRecyclerView 对象的列数，(recyclerView.layoutManager as StaggeredGridLayoutManager).spanCount 获取 recyclerView 对象的布局管理并强制转换为 StaggeredGridLayoutManager 类，再通过 StaggeredGridLayoutManager 类的 spanCount 属性获取列数。第 63 行获取显示的最后一个子视图的位置。由于已经将页脚子视图设置为独占一行，所以显示的最后一行是页脚子视图时只有一个子视图，也可以直接使用 positionArray[0]获取显示的最后一行第一个子视图的位置，就可以判断出是否是页脚子视图。第 64 行判断显示的最后一个子视图是否是最后一个子视图，如果是则继续加载更多的动态。第 65 行如果当前状态是加载状态，则直接返回，避免重复加载。第 71 ~ 75 行获取 IntArray 中的最大值。第 77 行 swiperefreshLayout 对象显示下拉刷新时用于刷新状态提示的子控件。由于加载 mRecyclerView 对象后，会自动调用 onScrolled(RecyclerView, Int, Int)方法首次加载数据，此时用户并没有下拉刷新，所以需要设置 isRefreshing 属性为 true。第 132 ~ 142 行 mHandle 对象加载线程对 UI 的更新，调用 DailyRecyclerViewAdapter 类的 changeStatus(Int)方法修改状态并更新 mRecyclerView 对象的子视图。

13.4 生物特征认证

生物特征认证是对指纹、人脸、虹膜等生物特征进行系统级的身份认证，可以代替密码用于解锁、动登录、支付等方面的验证。

3.4.1 Biometric 组件

1. BiometricManager 类

BiometricManager 类(androidx.biometric.BiometricManager)提供与生物特征相关的系统信息(常用法如表 13-5 所示)，用于检测是否可以使用生物特征进行身份验证。在 Android 10(API Level 29)及高版本且具有相关硬件支持的设备上录入指纹、人脸、虹膜等生物特征后才可以进行身份验证。

表 13-5 BiometricManager 类的常用方法

类型和修饰符	方 法
open Int	canAuthenticate() 检查用户是否可以使用生物特征进行身份验证
open Int	canAuthenticate(authenticators: Int) 检查用户是否可以使用符合给定要求的验证器进行身份验证
open static BiometricManager	from(@NonNull context: Context) 创建 BiometricManager 实例

2. BiometricPrompt 类

BiometricPrompt 类(androidx.biometric.BiometricPrompt)用于管理系统提供的生物特征提示框(常方法如表 13-6 所示)。当 App 位于后台时，将取消提示框，以保障安全性。

表 13-6 BiometricPrompt 类的常用方法

类型和修饰符	方 法
	<init>(@NonNull activity: FragmentActivity, @NonNull callback: BiometricPrompt.AuthenticationCallback) 主构造方法
	<init>(@NonNull fragment: Fragment, @NonNull callback: BiometricPrompt.AuthenticationCallback) 主构造方法
open Unit	authenticate(@NonNull info: BiometricPrompt.PromptInfo, @NonNull crypto: BiometricPrompt.CryptoObject) 显示生物特征认证的提示框
open Unit	authenticate(@NonNull info: BiometricPrompt.PromptInfo) 显示生物特征认证的提示框
open Unit	cancelAuthentication() 取消生物特征认证并关闭提示框

3. BiometricPrompt.AuthenticationCallback 类

BiometricPrompt.AuthenticationCallback 类(androidx.biometric.BiometricPrompt. AuthenticationCallback)于生物特征认证后回调，常用方法如表 13-7 所示。

4. BiometricPrompt.PromptInfo.Builder 类

BiometricPrompt.PromptInfo.Builder 类(androidx.biometric.BiometricPrompt.PromptInfo.Builder)用于建生物特征认证的提示框，常用方法如表 13-8 所示。

表 13-7　BiometricPrompt. AuthenticationCallback 类的常用方法

类型和修饰符	方　法
	<init>() 主构造方法
open Unit	onAuthenticationError(errorCode: Int, @NonNull errString: CharSequence) 生物特征认证遇到系统级错误时调用
open Unit	onAuthenticationFailed() 生物特征认证失败时调用
open Unit	onAuthenticationSucceeded(@NonNull result: BiometricPrompt.AuthenticationResult) 生物特征认证成功时调用

表 13-8　BiometricPrompt.PromptInfo.Builder 类的常用方法

类型和修饰符	方　法
	<init>() 主构造方法
open BiometricPrompt.PromptInfo	build() 创建 PromptInfo 对象
open BiometricPrompt.PromptInfo.Builder	setAllowedAuthenticators(allowedAuthenticators: Int) 设置生物特征提示调用的身份验证程序的类型，对用户进行身份验证
open BiometricPrompt.PromptInfo.Builder	setConfirmationRequired(confirmationRequired: Boolean) 设置识别之前是否需要明确的用户确认
open BiometricPrompt.PromptInfo.Builder	setDescription(@Nullable description: CharSequence?) 设置提示的说明
open BiometricPrompt.PromptInfo.Builder	setDeviceCredentialAllowed(deviceCredentialAllowed: Boolean) 设置是否可以使用设备 PIN、图案或密码替代生物特征进行身份验证 deviceCredentialAllowed 参数设置为 true 时，与 setNegativeButtonText (CharSequence) 方法必须二选一
open BiometricPrompt.PromptInfo.Builder	setNegativeButtonText(@NonNull negativeButtonText: CharSequence) 设置提示上否定按钮的文本
open BiometricPrompt.PromptInfo.Builder	setSubtitle(@Nullable subtitle: CharSequence?) 设置提示的副标题
open BiometricPrompt.PromptInfo.Builder	setTitle(@NonNull title: CharSequence) 设置提示的标题

13.4.2　实例工程：指纹支付

本实例演示了使用 BiometricManager 组件模拟支付时的指纹验证，单击“支付”按钮显示生物识别提示框(如图 13-6 所示)。

图 13-6　运行效果

1．录入模拟器的指纹

在模拟器中，依次打开【设置】→【安全】→【屏幕锁定】→【指纹+图案】，单击“下一步”按钮开始录入指纹，如图 13-7 所示。然后打开“Extended controls”对话框的“Fingerprint”选项卡，选择“Finger 1”选项，然后多次单击“TOUCH THE SENSOR”按钮完成指纹的录入，如图 13-8 所示。

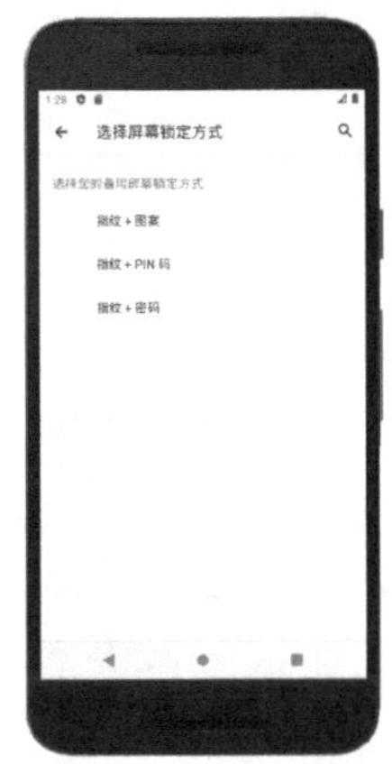

图 13-7　使用指纹解锁

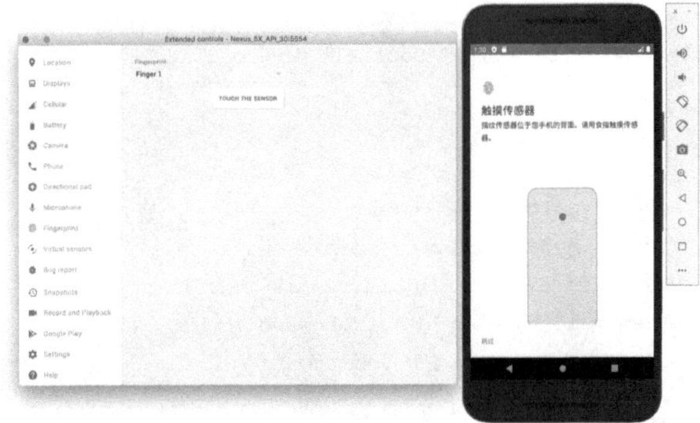

图 13-8　录入指纹

2．打开基础工程

打开“基础工程”文件夹中的 C1303 工程，该工程已经包含 MainActivity 类及资源文件。

3．添加依赖库

```
/app/build.gradle
33  implementation 'androidx.biometric:biometric:1.0.1'
```

第 33 行使用 implementation 关键字添加 Biometric 组件的依赖，版本为 1.0.1。下载依赖库后，自动更新 Gradle。

4．主界面的 Activity

```
/java/com/teachol/c1303/MainActivity.kt
class MainActivity : AppCompatActivity() {
    lateinit var executor: Executor
```

```
    lateinit var biometricPrompt: BiometricPrompt
    lateinit var promptInfo: BiometricPrompt.PromptInfo
    lateinit var resultTextView: TextView
    lateinit var resultImageView: ImageView
    lateinit var payButton: Button
    var authenticationCallback = object : BiometricPrompt.AuthenticationCallback() {
        //认证错误
        override fun onAuthenticationError(errorCode: Int, errString: CharSequence) {
            super.onAuthenticationError(errorCode, errString)
            errString.toString().toToast()
        }
        //认证成功
        override fun onAuthenticationSucceeded(result: BiometricPrompt.AuthenticationResult) {
            super.onAuthenticationSucceeded(result)
            resultTextView.text = "向小白魔成功支付一分钱"
            resultImageView.setImageResource(R.mipmap.img_succeeded)
        }
        //认证失败
        override fun onAuthenticationFailed() {
            super.onAuthenticationFailed()
            biometricPrompt.cancelAuthentication()
            resultTextView.text = "支付失败"
            resultImageView.setImageResource(R.mipmap.img_failed)
        }
    }
    override fun onCreate(savedInstanceState: Bundle?) {
        super.onCreate(savedInstanceState)
        setContentView(R.layout.activity_main)
        resultTextView = findViewById(R.id.result_text_view)
        resultImageView = findViewById(R.id.result_image_view)
        payButton = findViewById(R.id.pay_button)
        payButton.setOnClickListener {
            if (checkBiometric()) {
                resultTextView.text = "等待支付"
                resultImageView.setImageResource(R.mipmap.img_default)
                showbiometricPrompt()
            }
        }
    }
    //判断是否可以使用生物特征认证
    fun checkBiometric():Boolean{
        val biometricManager = BiometricManager.from(this)
        when (biometricManager.canAuthenticate()) {
            BiometricManager.BIOMETRIC_SUCCESS -> {
                "生物特征认证验证成功".toToast()
                return true
            }
            BiometricManager.BIOMETRIC_ERROR_NO_HARDWARE -> "无生物特征认证的硬件".toToast()
            BiometricManager.BIOMETRIC_ERROR_HW_UNAVAILABLE -> "生物特征认证的硬件不可用".toToast()
            BiometricManager.BIOMETRIC_ERROR_NONE_ENROLLED -> "未录入任何生物特征".toToast()
```

```
67              }
68              return false
69          }
70          //显示生物特征认证的提示框
71          fun showbiometricPrompt(){
72              executor = ContextCompat.getMainExecutor(this)
73              biometricPrompt = BiometricPrompt(this, executor, authenticationCallback)
74              //设置生物特征认证的提示框
75              promptInfo = BiometricPrompt.PromptInfo.Builder()
76                  .setTitle("指纹支付")
77                  .setSubtitle("您正在向小白魔支付 1 分钱")
78                  .setDeviceCredentialAllowed(true)
79                  .build()
80              biometricPrompt.authenticate(promptInfo) //进行生物特征认证
81          }
82          //显示提示信息
83          fun String.toToast() {
84              Toast.makeText(applicationContext, this, Toast.LENGTH_SHORT).show()
85          }
86      }
```

第 23～39 行重写认证回调对象的 3 个方法用于处理认证错误、认证成功和认证失败。认证失败后会反复提示指纹错误,并且不会关闭提示框,因此第 36 行调用 biometricPrompt.cancelAuthentication()方法取消认证并关闭提示框。第 59 行调用 biometricManager.canAuthenticate()方法返回检测结果，通过 BiometricManager 的常量判断检测结果。第 73 行创建生物特征认证的实例。第 75～79 行创建生物特征认证提示框的实例，setDeviceCredentialAllowed(true)设置可以使用设备 PIN、图案或密码替代生物特征进行认证，此时提示框的左下角会显示解锁设置的设备 PIN、图案或密码项作为文字按钮，单击后使用相应的方式替代验证。第 80 行显示生物特征认证的提示框进行认证。第 83～85 行声明扩展 String 类的扩展方法，用于显示提示信息，this 表示字符串实例。

13.5　感知生命周期

感知生命周期用于观察具有生命周期的组件的生命周期变化，控制资源释放，避免直接在生命周期方法中释放资源。拥有者(Lifecycle Owner)在生命周期事件发生后通知观察者对事件进行相应的处理。观察者(Lifecycle Observer)对象需要实现 LifecycleObserver 接口，该接口是一个空接口，不包含任何方法和属性。

13.5.1　Lifecycle 组件

1. State 枚举类

Lifecycle 组件是使用观察者模式进行设计的，观察者可以对 5 种事件进行处理。State 枚举类(androidx.lifecycle.Lifecycle.State)列举了这 5 种生命周期事件发生后的状态，如表 13-9 所示。拥有者和观察者的事件触发顺序并不是固定的,事件发生时先调用图 13～9 中离事件名称近的对象。onCreate、onStart 和 onResume 事件先调用拥有者的方法再调用观察者的方法，onPause、onStop 和 onDestroy 事件先调用观察者的方法再调用拥有者的方法。

表 13-9　State 枚举类的常量

常　量	说　明
CREATED	表示创建完成状态
DESTROYED	表示销毁后的状态
INITIALIZED	表示初始化后的状态
RESUMED	表示恢复后的状态
STARTED	表示开始后的状态

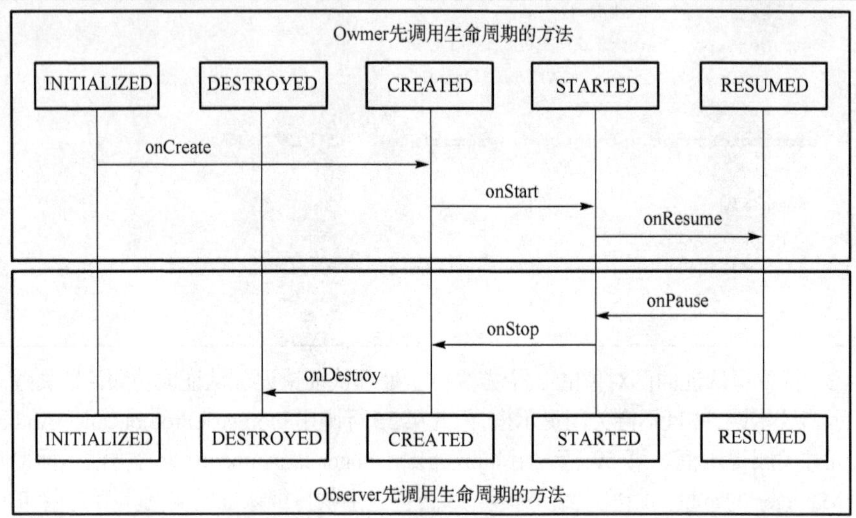

图 13-9　生命周期方法调用的顺序

提示：观察者模式

观察者模式（Observer Mode）是一种用于事件处理的设计模式。当事件发生或状态改变时，主动向观察者发出通知，通过观察者对事件进行相应的处理。

2．LifecycleOwner 接口

LifecycleOwner 接口（androidx.lifecycle.LifecycleOwner）用于获取及触发观察者的生命周期事件（方法如表 13-10 所示）。默认情况下，由于 AppCompatActivity 类通过父类已经实现了 LifecycleOwner 接口，所以无须添加依赖库。

表 13-10　LifecycleOwner 接口的方法

类型和修饰符	方　法
abstract Lifecycle	getLifecycle() 获取生命周期实例
suspend R	LifecycleOwner.whenCreated(block: suspend CoroutineScope.() -> T) 设置进入 Lifecycle.State.CREATED 状态后运行的代码块
suspend R	LifecycleOwner.whenResumed(block: suspend CoroutineScope.() -> T) 设置进入 Lifecycle.State.RESUMED 状态后运行的代码块
suspend R	LifecycleOwner.whenStarted(block: suspend CoroutineScope.() -> T) 设置进入 Lifecycle.State.STARTED 状态后运行的代码块

续表

类型和修饰符	方　法
suspend R	LifecycleOwner.withCreated (crossinline block: () -> R) 设置进入 Lifecycle.State.CREATED 状态后且保持该状态时运行的代码块
suspend R	LifecycleOwner.withResumed (crossinline block: () -> R) 设置进入 Lifecycle.State.RESUMED 状态后且保持该状态时运行的代码块
suspend R	LifecycleOwner.withStarted (crossinline block: () -> R) 设置进入 Lifecycle.State.STARTED 状态后且保持该状态时运行的代码块
suspend R	LifecycleOwner.withStateAtLeast (state: Lifecycle.State, crossinline block: () -> R) 设置进入指定状态后且保持该状态时运行的代码块

3. Lifecycle 类

Lifecycle 类(androidx.lifecycle.Lifecycle)用于存储 Activity 或 Fragment 的生命周期状态的信息(方法如表 13-11 所示)，在生命周期事件发生时调用观察对象的对应方法，以实现观察对象在生命周期事件发生时进行相应的处理。可以通过 LifecycleOwner.getLifecycle()方法获取 Lifecycle 实例，然后通过 addObserver(LifecycleObserver)方法添加观察者实例。

表 13-11　Lifecycle 类的方法

类型和修饰符	方　法
	<init>() 主构造方法
abstract Unit	addObserver (@NonNull observer: LifecycleObserver) 添加生命周期的观察对象
abstract Lifecycle.State	getCurrentState () 获取当前的生命周期状态
abstract Unit	removeObserver (@NonNull observer: LifecycleObserver) 移除生命周期的观察对象

4. LifecycleService 类

LifecycleService 类(androidx.lifecycle.LifecycleService)用于存储 Service 生命周期状态的信息(常用方法如表 13-12 所示)，是 android.app.Service 的子类。在生命周期事件发生时调用观察对象的对应方法，以实现观察对象在生命周期事件发生时应进行的处理。注意，需要添加“androidx.lifecycle:lifecycle-extensions”依赖库。

表 13-12　LifecycleService 类的常用方法

类型和修饰符	方　法
	<init>() 构造方法
open Lifecycle	getLifecycle () 获取生命周期实例
open IBinder?	onBind (@NonNull intent: Intent) 绑定后调用。返回 IBinder 实例与客户端进行数据通信。没有绑定到服务时，可以返回 null
open Unit	onCreate () 创建时调用
open Unit	onDestroy () 被销毁前调用
open Unit	onStartCommand (@Nullable intent: Intent?, flags: Int, startId: Int) 调用 Context.startService (Intent!) 方法启动服务时调用

5．ProcessLifecycleOwner 类

ProcessLifecycleOwner 类（androidx.lifecycle.ProcessLifecycleOwner）用于存储 App 进程的生命周期状态的信息，方法如表 13-13 所示。

表 13-13　ProcessLifecycleOwner 类的方法

类型和修饰符	方　法
open static LifecycleOwner	get() 获取整个 App 进程的生命周期的拥有者
open Lifecycle	getLifecycle() 获取生命周期实例

13.5.2　实例工程：改造使用 Camera2 类录制视频

本实例演示了使用 Lifecycle 组件对 C1104 工程进行改造，在 Camera 类内进行生命周期的处理。改造后功能性没有发生任何变化，只降低了功能的耦合性。

提示：耦合性

耦合性（Coupling），也称耦合度，是类或模块间依赖程度的度量，取决于控制关系、调用关系、数据传递关系。依赖程度越高，其耦合性越强，同时表明其独立性越差。降低耦合性，可以提高其独立性。软件工程中，使用耦合度和内聚度作为衡量类或模块独立程度的标准，即“高内聚低耦合”。

1．打开基础工程

打开“基础工程”文件夹中的 C1304 工程，该工程与 C1104 工程相同。

2．Camera 类

/java/com/teachol/c1304/Camera.kt

```
class Camera(val mContext: AppCompatActivity, val mPreviewView: PreviewTextureView, var mButton: Button, val mAuthority: String, var mPreviewActivity: Class<*>?):LifecycleObserver {
//Activity 恢复
    @OnLifecycleEvent(Lifecycle.Event.ON_RESUME)
    fun onResume() {
        Log.d(TAG, "onResume()")
        startBackgroundThread()
        //预览显示时打开摄像头，否则通过监听器监测打开摄像头
        if (mPreviewView.isAvailable) {
            openCamera(mPreviewView.width, mPreviewView.height)
        } else {
            mPreviewView.surfaceTextureListener = mSurfaceTextureListener
        }
    }
    //Activity 后台运行
    @OnLifecycleEvent(Lifecycle.Event.ON_PAUSE)
    fun onPause() {
        Log.d(TAG, "onPause()")
        if (mIsRecordingVideo) stopRecordingVideo(STOP_PAUSE)
        closeCamera()
```

```
111            stopBackgroundThread()
112        }
346        //请求权限
347        fun onRequestPermissionsResult(requestCode: Int) {
348            Log.d(TAG, "onRequestPermissionsResult()")
349            if(requestCode == Permissions.REQUEST_PERMISSIONS && !Permissions.hasPermissionsGranted(
350                    mContext,
351                    PERMISSIONS
352                ))
353                Permissions.requestPermissions(mContext, PERMISSIONS)
354        }
```

第 31 行声明实现 LifecycleObserver 接口，由于该接口为空接口，所以无须实现任何方法。第 94 行@OnLifecycleEvent(Lifecycle.Event.ON_RESUME) 注释表示第 95 ~ 104 行 onResume() 方法在 Activity 或 Fragment 进入前台调用 onResume() 方法后执行，使用注释后只要是合法的方法名就可以，无须只使用 onResume 作为方法名。第 106 行@OnLifecycleEvent(Lifecycle.Event.ON_PAUSE) 注释表示第 107 ~ 112 行 onPause() 方法在 Activity 或 Fragment 进入后台调用 onPause () 方法前执行。第 347 ~ 354 行处理请求权限的回调。

3. 主界面的 Activity

```
/java/com/teachol/c1304/MainActivity.kt
class MainActivity : AppCompatActivity() , View.OnClickListener{
    private val TAG = "MainActivity"
    private lateinit var mPreviewView: PreviewTextureView //预览拍摄画面
    private lateinit var mButton: Button //录制按钮
    private lateinit var mCamera:Camera
    override fun onCreate(savedInstanceState: Bundle?) {
        mPreviewView = findViewById(R.id.texture_view)
        mButton = findViewById(R.id.button)
        mButton.setOnClickListener(this)
        mCamera = Camera(this,mPreviewView,mButton,"com.teachol.c1303.fileprovider",PreviewActivity::class.java)
        getLifecycle().addObserver(mCamera)
    }
    override fun onResume() {
        Log.e(TAG,"onResume()")
        super.onResume()
    }
    override fun onPause() {
        Log.e(TAG,"onPause()")
        super.onPause()
    }
    //单击事件
    override fun onClick(v: View?) {
        if (mCamera.mIsRecordingVideo) mCamera.stopRecordingVideo(Camera.STOP_NORMAL)
        else mCamera.startRecordingVideo()
    }
    //请求权限回调
    override fun onRequestPermissionsResult(requestCode: Int, permissions: Array<String?>,
```

```
   grantResults: IntArray) {
40         mCamera.onRequestPermissionsResult(requestCode)
41     }
42 }
```

第 23 行将 mCamera 实例作为观察对象，使其能够处理生命周期事件。第 25 ~ 32 行 onResume() 方法和 onPause() 方法用于验证 MainActivity 和 Camera 生命周期的调用顺序，实际使用中可以不添加。第 39 ~ 41 行处理权限请求的回调，由于不属于生命周期事件，所以需要调用 Camera 类的 onRequestPermissionsResult(Int) 方法使 mCamera 实例能够处理回调的数据。

运行后，单击多任务键(底部正方形的虚拟键)后再单击该 App，使其从前台切换到后台再切换到前台显示。在 Logcat 窗口中，选择“Error”选项，可以观察到生命周期的调用顺序，如图 13-10 所示。MainActivity 先调用 OnResume() 方法，Camera 先调用 OnPause() 方法。

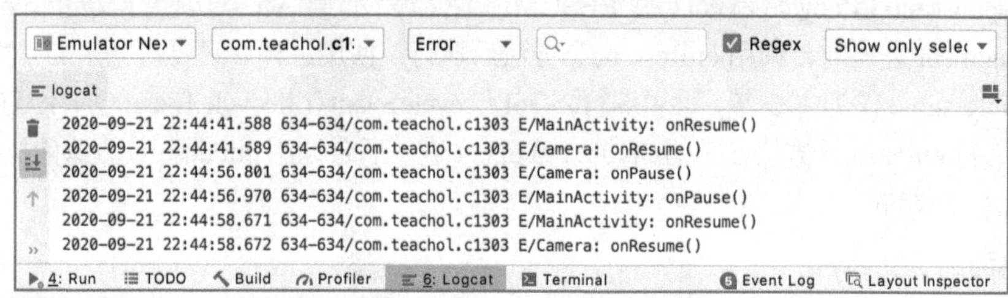

图 13-10 Logcat 窗口输出结果

13.6 视图模型

视图模型用于以感知生命周期的方式存储和管理界面相关的数据，在屏幕旋转等配置更改后能够继续还原数据，其生命周期与拥有 Lifecycle 的 Activity 或 Fragment 的生命周期一致。Activity 类的 onSaveInstanceState(Bundle) 方法与其相比，只能保存少量的、能支持序列化的数据。

13.6.1 ViewModel 组件

1. ViewModel 类

ViewModel 类(androidx.lifecycle.ViewModel) 由于作为父类创建视图模型的抽象类，所以无法直接实例化，除主构造函数外没有其他的公共方法。ViewModel 子类的实例不能引用视图、Lifecycle 或可能存储对 Activity 上下文引用的任何类，以避免出现异常。

提示：AndroidViewModel 类

AndroidViewModel 类是 ViewModel 类的子类，构造方法的唯一参数是 Application 类型。因为 Application 类是 Context 类的子类，所以支持 Application 类型的上下文。

2. ViewModelProvider 类

ViewModelProvider 类(androidx.lifecycle.ViewModelProvider) 为作用域提供视图模型，常用方法如表 13-14 所示。在实现 ViewModelStoreOwner 接口的组件(如 AppCompatActivity、Fragment 或 LifecycleService) 中，通过 ViewModelProvider 类获取 ViewModelProvider 实例。注意，需要添加“androidx.lifecycle:lifecycle-extensions”依赖库。

表 13-14　ViewModelProvider 类的常用方法

类型和修饰符	方　法
	<init>(@NonNull owner: ViewModelStoreOwner) 主构造方法
open T	get(@NonNull modelClass: Class<T>) 获取与 ViewModelProvider 关联的范围(通常是 AppCompatActivity 或 Fragment)中现有 ViewModel 或创建新的 ViewModel
open T	get(@NonNull key: String, @NonNull modelClass: Class<T>) 获取与 ViewModelProvider 关联的范围(通常是 AppCompatActivity 或 Fragment)中现有 ViewModel 或创建新的 ViewModel

13.6.2　实例工程：足球赛记分器

本实例演示了可以横屏或竖屏显示的足球赛记分器，使用 ViewModel 组件和 Activity 类的 onSaveInstanceState(Bundle)方法分别实现手机旋转后重建视图的数据恢复(如图 13-11 所示)。

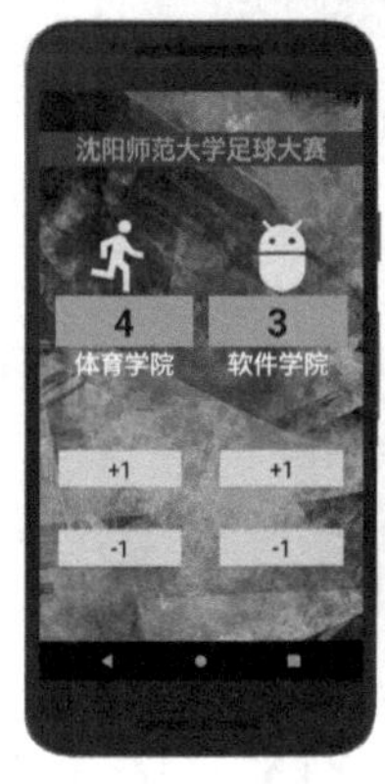

图 13-11　运行效果

1. 打开基础工程

打开“基础工程”文件夹中的 C1305 工程，该工程已经包含 MainActivity 类及资源文件，其中使用了内置的矢量资源，且已经添加了“androidx.lifecycle: lifecycle-extensions:2.2.0”依赖库。

2. 添加内置矢量资源

在“/res/drawable/”文件夹上单击右键，选择【New】→【Vector Asset】命令(如图 13-12 所示)。

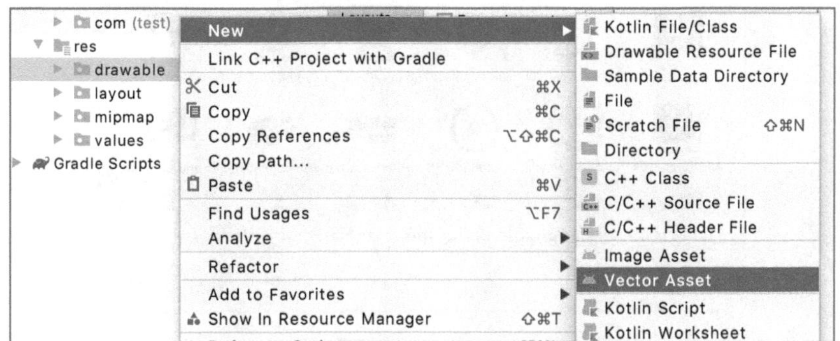

图 13-12　【New】→【Vector Asset】命令

在“Asset Studio”对话框中，设置名称、图标、尺寸和颜色(如图 13-13 所示)。单击“Clip Art”后的图标,弹出“Select Icon”对话框(如图 13-14 所示),选择所需的矢量图标,单击“OK”按钮。在“Asset Studio”对话框中，单击“Next”按钮，确认资源保存路径后，单击“Finish”按钮完成矢量资源的创建(如图 13-15 所示)。此步骤创建的矢量资源已经包含在基础工程中，这里仅演示创建方法。

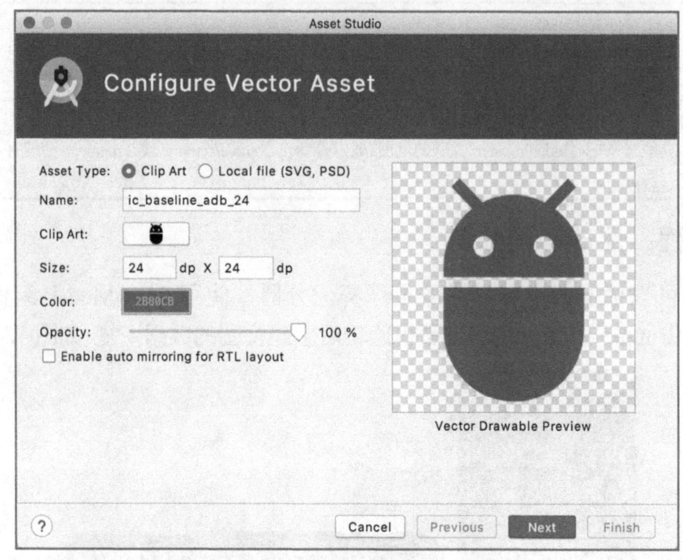

图 13-13 “Asset Studio”对话框

图 13-14 “Select Icon”对话框

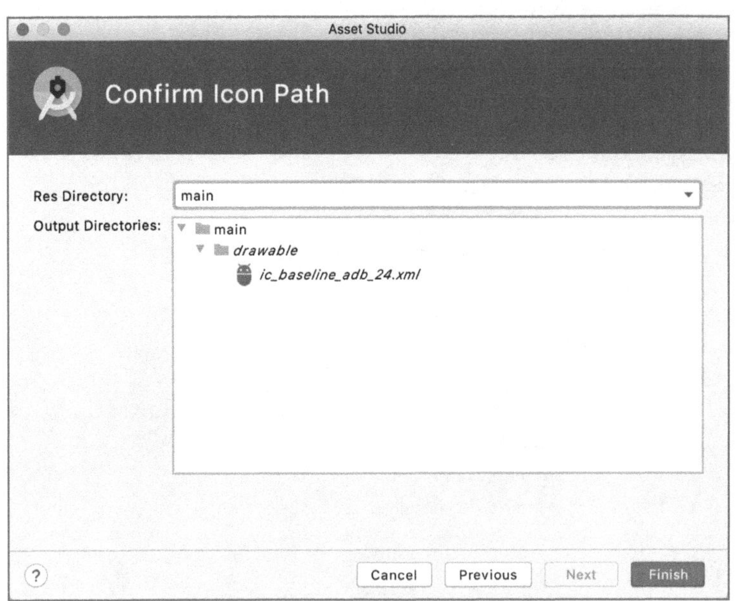

图 13-15　确认资源保存路径

3. CollegeViewModel 类

在“/java/com/teachol/c1305/”文件夹中，新建 CollegeViewModel.kt 文件，用于添加视图模型。

```
/java/com/teachol/c1305/CollegeViewModel.kt
03  import androidx.lifecycle.ViewModel
05  class CollegeViewModel: ViewModel() {
06      var score:Int = 0 //得分
07  }
```

第 03 行导入 ViewModel 类的命名空间。第 05 行声明 CollegeViewModel 类继承 ViewModel 类。第 06 行声明 score 变量保存得分。

4. 新建主界面的横向布局

打开 activity_main.xml 文件，单击“Orientation for Preview”按钮，在弹出的菜单中选择【Create Landscape Variation】命令，创建横向显示的布局文件(如图 13-16 所示)。此步骤创建的矢量资源已经包含在基础工程中，这里仅演示创建方法。

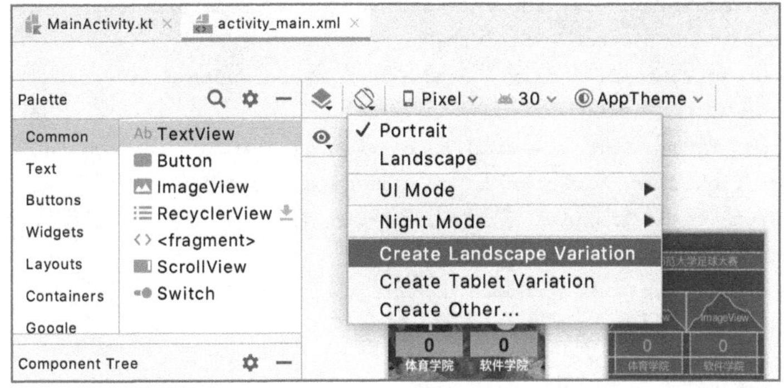

图 13-16　“Create Landscape Variation”命令

5. 主界面的 Activity

/java/com/teachol/c1305/MainActivity.kt

```
const val TAG = "MainActivity"
const val STATE_SCORE = "mScore"
class MainActivity : AppCompatActivity() {
    private var mScore: Int = 0 //软件学院得分
    override fun onCreate(savedInstanceState: Bundle?) {
        super.onCreate(savedInstanceState)
        setContentView(R.layout.activity_main)
        Log.e(TAG, "onCreate()")
        //设置全屏显示
        if (android.os.Build.VERSION.SDK_INT >= android.os.Build.VERSION_CODES.R) {
            window.insetsController?.also {
                it.hide(WindowInsets.Type.statusBars())
                it.hide(WindowInsets.Type.navigationBars())
            }
        } else {
            this.window.setFlags(WindowManager.LayoutParams.FLAG_FULLSCREEN, WindowManager.LayoutParams.FLAG_FULLSCREEN)
        }
        //视图模型：旋转后自动还原数据
        val teamTextView1: TextView = findViewById(R.id.teamTextView1)
        val addButton1: Button = findViewById(R.id.addButton1)
        val subtractButton1: Button = findViewById(R.id.subtractButton1)
        //通过 ViewModelProvider 得到 CollegeViewModel，如果 CollegeViewModel 不存在则创建
        val collegeViewModel: CollegeViewModel = ViewModelProvider(this).get(CollegeViewModel::class.java)
        teamTextView1.text = collegeViewModel.score.toString() //初始化体育学院的得分
        addButton1.setOnClickListener { //得分+1
            collegeViewModel.score ++
            teamTextView1.text = collegeViewModel.score.toString()
        }
        subtractButton1.setOnClickListener { //得分-1
            collegeViewModel.score--
            teamTextView1.text = collegeViewModel.score.toString()
        }
        //传统方式：无法在旋转后还原数据
        val teamTextView2: TextView = findViewById(R.id.teamTextView2)
        val addButton2: Button = findViewById(R.id.addButton2)
        val subtractButton2: Button = findViewById(R.id.subtractButton2)
        //从已保存状态恢复成员的值
        if (savedInstanceState != null) mScore = savedInstanceState.getInt(STATE_SCORE)
        teamTextView2.text = mScore.toString() //初始化软件学院的得分
        addButton2.setOnClickListener { //得分-1
            mScore ++
            teamTextView2.text = mScore.toString()
        }
        subtractButton2.setOnClickListener { //得分-1
            mScore --
```

```
57            teamTextView2.text = mScore.toString()
58        }
59    }
60    override fun onSaveInstanceState(savedInstanceState: Bundle) {
61        Log.e(TAG, "onSaveInstanceState()")
62        //保存软件学院的得分
63        savedInstanceState.putInt(STATE_SCORE, mScore)
64        //调用父类的 onSaveInstanceState(Bundle)方法，保存视图层次结构状态
65        super.onSaveInstanceState(savedInstanceState)
66    }
67 }
```

第 21 ~ 28 行根据版本号的不同使用两种方法设置全屏显示，这是因为第 27 行的方法已经在 API Level 30 时过期。第 34 行从当前的 AppCompatActivity 获取视图模型的实例，如果没有该实例则创建。第 35 行通过 collegeViewModel.score 获取保存的体育学院得分，将其转为 String 类型后显示在 teamTextView1 控件中。第 49 行当 savedInstanceState 参数不为 null（即重构布局）时，通过 savedInstanceState 参数提取软件学院的得分。第 60 ~ 66 行旋转屏幕时调用该方法，保存软件学院的得分，当重构布局时可用于还原数据。

运行工程后，单击模拟器左侧的向右旋转按钮将模拟器旋转后，在 Logcat 窗口中可以看到输出“onCreate()”和“onSaveInstanceState()”。再单击向左旋转按钮将模拟器方向还原，在 Logcat 窗口中可以看到输出“onCreate()”，如图 13-17 所示。

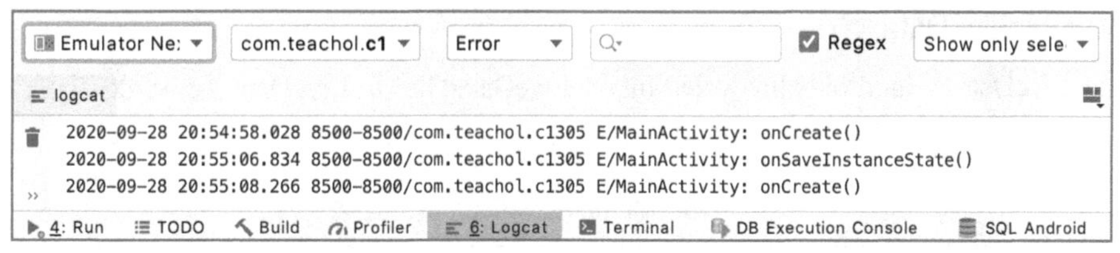

图 13-17　Logcat 窗口输出结果

13.7　实 时 数 据

实时数据用于以感知生命周期的方式更新处于活跃生命周期状态的组件数据，使数据始终保持最新状态，可用于 Activity、Fragment 或 Service 等具有生命周期的组件。如果活跃观察者（Active Observer）的生命周期处于 STARTED 或 RESUMED 状态，将发送更新的通知给观察者同步更新数据，而注册的非活跃观察者则不会收到更改通知。

13.7.1　LiveData 组件

1. LiveData 类

LiveData 类（androidx.lifecycle.LiveData<T>）是可观察的数据存储的抽象类（方法如表 13-15 所示），不能直接实例化。LiveData 具有生命周期感知能力，与实现 LifecycleOwner 接口的组件（如 AppCompatActivity、Fragment 或 LifecycleService）配合使用，可以确保 LiveData 仅更新处于活跃生命周期状态的应用组件观察者，并且在生命周期结束时自动移除。

表 13-15 LiveData 类的方法

类型和修饰符	方 法
	<init>(value: T) 泛型参数的主构造方法
	<init>() 主构造方法
open T?	getValue() 获取变量值
open Boolean	hasActiveObservers() 判断是否为活跃观察者
open Boolean	hasObservers() 判断是否为观察者
open Unit	observe(@NonNull owner: LifecycleOwner, @NonNull observer: Observer<in T>) 将观察者添加到生命周期拥有者的观察者列表中如果拥有者处于 DESTROYED 状态，该观察者将被自动移除
open Unit	observeForever(@NonNull observer: Observer<in T>) 将观察者添加到观察者列表中该观察者将接收所有事件，并且永远不会被自动删除可以调用 removeObserver (Observer)方法来停止观察
open Unit	removeObserver(@NonNull observer: Observer<in T>) 移除观察者
open Unit	removeObservers(@NonNull owner: LifecycleOwner) 移除与生命周期拥有者关联的所有观察者

2. MutableLiveData 类

MutableLiveData 类(androidx.lifecycle.MutableLiveData<T>)是 LiveData 类的子类(方法如表 13-16 所示)，可以直接实例化作为数据变化的被观察对象。

表 13-16 MutableLiveData 类的方法

类型和修饰符	方 法
	<init>(value: T) 泛型参数的主构造方法
	<init>() 主构造方法
open Unit	postValue(value: T) 更新数据，可以在线程中使用
open Unit	setValue(value: T) 更新数据，不可以在线程中使用

3. MediatorLiveData 类

MediatorLiveData 类(androidx.lifecycle.MediatorLiveData)是 MutableLiveData 类的子类，还能观察 LiveData 类的子类数据变化(方法如表 13-17 所示)。适合将多个 LiveData 数据合成为一个 MediatorLiveData 数据，可以针对每个 LiveData 数据添加一个观察者。

表 13-17 MediatorLiveData 类的方法

类型和修饰符	方 法
	<init>() 主构造方法

续表

类型和修饰符	方　　法
open Unit	addSource (@NonNull source: LiveData<S>, @NonNull onChanged: Observer<in S>) 添加观察数据变化的观察源，观察源必须是 LiveData 类的子类，通常是 MediatorLiveData 类实例。当观察源的数据改变时，调用 Observer<S>接口实例的 onChanged (T) 方法对数据的变化进行处理。必须先调用 observe (LifecycleOwner, Observer<in T>) 方法或 observeForever (Observer<in T>) 方法，否则无法观察到观察源的数据改变
open Unit	removeSource (@NonNull toRemote: LiveData<S>) 移除数据变化的观察源

4．Observer 接口

Observer 接口 (androidx.lifecycle.Observer) 是观察数据变化的接口，提供了 onChanged (T) 抽象方法
取数据的变化（方法如表 13-18 所示）。

表 13-18　Observer 类的方法

类型和修饰符	方　　法
abstract Unit	onChanged (t: T) 观察对象的数据改变时调用，可以观察到 MutableLiveData.postValue (T) 方法或 MutableLiveData.settValue (T) 方法对数据的改变

3.7.2　实例工程：联想搜索关键字

本实例演示了主流的搜索联想功能，根据输入的关键字通过后台网络搜索相关的关键字并显示来，单击“搜索”按钮或者单击搜索到的联想关键字，开启搜索结果（如图 13-18 所示）。为了减实例代码且突出核心功能，本实例直接在本地模拟后台网络获取数据，搜索结果中直接显示搜索的键字。

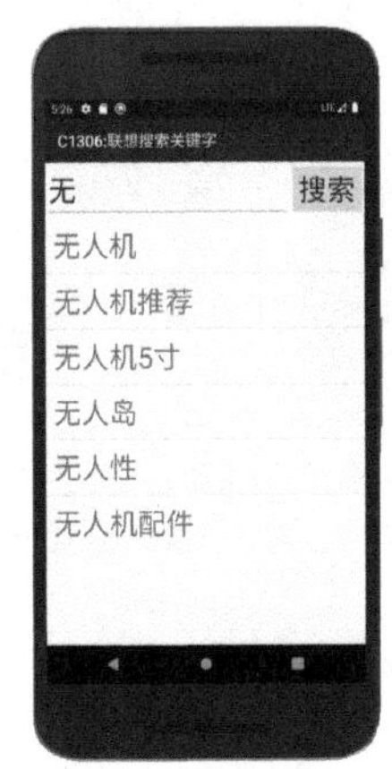

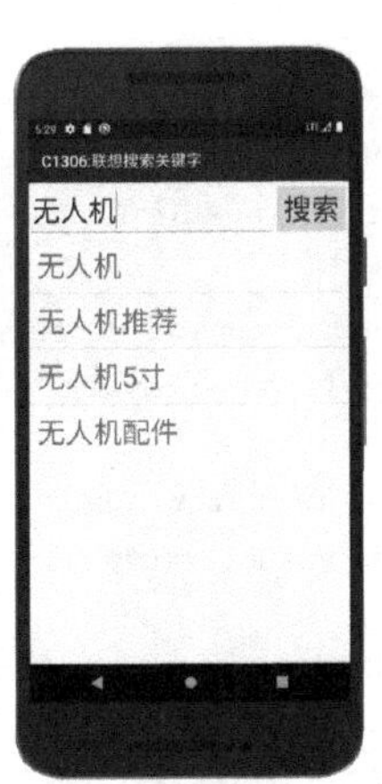

图 13-18　运行效果

1．打开基础工程

打开“基础工程”文件夹中的 C1306 工程，该工程已经包含 MainActivity、SearchResultActivity
及资源文件。

2．联想关键字子视图的布局

在“/res/layout/”文件夹中，新建 list_view_item.xml 文件，作为显示联想关键字的子视图布局。

/res/layout/list_view_item.xml

```
02 <TextView xmlns:android = "http://schemas.android.com/apk/res/android"
03    android:id = "@android:id/text1"
04    android:layout_width = "match_parent"
05    android:layout_height = "wrap_content"
06    android:padding = "10dp"
07    android:singleLine = "true"
08    android:textSize = "36sp" />
```

第 03 行 id 设置为@android:id/text1，可以自动匹配 ArrayAdapter 设置的数据。第 06 行设置内间距为 10dp，使其与分割线和屏幕边缘有一定的距离。

3．主界面的 Activity

/java/com/teachol/c1306/MainActivity.kt

```
const val MSG_SUCCESS = 1 //获取数据成功
const val MSG_NONDATA = 2 //无数据
class MainActivity : AppCompatActivity() {
   private val mMutableLiveData = MutableLiveData<String>() //保存关键字的 LiveData
   private val mKeys = mutableListOf("无人机", "无人机推荐", "无人机 5 寸", "无人岛", "无人性",
"无人机配件", "Android", "Android 11", "Android kotlin", "apple")
   private var mSearchKeys = mutableListOf<String>()//查询到的关键字
   private lateinit var mAdapter: ArrayAdapter<String>
   override fun onCreate(savedInstanceState: Bundle?) {
      super.onCreate(savedInstanceState)
      setContentView(R.layout.activity_main)
      val editText: EditText = findViewById(R.id.editText) //搜索框
      val button: Button = findViewById(R.id.button) //搜索按钮
      val listView: ListView = findViewById(R.id.listView) //联想关键字列表
      //观察 mutableLiveData 的数据变化
      mMutableLiveData.observe(this, object: Observer<String> {
         override fun onChanged(value: String?) {
            editText.setSelection(value.toString().length) //将光标移动到最后
            if (value.equals("")) mAdapter.clear() //为空时，清空适配器数据
            else SearchKeyThread(value!!).start() //不为空时，开启新线程搜索联想关键字
         }
      })
      //添加搜索内容变化事件的监听器
      editText.addTextChangedListener(object : TextWatcher {
         //输入文字产生变化时调用
         override fun onTextChanged(s: CharSequence?, start: Int, before: Int, count: Int) {
            //修改 MutableLiveData 的数据
            if (mMutableLiveData.value != s.toString()) mMutableLiveData.value = s.toString()
         }
      })
      //设置搜索按钮单击事件的监听器
      button.setOnClickListener {
         startSearchResultActivity(mMutableLiveData.value.toString())
```

```
        }
        //设置联想关键字的适配器
        mAdapter = ArrayAdapter(applicationContext, R.layout.list_view_item, mutableListOf<String>())
        listView.adapter = mAdapter //设置数据适配器
        //设置子视图的单击监听器
        listView.onItemClickListener = AdapterView.OnItemClickListener { _, _, position, _ ->
            editText.setText((listView.getChildAt(position) as TextView).text.toString(), TextView.BufferType.EDITABLE)
            startSearchResultActivity(mMutableLiveData.value.toString())
        }
    }
    //打开模拟搜索结果的 Activity
    private fun startSearchResultActivity(key:String){
        val intent = Intent(applicationContext, SearchResultActivity::class.java)
        intent.putExtra("key",key) //传递给 SearchResultActivity 的数据
        startActivity(intent)
    }
    //搜索联想关键字线程
    inner class SearchKeyThread(private val searchKey: String) : Thread() {
        override fun run() {
            mSearchKeys = getKeys(searchKey) //获取联想关键字
            if (mSearchKeys.size > 0) {
                mHandler.obtainMessage(MSG_SUCCESS).sendToTarget()
            } else {
                mHandler.obtainMessage(MSG_NONDATA).sendToTarget()
            }
        }
    }
    //模拟获取服务器端搜索联想关键字返回的结果
    fun getKeys(searchKey: String):MutableList<String>{
        val resultKeys = mutableListOf<String>() //查询到的联想关键字
        Thread.sleep(200)//模拟服务器端的延迟
        for (key in mKeys) {
            if (key.startsWith(searchKey, ignoreCase = true)) { //忽略英文字母大小写判断是否是联想关键字
                resultKeys.add(key.trim()) //保存查询到的关键字
            }
        }
        return resultKeys
    }
    //更新 UI
    private val mHandler: Handler = object : Handler(Looper.myLooper()!!) {
        override fun handleMessage(msg: Message) {
            when (msg.what) {
                MSG_SUCCESS -> {
                    mAdapter.clear() //清空联想关键字列表
                    mAdapter.addAll(mSearchKeys) //添加列表项数据
                    mAdapter.notifyDataSetChanged() //更新联想关键字视图
                }
                MSG_NONDATA -> {
```

```
100                    mAdapter.clear() //清空联想关键字列表
101                    mAdapter.notifyDataSetChanged() //更新联想关键字视图
102                }
103            }
104        }
105    }
106 }
```

第 19 行声明 mKeys 存储模拟网络数据库存储的关键字。第 31 ~ 37 行设置 mMutableLiveData 对象的观察者对象，观察者对象使用匿名对象生成，并重写其 onChanged(String?) 方法。当 mMutableLiveData 对象保存的数据改变时，开启新线程查询联想的关键字。第 39 ~ 48 行添加 editText 输入文本改变的监听器，必须重写 3 个方法，这里 afterTextChanged(Editable?) 方法和 beforeTextChanged(CharSequence?,Int,Int,Int) 方法虽然不需要使用，但是必须重写。onTextChanged(CharSequence?,Int,Int,Int) 方法用于输入的文本产生改变后，重新设置 mMutableLiveData 对象保存的数据，此时会触发 mMutableLiveData 对象的观察者并调用其 onChanged(String?) 方法。第 50 ~ 52 行设置单击“搜索”按钮后，开启 SearchResultActivity 并将搜索的关键字传递过去并显示出来。第 54 行设置显示联想关键字的 ListView 适配器，列表子视图的数据设置为空的 MutableList 对象。第 57 ~ 60 行设置子视图单击事件的监听器，单击子视图后将该联想关键字作为搜索关键字显示在搜索文本框中，开启 SearchResultActivity 并将搜索的关键字传递过去并显示出来。第 84 行忽略英文字母大小写判断模拟网络存储的关键字，是否是以搜索关键字开头的联想关键字。

第 14 章 “粉色辣椒”开发流程

“粉色辣椒”是一款开源社交类 App，由作者独立开发完成，后台采用 PHP+MYSQL 架构提供服务。该 App 面向所有开发者永久免费开放所有接口，包含注册、登录、发布动态、回复、关注、私信等常用的社交功能。希望该 App 成为一个远离广告、微商和网红的社交平台，回归社交的基本需求，更多地体现普通人的真实生活和社交需求。

14.1 项 目 介 绍

14.1.1 市场分析

目前，社交类 App 中用户浏览内容的来源分为三类：关注类、推送类和匿名类。关注类以用户主动关注或订阅获取内容为主；推送类以用户搜索、浏览、点赞、收藏、位置等喜好为主，关注或订阅内容为辅；匿名类以根据设定条件随机推送内容为主。

关注类需要主动关注其他用户，需要长期积累感兴趣的用户。使用推送类算法推送的内容严重同质化，导致用户浏览的内容会被限定在目前的喜好范围内，缺少随机推送未知喜好的内容。推送类还会对内容推送限流，向优质内容且高频发布内容的用户倾斜，导致强者愈强、弱者愈弱。匿名类随机性和不确定较强。

以某一类内容来源为主的 App 有明显的优势和劣势，融合多类内容来源的 App 没有明显的优势。

14.1.2 产品定位

“粉色辣椒”最初定位于面向 Android 学习者的教学实例，是一款基于位置推送内容的开源社交类 App，帮助用户找到周围感兴趣的人或兴趣相同的人只关注周围生活的普通人，希望能够通过网络使周围的陌生人走向现实生活。

“粉色辣椒”提供了开放平台，为每个开发者提供同等权限，希望汇集每个开发者的力量不断完善平台。同时，秉承网络契约精神，保障隐私和平等权利，对用户永久免费，不设置用户等级和收费用户。

14.1.3 产品展望

目前，“粉色辣椒”的平台功能只是一个简单的雏形，受租用服务器空间的限制，用户只能以图文形式发布内容。本平台重点发展以下几个方面。

- 共同开发：个人的创意和能力始终是有限的，汇集开发者创造不同界面或功能的客户端公用一个服务器端，为用户提供更多的选择。
- 共同管理：每个用户都可以申请成为管理者，参与制定管理条例，对举报的内容进行仲裁等。
- 信任参考：用户没有等级，用户之间可以将发布内容、身份认证、参与开发和管理等指标的评分作为信任参考。
- 非营利化：通过捐赠支持域名租用、服务器空间租用、组织活动等费用。

14.2 开发流程

下面先介绍企业开发 App 的流程。移动互联网产品是一项系统化的工程，开发过程不仅是 App 本身的开发，还涉及前期准备和后期运行的工作。标准化的开发流程包括商业需求、市场需求、产品需求、开发、测试、运维和运营等环节(如图 14-1 所示)；再根据运维环节和运营环节的反馈，返回到对应的环节对产品进行迭代。其中的开发环节，无论是个人开发还是团队开发都不能省略，其余环节可以根据实际情况进行精简或合并。Android 程序员的工作属于客户端开发环节。

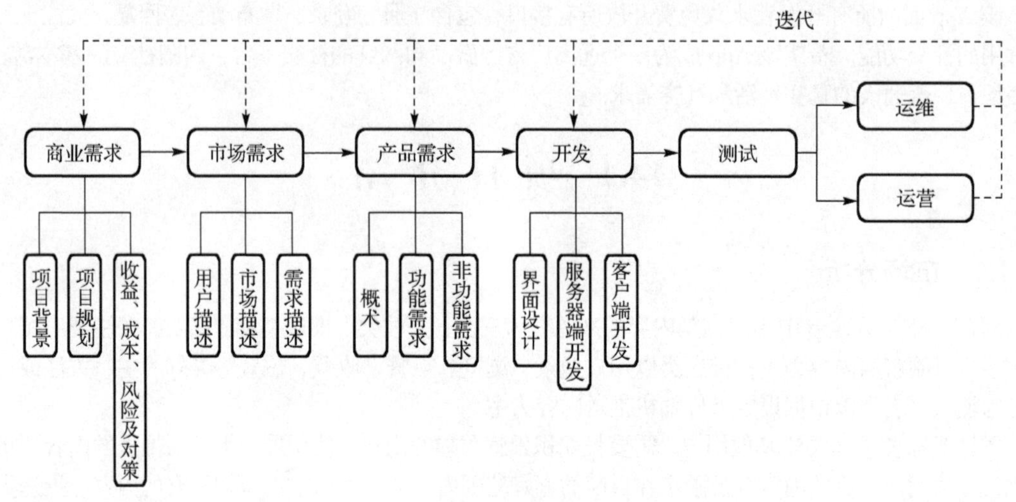

图 14-1　开发流程

14.3 开放平台介绍

“粉色辣椒”虽然现在只是用于学习者进行实践的社交平台，但是随着功能的不断完善，希望成为一个开发者共同开发和用户共同管理的平台。“粉色辣椒”开放平台永久免费开放，所有开发者的权限相同。由于水平和资源的限制，通过个人开发很难将该平台发展壮大，所以欢迎伙伴们加入平台接口和开源客户端的开发中，根据用户需求完善各项功能。

14.3.1 客户端框架结构

“粉色辣椒”的客户端包括欢迎、注册、登录、忘记密码、主页、发布动态、显示动态、回复动态、个人主页、发送私信、设置等模块，如图 14-2 所示，模块之间使用箭头连线表示跳转关系。其中，主页是使用 Tab 导航形式的容器，由附近、关注、偶遇、提醒和自己等 5 个 Fragment 构成，与主页之间使用直线连接。

14.3.2 开发者账号

“粉色辣椒”的 Android 客户端虽然是开源的，但是需要配置开发者账号才能调用服务器端的接口，这是为了确保服务器端的接口不被非法调用和免于遭受恶意攻击。“粉色辣椒”开放平台的网址为 http://www.weiju2014.com/pinkjiao/developer.html，采用实名制进行注册，界面如图 14-3 所示。

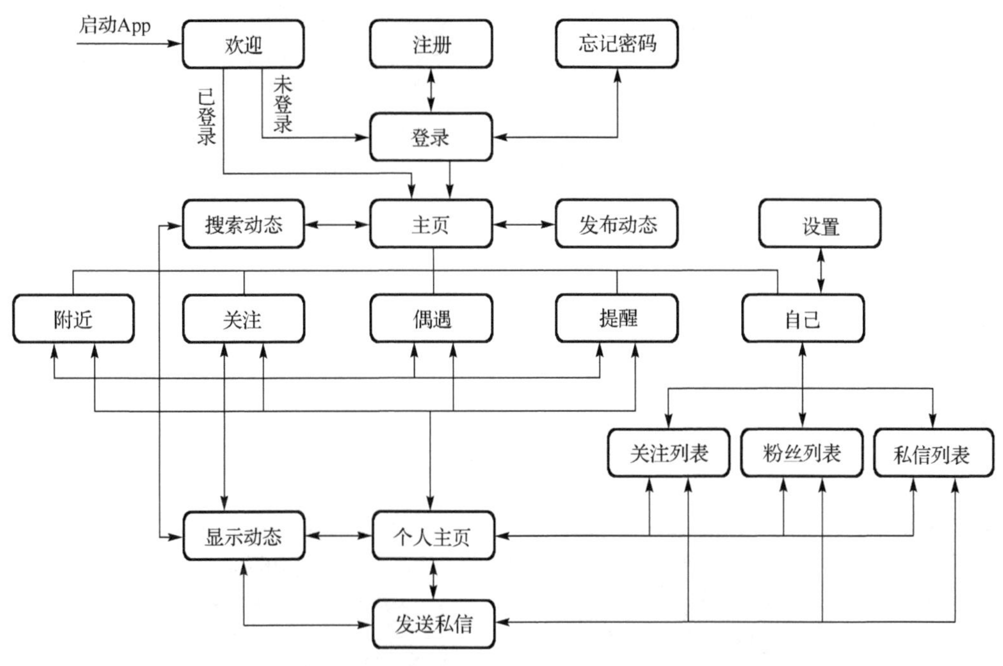

图 14-2 框架结构

图 14-3 开放平台界面

开发者注册完成后，登录开放平台。选择左侧导航栏中的“上传身份证”选项(如图 14-4 所示)，上传身份证照片必须与注册时的身份证信息一致，通过验证后可以使用开放平台。

提示：开放平台的实名制验证

实名制验证是为了保障服务器端接口不被恶意调用，平台不会对其进行公开和使用，并遵守法律法规进行妥善保存。注销开发者账号时，会在核实后销毁提交的身份证照片。

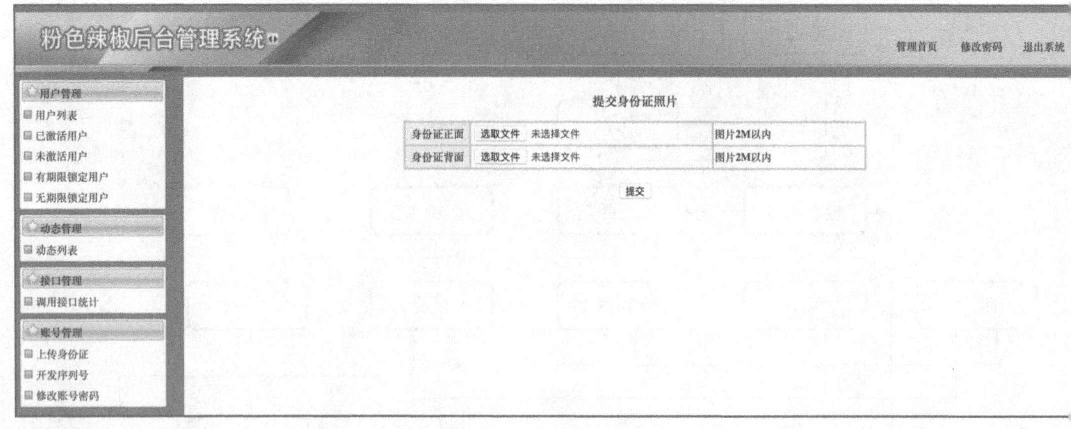

图 14-4　上传身份证

选择左侧导航栏中的“开发序列号”选项，可以查看开发序列号，如图 14-5 所示。在 Android 户端中设置开发者账号和开发序列号，可以调用“粉色辣椒”开放平台的接口。

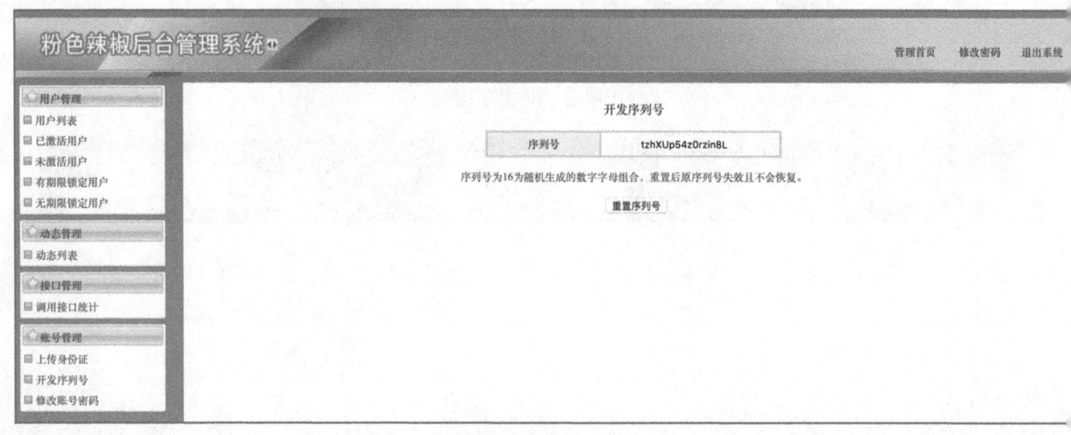

图 14-5　查看开发序列号

提示：开发序列号

开发序列号是使用开放平台的凭证，开发者应保密，以避免被盗用而导致封号。如果重置开发序列号，会重新生成 16 位的随机数字和字母组合的字符串，之前的序列号作废且无法恢复。

14.3.3　基础工程

打开“基础工程”文件夹中的 pinkjiao 工程，包含所有的基本资源文件、代码文件、权限配置 gradle 配置，且启用了视图绑定和数据绑定的功能。

其中，代码文件夹中包含 9 个自定义包，如图 14-6 所示。adapter 包用于存储适配器类及其接口 compat 包用于存储 Activity 的基类；fragment 包用于存储 Fragment 类及其接口；model 包用于存储据类；net 包用于保存访问 Http 的类及其接口；repository 包用于保存通过 Http 获取数据的存储类；u 包用于保存通用类；view 包包含 holder 包和 model 包，holder 包用于保存子视图的缓存类，model 用于保存视图模型类；widget 包用于保存自定义控件。

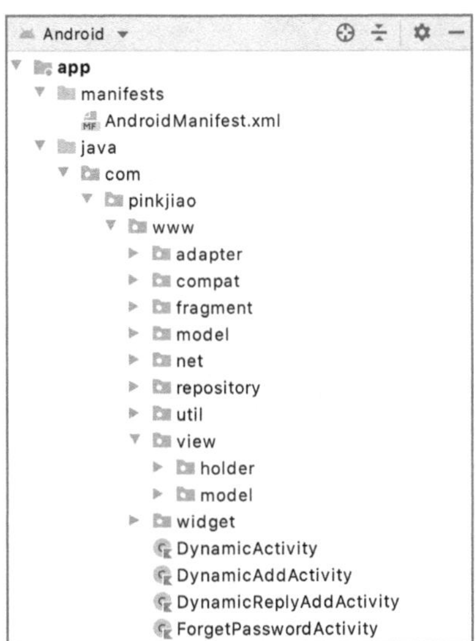

图 14-6　自定义包

14.3.4　设置服务器端

在 string.xml 文件中，设置服务器端的地址配置。虽然配置信息可以写在网络请求类中，但是为了便于统一管理，将所有需要配置的信息都放在 string.xml 文件中。

```
/res/values/strings/strings.xml
03 <string name = "app_domain">http://www.weiju2014.com/pinkjiao/</string>
04 <string name = "app_domain_folder">app_server</string>
05 <string name = "app_version">v1.0</string>
```

第 03 行 app_domain 标签设置域名。第 04 行 app_domain_folder 标签设置应用所在的文件夹。第 05 行 app_version 标签设置 App 的版本。三个标签的字符串连接起来就是服务器端接口所在的文件夹，即 http://www.weiju2014.com/pinkjiao/app_server/v1.0。

提示：app_domain 标签

app_domain 标签目前并没有使用 http://www.pinkjiao.com 域名地址，因为还没有通过域名备案，所以在已有域名下创建了一个文件夹临时提供服务。

14.3.5　设置开发者账号和开发序列号

在 string.xml 文件中，只需将属性值替换成注册的开发者账号和开发序列号即可，否则无法访问服务器端。

```
/res/values/strings/strings.xml
10 <string name = "developer_accountNumber">baizhe</string>
11 <string name = "developer_key">Xum23e1TYXeo3aQe</string>
```

第 10 行 developer_accountNumber 标签设置开发者账号，baizhe 是开发者账号。第 11 行 developer_key 标签设置开发序列号，Xum23e1TYXeo3aQe 是开发序列号。

14.4 启动图标

在“/res/”文件夹上单击右键，选择【New】→【Image Asset】命令(如图 14-7 所示)。

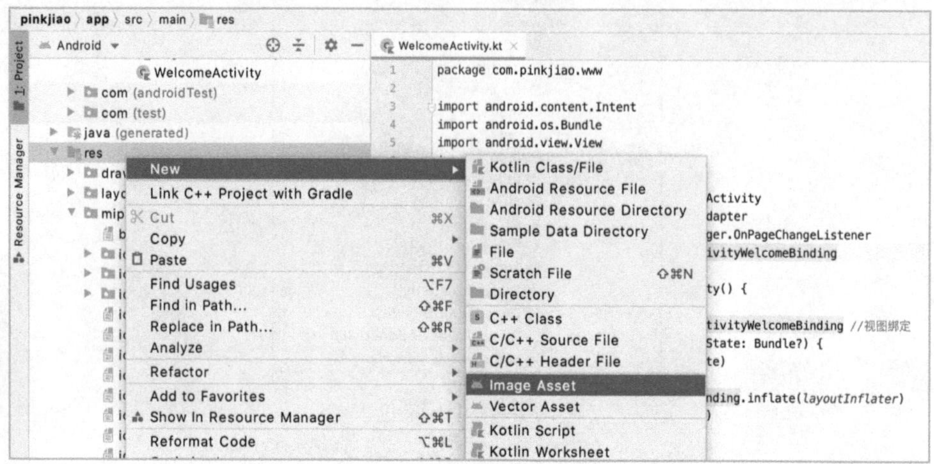

图 14-7 添加启动图标

在“Asset Studio”对话框中，“Path”选择启动图标文件的路径，右侧的预览窗口中会显示将要自动生成的不同图标样式效果，如图 14-8 所示。

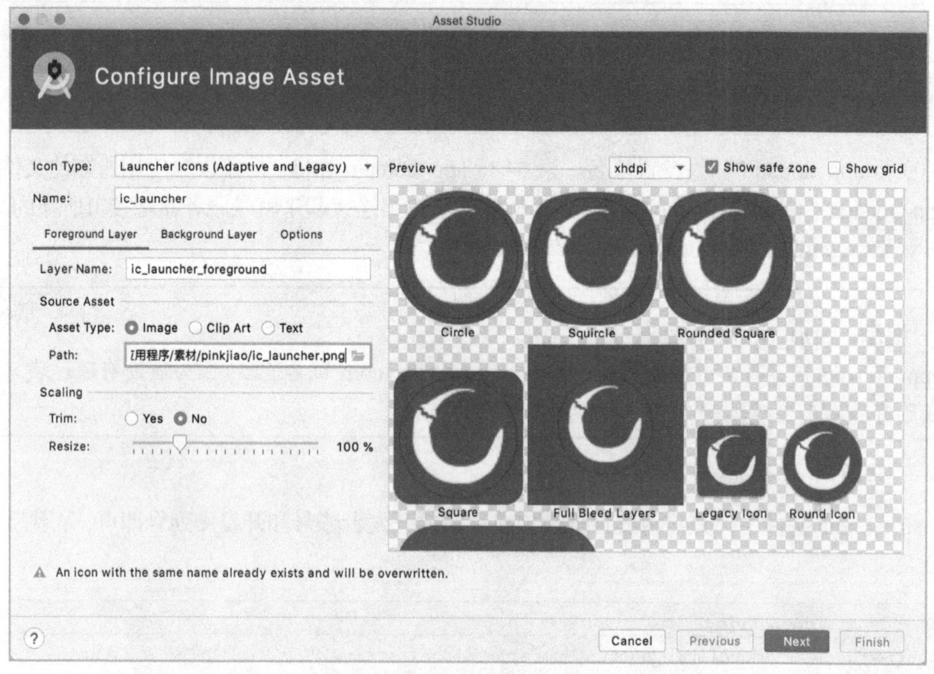

图 14-8 选择启动图标文件路径

单击“Next”按钮，由于图标文件已经存在，所以会提示确认覆盖，如图 14-9 所示。单击“Finish”按钮，覆盖原有的图标文件。

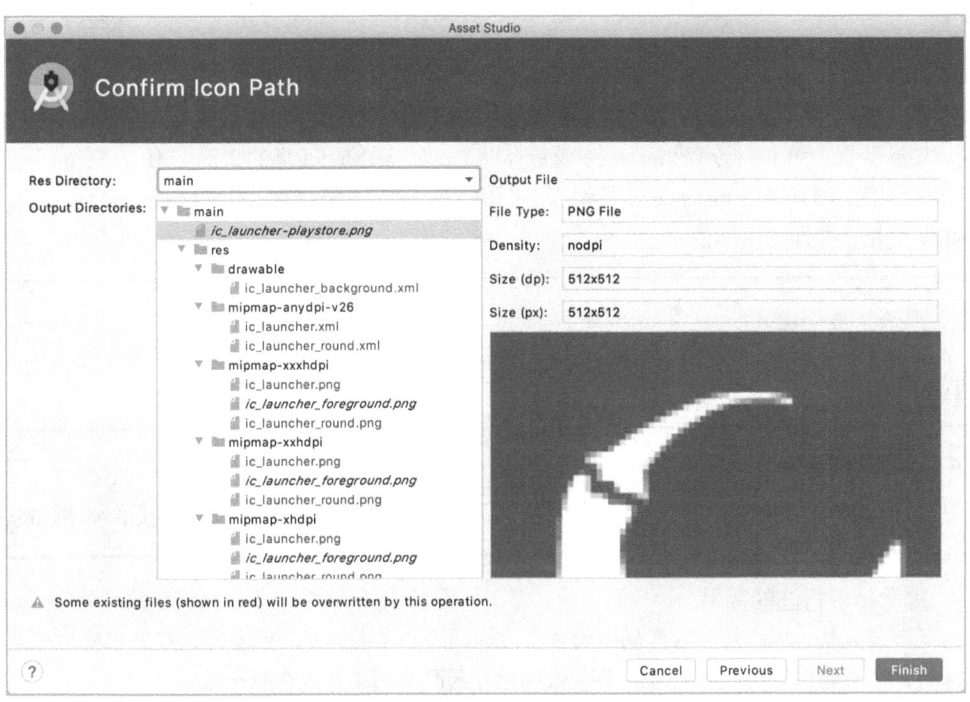

图 14-9 确认覆盖启动图标文件

14.5 数据模型

数据模型是数据特征的抽象，从抽象层次上描述系统的静态特征、动态行为和约束条件，为数据库系统的信息表示与操作提供一个抽象的框架。通常一个数据模型对应数据库中的一个数据表，数据模型的属性对应数据表中的一个字段。

“粉色辣椒”的后台数据库中包含 developer、user、dynamic、dynamic_reply、follow 和 message 等数据表，分别用于存储开发者、用户、动态、动态回复、关注和消息等数据。

14.5.1 开发者数据类

开发者数据类用于从 strings.xml 文件中获取开发者账号和开发序列号，通过 getDeveloper(Context) 方法获取 DeveloperModel 类的单例。

```
/java/com/pinkjiao/www/model/DeveloperModel.kt
class DeveloperModel(val developerAccountNumber:String, val developerKey:String) {
    companion object{
        private var instance: DeveloperModel? = null
        fun getDeveloper(context: Context):DeveloperModel{
            instance?.let { return it } //缓存 ContactsDatabase 实例存在时直接返回
            val developerAccountNumber:String = context.getString(R.string.developer_accountNumber)
            val developerKey:String =  context.getString(R.string.developer_key)
            instance = DeveloperModel(developerAccountNumber,developerKey)
            return instance!!
        }
```

```
16        }
17    }
```

第 06 行通过主构造方法声明存储开发者账号和开发序列号的 developerAccountNumber 变量和 developerKey 变量。第 08 行声明 instance 变量存储 DeveloperModel 类的实例。第 10 行 instance 变量如果已经实例化，则直接将 instance 变量作为返回值。第 11 ~ 14 行创建一个 DeveloperModel 类实例赋给 instance 变量，并作为返回值。

提示：单例模式

单例模式是类只有一个实例的设计模式。单例模式的类不直接通过构造方法进行实例化，而是通过静态方法返回该类的实例。

14.5.2 用户数据类

用户数据类用于存储用户的数据，同时集成了使用共享偏好设置将其保存、获取和清除的方法。

提示：多行注释简写

为了节省篇幅，从此处开始代码中的多行注释简写成单行模式，配套的工程源代码中依然使用多行注释。简写的注释前没有行号，其余代码行号与配套的工程源代码一致。

/java/com/pinkjiao/www/model/UserModel.kt

```
class UserModel {
    var accountID: String = "" //用户账号 ID
    var accountNumber: String = "" //用户账号
    var phone: String = "" //用户手机
    var name: String = "" //用户姓名
    var password: String = "" //用户密码
    var gender: String = "" //用户性别
    var birthday: String = "" //用户生日
    var nationalIdNumber: String = "" //用户身份证号
    var faceImage: String = "" //用户头像
    var registrationTime: String = "" //用户注册时间
    var followFlag:String = "" //关注标记
    var followNum:Int = 0 //关注数量
    var followerNum:Int = 0 //粉丝数量
    var dynamicNum:Int = 0 //动态数量
    var sendMessageNum:Int = 0 //发送私信数量
    var receiveMessageNum:Int = 0 //接收私信数量
    var nationalIDImageCheck:Int = 0 //身份证验证状态
    companion object{
        //设置共享偏好设置。Context 参数为上下文，userModel 参数为用户模型，返回值类型为 Boolean
        fun setSharedPreferences(context:Context,userModel:UserModel):Boolean{
            val sp: SharedPreferences = context.getSharedPreferences("User", Context.MODE_PRIVATE)
            val editorSP = sp.edit()
            editorSP.putString("userAccountID", userModel.accountID)
            editorSP.putString("userPassword", userModel.password)
            editorSP.putString("userFaceImage", userModel.faceImage)
            editorSP.putInt("userNationalIDImageCheck", userModel.nationalIDImageCheck)
```

```
38              return editorSP.commit()
39          }
            //获取共享偏好设置。Context 参数为上下文，userModel 参数为用户模型，返回值类型为 Boolean
46          fun getSharedPreferences(context:Context,userModel:UserModel):Boolean{
47              val sp: SharedPreferences = context.getSharedPreferences("User", Context.MODE_PRIVATE)
48              userModel.accountID = sp.getString("userAccountID", "").toString()
49              userModel.password = sp.getString("userPassword", "").toString()
50              userModel.faceImage = sp.getString("userFaceImage", "").toString()
51              userModel.nationalIDImageCheck = sp.getInt("userNationalIDImageCheck", 0)
52              return !(userModel.accountID == "" || userModel.password == "" || userModel.
    faceImage == "" || userModel.nationalIDImageCheck == 0)
53          }
            //清空共享偏好设置。context 参数为上下文，返回值类型为 Boolean
59          fun clearSharedPreferences(context:Context):Boolean{
60              val sp: SharedPreferences = context.getSharedPreferences("User", Context.MODE_PRIVATE)
61              return sp.edit().clear().commit()
62          }
63      }
64  }
```

第 07 ~ 23 行声明用户的属性，与服务器端数据库中的 user 表字段对应。第 31 ~ 39 行在本地保存登录用户的数据，保存成功时返回 true。第 46 ~ 53 行获取本地保存的用户登录数据，获取成功时返回 true。第 59 ~ 62 行清除登录用户的数据，在退出登录时调用，清除成功时返回 true。

14.5.3 动态数据类

动态数据类用于存储动态的数据，实现 Parcelable 接口对数据进行序列化，以便在不同的 Activity 之间传递 DynamicModel 实例。

提示：序列化

序列化的目的是永久保存对象数据，将对象数据保存在文件中或者跨进程和网络间进行对象传递。在 Android 中可以使用 Parcelable 接口或 Serializable 接口进行序列化。Parcelable 接口声明后需要重写一些方法，而 Serializable 接口只需声明即可。Parcelable 接口虽然使用起来烦琐一些，但是性能更高。Parcelable 接口是在内存中直接进行读写的，而 Serializable 接口是通过使用 I/O 流的形式将数据读写入在硬盘上的。

```
/java/com/pinkjiao/www/model/DynamicModel.kt
class DynamicModel() : Parcelable {
    var userName: String = "" //姓名
    var userAccountID: String = "" //昵称
    var userFaceImage: String = "" //头像
    var id: Int = 0 //动态 ID
    var content: String = "" //动态内容
    var images = arrayOfNulls<String>(4) //图片
    var label: String = "" //动态内容
    var createTime: String = "" //发布时间
    var replyPrivateNum: Int = 0 //私回数量
    var replyNum: Int = 0 //回复数量
```

```
    //次构造方法：读取序列化数据。parcel 参数为序列化数据
21  constructor(parcel: Parcel) : this() {
22      userName = parcel.readString()!!
23      userAccountID = parcel.readString()!!
24      userFaceImage = parcel.readString()!!
25      id = parcel.readInt()
26      content = parcel.readString()!!
27      images = parcel.createStringArray()!!
28      label = parcel.readString()!!
29      createTime = parcel.readString()!!
30      replyPrivateNum = parcel.readInt()
31      replyNum = parcel.readInt()
32  }
    //写入序列化数据。Parcel 参数为序列化数据，flags 参数为标记
38  override fun writeToParcel(parcel: Parcel, flags: Int) {
39      parcel.writeString(userName)
40      parcel.writeString(userAccountID)
41      parcel.writeString(userFaceImage)
42      parcel.writeInt(id)
43      parcel.writeString(content)
44      parcel.writeStringArray(images)
45      parcel.writeString(label)
46      parcel.writeString(createTime)
47      parcel.writeInt(replyPrivateNum)
48      parcel.writeInt(replyNum)
49  }
    //描述此 Parcelable 实例的封送表示中包含特殊对象的种类
54  override fun describeContents(): Int {
55      return 0 //位掩码，指示由此 Parcelable 对象实例封送的特殊对象类型集
56  }
    //读取接口：从 Parcel 中构造一个实现了 Parcelable 的类的实例处理
60  companion object CREATOR : Parcelable.Creator<DynamicModel> {
        //从 Parcel 中读取数据。parcel 参数为序列化数据，返回值类型为 DynamicModel
66      override fun createFromParcel(parcel: Parcel): DynamicModel {
67          return DynamicModel(parcel)
68      }
        //创建新数组：供外部类反序列化本类数组使用。Size 参数为数组尺寸，返回值类型为 Array<DynamicModel?>
74      override fun newArray(size: Int): Array<DynamicModel?> {
75          return arrayOfNulls(size)
76      }
77  }
78 }
```

第 07 ~ 16 行声明动态属性，属性与服务器端数据库中 danymic 数据表的字段对应。第 21 ~ 32 行通过次构造方法从 Parcel 对象读取 DynamicModel 类的属性。第 38 ~ 49 行将 DynamicModel 类的属性写入 Parcel 对象中。第 54 ~ 56 行设置是否返回文件描述，如果需要 writeToParcel(Parcel,int)方法输出时包含文件描述符，则返回值设置为 Parcelable.CONTENTS_FILE_DESCRIPTOR。第 60 ~ 77 行 CREATOR 提供对外的读取，实现 Creator<T>接口。因为实现类的类型是未知的，所以继承类名通过泛型参数传入。重写接口的 createFromParcel(Parcel)方法和 newArray(Int)方法分别返回单个和数组实例。

14.5.4 动态回复数据类

动态回复数据类用于存储用户回复动态的数据，声明的属性与服务器端数据库中 danymic_reply 数据表的字段对应。

```
/java/com/pinkjiao/www/model/DynamicReplyModel.kt
class DynamicReplyModel {
    var userName: String = "" //用户姓名
    var userAccountID: String = "" //用户账号 ID
    var userFaceImage: String = "" //用户头像
    var id: Int = 0 //回复 ID
    var content: String = "" //回复内容
    var createTime: String = "" //发布时间
    var dynamicID: Int = 0 //动态 ID
    var repliedDynamicReplyID: Int = 0 //被回复的 ID
    var repliedUserName: String = "" //被回复的作者姓名
    var repliedUserAccountID: String = "" //被回复的作者账号 ID
    var privateFlag: Int = 0 //私回标记
}
```

14.5.5 关注数据类

关注数据类用于存储关注和粉丝的数据，关注和粉丝的关注模型数据是反向关系，声明的属性与服务器端数据库中 follow 数据表的字段对应。

```
/java/com/pinkjiao/www/model/FollowModel.kt
class FollowModel {
    var userAccountID: String = "" //用户账号
    var userName: String = "" //用户昵称
    var userFaceImage: String = "" //用户头像
    var followBecomeTime: String = "" //关注时间
}
```

14.5.6 消息数据类

消息数据类用于存储提醒和私信的数据，声明的属性与服务器端数据库中 message 数据表的字段对应。由于私信也需要进行提醒，且私信只包含 255 个以内的字符，所以将提醒和私信合并成统一的数据结构。

```
/java/com/pinkjiao/www/model/MessageModel.kt
class MessageModel {
    var messageID: String = "" //私信 ID
    var messageType: String = "" //私信类型
    var messageTypeParam: String = "" //私信类型的附带参数: 0 系统信息; 1 私信; 2 动态回复; 3 动态私回; 4 关注
    var senderUserAccountID: String = "" //发送者账号 ID
    var senderUserName: String = "" //发送者账号昵称
    var senderUserFaceImage: String = "" //发送者头像
    var receiverUserAccountID: String = "" //接收者账号 ID
    var receiverUserName: String = "" //接收者账号昵称
    var receiverUserFaceImage: String = "" //接收者头像
```

```
    var messageContent: String = "" //私信内容
    var messageCreateTime: String = "" //发送时间
    var messageReadFlag: String = "" //接收者阅读标记
}
```

14.6 欢迎模块

欢迎模块已经包含在基础工程中，与 C0707 工程的实现效果基本相同，读者可以自行查看代码，运行效果如图 14-10 所示。此处去掉了滚动状态的显示，使用了视图绑定调用控件，将 WelcomePagerAdapter 类放置在 WelcomeActivity 类内部。单击“开始”按钮后，进入登录模块。

图 14-10　欢迎模块的运行效果

14.7 注册模块组

注册模块组主要包含注册账号(RegActivity)、忘记密码(ForgetPasswordActivity)和登录(LoginActivity)等三个 Activity，登录的 Activity 连接注册账号的 Activity 和忘记密码的 Activity，运行效果如图 14-11 所示。

图 14-11　注册模块组的运行效果

4.7.1 注册账号模块

注册账号模块有三个重点：验证输入的注册信息、向服务器端发送注册信息和处理服务器端返回信息，运行效果如图 14-12 所示。其中，由于没有购买验证码短信的服务，此处直接将验证码返回客户端，验证过程与手机接收验证码是一样的。

图 14-12　注册账号模块的运行效果

> **提示：手机验证码服务**
>
> 各大云服务平台几乎都提供了收费的手机验证码服务，提供相应的接口发送验证码，包括短信和语音两种形式。

1. 注册信息验证类

util 包内的 Validator 类是一个对象类，用于验证注册信息的规范性。客户端对注册信息进行验证为了减少不规范信息发送到服务器端，以确保注册信息的规范性。即使客户端对注册信息进行了验证，服务器端也会再进行一次验证，防止黑客发送不规范的注册信息或使用开放平台的非官方客户端验证发送的注册信息。

```
/java/com/pinkjiao/www/util/Validator.kt
//账号的有效性验证。accountNumber 参数为账号，返回值类型为 Boolean
fun isAccountNumber(accountNumber: String): Boolean {
    val regex = "^([a-zA-Z0~9])[a-zA-Z0~9_]{5,11}$" //正则表达式
    return Pattern.matches(regex, accountNumber)
}
//手机号的有效性验证。phoneNumber 参数为手机号，返回值类型为 Boolean
fun isPhoneNumber(phoneNumber: String): Boolean {
    val regex = "^(1[3~9][0~9])[0~9]{8}$" //正则表达式
    return Pattern.matches(regex, phoneNumber)
}
//姓名的有效性验证。name 参数为姓名，返回值类型为 Boolean
fun isName(name: String): Boolean {
    val regex = "^(?:[\\u4e00-\\u9fa5]+)(?:·[\\u4e00-\\u9fa5]+)*\$|^[a-zA-Z0~9]+\\s?
[.·\\-()a-zA-Z]*[a-zA-Z]+\$"
 //正则表达式
```

```
39        return Pattern.matches(regex, name)
    }
46  //密码的有效验证。password 参数为密码，返回值类型为 Boolean
47  fun isPassword(password: String): Boolean {
48      val regex = "^[a-zA-Z0~9]{6,12}$" //正则表达式
49      return Pattern.matches(regex, password)
    }
```

第 16 ~ 19 行判断账号名是否规范，账号名是以数字或字母开头的 6 ~ 12 位的由数字、字母或下画线组成的字符串。第 26 ~ 29 行判断手机号是否规范，手机号必须是以 13 ~ 19 开头的 13 位数字。第 36 ~ 39 行判断姓名是否规范，姓名是由中文或者英文组成的 20 个字符以内的字符串，姓名中可以包含·符号和空格，\u4e00-\u9fa5 表示所有中文字符。第 47 ~ 49 行判断密码是否符合规范，密码是由数字和字母组成的 6 ~ 12 位字符串。

提示：正则表达式

正则表达式(Regular Expression)是描述一种字符串匹配模式(Pattern)的表达式，可以用来检查一个字符串是否含有某种子字符串、将匹配的子字符串替换或者从某个字符串中取出符合某个条件的子字符串等。

- [ABC]：匹配 ABC 字符串，如[ao]匹配字符串“are you ok”中所有的 a 字母和 o 字母。
- [^ABC]：匹配除 ABC 字符串外的所有字符。
- [A-Z]：匹配所有大写字母。
- [a-z]：匹配所有小写字母。
- ( )：标记一个子表达式的开始和结束位置。
- *：匹配前面的子表达式零次或多次。匹配*字符时使用*。
- +：匹配前面的子表达式一次或多次。匹配+字符时使用\+。
- .：匹配除换行符\n 外的任何单字符。匹配.字符时使用\.。
- ?：匹配前面的子表达式零次或一次，或指明一个非贪婪限定符。匹配?字符时使用\?。
- $：匹配输入字符串的结尾位置。匹配$字符本身时使用\$。
- ^：匹配输入字符串的开始位置。当该符号在方括号表达式中使用时，表示不接收该方括号表达式中的字符集合。匹配^字符时使用\^。
- |：指明两项之间的一个选择。匹配|字符时使用\|。
- {n}：匹配 n 次，n 为非负整数。例如，o{2}会匹配 food 中的两个 o，不能匹配 job 中的单个 o。
- {n,}：匹配至少 n 次，n 是一个非负整数。
- {n,m}：至少匹配 n 次且最多匹配 m 次，m 和 n 均为非负整数，且 n<= m。例如，o{1,3}将匹配 joooooob 中的前三个 o；o{0,1}等价于 o?。
- \：将下一个字符标记为或特殊字符、或原义字符、或向后引用、或八进制转义符。例如，n 匹配 n 字符；\n 匹配换行符；\\匹配\字符；\('匹配(符号。
- \cx：匹配由 x 指明的控制字符。例如，\cM 匹配 Control-M 或回车符。x 的值必须为 A ~ Z 或 a ~ z 之一；否则，将 c 视为一个原义的 c 字符。
- \f：匹配一个换页符，等价于\x0c 和\cL。
- \n：匹配一个换行符，等价于\x0a 和\cJ。
- \r：匹配一个回车符，等价于\x0d 和\cM。

- \s: 匹配所有空白符，包括换行，等价于[\f\n\r\t\v]。
- \S: 匹配所有非空白符，包括换行，等价于[^\f\n\r\t\v]。
- \t: 匹配一个制表符，等价于\x09 和\cI。
- \v: 匹配一个垂直制表符，等价于\x0b 和\cK。
- \w: 匹配所有的字母、数字和下画线，等价于[A-Za-z0~9_]。

/java/com/pinkjiao/www/util/Validator.kt

```
//身份证的有效性验证。nationalIdNumber 参数为身份证号，返回值类型为 Boolean
fun isNationalIdNumber(nationalIdNumber: String): Boolean {
    val valCodeArr = arrayOf("1", "0", "X", "9", "8", "7", "6", "5", "4", "3", "2")
    val wi = arrayOf("7","9","10","5","8","4","2","1","6","3","7","9","10","5","8","4","2")
    var ai: String //前 17 位身份证号
    //验证长度是否为 18 位
    if (nationalIdNumber.length != 18) {
        return false
    } else {
        ai = nationalIdNumber.substring(0, 17)
    }
    //验证前 17 位是否为数字
    if (!isNumeric(ai)) {
        return false //18 位号码除最后一位外，都应为数字。
    }
    //验证出生年月是否有效
    val strYear = ai.substring(6, 10) //年
    val strMonth = ai.substring(10, 12) //月
    val strDay = ai.substring(12, 14) //日
    if (!isDate("$strYear-$strMonth-$strDay")) {
        return false //身份证生日无效
    }
    val gc = GregorianCalendar()
    val s = SimpleDateFormat("yyyy-MM-dd", Locale.ENGLISH)
    val dateStr = "$strYear-$strMonth-$strDay"
    val pos = ParsePosition(0)
    try {
        if (gc[Calendar.YEAR] - strYear.toInt() > 150 || gc.time.time - s.parse(dateStr, pos)!!.time < 0) {
            return false //身份证生日不在有效范围内
        }
    } catch (e: NumberFormatException) {
        e.printStackTrace()
    } catch (e: ParseException) {
        e.printStackTrace()
    }
    if (strMonth.toInt() > 12 || strMonth.toInt() == 0) {
        return false //身份证月份无效
    }
    if (strDay.toInt() > 31 || strDay.toInt() == 0) {
        return false //身份证日期无效
    }
```

```
    //验证地区编码是否有效
    val h = getAreaCode()
    if (h[ai.substring(0, 2)] == null) {
        return false //身份证地区编码错误
    }
    //验证最后一位
    var totalmulAiWi = 0
    for (i in 0..16) {
        totalmulAiWi += ai[i].toString().toInt() * wi[i].toInt()
    }
    val modValue = totalmulAiWi % 11
    val strVerifyCode = valCodeArr[modValue]
    ai += strVerifyCode
    if (nationalIdNumber.length == 18) {
        if (ai != nationalIdNumber) {
            return false //身份证无效，不是合法的身份证号码
        }
    } else {
        return true
    }
    return true
}
//获取地区编码。返回值类型为Hashtable<*, *>
private fun getAreaCode(): Hashtable<*, *> {
    val hashtable = Hashtable<String, String>()
    hashtable["11"] = "北京"
    hashtable["12"] = "天津"
    hashtable["13"] = "河北"
    //省略了其余省份、自治区、直辖市和特别行政区的代码
    return hashtable
}
//判断字符串是否为数字。str参数为字符串，返回值类型为Boolean
private fun isNumeric(str: String): Boolean {
    val pattern = Pattern.compile("[0~9]*")
    val isNum = pattern.matcher(str)
    return isNum.matches()
}
//判断字符串是否为日期格式。strDate参数为字符串，返回值类型为Boolean
private fun isDate(strDate: String): Boolean {
    var convertSuccess = true
    //指定日期格式为四位年/两位月份/两位日期，注意yyyy/MM/dd区分英文大小写
    val format = SimpleDateFormat("yyyy-MM-dd", Locale.ENGLISH)
    try {
        //设置lenient为false. 否则SimpleDateFormat会比较宽松地验证日期，如2021/02/29会被接收，并转换成2021/03/01
        format.isLenient = false
        format.parse(strDate)
    } catch (e: ParseException) {
        //如果有throw java.text.ParseException或者NullPointerException，说明格式不对
        convertSuccess = false
```

```
190        }
191        return convertSuccess
192    }
    //判断字符串是否只包含字母、数字和汉字。str参数为字符串，minLength参数为最小长度，maxLength参数为最大长度，返回值类型为Boolean
201    fun isNoSymbolString(str: String, minLength: Int, maxLength: Int): Boolean {
202        val regEx = "^[\\u4e00-\\u9fa5a-zA-Z0~9]{$minLength,$maxLength}$"
203        return Pattern.matches(regEx, str)
204    }
```

第 56 ~ 117 行判断身份证号是否符合规范，身份证有 18 位，前 2 位是省级行政区（包括省、自治区、直辖市和特区）的编号，第 3、4 位数表示地市级行政区的编号，第 5、6 位数表示区级行政区的编号，第 7 ~ 14 位表示出生年月日，第 15、16 位表示所在地派出所的编号，第 17 位表示性别（奇数为男性），最后 1 位表示校验码。第 123 ~ 161 行使用 Hashtable<String, String>() 类型保存所有的省级行政区编号和名称，作为 getAreaCode() 方法的返回值。第 98 行 h 是保存省级行政区编号和名称的哈希表，h[ai.substring(0, 2)]调用身份证号前两位为键名的键值，如果为 null 则表示身份证号前两位是错误的省级行政区编号。第 179 ~ 192 行 isDate(strDate:String) 方法用于判断 strDate 字符串所表示的日期是否为合法日期，将 strDate 字符串转换为日期时，如果没有抛出异常则表示转换成功，意味着 strDate 字符串表示的是合法日期。第 201 ~ 204 行 isNoSymbolString(str:String, minLength:Int,maxLength:Int) 方法用于判断 str 字符串的长度是否在 minLength 至 maxLength 长度范围内，且只包含汉字、数字或字母。

图 14-13 选择日期对话框

2. 日期选取对话框类

DataPickerFragment 类继承 DialogFragment 类，并实现了 OnDateSetListener 接口，用于实现选择日期的对话框，运行效果如图 14-13 所示。

```
/java/com/pinkjiao/www/fragment/DataPickerFragment.kt
12  class DataPickerFragment : DialogFragment(), OnDateSetListener {
13      //设置对话框方法
14      override fun onCreateDialog(savedInstanceState: Bundle?): Dialog {
15          val c: Calendar = Calendar.getInstance() //调用getInstance
16          val year: Int = c.get(Calendar.YEAR) - 18 //得到年
17          val month = 0 //月
18          val day = 1 //日
19          return DatePickerDialog(requireActivity(), this, year, month, day)
20      }
        //设置onDateSet方法。dataPicker参数为日期选择器，year参数为年，month参数为月，day参数为日
28      override fun onDateSet(datePicker: DatePicker, year: Int, month: Int, day: Int) {
29          val activity: RegActivity = activity as RegActivity
30          activity.processDatePickerResult(year, month, day)
31      }
32  }
```

第 14 ~ 20 行重写 onCreateDialog(savedInstanceState: Bundle?) 方法，将默认显示的日期设置为当前年的 18 年前的 1 月 1 日。DatePickerDialog() 方法的参数中，月份从 0 开始，每月的第几天从 1 开始。因此，设置 month 变量为 0，表示 1 月份；设置 day 变量为 1，表示每个月的第一天。第 28 ~ 31

行重写 onDateSet(datePicker:DatePicker,year:Int,month:Int,day:Int) 方法，将选取的日期通过调用 RegActivity 类的 processDatePickerResult(year:Int,month:Int,day:Int)方法进行回传。

3. 修改状态栏颜色模式类

工程设置默认样式为 NoActionBar，状态栏和导航设置为透明。此时，Activity 会占满屏幕且与状态栏的文字叠加。因此，不但需要将显示状态栏文字的区域空出来，防止文字与 Activity 的内容不当叠加；还需要根据状态栏区域的背景颜色修改状态栏的颜色模式，防止状态栏文字颜色与 Activit 重叠区域的颜色过于接近影响用户观看状态栏。

```
/res/values/themes/themes.xml
<style name = "Theme.pinkjiao" parent = "Theme.MaterialComponents.Light.NoActionBar">
    <item name = "android:windowTranslucentStatus">false</item>
    <item name = "android:windowTranslucentNavigation">true</item>
</style>
```

BaseActivity 类继承 AppCompatActivity 类，作为工程中需要改变状态栏颜色模式的 Activity 的基类，设置状态颜色模式后的运行效果如图 14-14 所示。

图 14-14　深色模式和浅色模式的状态栏运行效果

```
/java/com/pinkjiao/www/compat/BaseActivity.kt
open class BaseActivity: AppCompatActivity() {
    companion object{
        const val STATUS_BAR_LIGHT = true //浅色模式
        const val STATUS_BAR_DARK = false //深色模式
    }
    //设置状态栏颜色模式。Status 参数为状态
    fun setStatusBar(status: Boolean){
        if (status) {
            if (Build.VERSION.SDK_INT >= Build.VERSION_CODES.R) {
                window.insetsController?.setSystemBarsAppearance(
                    WindowInsetsController.APPEARANCE_LIGHT_STATUS_BARS,
                    WindowInsetsController.APPEARANCE_LIGHT_STATUS_BARS)
            } else {
                window.decorView.systemUiVisibility = View.SYSTEM_UI_FLAG_LIGHT_STATUS_BAR
            }
```

```
27          } else {
28              if (Build.VERSION.SDK_INT >= Build.VERSION_CODES.R) {
29                  window.insetsController?.setSystemBarsAppearance(
30                      WindowInsetsController.APPEARANCE_LIGHT_STATUS_BARS, 0)
31              } else {
32                  window.decorView.systemUiVisibility = View.SYSTEM_UI_FLAG_LAYOUT_FULLSCREEN or
View.SYSTEM_UI_FLAG_VISIBLE
33              }
34          }
35      }
36  }
```

第 11、12 行定义两个静态常量区分状态栏的颜色模式。第 21、23 行对 Android API Level 30 及以上版本的设备设置浅色模式的状态栏。第 25 行对其他版本的设备设置浅色模式的状态栏。第 29、30 行对 Android API Level 30 及以上版本的设备设置深色模式的状态栏。第 32 行对其他版本的设备设置深色模式的状态栏。

4. 隐藏虚拟键盘类

HideKeyboardActivity 类实现单击空白区域隐藏虚拟键盘的功能，大部分情况下还需要使用 BaseActivity 类的功能，而 Kotlin 只能单继承，所以继承了 BaseActivity 类。

```
/java/com/pinkjiao/www/compat/HideKeyboardActivity.kt
06  open class HideKeyboardActivity : BaseActivity() {
        //调度触摸事件：单击时隐藏软键盘。event 参数为运动事件，返回值类型为 Boolean
13      override fun dispatchTouchEvent(event: MotionEvent): Boolean {
14          if (event.action == MotionEvent.ACTION_DOWN && Keyboard.isShouldHide(currentFocus, event)) {
15              val view = currentFocus
16              Keyboard.hide(this, currentFocus?.windowToken)
17              view!!.clearFocus()
18          }
19          return super.dispatchTouchEvent(event)
20      }
21  }
```

第 13 行重写 dispatchTouchEvent(event: MotionEvent)方法，处理触摸事件分发。第 14 行 event.action 表示当前的事件，MotionEvent.ACTION_DOWN 表示单击时按下的事件。Keyboard 是自定义的对象类，用于隐藏判断虚拟键盘显示状态和虚拟键盘。调用 Keyboard.isShouldHide(View?,MotionEvent)方法判断虚拟键盘是否是显示状态。第 15 行获取当前获取焦点的控件。第 16 行 Keyboard.hide(Context,IBinder?)方法隐藏虚拟键盘。第 17 行取消当前控件的焦点。第 19 行调用 super.dispatchTouchEvent(Event)方法将事件继续向下分发，并将返回值作为 dispatchTouchEvent(event: MotionEvent)方法的返回值。

5. 网络请求类

本实例中会多次访问服务器端的 Http 地址，每次访问都需要提交开发者账号和开发序列号，用户登录后每次访问都需要提交登录用户的账号 ID 和使用 md5 算法加密的密码，下面介绍封装这些功能的自定义类。自定义 RequestUrl 类用于获取服务器端域名的 Url 及访问请求页面所在文件夹的 Url；自定义 GetRequest 类将开发者账号和序列号及登录用户账号 ID 和使用 MD5 算法加密的密码合成为完整的 Get 请求 Url；自定义 PostRequest 类用于保存 Post 请求所需的数据。

```
/java/com/pinkjiao/www/net/Resquest.kt
08 class RequestUrl {
09     companion object{
        //获取服务器端页面所在的文件夹 Url 路径。Context 参数为上下文，返回值类型为 String
15         fun getBaseUrl(context: Context): String {
16             val domain:String =  getDomainUrl(context)
17             val domainFolder:String =  context.getString(R.string.app_domain_folder)
18             val version:String =  context.getString(R.string.app_version)
19             return "$domain$domainFolder/$version/"
20         }
        //获取服务器端 App 服务所在的文件夹 Url 路径。Context 参数为上下文，返回值类型为 String
26         fun getDomainUrl(context: Context): String {
27             return context.getString(R.string.app_domain)
28         }
29     }
30 }
//Get 请求类：合成包含 Get 方法参数的 Url。baseUrl 参数为服务器端页面所在的文件夹 Url，page 参数为页面名称，developer 参数为开发者，user 参数为当前登录用户，parameter 参数为其他参数
39 class GetRequest(var baseUrl: String, page: String, developer: DeveloperModel, user: UserModel, var parameter:String){
40     var url:String
41     init {
42         val developerParameter = "developerAccountNumber =" + developer.developerAccountNumber + "&developerKey=" + developer.developerKey //开发者账号和密码
43         val userParameter = "&currentUserAccountID =" + user.accountID + "&currentUserPassword =" + user.password //用户账号和密码
44         url = "$baseUrl$page?$developerParameter$userParameter$parameter" //完整的 Get 请求 Url
45     }
46 }
//Post 请求类：包含 Post 请求所需的所有数据。baseUrl 参数为服务器端页面所在的文件夹 Url，page 参数为页面名称，developer 参数为开发者，user 参数为当前登录用户，parameter 参数为其他参数，files 参数为文件
56 class PostRequest(var baseUrl: String, page: String, developer:DeveloperModel, user:UserModel, val parameter: MutableMap<String, String>, val files: Map<String, String>?){
57     var url:String
58     init {
59         //开发者账号和密码
60         parameter["developerAccountNumber"] = developer.developerAccountNumber
61         parameter["developerKey"] = developer.developerKey
62         //用户账号和密码
63         parameter["currentUserAccountID"] = user.accountID
64         parameter["currentUserPassword"] = user.password
65         url = "$baseUrl$page" //Post 请求的 Url
66     }
67 }
```

第 19 行将其上三个字符串合成起来成为完整的服务器端页面所在的路径，合成后的路径为 http://www.pinkjiao.com/app_server/v1.0/。第 41 ~ 45 行 init()初始化块中合成完整的 Get 请求 Url，第 56 ~ 67 行将开发者和当前登录用户的账号和密码保存在 parameter 变量中，并合成 Post 请求的 Url。

6. 注册账号的服务器端接口页面

注册账号的服务器端接口页面为 reg.php，通过网络请求类合成后的完整访问地址为 http://www.pinkjiao.com/app_server/v1.0/reg.php，由于通过网络请求类简化了服务器端接口页面的设置，**后面所提到的服务器端接口页面均为页面名称，需要通过网络请求类合成的完整访问地址。**该接口页面使用 Get 方式接收数据，所需参数如表 14-1 所示。验证注册验证码的服务器端接口页面为 regCheckCode.php，所需参数如表 14-2 所示。

表 14-1 reg.php 接收的参数

参　数	说　明
developerAccountNumber	开发者账号
developerKey	开发序列号
accountNumber	注册用户的账号
phone	注册用户的电话号码
name	注册用户的姓名
password	注册用户的 MD5 算法加密后密码
gender	注册用户性别(0 表示女性，1 表示男性)
birthday	注册用户的生日
nationalIdNumber	注册用户的身份证号

表 14-2 regCheckCode.php 接收的参数

参　数	说　明
developerAccountNumber	开发者账号
developerKey	开发序列号
accountNumber	注册用户的账号
password	注册用户的 MD5 算法加密后密码
registrationCode	验证码

注册成功时，返回的 JSON 数据结构为{"state":"ok","code":验证码}，验证码是服务器端生成的 5 位随机数。验证验证码成功时，返回的 JSON 数据为{"state":"ok"}。注册和验证验证码失败时，返回的 JSON 数据结构为{"state":"错误码","info":"错误提示信息"}，如表 14-3 所示。

表 14-3 返回的错误码

错 误 码	错误提示信息
error501	用户名包含禁用字符串
error502	账号已存在
error503	复制头像失败
error504	验证码错误
error505	数据格式不符合规范

7. 注册账号的 Activity 类

RegActivity 类继承自定义的 HideKeyboardActivity 类，实现注册账号的功能。

```
/java/com/pinkjiao/www/RegActivity.kt
42  //判断账号
43  accountNumberEditText.setOnFocusChangeListener { _, hasFocus ->
44      if (!hasFocus && !Validator.isAccountNumber(accountNumberEditText.text.toString()) &&
    accountNumberEditText.text.toString() != "") {
45          accountNumberEditText.setText("")
46          accountNumberEditText.hint = getString(R.string.reg_account_number_hint_error)
47          accountNumberEditText.setHintTextColor(errorHintTextColor)
48      } else {
49          accountNumberEditText.hint = getString(R.string.reg_account_number_hint)
50          accountNumberEditText.setHintTextColor(hintTextColor)
51      }
52  }
53  accountNumberEditText.addTextChangedListener(
54      object : TextWatcher {
57          override fun onTextChanged(s: CharSequence, start: Int, before: Int, count: Int) {
58              if (Validator.isAccountNumber(accountNumberEditText.text.toString())) {
59                  accountNumberOkImageView.visibility = View.VISIBLE
60              } else {
61                  accountNumberOkImageView.visibility = View.INVISIBLE
62              }
63          }
64      }
65  )
```

输入注册信息的控件比较多，这里仅通过输入账号的 accountNumberEditText 控件进行讲解。43～52 行 setOnFocusChangeListener(OnFocusChangeListener)方法在改变焦点时验证输入的内容是符合规范。如果获取焦点时，输入内容不为空且不符合规范，则清空输入内容并以红色显示提示文字否则还原提示文字及其颜色。第 53～65 行 addTextChangedListener(TextWatcher)方法监听输入的文内容，其中只使用重写的 onTextChanged(CharSequence,Int,Int,Int)方法在输入内容改变时验证输入的容，如果符合规范，则显示后面的 accountNumberOkImageView 控件(绿色圆点)。

```
/java/com/pinkjiao/www/RegActivity.kt
private fun regSend(){
    val baseUrl = RequestUrl.getBaseUrl(context)
    val page = "reg.php"
    val parameter = "&accountNumber = " + userModel.accountNumber + "&phone = " + userModel.phone
+ "&name = " + userModel.name + "&password = " + userModel.password + "&gender = " + userModel.gender
+ "&birthday = " + userModel.birthday + "&nationalIdNumber = " + userModel.nationalIdNumber
    val getRequest = GetRequest(baseUrl, page, DeveloperModel.getDeveloper(context), UserModel(),
parameter)
    Http.get(getRequest.url, object : Http.GetResponseListener {
        override fun onCreate() {
            //显示等待对话框
            LoadingProgressDialog.showProgressDialog(context, getString(R.string.reg_dialog_tip))
        }
        override fun onResponse(res: String) {
            val jsonParser = JSONTokener(JSON.cleanBOM(res))
            val jsonObject = jsonParser.nextValue() as JSONObject
            if (jsonObject.getString("state") == "ok") { //注册成功 锁定注册控件
```

```
258                 with(viewBinding) {
260                     regInfoLinearLayout.visibility = View.GONE //隐藏注册控件的布局
262                     regCodeLinearLayout.visibility = View.VISIBLE //显示输入注册码的布局
263                     //显示注册码(替代发送到手机的注册码)
264                     regSendTextView.text = regSendTextView.text.toString().plus(":" + jsonObject.getString("code"))
265                 }
266             } else {
267                 when {
268                     jsonObject.getString("state") == "error502" -> {
269                         Toast.makeText(context, resources.getString(R.string.toast_reg_account_error), Toast.LENGTH_LONG).show()
270                     }
271                     jsonObject.getString("state") == "error505" -> {
272                         Toast.makeText(context, resources.getString(R.string.toast_reg_check_error), Toast.LENGTH_LONG).show()
273                     }
274                     else -> {
275                         Toast.makeText(context, resources.getString(R.string.toast_reg_error), Toast.LENGTH_LONG).show()
276                     }
277                 }
278             }
280             LoadingProgressDialog.dismiss() //隐藏等待对话框
281         }
282         override fun onFailure(res: String) {
283             Toast.makeText(context, resources.getString(R.string.toast_net_error), Toast.LENGTH_LONG).show()
285             LoadingProgressDialog.dismiss() //隐藏等待对话框
286         }
287     })
288 }
```

第 244 行获取访问页面的域名和所在的文件夹路径的字符串，赋给 baseUrl 变量。第 245 行将要访问的页面名称，赋给 page 变量。baseUrl 变量和 page 变量连接起来就是访问的服务器端地址。第 246 行 parameter 变量用于保存使用 get 方法向服务器端传递的参数。第 247 行初始化 getRequest 对象，合成完整的注册用户的 Url。第 248 行通过自定义的 Http 类发送 get 请求，其中 getRequest.url 是合成后的完整 Url 地址，Http.GetResponseListener 是回调 Http 请求的自定义接口。第 249 ~ 252 行是发送 Http 请求前的回调方法，显示等待对话框，其中 LoadingProgressDialog 类是自定义的等待对话框类。第 253 ~ 281 行是处理服务器端返回结果的回调方法，如果注册信息符合规范，则隐藏输入注册信息的控件所在的布局，并显示输入验证码的布局；否则显示服务器端返回的错误信息。第 282 ~ 286 行处理网络请求失败后的界面反馈，显示网络错误的信息，并隐藏等待对话框。

提示：发送验证码数据

发送验证码数据的 checkCodeSend() 方法与发送注册的 regSend() 方法，二者实现的方式类似，这里就不进行详细分析了，读者可以自行查看源代码或者参考忘记密码的发送验证码内容。

```
/java/com/pinkjiao/www/RegActivity.kt
    //显示日期选择对话框
328 private fun showDatePicker() {
329     val newFragment: DialogFragment = DataPickerFragment() //实例化
330     newFragment.show(supportFragmentManager, "") //显示出来
331 }
    //获取选择日期，year参数为年，month参数为月，day参数为日
339 fun processDatePickerResult(year: Int, month: Int, day: Int) {
340     val date = "${year}-${month + 1}-${day}" //月从0开始，所以要加1
341     with(viewBinding) {
342         birthdayEditText.setText(date)
343         birthdayOkImageView.visibility = View.VISIBLE
344     }
345 }
```

第 328～331 行显示日期选择对话框。第 339～345 行获取选择的出生日期并在 birthdayEditText 控件中显示出来，日期选择对话框中选择日期后单击“确定”按钮后调用该方法。

14.7.2 找回密码模块

找回密码模块通过两个步骤找回密码：输入的注册信息和输入接收到的验证码，运行效果如图 14-15 所示。由于输入和发送找回密码数据的方法与注册类似，所以下面主要介绍发送验证码。

图 14-15 找回密码模块的运行效果

1. 找回密码的服务器端接口页面

找回密码的服务器端接口页面为 forgetPassword.php，使用 Get 方式接收数据，所需参数如表 14-4 所示。验证验证码的服务器端接口页面为 forgetPasswordCheckCode.php，使用 Get 方式接收数据，所需参数如表 14-5 所示。

表 14-4 forgetPassword.php 接收的参数

参 数	说 明
developerAccountNumber	开发者账号
developerKey	开发序列号
accountNumber	注册用户的账号

续表

参　数	说　明
phone	注册用户的电话
name	注册用户的姓名
nationalIdNumber	注册用户的身份证号

表 14-5　forgetPasswordCheckCode.php 接收的参数

参　数	说　明
developerAccountNumber	开发者账号
developerKey	开发序列号
accountNumber	注册用户的账号
phone	注册用户的电话号码
name	注册用户的姓名
nationalIdNumber	注册用户的身份证号
password	注册用户的 MD5 算法加密后的新密码
forgetPasswordCode	验证码

找回密码成功时，返回的 JSON 数据结构为{"state":"ok","code":验证码}。验证验证码成功时，返回的 JSON 数据为{"state":"ok"}。找回密码和验证验证码失败时，返回的 JSON 数据结构为{"state":"错误码","info":"错误提示信息"}，如表 14-6 所示。

表 14-6　返回的错误码

错 误 码	错误提示信息
error701	密码错误
error702	验证码错误

2．找回密码的 Activity

ForgetPasswordActivity 类继承自定义的 HideKeyboardActivity 类，实现忘记密码时，通过验证注册信息找回注册账号密码的功能。

```
/java/com/pinkjiao/www/ForgetPasswordActivity.kt
private fun checkCode(){
    val baseUrl = RequestUrl.getBaseUrl(context)
    val page = "forgetPasswordCheckCode.php"
    val parameter = "&accountNumber = " + userModel.accountNumber + "&phone = " + userModel.phone +
        "&name = " + userModel.name + "&nationalIdNumber = " + userModel.nationalIdNumber +
        "&password = " + userModel.password + "&forgetPasswordCode = " +
         viewBinding.codeEditText.text.toString()
    val getRequest = GetRequest(baseUrl, page, DeveloperModel.getDeveloper(context),
UserModel(),parameter)
    Http.get(getRequest.url, object : Http.GetResponseListener {
        override fun onCreate() {
            //显示等待对话框
            LoadingProgressDialog.showProgressDialog(context, getString(R.string.reg_dialog_tip))
        }
```

```
271         override fun onResponse(res: String) {
272             val jsonParser = JSONTokener(JSON.cleanBOM(res))
273             val jsonObject = jsonParser.nextValue() as JSONObject
274             if (jsonObject.getString("state") == "ok") { //找回密码成功，锁定注册控件
276                 with(viewBinding) {
277                     forgetPasswordCodeLinearLayout.visibility = View.GONE
278                     forgetPasswordSuccessLinearLayout.visibility = View.VISIBLE
279                 }
280             } else {
281                 Toast.makeText(context, resources.getString(R.string.toast_forget_password_check_code_error), Toast.LENGTH_LONG).show()
282             }
284             LoadingProgressDialog.dismiss() //隐藏等待对话框
285         }
286         override fun onFailure(res: String) {
287             Toast.makeText(context, resources.getString(R.string.toast_net_error), Toast.LENGTH_LONG).show()
289             LoadingProgressDialog.dismiss() //隐藏等待对话框
290         }
291     })
292 }
```

第 262 行获取访问页面的域名和所在的文件夹路径的字符串，赋给 baseUrl 变量。第 263 行将要访问的页面名称，赋给 page 变量。第 264 行 parameter 变量保存使用 get 方法向服务器端传递的参数。第 265 行初始化 getRequest 对象，合成完整的验证验证码的 Url。第 266 行调用 Http 类 get 静态方法发送验证验证码的请求。第 267～270 行是发送 Http 请求前的回调方法，显示等待对话框。第 271～285 行是处理服务器端返回结果的回调方法，如果验证成功，则隐藏输入验证码的布局，显示验证成功的布局；否则显示服务器端返回的错误信息。第 286～290 行处理网络请求失败后的界面反馈，显示网络错误的信息，并隐藏等待对话框。

图 14-16　登录模块的运行效果

14.7.3　登录模块

登录模块的 Activity 是注册模块的中心，连接注册账号和找回密码的 Activity，主要实现的是登录功能，如图 14-16 所示。

1. 登录的服务器端接口页面

登录的服务器端接口页面为 login.php，使用 Get 方式接收数据，所需参数如表 14-7 所示。

表 14-7　login.php 接收的参数

参　数	说　明
developerAccountNumber	开发者账号
developerKey	开发序列号
accountNumber	登录用户的账号
password	登录用户的 MD5 算法加密后密码

登录成功时，返回的 JSON 数据结构为{"state":"ok", 'user' = > {登录用户数据的 JSON 数组}}。登录用户数据的 JSON 数组的键名如表 14-8 所示。

登录成功时返回的 JSON 数据样例


```
{"state":"ok","user":{"userAccountID":"2181821129","userName":"小飞飞","userFaceImage":"common\/face\/
202101\/2181821129.jpg","userNationalIDImageCheck":"0"}}
```


表 14-8　登录用户数据的 JSON 数组的键名

键　名	说　明
userAccountID	登录用户的账号 ID
userName	登录用户的姓名
userFaceImage	登录用户的头像图片地址
userNationalIDImageCheck	登录用户的身份证验证状态

登录失败时，返回的 JSON 数据结构为{"state":"错误码","info":"错误提示信息"}，如表 14-9 所示。

表 14-9　返回的错误码

错 误 码	错误提示信息
error601	账号错误
error602	账号未激活
error603	账号有限期锁定
error604	账号无限期锁定

2. 登录的 Activity 类

LoginActivity 类继承自定义的 HideKeyboardActivity 类，实现登录的功能，同时可以通过对应按钮启动注册账号和找回密码的 Activity。

/java/com/pinkjiao/www/LoginActivity.kt

```
private fun userLogin(){
    val baseUrl = RequestUrl.getBaseUrl(context)
    val domainUrl = RequestUrl.getDomainUrl(context)
    val page = "login.php"
    val parameter = "&accountNumber = " + loginUserModel.accountNumber + "&password = " +loginUserModel.password
    val getRequest = GetRequest(baseUrl, page, DeveloperModel.getDeveloper(context), UserModel(),parameter)
    Http.get(getRequest.url, object : Http.GetResponseListener {
        override fun onCreate() {
            //显示等待对话框
            LoadingProgressDialog.showProgressDialog(context, getString(R.string.login_dialog_tip))
        }
        override fun onResponse(res: String) {
            val jsonParser = JSONTokener(JSON.cleanBOM(res))
            val jsonObject = jsonParser.nextValue() as JSONObject
            if (jsonObject.getString("state") == "ok") {
                val userJSON = JSONTokener(jsonObject.getString("user")).nextValue() as JSONObject
                loginUserModel.accountID = userJSON.getString("userAccountID")
                loginUserModel.name = userJSON.getString("userName")
                loginUserModel.faceImage = Image.getUrl(domainUrl, userJSON.getString("userFaceImage"))
                loginUserModel.nationalIDImageCheck = userJSON.getInt("userNationalIDImageCheck")
```

```
75                If (UserModel.setSharedPreferences(context,loginUserModel)) {
76                    startActivity(Intent(context, MainActivity::class.java))
77                    this@LoginActivity.finish()
78                } else {
79                    Toast.makeText(context, resources.getString(R.string.toast_set_shared_
    preferences_error), Toast.LENGTH_LONG).show()
80                }
81            } else {
82                Toast.makeText(context, resources.getString(R.string.toast_login_error), Toast.
    LENGTH_LONG).show()
83            }
85            LoadingProgressDialog.dismiss() //隐藏等待对话框
86        }
87        override fun onFailure(res: String) {
88            Toast.makeText(context, resources.getString(R.string.toast_net_error), Toast.
    LENGTH_LONG).show()
90            LoadingProgressDialog.dismiss() //隐藏等待对话框
91        }
92    })
93 }
```

第 62 ~ 65 行是发送 Http 请求前的回调方法，显示等待对话框。第 66 ~ 86 行处理请求返回的 JSON 字符串的响应数据。第 69 行判断返回 JSON 字符串的 state 键值是否为“ok”字符串。第 70 行读取 user 键名存储的登录用户数据的 JSON 数组。第 71 ~ 74 行解析 JSON 数组获取登录用户数据，第 73 行调用 Image.getUrl(domain:String,path:String,quality:String = originalQuality) 方法，获取头像图片的完整路径。第 75 ~ 80 行本地化保存用户的登录信息，保存成功后启动 MainActivity，否则显示错误提示。第 87 ~ 91 行处理网络请求失败后的界面反馈。

Image 类提供了获取图片实际路径的 getUrl(domain:String,path:String,quality:String = originalQuality 方法，头像图片可以获取原始质量和低质量压缩图片的真实路径，动态图片的可以获取原始质量和高中低三种质量压缩图片的真实路径。

```
/java/com/pinkjiao/www/Util/Image.kt
21  private const val originalQuality = "" //原始质量图片
22  const val highQuality = "_H_" //高质量压缩图片
23  const val mediumQuality = "_M_" //中质量压缩图片
24  const val smallQuality = "_S_" //低质量压缩图片
    //获取图片的 Url 路径，domain 参数为域名，path 参数为图片路径，quality 参数为图片质量
194 fun getUrl(domain:String,path:String,quality:String = originalQuality):String{
195     var url = path
196     if(path != "") {
197         val list = path.split("/").toMutableList()
198         url = domain + path.replace(list[list.size-1],quality + list[list.size-1])
199     }
200     return url
201 }
```

第 21 ~ 24 行声明 4 种不同质量的常量，由于获取不同质量使用字符串替换的方式，所以原始量图片的常量使用空字符串。第 197 行对图片路径使用“/”字符进行分割，并保存到 MutableList

型的 list 变量。第 198 行将图片名称前添加图片质量的前缀进行替换并添加到域名后，合成图片的真实路径。

14.8 首页模块组 1

首页模块组 1 包括首页模块（MainActivity）、附近模块（VicinityFragment）、动态列表模块和关注模块（FollowFragment）。

14.8.1 首页模块

首页模块用于显示附近、关注、偶遇、提醒和自己等 5 个 Fragment，如图 14-17 所示。

图 14-17 首页模块的运行效果

1. 显示提醒信息数量的接口

TabMenu 接口通过首页模块的 Fragment 调用，用于在底部的 Tab 按钮中显示提醒信息数量。

```
/java/com/pinkjiao/www/fragment/TabMenu.kt
interface TabMenu {
    fun setTabNum(index:Int,num:Int)
}
```

第 04 行定义 setTabNum(index:Int,num:Int)方法，index 参数表示 Tab 按钮的序号，num 参数表显示的信息数量。

2. 首页的 Activity 类

MainActivity 类用于显示底部导航的 Tab 按钮及其对应的 Fragment，单击底部的 Tab 按钮会切显示不同的 Fragment，并调用 setStatusBar()方法修改状态栏的颜色模式。

```
/java/com/pinkjiao/www/MainActivity.kt
class MainActivity : BaseActivity(), TabMenu {
    private lateinit var fragmentManager: FragmentManager
    private val tabTextView = arrayOfNulls<TextView>(5) //Tab 按钮中显示文字的控件
    private val tabNumTextView = arrayOfNulls<TextView>(5) //Tab 按钮中显示数字提示的控件
    private val tabTextViewID = intArrayOf(R.id.vicinityTabTextView, R.id.followTabTextView, R.id.meetTabTextView, R.id.remindTabTextView, R.id.tab_mine) //Tab 的控件 id
    private val tabNumTextViewID = intArrayOf(R.id.vicinityTabNumTextView, R.id.followTabNumTextView, R.id.meetTabNumTextView, R.id.remindTabNumTextView, R.id.tab_mine_num)
    private lateinit var viewBinding: ActivityMainBinding //视图绑定
    private var currentTab = 0
    override fun onCreate(savedInstanceState: Bundle?) {
        super.onCreate(savedInstanceState)
        viewBinding = ActivityMainBinding.inflate(layoutInflater)
        setContentView(viewBinding.root)
        //初始化分页数组
        val fragment: Array<RefreshFragment>= arrayOf(VicinityFragment.newFragment("附近", this), FollowFragment.newFragment("关注",this), MeetFragment.newFragment("偶遇",this), RemindFragment.newFragment("提醒",this), MineFragment.newFragment(this))
        //初始化控件
        for (i in tabTextView.indices) {
            tabTextView[i] = findViewById(tabTextViewID[i])
            tabTextView[i]!!.tag = i
            tabTextView[i]!!.setOnClickListener { v ->
                If (currentTab == i) {
                    fragment[i].refresh() //更新当前 Fragment 显示的内容
                } else {
                    (fragment[i] as StatusBar).setStatusBar()
                    currenttab = i
                    showTabFragment(v.tag as Int, fragment) //显示 fragment
                }
            }
            tabnumtextview[i] = findViewById(tabnumtextviewid[i])
        }
        fragmentmanager = supportFragmentManager //获取 FragmentManager 实例
        showTabFragment(0,fragment) //设置默认显示的 Fragment
    }
    //隐藏 Tab 的圆点数字，index 参数为序号
    Private Fun hideTabNum(index: Int) {
        TABNUMTEXTVIEW[index]!!.visibility = View.GONE
    }
    //显示 Fragment，index 参数为序号，Tab 参数为对应的 Fragment 数组
    private fun showTabFragment(index: Int,fragment:Array<RefreshFragment>) {
```

```
61          //创建事务
62          val fragmentTransaction: FragmentTransaction = fragmentManager.beginTransaction()
64          if (fragmentManager.fragments.isEmpty()) { //判断是否已经添加 Fragment
65              for (i in tabTextView.indices) {//添加 Fragment
66                  fragmentTransaction.add(R.id.fragment, fragment[i])
67                  fragmentTransaction.show(fragment[i])
68              }
69          }
71          for (i in tabTextView.indices) { //还原状态
72              fragmentTransaction.hide(fragment[i]) //隐藏 Fragment
73              tabTextView[i]!!.isSelected = false //取消 Tab 选中状态
74          }
75          tabTextView[index]!!.isSelected = true //设置 Tab 选中状态
76          hideTabNum(index) //隐藏 tab 圆点
77          fragmentTransaction.show(fragment[index]) //显示 Fragment
78          fragmentTransaction.commit() //提交事务
79      }
        //设置底部 Tab 显示的数字提示，index 参数为序号，num 参数为数字
85      override fun setTabNum(index: Int, num: Int) {
86          tabNumTextView[index]!!.text = num.toString()
87          tabNumTextView[index]!!.visibility = View.VISIBLE
88      }
89  }
```

第 27 行通过静态方法实例化 5 个 Fragment，保存在 fragment 数组变量中。这些 Fragment 都继承自 RefreshFragment 抽象类，该抽象类后续介绍。第 32 ~ 40 行添加底部 Tab 按钮的单击事件，如果单击当前显示 Fragment 对应的 Tab 按钮，则调用 refresh() 方法刷新 Fragment 显示的数据。第 52 ~ 54 行 hideTabNum(index:Int) 方法用于隐藏 Tab 按钮上的提醒数字。第 60 ~ 79 行 showTabFragment (index:Int,fragment:Array<RefreshFragment>) 方法用于设置当前显示的 Fragment，并切换显示的 Tab 按钮。第 85 ~ 88 行 setTabNum(index: Int, num: Int) 方法用于设置底部 Tab 显示的数字提示。

14.8.2 附近模块

附近模块显示在指定范围内发布的动态，如图 14-18 所示。由于目前用户量不多，限定范围后可能会没有附近发布的动态，所以暂时显示的是平台所有发布的动态，根据发布时间进行排序。

图 14-18 附近模块的运行效果

1. 刷新 Fragment 的抽象类

RefreshFragment 类是一个抽象类，声明 refresh() 方法。该方法是抽象方法，用于刷新 Fragment 的显示内容。MainActivity 类中显示 Fragment 都继承了该类，因此可以通过单击 Tab 按钮对其内容进行刷新。

```
/java/com/pinkjiao/www/fragment/RefreshFragment.kt
abstract class RefreshFragment : Fragment() {
    abstract fun refresh()
}
```

第 05 行声明 RefreshFragment 为抽象类，并继承 Fragment 类。第 06 行声明 refresh()方法为抽象方法。

2. 设置状态栏颜色的接口

StatusBar 接口用于设置状态栏的颜色模式，由于 Fragment 和 Activity 都需要修改状态栏的颜色模式，所以需要使用接口，而不是类。

```
/java/com/pinkjiao/www/fragment/StatusBar.kt
05 interface StatusBar {
06     fun setStatusBar()
07 }
```

第 05 行声明 StatusBar 接口。第 06 行声明 setStatusBar()方法，用于设置状态栏的颜色模式。

3. 附近的 Fragment 类

VicinityFragment 类用于显示动态列表的 DynamicListFragment 类实例，继承 RefreshFragment 类，实现 StatusBar 接口。VicinityFragment 类并不直接作为显示动态列表的容器，而是作为 DynamicListFragment 类实例的容器。

```
/java/com/pinkjiao/www/fragment/VicinityFragment.kt
//附近的动态。Title 参数为标题，baseActivity 参数为所属的 Activity
class VicinityFragment(private val title:String, private val baseActivity: BaseActivity) : RefreshFragment(),StatusBar {
    companion object{
        lateinit var fragment:VicinityFragment
        fun newFragment(title:String, parentActivity: BaseActivity):VicinityFragment{
            fragment = VicinityFragment(title,parentActivity)
            return fragment
        }
    }
    private lateinit var viewBinding: FragmentVicinityBinding //视图绑定
    private lateinit var dynamicListFragment: DynamicListFragment
    override fun onCreateView(inflater: LayoutInflater, container: ViewGroup?, savedInstanceState: Bundle?): View {
        viewBinding = FragmentVicinityBinding.inflate(inflater, container, false)//实例化ViewBinding
        viewBinding.titleTextView.text = title //标题
        viewBinding.searchImageView.setOnClickListener { //搜索按钮
            val intent = Intent(context, SearchDynamicActivity::class.java)
            startActivity(intent)
        }
        viewBinding.addImageView.setOnClickListener { //添加按钮
            val intent = Intent(context, DynamicAddActivity::class.java)
            startActivity(intent)
        }
        //添加显示动态的 Fragment
        dynamicListFragment = DynamicListFragment.newInstance("dynamicList.php","",2)
        val transaction = childFragmentManager.beginTransaction()
        transaction.add(R.id.dynamicListFragment, dynamicListFragment).commit()
        return viewBinding.root
    }
    override fun refresh(){ //刷新 Fragment
```

```
54            dynamicListFragment.refresh()
55        }
59        override fun setStatusBar() { //设置状态栏颜色模式
60            baseActivity.setStatusBar(BaseActivity.STATUS_BAR_DARK)
61        }
62    }
```

第 19 行声明 VicinityFragment 类，虽然 baseActivity 参数始终传递的都是 MainActivity 类实例，但是由于需要调用 BaseActivity 类的 setStatusBar(Boolean)方法，**所以 baseActivity 参数的数据类型声明为 BaseActivity 类，而不是直接声明为 MainActivity 类。这样会降低代码的耦合性，重用代码时更加便利**。第 22 ~ 25 行使用静态方法实例化 VicinityFragment 类，避免重构时出现异常。第 47 行将 dynamicListFragment 添加到 dynamicListFragment 布局中并显示出来。第 53 ~ 55 行重写 RefreshFragment 类的 refresh()抽象方法，调用 dynamicListFragment 的 refresh()方法更新显示的内容。第 59 ~ 61 行重写 StatusBar 接口的 setStatusBar()方法，调用所属容器 BaseActivity 类实例的 setStatusBar(Boolean)方法设置状态栏的颜色模式。

14.8.3 动态列表模块

由于动态列表多处使用，所以单独创建 DynamicListFragment 类，根据传递的参数不同显示不同筛选条件的动态列表。

1. 动态列表的服务器端接口页面

动态列表的服务器端接口页面包括 dynamicList.php、searchDynamicList.php 和 userDynamicList.php，都使用 Get 方式接收数据。dynamicList.php 用于获取附近发布或关注用户发布的动态列表，所需参数如表 14-10 所示。searchDynamicList.php 用于获取根据关键字搜索到的动态列表，所需参数如表 14-11 所示。userDynamicList.php 用于获取指定用户发布的动态列表，所需参数如表 14-12 所示。

表 14-10 dynamicList.php 接收的参数

参 数	说 明
developerAccountNumber	开发者账号
developerKey	开发序列号
currentUserAccountID	登录用户的账号
currentUserPassword	登录用户的 MD5 算法加密后密码
start	开始加载的序号
num	加载数量

表 14-11 searchDynamicList.php 接收的参数

参 数	说 明
developerAccountNumber	开发者账号
developerKey	开发序列号
currentUserAccountID	登录用户的账号
currentUserPassword	登录用户的 MD5 算法加密后密码
searchKey	搜索关键字
start	开始加载的序号
num	加载数量

表 14-12　userDynamicList.php 接收的参数

参　数	说　明
developerAccountNumber	开发者账号
developerKey	开发序列号
currentUserAccountID	登录用户的账号
currentUserPassword	登录用户的 MD5 算法加密后密码
userAccountID	被搜索的用户账号 ID
start	开始加载的序号
num	加载数量

获取动态列表数据成功时，返回的 JSON 数据结构为{"state":"ok","dynamicList":[动态列表数据的 JSON 数组]}，动态列表数据的 JSON 数组的键名如表 14-13 所示。没有动态列表数据时，返回的 JSON 数据结构为{"state":"ok","dynamicList":[]}。

表 14-13　动态列表数据的 JSON 数组的键名

键　名	说　明
userName	用户姓名
userAccountID	用户账号 ID
userFaceImage	用户头像图片地址
dynamicID	动态 ID
dynamicContent	动态内容
dynamicImage0	第一张动态图片地址
dynamicImage1	第二张动态图片地址
dynamicImage2	第三张动态图片地址
dynamicImage3	第四张动态图片地址
dynamicLabel	动态标签
dynamicCreateTime	动态发布时间
dynamicReplyNum	动态的回复数量
dynamicReplyPrivateNum	动态的私回数量

动态列表的 JSON 数据样例(服务器端实际返回单行形式的 JSON 数据)

```
{
    "state":"ok",
    "dynamicList":[
        {
            "dynamicID":"60",
            "dynamicContent":"蒙餐里奇奇怪怪的食物",
            "userName":"不可信其无",
            "userAccountID":"98323430",
            "userFaceImage":"common\/face\/202101\/538221885014.jpg",
            "dynamicImage0":"common\/dynamic\/202101\/20210126204154_960.jpg",
            "dynamicImage1":"common\/dynamic\/202101\/20210126204155_415.jpg",
            "dynamicImage2":"",
            "dynamicImage3":"",
```

```
        "dynamicLabel":"蒙餐",
        "dynamicCreateTime":"2021-01-26 20:41:54",
        "dynamicReplyNum":"2",
        "dynamicReplyPrivateNum":"0"
    },
    {
        "dynamicID":"41",
        "dynamicContent":"东北洗浴中心，冬天最热闹的地方莫非于此了吧。",
        "userName":"璐璐",
        "userAccountID":"4294967295",
        "userFaceImage":"common\/face\/202101\/7252121129.jpg",
        "dynamicImage0":"common\/dynamic\/202101\/20210123165331_779.jpg",
        "dynamicImage1":"common\/dynamic\/202101\/20210123165332_585.jpg",
        "dynamicImage2":"common\/dynamic\/202101\/20210123165332_501.jpg",
        "dynamicImage3":"common\/dynamic\/202101\/20210123165332_260.jpg",
        "dynamicLabel":"澡堂",
        "dynamicCreateTime":"2021-01-26 16:53:31",
        "dynamicReplyNum":"1",
        "dynamicReplyPrivateNum":"0"
    }
  ]
}
```

2. 自动截取内容的文本控件

AutoCutTextView 控件是根据行数自动截取文本内容的文本控件，继承 AppCompatTextView 类。对动态内容进行截取前后的效果对比如图 14-19 所示。当显示动态列表时，如果将动态内容全部显示出来，会影响界面效果，也不便于快速预览。因此需要对内容进行截取，因为字号和中英文字符所占宽度不同，所以根据行数进行截取会更加美观。但这需要获取控件的宽度，因此通过 AutoCutTextView 控件实现该功能更加便捷。

图 14-19　截取与未截取的动态内容对比

```
/java/com/pinkjiao/www/widget/AutoCutTextView.kt
class AutoCutTextView : androidx.appcompat.widget.AppCompatTextView {
    var textRowNum = 5 //截取的行数
    private var cacheText = "" //缓存文本
```

```
11      constructor(context: Context) : super(context)
12      constructor(context: Context, attrs: AttributeSet) : super(context, attrs)
13      var text: String = ""
14          set(value) {
15              cacheText = value
16              field = ""
17          }
18      override fun onDraw(canvas: Canvas?) { //用于获取单个文字的宽度
20          val paint = Paint()
21          paint.textSize = textSize
22          val strWidth: Float = paint.measureText("字")
24          val strNumPerRow: Int = (width / strWidth).toInt() //计算每行的文字数量
26          if (cacheText.length > strNumPerRow * textRowNum) { //如果字符串长度大于两行则截取字符串
27              cacheText = cacheText.substring(0, strNumPerRow * textRowNum)
28          }
29          if (text != cacheText) { setText(cacheText) }
30          super.onDraw(canvas)
31      }
32  }
```

第 09 行 textRowNum 属性用于设置截取的行数，默认值为 5。第 11、12 行通过次构造方法调用父类的次构造方法，实现初始化。第 14 ~ 17 行通过 text 属性的 set()方法将设置的文本内容赋给 cacheText 变量进行缓存，不直接赋给 text 属性。第 18 ~ 31 行绘制控件时，计算单个汉字的宽度，然后计算出每行需要多少个汉字。如果 cacheText 变量的长度大于截取行数的汉字数量总和，则对 cacheText 变量截取至截取行数的汉字数量总和，作为 text 属性值；否则将 cacheText 变量作为 text 属性值。

```
/res/layout/item_dynamic_list_span_count2.xml
61  <com.pinkjiao.www.widget.AutoCutTextView
62      android:id = "@+id/contentTextView"
63      android:layout_width = "match_parent"
64      android:layout_height = "wrap_content"
65      android:layout_margin = "5dp"
66      android:textColor = "@color/black"
67      tools:text = "内容" />
```

第 61 行使用自定义控件标签时，需要使用包含命名空间的完整控件名称。第 62 ~ 66 行的 4 个属性是继承 AppCompatTextView 类的属性。

3. 动态列表的适配器类

DynamicListAdapter 类用于显示动态列表数据的 RecyclerView 控件的数据适配，可以适配单列和双列两种形式显示的数据。为了便于理解，下面将 DynamicListAdapter 类的主体结构和数据绑定子视图控件分开介绍。

```
/java/com/pinkjiao/www/adapter/DynamicListAdapter.kt
22  class DynamicListAdapter(
23      private val dynamicModelList: List<DynamicModel>, //动态数据模型的列表
24      private val userAccountID:String = "", //用户账号 ID
25      private val spanCount: Int = 1 //列数
26  ) : RecyclerView.Adapter<RecyclerView.ViewHolder>(){
```

```
    private val item1: Int = R.layout.item_dynamic_list_span_count1 //单列显示的动态子视图布局资源
    private val item2: Int = R.layout.item_dynamic_list_span_count2 //双列显示的动态子视图布局资源
    private val footer: Int = R.layout.footer_dynamic_list //页脚子视图布局资源
    private var context: Context? = null //获取子视图数量
    var currentState = "" //当前状态
    companion object{
        const val TYPE_ITEM = 0 //动态子视图类型
        const val TYPE_FOOTER = 1 //页脚子视图类型
        const val STATUS_REFRESHING = "上拉刷新状态"
        const val STATUS_LOADING_MORE = "正在加载状态"
        const val STATUS_FINISH = "加载完成状态"
        const val STATUS_NET_ERROR = "网络错误状态"
        const val STATUS_NON_DATA = "无数据状态"
        const val STATUS_ALL_DATA = "已经加载全部数据状态"
        const val STATUS_MAX_COUNT = "达到加载上限状态"
    }
    override fun onCreateViewHolder(parent: ViewGroup, viewType: Int): RecyclerView.ViewHolder {
        context = parent.context
        val holder: RecyclerView.ViewHolder
        //返回缓存子视图的 ViewHolder
        if (viewType == TYPE_ITEM) {
            holder = if(spanCount == 1) {
                ItemViewHolder1(LayoutInflater.from(parent.context).inflate(item1, parent, false))
            } else {
                ItemViewHolder2(LayoutInflater.from(parent.context).inflate(item2, parent, false))
            }
        } else {
            holder = FooterViewHolder(LayoutInflater.from(parent.context).inflate(footer, parent, false)).also {
                (it.linearLayout.layoutParams as StaggeredGridLayoutManager.LayoutParams).isFullSpan = true
            }
        }
        return holder
    }
    override fun onBindViewHolder(holder: RecyclerView.ViewHolder, position: Int) {
        //数据绑定子视图控件，稍后讲解
    }
    override fun getItemCount(): Int {  //获取子视图数量
        return dynamicModelList.size + 1//增加一个子视图用于显示加载提示子视图
    }
    override fun getItemViewType(position: Int): Int { //用于获取子视图类型
        return if (position + 1 == itemCount) TYPE_FOOTER else TYPE_ITEM
    }
    fun changeStatus(state: String) { //改变状态
        currentState = state
        if (state != STATUS_LOADING_MORE) {
            notifyDataSetChanged() //通知数据集改变，刷新 RecyclerView
        }
    }
    //单列显示的动态子视图 ViewHolder
    open class ItemViewHolder1(view: View) : RecyclerView.ViewHolder(view) {
```

```
234         var itemLinearLayout: LinearLayout = view.findViewById(R.id.itemLinearLayout)
235         var faceImageView: ImageView = view.findViewById(R.id.faceImageView)
236         var nameTextView: TextView = view.findViewById(R.id.nameTextView)
237         var contentTextView: TextView = view.findViewById(R.id.contentTextView)
238         var imagePagerView: ImagePagerView = view.findViewById(R.id.imagePagerView)
239         var creatTimeTextView: TextView = view.findViewById(R.id.createTimeTextView)
240         var replyPrivateNumTextView: TextView = view.findViewById(R.id.replyPrivateNumTextView)
241         var replyNumTextView: TextView = view.findViewById(R.id.replyNumTextView)
242     }
        //双列显示的动态子视图 ViewHolder
247     open class ItemViewHolder2(view: View) : RecyclerView.ViewHolder(view) {
            //与单列显示的动态子视图类似，详细代码查看工程文件
256     }
        //页脚子视图 ViewHolder
260     open class FooterViewHolder(view: View) : RecyclerView.ViewHolder(view) {
261         var linearLayout: LinearLayout = view.findViewById(R.id.linearLayout)
262         var progressBar: ProgressBar = view.findViewById(R.id.progressBar)
263         var footerTextView: TextView = view.findViewById(R.id.footerTextView)
264     }
265 }
```

第 35、36 行声明两种不同类型子视图的静态常量。第 37 ~ 43 行声明动态列表状态的静态常量。第 49 行判断显示的子视图是动态子视图还是页脚子视图。第 50 ~ 54 行根据 RecyclerView 控件显示的列数初始化单列子视图或双列子视图的 ViewHolder。第 57 行设置页脚子视图独占一行。第 62 ~ 203 行数据绑定子视图的代码，稍后介绍。第 209 行动态列表的长度+1 作为子视图的数量，增加的一个子视图就是页脚子视图。第 217 行如果当前位置+1（即 position+1）等于子视图数量 itemCount（即 getItemCount()方法的返回值）。第 233 ~ 228 行 changeStatus(state: String) 方法供外部加载数据后调用，更新子视图显示的内容。第 233 ~ 242 行是缓存单列显示子视图的缓存视图类。第 260 ~ 264 行是缓存页脚子视图的缓存视图类。

动态子视图分为单列和双列两种主要形式，有三种显示效果，如图 14-20 所示。单列动态子视图的图片可以左右滑动切换显示。双列动态子视图又分为两种效果：当传递的用户账号 ID 为空字符串时，顶部显示用户头像和用户姓名，隐藏发布时间；否则，顶部显示发布时间，隐藏用户头像和用户姓名。

图 14-20　动态子视图的三种显示效果

```
/java/com/pinkjiao/www/adapter/DynamicListAdapter.kt
//为缓存子视图的 ViewHolder 匹配数据
if (holder is ItemViewHolder1) { //单列动态子视图
    holder.itemLinearLayout.setOnClickListener { //单击子视图启动显示动态内容的 Activity
        val intent = Intent(context, DynamicActivity::class.java)
        intent.putExtra(DynamicActivity.EXTRA_DYNAMIC_MODEL, dynamicModelList[position])
        context!!.startActivity(intent)
    }
    //头像
    if (dynamicModelList[position].userFaceImage != "") {
        Image.loadFromUrl(holder.faceImageView, dynamicModelList[position].userFaceImage, CacheDir.getFace(context!!)) //加载头像图片
        holder.faceImageView.setOnClickListener { //单击头像启动个人主页的 Activity
            val intent = Intent(context, UserHomeActivity::class.java)
            intent.putExtra(UserHomeActivity.EXTRA_USER_ACCOUNT_ID, dynamicModelList[position].userAccountID)
            intent.putExtra(UserHomeActivity.EXTRA_USER_NAME, dynamicModelList[position].userName)
            context!!.startActivity(intent)
        }
    }
    //图片
    if (dynamicModelList[position].images.isNotEmpty()) {
        holder.imagePagerView.visibility = View.VISIBLE
        holder.imagePagerView.setAdapter(dynamicModelList[position].images) //设置显示图片的适配器
    } else { //加载提示子视图
        holder.imagePagerView.visibility = View.GONE
    }
    //私回
    holder.replyPrivateNumTextView.text = "私回[${dynamicModelList[position].replyPrivateNum}条]"
    holder.replyPrivateNumTextView.setOnClickListener { //单击启动回复的 Activity
        val intent = Intent(context, DynamicReplyAddActivity::class.java)
        intent.putExtra(DynamicReplyAddActivity.EXTRA_PRIVATE_FLAG, 1) //1 表示私回
        intent.putExtra(DynamicReplyAddActivity.EXTRA_DYNAMIC_ID, dynamicModelList[position].id)
        context!!.startActivity(intent)
    }
    //回复
    holder.replyNumTextView.text = "回复[${dynamicModelList[position].replyNum}条]"
    holder.replyNumTextView.setOnClickListener { //单击启动回复的 Activity
        val intent = Intent(context, DynamicReplyAddActivity::class.java)
        intent.putExtra(DynamicReplyAddActivity.EXTRA_PRIVATE_FLAG, 0) //0 表示普通回复
        intent.putExtra(DynamicReplyAddActivity.EXTRA_DYNAMIC_ID, dynamicModelList[position].id)
        context!!.startActivity(intent)
    }
} else if (holder is ItemViewHolder2) { //双列动态子视图
    //数据绑定子视图
} else if (holder is FooterViewHolder) { //页脚子视图
    when (currentState) { //根据当前状态显示页脚子视图的内容
        STATUS_FINISH -> holder.linearLayout.visibility = View.GONE
        STATUS_REFRESHING -> holder.linearLayout.visibility = View.GONE
    }
}
```

第 64 行判断是不是单列动态子视图。第 74 ~ 77 行启动用户个人主页的 Activity，intent 对象传递两个参数，分别是用户账号 ID 和用户姓名。**只通过用户账号 ID 就可以搜索到用户，额外增加的用户姓名是为了防止黑客通过自动生成的用户账号 ID 遍历所有用户的数据。**第 114 行判断是不是双列动态子视图。第 172 行判断是不是页脚子视图。第 173 ~ 201 行根据动态列表的状态显示页脚子视图。

4. 滚动监听类

由于每个列表都需要实现滚动到底部自动加载数据的功能，所以需要在 RecyclerView.OnScrollListener 类的 onScrolled(RecyclerView,Int,Int) 方法中判断是否已经滚动到最后一个子视图。使用 OnScrollListener 类继承 RecyclerView.OnScrollListener 类，添加获取最后一个子视图位置的方法。

```
/java/com/pinkjiao/www/adapter/OnScrollListener.kt
05  open class OnScrollListener: RecyclerView.OnScrollListener() {
        //获取最后一个子视图的位置。positionArray 参数为最后一行子视图位置数组
11      fun findMax(positionArray: IntArray): Int {
12          var max = positionArray[0]
13          for (value: Int in positionArray) if (value > max) max = value
14          return max
15      }
16  }
```

第 12 行暂时将 positionArray 数组的第一个元素作为最大值作为 max 变量的初始值。第 13 行遍历 positionArray 数组，找出数组的最大值赋给 max 变量。第 14 行将 max 变量作为返回值。

5. 动态列表的 Fragment

动态列表的 Fragment 会在附近、关注、自己、搜索动态和个人主页中被调用，用于显示不同筛选条件的动态列表。DynamicListFragment 类代码量比较大，以下对代码进行拆分讲解，首先了解该类的代码结构。

```
/java/com/pinkjiao/www/fragment/DynamicListFragment.kt
class DynamicListFragment(
    private var serverPage: String, //提供服务的页面
    private var userAccountID: String, //用户账号 ID
    private var spanCount: Int = 1, //显示的列数
    private var autoLoadEnable: Boolean, //自动加载
    private val swipeRefreshLayoutEnable: Boolean = true //下拉刷新是否启用
) : RefreshFragment() {
    companion object {
        //新建 DynamicListFragment 实例
        fun newInstance(page: String, userAccountID: String, spanCount: Int = 1, autoLoadEnable: Boolean = true, swipeRefreshLayoutEnable: Boolean = true): DynamicListFragment {
            return DynamicListFragment(page, userAccountID, spanCount,autoLoadEnable, swipeRefreshLayoutEnable)
        }
    }
    private var maxNum: Int = 5000//最大加载数据量
    private var pageNum: Int = 100//每次加载的数据量
    private lateinit var viewBinding: FragmentDynamicListBinding //视图绑定
    private lateinit var adapter: DynamicListAdapter //适配器
    private val dynamicModelList = ArrayList<DynamicModel>() //动态模型数组
```

```
54      private val userModel = UserModel() //用户模型
55      private var startIndex = 0 //开始加载的序号
56      private var searchKey: String = "" //搜索关键字
57      override fun onCreateView(inflater: LayoutInflater, container: ViewGroup?, savedInstanceState:
    Bundle?): View {
58          //从共享偏好设置获取用户数据
59          if (!UserModel.getSharedPreferences(requireContext(), userModel)) {
60              startActivity(Intent(context, LoginActivity::class.java))
61              Toast.makeText(context,  getString(R.string.toast_get_shared_preferences_error),
    Toast.LENGTH_LONG).show()
62          }
63          initView(inflater,container)//初始化视图控件
64          return viewBinding.root
65      }
        //刷新数据
116     override fun refresh(){
117         startIndex = 0
118         loadDynamicListData(startIndex, pageNum)
119     }
        //加载动态列表数据，star参数为开始序号，num参数为加载数量
125     private fun loadDynamicListData(start: Int, num: Int) {
126         if (start == 0) {
127             adapter.changeStatus(FooterViewHolder.STATUS_REFRESHING) //刷新状态
128         } else {
129             adapter.changeStatus(FooterViewHolder.STATUS_LOADING_MORE)//设置加载状态
130         }
131         getDynamicListData(start, num)
132         viewBinding.swipeRefreshLayout.isRefreshing = swipeRefreshLayoutEnable
133     }
        //从服务器端通过Get方式获取动态列表数据，start参数为开始序号，num参数为加载数量
139     private fun getDynamicListData(start: Int = 0, num: Int = pageNum) {
156     }
        //处理Http响应数据，res参数为返回的JSON数据
161     private fun httpResponseDataHandle(res:String){
201     }
        //搜索动态，searchKey参数为搜索关键字，autoLoadEnable参数为是否自动加载
207     fun searchDynamic(searchKey: String, autoLoadEnable: Boolean = true) {
208         this@DynamicListFragment.searchKey = searchKey
209         loadDynamicListData(0, pageNum)
210         this@DynamicListFragment.autoLoadEnable = autoLoadEnable
211     }
212 }
```

第 28 ~ 34 行声明 DynamicListFragment 类继承 RefreshFragment 类，通过主构造方法声明 5 个属性，作为查询动态列表数据的筛选条件。第 45 ~ 47 行是实例化 DynamicListFragment 类的静态方法。第 57 ~ 65 行实现创建视图后对控件进行初始化。第 116 ~ 119 行重写 RefreshFragment 类的 refresh() 抽象方法，重新加载并刷新列表数据。第 125 ~ 133 行加载动态列表数据及视图控件状态的修改。第 139 ~ 156 行通过 Get 请求获取动态列表的数据。第 161 ~ 201 行实现对 Get 请求返回的 JSON 数据的处理。第 207 ~ 211 行实现指定关键字搜索符合条件的动态，并以列表的形式返回数据。

下面介绍 initView(LayoutInflater, ViewGroup?) 方法、getDynamicListData(Int,Int) 方法和 httpResponseDataHandle(String) 方法。

```
/java/com/pinkjiao/www/fragment/DynamicListFragment.kt
66  private fun initView(inflater: LayoutInflater, container: ViewGroup?){
68      adapter = DynamicListAdapter(dynamicModelList, userAccountID, spanCount) //实例化适配器
70      viewBinding = FragmentDynamicListBinding.inflate(inflater, container, false) //实例化视图绑定
72      viewBinding.swipeRefreshLayout.setOnRefreshListener { //滑动刷新布局的刷新事件
73          startIndex = 0
74          loadDynamicListData(startIndex, pageNum)
75      }
77      viewBinding.swipeRefreshLayout.isEnabled = swipeRefreshLayoutEnable //设置下拉刷新状态
78      //设置瀑布流布局的纵向列
79      val layoutManager = StaggeredGridLayoutManager(spanCount, StaggeredGridLayoutManager.VERTICAL)
80      //设置回收视图
81      viewBinding.recyclerView.layoutManager = layoutManager
82      viewBinding.recyclerView.adapter = adapter
83      viewBinding.recyclerView.setItemViewCacheSize(10) //设置缓存子视图的数量
84      viewBinding.recyclerView.addOnScrollListener(object : OnScrollListener() { //添加滚动监听器
87          override fun onScrolled(recyclerView: RecyclerView, dx: Int, dy: Int) { //重写滚动完成事件
88              super.onScrolled(recyclerView, dx, dy)
89              if (autoLoadEnable) {
91                  val positionArray = IntArray((recyclerView.layoutManager as StaggeredGridLayoutManager).
    spanCount) //存储最后一行子视图位置的数组
93                  layoutManager.findLastVisibleItemPositions(positionArray) //获取最后一行子视图位置的数组
94                  //判断显示的最后一个子视图是否是最后一个视图
95                  if (findMax(positionArray) + 1 == adapter.itemCount) {
96                      //下拉刷新
97                      if (adapter.currentState == FooterViewHolder.STATUS_REFRESHING) return
98                      //避免重复加载
99                      if (adapter.currentState == FooterViewHolder.STATUS_LOADING_MORE) return
100                     //已加载所有数据
101                     if (adapter.currentState == FooterViewHolder.STATUS_ALL_DATA) return
102                     //无数据不加载
103                     if (adapter.currentState == FooterViewHolder.STATUS_NON_DATA) return
104                     //设置达到加载上限状态
105                     if (adapter.currentState == FooterViewHolder.STATUS_MAX_COUNT) return
106                     //获取动态列表数据
107                     loadDynamicListData(startIndex, pageNum)
108                 }
109             }
110         }
111     })
112 }
```

第 68 行实例化适配器。第 70 行通过绑定类生成实例化 viewBinding 变量。第 72～75 行设置 swipeRefreshLayout 布局的滑动刷新事件，将 startIndex 属性重置为 0，从第一条数据开始重新加载动态列表。第 83 行设置缓存子视图的数量，如果连续出现没有图片的动态，标准尺寸的手机屏幕上显示的数量不会超过 10 个，略微增加一些缓存子视图的数量会提升加载时的流畅度，同时也会略微增加

些系统资源的消耗。第 84 ~ 111 行添加 recyclerView 控件的滚动监听器，autoLoadEnable 属性设置为 true 时，滚动到底部自动加载数据。第 95 行调用的 findMax(IntArray)方法是 OnScrollListener()自定义类内的方法。

```
/java/com/pinkjiao/www/fragment/DynamicListFragment.kt
139 private fun getDynamicListData(start: Int = 0, num: Int = pageNum) {
140     val baseUrl = RequestUrl.getBaseUrl(requireContext())
141     val parameter = "&userAccountID = $userAccountID&searchKey = $searchKey&start = $start&num = $num"
142     val getRequest = GetRequest(baseUrl, serverPage, DeveloperModel.getDeveloper(requireContext()), userModel, parameter)
143     Http.get(getRequest.url, object : Http.ResponseListener {
144         override fun onResponse(res: String) {
145             if (startIndex == 0) { dynamicModelList.clear() }
146             httpResponseDataHandle(res) //处理返回的 JSON 数据
147         }
148         override fun onFailure(res: String) {
149             adapter.changeStatus(FooterViewHolder.STATUS_NET_ERROR)
150             viewBinding.swipeRefreshLayout.isRefreshing = false //关闭刷新状态
151             Toast.makeText(context, res, Toast.LENGTH_LONG).show()
152         }
153     })
154 }
```

第 140 行获取服务器端页面所在的文件夹 Url 路径。第 141 行合成查询条件的 Get 参数。第 142 行实例化 Get 请求对象。第 143 ~ 153 行发送 Get 请求并处理请求结果的回调。第 144 ~ 147 行重写 Http.ResponseListener 接口的 onResponse(String)方法，处理成功获取的动态列表数据。第 148 ~ 152 行处理网络请求失败后的界面反馈。

```
/java/com/pinkjiao/www/fragment/DynamicListFragment.kt
private fun httpResponseDataHandle(res:String){
    val jsonParser = JSONTokener(JSON.cleanBOM(res))
    val jsonObject = jsonParser.nextValue() as JSONObject
    if (jsonObject.getString("state") == "ok") {
        val dynamicList = JSONArray(jsonObject.getString("dynamicList"))
        for (i in 0 until dynamicList.length()) { //遍历 JSON 数组数据
            val jo = dynamicList.opt(i) as JSONObject
            val dynamicModel = DynamicModel()
            dynamicModel.userName = jo.getString("userName")
            dynamicModel.userAccountID = jo.getString("userAccountID")
            dynamicModel.userFaceImage = Image.getUrl(getString(R.string.app_domain), jo.getString("userFaceImage"), Image.smallQuality)
            dynamicModel.id = jo.getInt("dynamicID")
            dynamicModel.content = jo.getString("dynamicContent")
            dynamicModel.images[0] = Image.getUrl(getString(R.string.app_domain), jo.getString("dynamicImage0"), Image.highQuality)
            dynamicModel.images[1] = Image.getUrl(getString(R.string.app_domain), jo.getString("dynamicImage1"), Image.highQuality)
            dynamicModel.images[2] = Image.getUrl(getString(R.string.app_domain), jo.getString("dynamicImage2"), Image.highQuality)
            dynamicModel.images[3] = Image.getUrl(getString(R.string.app_domain), jo.getString
```

```
    ("dynamicImage3"), Image.highQuality)
178            dynamicModel.images = Common.removeEmptyArrayElement(dynamicModel.images)
179            dynamicModel.replyNum = jo.getInt("dynamicReplyNum")
180            dynamicModel.replyPrivateNum = jo.getInt("dynamicReplyPrivateNum")
181            dynamicModel.createTime = jo.getString("dynamicCreateTime")
182            dynamicModel.label = jo.getString("dynamicLabel")
183            dynamicModelList.add(dynamicModel)
184        }
185        when {
186            //刷新后滚动到顶部
187            startIndex == 0 ->{ viewBinding.recyclerView.scrollToPosition(0) }
188            //返回的数据少于加载的数量，设置底部视图的状态
189            dynamicList.length() < pageNum -> {
190                adapter.changeStatus(FooterViewHolder.STATUS_ALL_DATA)
191            }
192            //再次加载的总数大于最大加载数据量，设置底部视图的状态
193            startIndex + dynamicList.length() + pageNum > maxNum -> {
194                adapter.changeStatus(FooterViewHolder.STATUS_MAX_COUNT)
195            }
196            //其他情况，设置下次开始加载的序号
197            else -> { startIndex += dynamicList.length() }
198        }
199    } else { adapter.changeStatus(FooterViewHolder.STATUS_NON_DATA) }
200    viewBinding.swipeRefreshLayout.isRefreshing = false //关闭刷新状态
201 }
```

第 164 行读取 JSON 数据中的 state 键值，并判断键值是否为 "ok"。第 165 行读取 JSON 数据中的 dynamicList 键值。第 166 ~ 184 行遍历 dynamicList 键值以数组形式存储的动态数据，每条动态数据存储在一个 DynamicModel 对象中，然后将存储每条动态数据的 DynamicModel 对象添加到 dynamicModelList 变量。

14.8.4 关注模块

关注模块与附近模块的实现方法基本相同，主要区别在于服务器端接口和子视图显示方式的不同，如图 14-21 所示。关注以单列子视图显示动态列表的数据，而附近使用双列子视图显示动态列表的数据。单列子视图使用自定义的 ImagePagerView 控件显示图片，可以实现多张图片循环滑动的效果。

图 14-21 关注模块的运行效果

1. 图片分页滑动控件

ImagePagerView 控件是自定义的图片分页滑动控件，继承 FrameLayout 类，显示效果如图 14-2 所示。由于篇幅限制，此处不介绍该控件，读者可以自行查看项目中的源代码，其中包含详细的注释 下面是在动态列表的单列子视图布局文件中使用该自定义控件的代码。

```
/res/layout/item_dynamic_list_span_count1.xml
<com.pinkjiao.www.widget.ImagePagerView
    android:id = "@+id/imagePagerView"
    android:layout_width = "match_parent"
```

```
53     android:layout_height = "400dp"
54     android:layout_margin = "5dp"
55     android:background = "@color/bg_gray" />
```

第 50 行使用自定义控件标签时，需要使用包含命名空间的完整控件名称。第 51 ~ 55 行 4 个属性是继承 FrameLayout 类的属性。

图 14-22　ImagePagerView 控件的显示效果

2. 关注的 Fragment 类

FollowFragment 类与 VicinityFragment 类的功能基本相同，调用的服务器端接口的参数不同，调用 DynamicListFragment 类实例使用单列形式显示子视图。

```
/java/com/pinkjiao/www/fragment/FollowFragment.kt
45 val loginUserModel = UserModel()
46 //从共享偏好设置获取用户数据
47 if (UserModel.getSharedPreferences(requireContext(),loginUserModel)) {
48     //添加显示动态的 Fragment
49     dynamicListFragment = DynamicListFragment.newInstance("followDynamicList.php",
   loginUserModel.accountID,1)
50     val transaction = childFragmentManager.beginTransaction()
51     transaction.add(R.id.dynamicListFragment, dynamicListFragment).commit()
52 }
```

第 45 行实例化 loginUserModel 变量。第 47 ~ 52 行获取本地存储的登录用户数据，如果获取成功，则实例化显示个人动态的 Fragment，并显示到 dynamicListFragment 布局中。

14.9　发布动态模块

发布动态模块将发布的动态数据发送到服务器端存储起来，动态数据包括内容、图片（可选）和标签（可选），如图 14-23 所示。

14.9.1　发布动态的服务器端接口页面

发布动态的服务器端接口页面为 dynamicAdd.php，使用 Post 方式接收数据，所需参数如表 14-14 所示。

图 14-23　发布动态的运行效果

表 14-14　dynamicAdd.php 接收的参数

参　数	说　明
developerAccountNumber	开发者账号
developerKey	开发序列号
currentUserAccountID	登录用户的账号
currentUserPassword	登录用户的 MD5 算法加密后密码
dynamicContent	动态的内容
uploadInput	动态的图片（可选）
dynamicLabel	动态的标签（可选）

发布动态成功时，返回的 JSON 数据结构为{"state":"ok"}。发布动态失败时，返回的 JSON 数据结构为{"state":"错误码","info":"错误提示信息"}，如表 14-15 所示。

表 14-15　返回的错误码

错　误　码	错误提示信息
error804	发布动态失败
error805	动态数据有误

14.9.2　过滤类

Input 类用于过滤 EditText 控件输入的内容，与监听事件过滤输入内容的方式相比，耦合度更低，代码更加简洁，所以单纯过滤输入内容更加适合。

```
/java/com/pinkjiao/www/util/Input.kt
object Input {
    fun contentFilter():InputFilter{ //内容过滤器。返回值类型为 InputFilter
        return InputFilter { source, _, _, dest, dstart, _ ->
            val placeStart = dest.subSequence(0, dstart).toString() //获取输入位置之前的字符
            //判断是否换行
            if (source.toString().contains("\n")) {
                if (placeStart == "") {
                    return@InputFilter "" //过滤输入的内容
                } else if (placeStart.substring(placeStart.length - 1, placeStart.length) == "\n") {
                    return@InputFilter "" //过滤输入的内容
                }
            }
            null //不过滤输入的内容
        }
    }
    fun labelFilter(maxLength:Int):InputFilter{ //标签过滤器。返回值类型为 InputFilter
        return InputFilter { source, _, _, dest, dstart, dend ->
            val placeStart = dest.subSequence(0, dstart) //获取输入位置之前的字符
            val placeEnd = dest.subSequence(dend, dest.length) //获取输入位置之后的字符
            val newString = placeStart.toString() + source.toString() + placeEnd.toString()
```

```
35            if (!Validator.isNoSymbolString(newString, 1, maxLength)) {
36                return@InputFilter "" //过滤输入的内容
37            }
38            null //不过滤输入的内容
39        }
40    }
41 }
```

第 15 行判断输入的字符串(粘贴时涉及多字符的字符串)中是否包含换行符。第 16 ~ 20 行如果当输入字符串之前的字符为空字符，则过滤输入的字符串；如果输入字符串的最后一个字符为换行符，过滤输入的字符串。第 34 行将当前输入的字符串与当前输入位置之前的字符串和当前输入位置之后字符串连接起来，赋给 newString 变量。第 35 ~ 37 行通过 Validator.isNoSymbolString(String,Int,Int)去判断 newString 变量是否在 maxLength 的长度范围内且只包含数字、字母和汉字，否则进行过滤。

.9.3 发布动态的 Activity

发布动态的 Activity 调用系统图库选取图片，最多可以选择 4 幅图片。目前，基本上所有主流 App 使用 ContentProvider 组件获取系统相册的数据开发自定义的图片选择器，可以实现限定选择数量、寸质量、图片视频筛选等功能，甚至可以为图片添加标签、进行图像处理等，如图 14-24 所示。因篇幅所限，所以本项目暂时没有使用自定义的图片选择器。

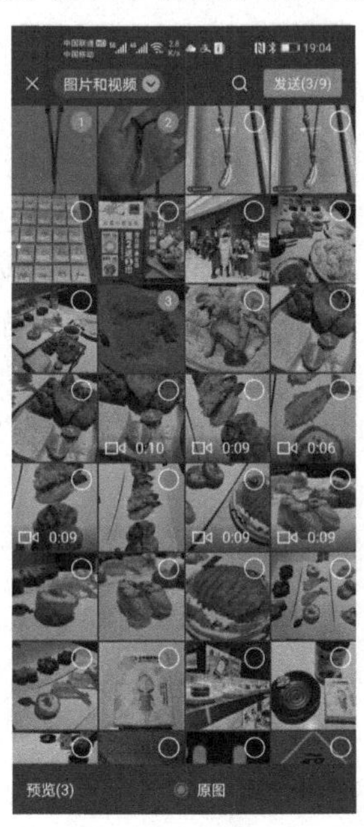

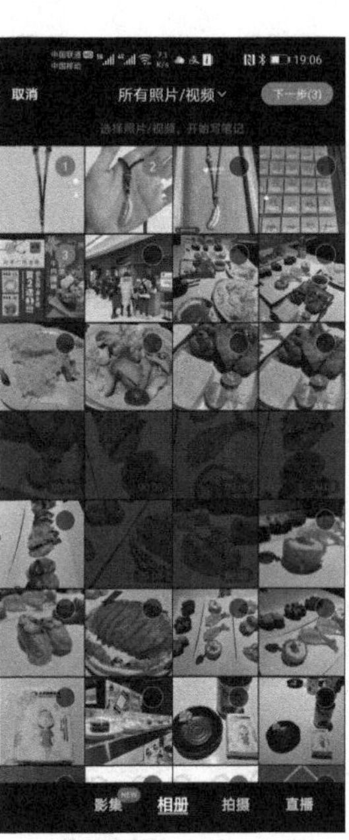

图 14-24 主流 App 的图片选择器

DynamicAddDynamic 类代码量比较大，所以对代码进行拆分介绍，首先了解该类的代码结构。

```
java/com/pinkjiao/www/DynamicAddActivity.kt
class DynamicAddActivity : HideKeyboardActivity()  {
    private lateinit var viewBinding:ActivityDynamicAddBinding //视图绑定
    private lateinit var selectImageView: ImageView //当前选择图片的 ImageView
    private var context: Context = this //上下文
    private val imageViewCount = 4 //表示上传图片的最大数量
```

```
29      private var imagesPath =  arrayOfNulls<String>(imageViewCount) //上传图片的文件路径数组
30      private var imagesDegree = IntArray(imageViewCount) //上传图片的旋转角度数组
31      private var imagesUri = arrayOfNulls<Uri>(imageViewCount) //上传图片的 Uri 数组
32      private var imageViewSelectedIndex = 0 //最后选择图片的 ImageView 序号
33      private var cacheImagesPrefixName: String = "postImage" //缓存图片文件的前缀名
34      private var cacheImageMaxHeight = 1920 //缓存图片最大高度
35      private var cacheImageMaxWidth = 1280 //缓存图片最大宽度
36      private var thumbnailHeight = 80 //缩略图高度
37      private var thumbnailWidth = 80 //缩略图宽度
38      private var imageViewWeight = 0 //上传图片按钮的尺寸权重
43      override fun onWindowFocusChanged(hasFocus: Boolean) { //窗口焦点改变事件
59  }
60      override fun onCreate(savedInstanceState: Bundle?) {
103 }
        //选取图片。selectIndex 参数为图片控件序号，imageView 参数为图片控件
109     private fun selectImage(selectIndex: Int, imageView: ImageView){
121     }
        //发布动态。userModel 参数为当前用户数据，parameter 参数为动态参数，postFiles 参数为动态的图片附件
128     private fun send(userModel:UserModel,parameter: MutableMap<String, String>, postFiles
    Map<String, String>?){
158     }
        //回调。requestCode 参数为请求码，resultCode 参数为结果码，intent 参数为 intent 对象
165     override fun onActivityResult(requestCode: Int, resultCode: Int, intent: Intent?) {
176     }
177 }
```

第 25 ~ 38 行声明所需的属性，并初始化一些配置数据。第 43 ~ 59 行 onWindowFocusChan (Boolean)方法用于实现选择图片的 4 个 ImageView 控件尺寸自动匹配屏幕的宽度。第 60 ~ 103 onCreate(Bundle?)方法用于初始化控件。第 109 ~ 121 行 selectImage(Int,ImageView)方法用于调用系图库选取图片。第 128 ~ 158 行 send(UserModel,MutableMap<String,String>,Map<String,String>?)方法于发送动态数据并处理响应结果。第 165 ~ 176 行 onActivityResult(Int,Int,Intent?)方法用于处理通过统图册选择图片后的回调。

/java/com/pinkjiao/www/DynamicAddActivity.kt

```
43  override fun onWindowFocusChanged(hasFocus: Boolean) {
44      super.onWindowFocusChanged(hasFocus)
45      if (hasFocus) {
46          //根据容器宽度自动计算每个控件的宽度和高度
47          val measuredWidth: Int = viewBinding.imageLinearLayout.measuredWidth
48          imageViewWeight = (measuredWidth - Display.dp2px(context, 2.5f * 8 + 10f))/4
49          with (viewBinding) {
50              val margin = Display.dp2px(context, 2.5f)
51              val params = LinearLayout.LayoutParams(imageViewWeight, imageViewWeight)
52              params.setMargins(margin, margin, margin, margin)
53              dynamicImageView0.layoutParams = params
54              dynamicImageView1.layoutParams = params
55              dynamicImageView2.layoutParams = params
56              dynamicImageView3.layoutParams = params
57          }
58      }
59  }
```

第 47 行获取图片(默认显示+图标的图片)所在布局的宽度。第 48 行计算每幅图片的宽度匹配所在布局的宽度。第 49 ~ 57 行设置每幅图片的宽度、高度和间距。

```
/java/com/pinkjiao/www/DynamicAddActivity.kt
60  override fun onCreate(savedInstanceState: Bundle?) {
61      super.onCreate(savedInstanceState)
62      viewBinding = ActivityDynamicAddBinding.inflate(layoutInflater)
63      setContentView(viewBinding.root)
64      with (viewBinding) {
65          cancelTextView.setOnClickListener { this@DynamicAddActivity.finish() }
66          publishTextView.setOnClickListener {
67              val loginUserModel = UserModel()
68              if (!UserModel.getSharedPreferences(context, loginUserModel)){ //从共享偏好设置获取用户数据
69                  startActivity(Intent(context, LoginActivity::class.java))
70                  Toast.makeText(context, getString(R.string.toast_get_shared_preferences_
    error), Toast.LENGTH_LONG).show()
71              } else if (contentEditText.text.toString() != "") { //发布动态
73                  Keyboard.hide(context, currentFocus?.windowToken) //关闭软键盘
75                  val parameter: MutableMap<String, String>= HashMap() //post 参数
76                  parameter["dynamicContent"] = contentEditText.text.toString()
77                  parameter["dynamicLabel"] = labelEditText.text.toString()
78                  //post 文件
79                  val postFileUri:Array<Uri?> = Common.removeEmptyArrayElement(imagesUri)//清空
    Uri 数组中的空元素
80                  val postFiles: MutableMap<String, String>
81                  if (postFileUri.isNotEmpty()) {
83                      LoadingProgressDialog.showProgressDialog(context, getString
    (R.string.dynamic_add_dialog_tip)) //显示等待对话框
84                      postFiles = Image.cacheImage(contentResolver, "uploadInput",
85                              CacheDir.getImage(context), cacheImagesPrefixName,
86                              cacheImageMaxHeight, cacheImageMaxWidth, postFileUri, imagesDegree)
87                  } else { postFiles = mutableMapOf() }
88                  send(loginUserModel, parameter, postFiles)
89              } else {
90                  Toast.makeText(context, getString(R.string.dynamic_add_content_hint), Toast.
    LENGTH_LONG).show()
91              }
92          }
93          dynamicImageView0.setOnClickListener { selectImage(0, dynamicImageView0) }
94          dynamicImageView1.setOnClickListener { selectImage(1, dynamicImageView1) }
95          dynamicImageView2.setOnClickListener { selectImage(2, dynamicImageView2) }
96          dynamicImageView3.setOnClickListener { selectImage(3, dynamicImageView3) }
98          contentEditText.filters = arrayOf(Input.contentFilter()) //设置内容过滤器
99          labelEditText.filters = arrayOf(Input.labelFilter(8)) //设置标签过滤器
101         Keyboard.show(this@DynamicAddActivity, contentEditText) //显示软键盘
102     }
103 }
```

第 68 行读取登录用户存储在本地的数据，后面将其作为发布动态的 Post 参数。第 75 ~ 77 行将内容控件和标签控件输入的内容合成 Post 参数。第 79 行清空 imagesUri 数组中的空元素，因为每幅图片

都是单独调用系统图库进行选择的，允许按照任意顺序选择，也可以不选择，所以会导致空元素的存在。第 84 ~ 86 行将选择的图片在本地进行压缩后缓存，并将上传图片对应的 Post 参数名和保存路径保存在 MutableMap<String, String>对象，赋给 postFiles 变量。第 98 行设置 contentEditText 控件的过滤器，过滤连续的回车输入。第 99 行设置 labelEditText 控件的过滤器，将输入字符的数量控制在 8 个以内。

```
/java/com/pinkjiao/www/DynamicAddActivity.kt
109 private fun selectImage(selectIndex: Int, imageView: ImageView){
110     imageViewSelectedIndex = selectIndex
111     selectImageView = imageView
112     if (imagesPath[imageViewSelectedIndex] == null) {
113         val intent = Intent(Intent.ACTION_PICK)
114         intent.type = "image/*"
115         startActivityForResult(intent, RESULT_CANCELED)
116     } else {
117         imagesPath[imageViewSelectedIndex] = null
118         imagesUri[imageViewSelectedIndex] = null
119         selectImageView.setImageDrawable(ContextCompat.getDrawable(context,
R.drawable.ic_baseline_add_50))
120     }
121 }
```

第 112 行判断当前单击的图片控件是否选择了图片。第 113 ~ 115 行通过 intent 对象启动系统图库选择图片。第 117 ~ 119 行将单击位置的图片路径和 Uri 清空，并设置为默认图片（即+图标图片）。

```
/java/com/pinkjiao/www/DynamicAddActivity.kt
private fun send(userModel:UserModel,parameter: MutableMap<String, String>, postFiles:
Map<String, String>?){
    val baseUrl = RequestUrl.getBaseUrl(context)
    val page = "dynamicAdd.php"
    val postRequest = PostRequest(baseUrl, page, DeveloperModel.getDeveloper(context),
userModel, parameter, postFiles)
    //post 方式上传
    Http.post(postRequest.url, postRequest.parameter, postRequest.files, object :
Http.PostResponseListener {
        override fun onCreate() {
            LoadingProgressDialog.dismiss() //隐藏等待对话框
            LoadingProgressDialog.showProgressDialog(context, getString(R.string.dynamic_
add_dialog_tip)) //显示等待对话框
        }
        override fun onResponse(res: String) {
            LoadingProgressDialog.dismiss() //隐藏等待对话框
            val jsonParser = JSONTokener(JSON.cleanBOM(res))
            val jsonObject = jsonParser.nextValue() as JSONObject
            if (jsonObject.getString("state") == "ok") {
                finish()
                Toast.makeText(context, getString(R.string.toast_dynamic_add_publish_
success), Toast.LENGTH_LONG).show()
            } else {
                Toast.makeText(context, getString(R.string.toast_dynamic_add_publish_error),
```

```
    Toast.LENGTH_LONG).show()
150         }
151     }
152     override fun onFailure(res: String) {
153         Toast.makeText(context, getString(R.string.toast_net_error), Toast.LENGTH_LONG).show()
155         LoadingProgressDialog.dismiss() //隐藏等待对话框
156     }
157   })
158 }
```

第 129 行获取服务器端页面所在的文件夹 Url 路径。第 130 行设置发布动态的服务器端接口页面。第 131 行实例化发布动态的 post 请求对象。第 134 ~ 139 行是发送 http 请求前的回调方法，显示等待对话框。第 140 ~ 151 行重写 Http.PostResponseListener 接口的 onResponse(String) 方法，处理服务器端接口页面返回的 JSON 数据。如果发布成功，则关闭当前 Activity，显示发布成功的提示信息；否则显示发布失败的提示信息。第 152 ~ 155 行处理网络请求失败的界面反馈。

```
/java/com/pinkjiao/www/DynamicAddActivity.kt
165 override fun onActivityResult(requestCode: Int, resultCode: Int, intent: Intent?) {
166     super.onActivityResult(requestCode, resultCode, intent)
167     if (resultCode == RESULT_OK && requestCode == RESULT_CANCELED) {
168         val uri = intent!!.data!!
169         imagesUri[imageViewSelectedIndex] = uri
170         imagesDegree[imageViewSelectedIndex] = Image.getImageOrientation(contentResolver, uri)
171         imagesPath[imageViewSelectedIndex] = Common.uriToPath(contentResolver, uri)
173         val thumbnailBitmap = Image.loadThumbnail (contentResolver, uri, thumbnailHeight,
thumbnailWidth) //获取选取图片的缩略图
174         selectImageView.setImageBitmap(thumbnailBitmap) //显示缩略图
175     }
176 }
```

第 168 行获取调用系统图库选择图片后返回的 uri。第 169 行将 uri 保存到对应的 imagesUri 数组元素。第 170 行获取选择图片的旋转数据保存到对应的 imagesDegree 数组元素。第 171 行通过 uri 获取选择图片的物理路径保存到对应的 imagesPath 数组元素。第 173 行通过 uri 获取选择图片指定尺寸的缩略图。第 174 行将缩略图显示到对应的 ImageView 控件中。

14.10　MVVM 模式

MVVM(Model-View-ViewModel，模型-视图-视图模型)模式的主要目的是降低耦合度，使逻辑更清晰，统一处理 UI 操作，便于修改视图。Model 将数据进行模型化处理，通过 ViewModel 将 Model 与 View 相分离。同时 Model 与 ViewModel 之间通过观察者模式实现数据的双向绑定(也可以是单向绑定)。

提示：MVC 模式

MVC 是模型(model)、视图(view)、控制器(controller)的缩写。MVC 模式以将业务逻辑、数据、界面分离的方式组织代码。

14.10.1 逻辑关系

在本工程中使用 MVVM 模式时，ViewModel 功能通过一个 ViewModel 类实现，Model 和 View 功能的实现都涉及多个非继承关系的类。因此将 MVVM 模式归纳为三层结构较为适合，如图 14-25 所示。

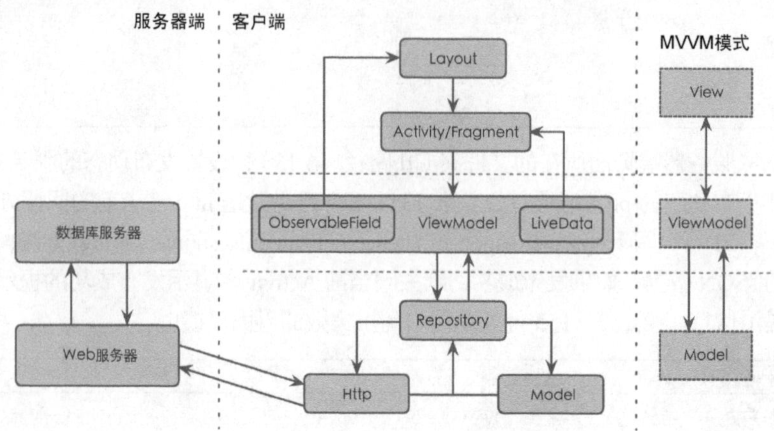

图 14-25 本工程的 MVVM 模式

View 层负责界面数据的展示，与用户进行交互。通常由 Activity 类或 Fragment 类与 Layout 布局组成。Layout 布局中的控件属性可以和 ViewModel 类的 ObserableField 属性绑定。

ViewModel 层负责业务逻辑，通常每个视图对应一个 ViewModel 类。从 Model 层获取数据，进行业务处理，然后通过 ObserableField 属性向 Layout 布局中的控件属性单向传递数据，通过 LiveData 属性向 Activity 类或 Fragment 类单向传递数据。

Model 层负责处理数据的加载或者存储，包括 Repository 类（存储类）、Http 类（网络通信类）、Model 类（保存数据类）。Repository 类通过 Http 类获取服务器端的数据，然后保存到 Model 类中，需要时再从 Model 类中调用。

14.10.2 优势和劣势

凡事必有利弊，MVVM 模式也是一样的。需要根据项目实际情况判断是否使用 MVVM 模式，越复杂的项目越能体现出该模式的优势。

MVVM 模式的优势：由于降低了耦合度，所以减少了 Activity 类和 Fragment 类中的代码量，逻辑结构更清晰；双向数据绑定可以使 ViewModel 专注业务逻辑的处理，而不需要处理 UI。

MVVM 模式的劣势：Android Studio 对数据绑定的 Bug 调试还不完善；双向数据绑定和 Model 会占用更多的系统资源；逻辑结构的增加看似逻辑清晰，但增加了理解的难度，增加了学习成本。

14.11 首页模块组 2

首页模块组 2 包括偶遇模块（MeetActivity）、提醒模块（RemindFragment）和自己模块（MineFragment），从此开始使用 MVVM 模式进行程序编写。

14.11.1 偶遇模块

偶遇模块会随机显示一条动态，如图 14-26 所示。服务器端完全随机生成动态的 ID，不受推荐法、发布动态的时间和发布动态的定位限制。显示偶遇后，再次单击偶遇的 Tab 按钮会重新随机显示一条动态。本模块涉及的类较多，核心类之间的调用关系如图 14-27 所示。

图 14-26 偶遇模块的运行效果

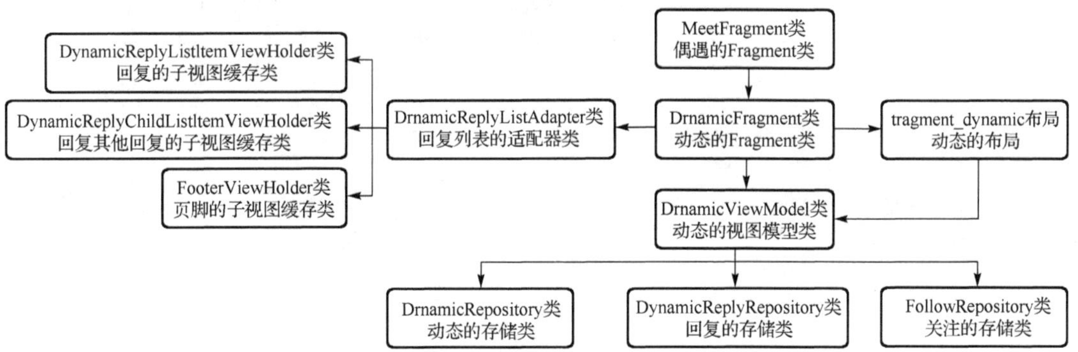

图 14-27 偶遇模块核心类之间的调用关系

1．动态的存储类

DynamicRepository 类用于存储(包括添加、查询、更新、删除数据等)动态数据。下面先介绍该类主体结构，再介绍存储数据的方法。

```
/java/com/pinkjiao/www/repository/DynamicRepository.kt
class DynamicRepository {
    //获取动态
    fun getDynamic(domainUrl:String, baseUrl:String, developerModel:DeveloperModel, loginUserModel:
UserModel, dynamicID:String?, responseListener : ResponseListener){
    }
    //获取关键字
    fun getKeys(baseUrl:String, developerModel:DeveloperModel, loginUserModel:UserModel,
searchKey:String?, responseListener : KeysResponseListener){
    }
    //删除动态
    fun deleteDynamic(baseUrl:String, developerModel:DeveloperModel, loginUserModel:UserModel,
dynamicID:Int, responseListener : DeleteResponseListener){
    }
```

```
149      interface ResponseListener {
            //Http 请求的回调。Res 参数为响应结果，dynamicModel 参数为动态数据，followFlag 参数为是否关注动态作者
156         fun onCallBack(res: String, dynamicModel: DynamicModel?, followFlag:String = "0")
157      }
158      interface KeysResponseListener {
            //搜索字符串的 Http 请求的回调。Res 参数为响应结果，keyList 参数为关键字集合
164         fun onCallBack(res: String, keyList: MutableMap<String,String>?)
165      }
166      interface DeleteResponseListener {
            //删除字符串的 Http 请求的回调。Res 参数为响应结果。
171         fun onCallBack(res: String)
172      }
173  }
```

第 149 ~ 172 行声明三种类型的监听接口，且都声明了 onCallBack()方法，但所接收的参数有所区别。res 参数是 String 类型，用于接收响应的结果类型；其余参数用于接收服务器端获取的数据。

/java/com/pinkjiao/www/repository/DynamicRepository.kt

```
//获取动态。domainUrl 参数为服务器端 App 服务所在的文件夹 Url 路径，baseUrl 参数为服务器端页面所在的文件夹 Url 路径，developerModel 参数为开发者，loginUserModel 参数为登录用户，dynamicID 参数为动态 ID，responseListener 参数为监听器接口
fun getDynamic(domainUrl:String, baseUrl:String, developerModel:DeveloperModel, loginUserModel:
UserModel, dynamicID:String?, responseListener : ResponseListener){
    val page = "dynamic.php" //接口页面
    var parameter = ""
    if (dynamicID != null) {
        parameter = "&dynamicID=$dynamicID" //get 参数
    }
    //get 请求对象
    val getRequest = GetRequest(baseUrl, page, developerModel, loginUserModel, parameter)
    //get 方法获取动态数据
    Http.get(getRequest.url, object : Http.ResponseListener {
        override fun onResponse(res: String) {
            val jsonParser = JSONTokener(JSON.cleanBOM(res))
            val jsonObject = jsonParser.nextValue() as JSONObject
            if (jsonObject.getString("state") == "ok") {
                val followFlag = jsonObject.getString("followFlag")
                val dynamicJSONObject = jsonObject.getJSONObject("dynamic")
                val dynamicModel = DynamicModel()
                dynamicModel.userName = dynamicJSONObject.getString("userName")
                dynamicModel.userAccountID = dynamicJSONObject.getString("userAccountID")
                dynamicModel.userFaceImage = Image.getUrl(domainUrl, dynamicJSONObject.
getString("userFaceImage"), Image.smallQuality)
                //赋值 dynamicModel 变量其他属性的方法类似，此处省略，可在工程文件中查看。
                responseListener.onCallBack(Http.RESPONSE_SUCCESS,dynamicModel,followFlag)
            } else {
                responseListener.onCallBack(jsonObject.getString("info"), null)
            }
        }
        override fun onFailure(res: String) {
            responseListener.onCallBack(res, null)
```

```
82           }
83       })
84   }
```

第 29 行设置服务器端接口页面。第 31 ~ 33 行判断是否获取指定 ID 的动态，如果 dynamicID 参数为 null，表示随机获取一条动态，实现偶遇的功能。第 35 行实例化 Get 请求对象。第 37 行发送获取动态的请求。第 38 ~ 79 行重写 onResponse()方法，解析请求返回的 JSON 数据。如果获取动态成功，通过 onCallBack()方法将动态数据和登录用户是否关注该动态的发布者回调，否则通过 onCallBack()方法回调错误信息。第 80 ~ 83 行重写 onFailure()方法，调用 onCallBack()方法回调网络错误信息。

/java/com/pinkjiao/www/repository/DynamicRepository.kt

```
    //获取关键字的相关关键字。baseUrl 参数为服务器端页面所在的文件夹 Url 路径，developerModel 参数为开发者，
loginUserModel 参数为登录用户，searchKey 参数为被搜索的关键字核心词，responseListener 参数为监听器接口
93  fun  getKeys(baseUrl:String,  developerModel:DeveloperModel,  loginUserModel:UserModel,
searchKey:String?, responseListener : KeysResponseListener){
94      val page = "searchDynamicKeys.php" //接口页面
95      val parameter = "&searchKey=$searchKey" //get 参数
96      //get 请求对象
97      val getRequest = GetRequest(baseUrl, page, developerModel, loginUserModel, parameter)
98      //get 方法获取关键字
99      Http.get(getRequest.url, object : Http.ResponseListener {
100         override fun onResponse(res: String) {
101             val jsonParser = JSONTokener(JSON.cleanBOM(res))
102             val jsonObject = jsonParser.nextValue() as JSONObject
103             if (jsonObject.getString("state") == "ok") {
104                 val keyList = mutableMapOf<String,String>()
105                 val dynamicJSON = JSONArray(jsonObject.getString("searchKeyList"))
106                 for (i in 0 until dynamicJSON.length()) { //遍历数据
107                     val jo = dynamicJSON.opt(i) as JSONObject
108                     keyList[jo.getString("searchKey")] = jo.getString("count")
109                 }
110                 responseListener.onCallBack(Http.RESPONSE_SUCCESS, keyList)
111             } else {
112                 responseListener.onCallBack(jsonObject.getString("info"), null)
113             }
114         }
115         override fun onFailure(res: String) {
116             responseListener.onCallBack(res, null)
117         }
118     })
119 }
```

第 94 行设置服务器端接口页面。第 99 行发送查询关键字的相关动态关键字的请求，即搜索动态输入关键字后查询到的相关关键字及包含该关键字的动态数量，便于用户快速选择。第 100 ~ 114 重写 onResponse()方法，解析请求返回的 JSON 数据。如果获取动态成功，遍历返回的 JSON 数据 searchKeyList 元素保存的 JSON 数组，使用 keyList 变量的键名保存相关关键字，keyList 变量键值存包含相关关键字的动态数量，然后通过 onCallBack()方法将 keyList 变量回调。如果获取动态失败，通过 onCallBack 方法回调错误信息。第 115 ~ 117 行重写 onFailure()方法，调用 onCallBack()方法调网络错误信息。

```
/java/com/pinkjiao/www/repository/DynamicRepository.kt
    //删除动态。baseUrl 参数为服务器端页面所在的文件夹 Url 路径，developerModel 参数为开发者，loginUserModel 参数为登录用户，dynamicID 参数为动态 ID，responseListener 参数为监听器接口
128 fun deleteDynamic(baseUrl:String, developerModel:DeveloperModel, loginUserModel:UserModel, dynamicID:Int, responseListener : DeleteResponseListener){
129     val page = "dynamicDelete.php"
130     val parameter = "&dynamicID=$dynamicID" //get 参数
131     //get 请求对象
132     val getRequest = GetRequest(baseUrl, page, developerModel, loginUserModel, parameter)
133     //get 方法删除动态数据
134     Http.get(getRequest.url, object : Http.GetResponseListener {
135         override fun onCreate() {
136             responseListener.onCallBack(Http.CREATE)
137         }
138         override fun onResponse(res: String) {
139             val jsonParser = JSONTokener(JSON.cleanBOM(res))
140             val jsonObject = jsonParser.nextValue() as JSONObject
141             if (jsonObject.getString("state") == "ok") {
142                 responseListener.onCallBack(Http.RESPONSE_SUCCESS)
143             } else {
144                 responseListener.onCallBack(jsonObject.getString("info"))
145             }
146         }
147         override fun onFailure(res: String) {
148             responseListener.onCallBack(res)
149         }
150     })
151 }
```

第 129 行设置服务器端接口页面。第 134 行发送删除动态的请求。第 135 ~ 137 行重写 onCreate() 方法，调用 onCallBack() 方法回调请求创建的状态。第 135 ~ 146 行重写 onResponse() 方法，解析请求返回的 JSON 数据，回调删除结果。第 147 ~ 149 行重写 onFailure() 方法，调用 onCallBack() 方法回调网络错误信息。

2. 回复的存储类

DynamicReplyRepository 类用于添加和查询动态的回复数据。下面先介绍该类主体结构，再介绍存储数据的方法。

```
/java/com/pinkjiao/www/repository/DynamicReplyRepository.kt
class DynamicReplyRepository {
    //添加回复
    fun reply(baseUrl: String, developerModel: DeveloperModel, loginUserModel: UserModel, parameter: MutableMap<String, String>, responseListener:ResponseListener){
    }
    //获取回复列表数据
    fun getDynamicReplyList(domainUrl:String, baseUrl: String, developerModel: DeveloperModel, loginUserModel: UserModel,start:Int,num:Int,dynamicReplyDynamicID:Int, responseListener:ResponseListener) {
    }
    interface ResponseListener {
```

```
        //Http 请求的回调。res 参数为响应结果，count 参数为回复列表数据的数量，dynamicReplyModelList 参数为回复列表数据
105         fun onCallBack(res: String, count:Int = 0, dynamicReplyModelList:ArrayList<DynamicReplyModel>? = null)
106     }
107 }
```

第 25 ~ 47 行添加动态的回复。第 58 ~ 97 行获取指定动态的回复列表数据。第 98 ~ 106 行声明响应请求的监听接口，声明了 onCallBack() 方法，处理响应数据。

```
/java/com/pinkjiao/www/repository/DynamicReplyRepository.kt
    //添加回复。baseUrl 参数为服务器端页面所在的文件夹 Url 路径，developerModel 参数为开发者，loginUserModel 参数为登录用户，parameter 参数为回复数据，responseListener 参数为监听器接口
25  fun reply(baseUrl: String, developerModel: DeveloperModel, loginUserModel: UserModel, parameter: MutableMap<String, String>, responseListener:ResponseListener){
26      val page = "dynamicReplyAdd.php"
27      //post 请求对象
28      val postRequest = PostRequest(baseUrl, page, developerModel, loginUserModel, parameter, null)
29      //post 方式上传
30      Http.post(postRequest.url, postRequest.parameter, postRequest.files, object : Http.PostResponseListener {
31          override fun onCreate() {
32              responseListener.onCallBack(Http.CREATE)
33          }
34          override fun onResponse(res: String) {
35              val jsonParser = JSONTokener(JSON.cleanBOM(res))
36              val jsonObject = jsonParser.nextValue() as JSONObject
37              if (jsonObject.getString("state") == "ok") {
38                  responseListener.onCallBack(Http.RESPONSE_SUCCESS)
39              } else {
40                  responseListener.onCallBack(jsonObject.getString("info"))
41              }
42          }
43          override fun onFailure(res: String) {
44              responseListener.onCallBack(res)
45          }
46      })
47  }
```

第 26 行设置服务器端接口页面。第 28 行实例化 Post 请求对象。第 29 行发送添加动态回复的请求。第 31 ~ 33 行重写 onCreate() 方法，调用 onCallBack() 方法回调请求创建的状态。第 34 ~ 42 行重写 onResponse() 方法，解析请求返回的 JSON 数据，并通过 onCallBack() 方法进行回调。第 43 ~ 46 行重写 onFailure() 方法，调用 onCallBack() 方法回调网络错误信息。

```
/java/com/pinkjiao/www/repository/DynamicReplyRepository.kt
    //获取回复列表。baseUrl 参数为服务器端页面所在的文件夹 Url 路径，developerModel 参数为开发者，loginUserModel 参数为登录用户，start 参数为开始加载的序号，num 参数为数量，dynamicReplyDynamicID 参数为动态 ID，responseListener 参数为监听器接口
58  fun getDynamicReplyList(domainUrl:String, baseUrl: String, developerModel: DeveloperModel, loginUserModel: UserModel,start:Int,num:Int,dynamicReplyDynamicID:Int, responseListener:
```

```
   ResponseListener) {
59     val page = "dynamicReplyList.php"
60     val parameter = "&dynamicReplyDynamicID=$dynamicReplyDynamicID&start=$start&num=$num"
61     val getRequest = GetRequest(baseUrl, page, developerModel, loginUserModel, parameter)
62     Http.get(getRequest.url, object : Http.ResponseListener {
63         override fun onResponse(res: String) {
64             val dynamicReplyModelList = ArrayList<DynamicReplyModel>() //动态回复模型数组
65             val jsonParser = JSONTokener(JSON.cleanBOM(res))
66             val jsonObject = jsonParser.nextValue() as JSONObject
67             val count = jsonObject.getInt("count")
68             if (jsonObject.getString("state") == "ok" && count > 0) {
69                 val dynamicJSON = JSONArray(jsonObject.getString("dynamicReplyList"))
70                 for (i in 0 until dynamicJSON.length()) { //遍历数据
71                     val jo = dynamicJSON.opt(i) as JSONObject
72                     val dynamicReplyModel = DynamicReplyModel()
73                     dynamicReplyModel.userName = jo.getString("userName")
74                     dynamicReplyModel.userAccountID = jo.getString("userAccountID")
75                     dynamicReplyModel.userFaceImage = Image.getUrl(domainUrl, jo.getString
   ("userFaceImage"), Image.smallQuality)
76                     dynamicReplyModel.id = jo.getInt("dynamicReplyID")
77                     dynamicReplyModel.content = jo.getString("dynamicReplyContent")
78                     dynamicReplyModel.createTime = jo.getString("dynamicReplyCreateTime")
79                     dynamicReplyModel.dynamicID = jo.getInt("dynamicReplyDynamicID")
80                     dynamicReplyModel.repliedDynamicReplyID = jo.getInt
   ("dynamicReplyRepliedDynamicReplyID")
81                     dynamicReplyModel.privateFlag = jo.getInt("dynamicReplyPrivateFlag")
82                     dynamicReplyModel.repliedUserName = jo.getString("repliedUserName")
83                     dynamicReplyModel.repliedUserAccountID = jo.getString("repliedUserAccountID")
84                     dynamicReplyModelList.add(dynamicReplyModel)
85                 }
86                 responseListener.onCallBack(
                       FooterViewHolder.STATUS_FINISH,count,dynamicReplyModelList)
87             } else if (count == 0) {
88                 responseListener.onCallBack(
                       FooterViewHolder.STATUS_NON_DATA,count,dynamicReplyModelList)
89             } else {
90                 responseListener.onCallBack(
                       FooterViewHolder.STATUS_NET_ERROR,count,dynamicReplyModelList)
91             }
92         }
93         override fun onFailure(res: String) {
94             responseListener.onCallBack(res)
95         }
96     })
97 }
```

第 59 行设置服务器端接口页面。第 61 行实例化 Get 请求对象。第 62 行发送获取动态的回复列表的请求。第 63～92 行重写 onResponse()方法，解析请求返回的 JSON 数据，并通过 onCallBack(方法将页脚子视图状态、回复列表数据数量和回复列表数据回调。第 93～95 行重写 onFailure()方法调用 onCallBack()方法回调网络错误信息。

3. 关注的存储类

FollowRepository 类用于存储(包括添加、查询、删除数据等)关注和粉丝数据，关注和粉丝是互反关系的数据。例如，A 用户关注了 B 用户，数据库中插入一个 A 用户关注 B 用户的数据，A 用户获取关注列表时会查询到该数据，而 B 用户获取粉丝列表时也会查询到该数据。下面先介绍该类主体结构，再介绍存储数据的方法。

/java/com/pinkjiao/www/repository/FollowRepository.kt

```
14  class FollowRepository {
        //添加关注
23      fun addFollow(baseUrl:String, developerModel:DeveloperModel, loginUserModel:UserModel,
    followAccountID:String,responseListener: ResponseListener){
43      }
        //取消关注
52      fun removeFollow(baseUrl:String, developerModel:DeveloperModel, loginUserModel:UserModel,
    followAccountID:String,responseListener: ResponseListener){
72      }
        //获取关注列表数据
83      fun getFollowList(domain:String, baseUrl: String, developerModel: DeveloperModel,
    loginUserModel: UserModel, start:Int, num:Int, responseListener: ResponseListener) {
84          val page = "followList.php" //接口页面
85          getList(page, domain, baseUrl, developerModel, loginUserModel, start, num,
    responseListener)
86      }
        //获取粉丝列表数据
97      fun getFollowerList(domain:String, baseUrl: String, developerModel: DeveloperModel,
    loginUserModel: UserModel, start:Int, num:Int, responseListener: ResponseListener) {
98          val page = "followerList.php" //接口页面
99          getList(page, domain, baseUrl, developerModel, loginUserModel, start, num,
    responseListener)
100     }
        //获取列表数据
112     private fun getList(page:String, domain:String, baseUrl: String, developerModel:
    DeveloperModel, loginUserModel: UserModel, start:Int, num:Int, responseListener:
    ResponseListener){
141     }
142     interface ResponseListener {
            //Http 请求的回调。res 参数为响应结果，followModelList 参数为关注列表数据
148         fun onCallBack(res: String,followModelList:ArrayList<FollowModel>?)
149     }
150 }
```

第 83 ~ 86 行获取关注列表和第 97 ~ 100 行获取粉丝列表，都调用 getList 方法处理返回的数据，因为获取列表成功时返回的数据类型都是 ArrayList<FollowModel>。第 142 ~ 149 行声明响应请求的监听接口，且声明 onCallBack() 方法接收响应数据。res 参数是 String 类型，接收响应的结果类型；followModelList 参数是可空的 ArrayList<FollowModel>类型，接收关注或粉丝的列表数据。

/java/com/pinkjiao/www/repository/FollowRepository.kt

```
    //添加关注。baseUrl 参数为基础路径，developerModel 参数为开发者，loginUserModel 参数为登录用户，
followAccountID 参数为关注用户账号 ID，responseListener 参数为响应监听器
```

```
23  fun  addFollow(baseUrl:String,  developerModel:DeveloperModel,  loginUserModel:UserModel,
followAccountID:String,responseListener: ResponseListener){
24      val page = "followAdd.php" //接口页面
25      val parameter = "&followUserAccountID=$followAccountID" //get 参数
26      //get 请求对象
27      val getRequest = GetRequest(baseUrl, page, developerModel, loginUserModel, parameter)
28      //get 方法添加关注
29      Http.get(getRequest.url, object : Http.ResponseListener {
30          override fun onResponse(res: String) {
31              val jsonParser = JSONTokener(JSON.cleanBOM(res))
32              val jsonObject = jsonParser.nextValue() as JSONObject
33              if (jsonObject.getString("state") == "ok") {
34                  responseListener.onCallBack(Http.RESPONSE_SUCCESS, null)
35              } else {
36                  responseListener.onCallBack(jsonObject.getString("info"), null)
37              }
38          }
39          override fun onFailure(res: String) {
40              responseListener.onCallBack(res, null)
41          }
42      })
43  }
```

第 24 行设置服务器端接口页面。第 25 行合成 Get 请求参数。第 27 行实例化 Get 请求对象。第 29 行发送添加关注的请求。第 30 ~ 38 行重写 onResponse()方法，解析请求返回的 JSON 数据。如果添加关注成功，通过 onCallBack()方法回调成功信息，否则通过 onCallBack()方法回调错误信息。第 39 ~ 41 行重写 onFailure()方法，调用 onCallBack()方法回调网络错误信息。

```
/java/com/pinkjiao/www/repository/FollowRepository.kt
//取消关注。baseUrl 参数为基础路径，developerModel 参数为开发者，loginUserModel 参数为登录用户，
followAccountID 参数为关注用户账号 ID，responseListener 参数为响应监听器
fun removeFollow(baseUrl:String, developerModel:DeveloperModel, loginUserModel:UserModel,
followAccountID:String,responseListener: ResponseListener){
    val page = "followRemove.php" //接口页面
    val parameter = "&followUserAccountID=$followAccountID" //get 参数
    //get 请求对象
    val getRequest = GetRequest(baseUrl, page, developerModel, loginUserModel, parameter)
    //get 方法移除关注
    Http.get(getRequest.url, object : Http.ResponseListener {
        override fun onResponse(res: String) {
            val jsonParser = JSONTokener(JSON.cleanBOM(res))
            val jsonObject = jsonParser.nextValue() as JSONObject
            if (jsonObject.getString("state") == "ok") {
                responseListener.onCallBack(Http.RESPONSE_SUCCESS, null)
            } else {
                responseListener.onCallBack(jsonObject.getString("info"), null)
            }
        }
        override fun onFailure(res: String) {
            responseListener.onCallBack(res, null)
```

```
0         }
1     })
2  }
```

第 53 行设置服务器端接口页面。第 54 行合成 Get 请求参数。第 56 行实例化 Get 请求对象。第行发送移除关注的请求。第 59 ~ 67 行重写 onResponse()方法，解析请求返回的 JSON 数据。如果除关注成功，通过 onCallBack()方法回调成功信息，否则通过 onCallBack()方法回调错误信息。第 ~ 70 行重写 onFailure()方法，调用 onCallBack()方法回调网络错误信息。

java/com/pinkjiao/www/repository/FollowRepository.kt

```
   //获取列表数据。page 参数为接口页面，domain 参数为域名，baseUrl 参数为基础路径，developerModel 参数为开发者，
   loginUserModel 参数为登录用户，start 参数为开始序号，num 参数为数量，responseListener 参数为响应监听器
12 private fun getList(page:String, domain:String, baseUrl: String, developerModel: DeveloperModel,
   loginUserModel: UserModel, start:Int, num:Int, responseListener: ResponseListener){
13     val parameter = "&start = $start&num = $num" //get 参数
14     val getRequest = GetRequest(baseUrl, page, developerModel, loginUserModel, parameter)
15     //get 方法获取关注列表
16     Http.get(getRequest.url, object : Http.ResponseListener {
17         override fun onResponse(res: String) {
18             val followModelList = ArrayList<FollowModel>() //动态回复模型数组
19             val jsonParser = JSONTokener(JSON.cleanBOM(res))
20             val jsonObject = jsonParser.nextValue() as JSONObject
21             if (jsonObject.getString("state") == "ok") {
22                 val followJSON = JSONArray(jsonObject.getString("followList"))
23                 for (i in 0 until followJSON.length()) { //遍历数据
24                     val jo = followJSON.opt(i) as JSONObject
25                     val followModel = FollowModel()
26                     followModel.userAccountID = jo.getString("userAccountID")
27                     followModel.userName = jo.getString("userName")
28                     followModel.userFaceImage = Image.getUrl(domain, jo.getString
   ("userFaceImage"), Image.smallQuality)
29                     followModel.followBecomeTime = jo.getString("followBecomeTime")
30                     followModelList.add(followModel)
31                 }
32                 responseListener.onCallBack(Http.RESPONSE_SUCCESS,followModelList)
33             } else {
34                 responseListener.onCallBack(jsonObject.getString("info"),null)
35             }
36         }
37         override fun onFailure(res: String) {
38             responseListener.onCallBack(res,null)
39         }
40     })
41 }
```

第 113 行合成 Get 请求参数。第 114 行实例化 Get 请求对象，接口页面通过 page 参数传递。第行发送请求。第 117 ~ 136 行重写 onResponse()方法，解析请求返回的 JSON 数据。如果获取用户力，通过 onCallBack()方法回调成功信息及用户列表数据，否则通过 onCallBack()方法回调错误信第 137 ~ 139 行重写 onFailure()方法，调用 onCallBack()方法回调网络错误信息。

4. 动态的视图模型类

DynamicViewModel 类用于为显示动态的视图提供绑定数据及数据处理。下面先介绍该类主体结构，再介绍数据处理的方法。

```
/java/com/pinkjiao/www/view/model/DynamicViewModel.kt
//动态视图模型。baseUrl 参数为基础 Url，developerModel 参数为开发者，loginUserModel 参数为当前登录用户
class DynamicViewModel(var baseUrl:String, var developerModel: DeveloperModel, private var loginUserModel: UserModel): ViewModel() {
    companion object{
        const val STATE_GET_DYNAMIC_SUCCESS = "获取动态数据成功"
        const val STATE_GET_DYNAMIC_FAILURE = "获取动态数据失败"
        const val STATE_ADD_FOLLOW_SUCCESS = "添加关注成功"
        const val STATE_ADD_FOLLOW_FAILURE = "添加关注失败"
        const val STATE_REMOVE_FOLLOW_SUCCESS = "取消关注成功"
        const val STATE_REMOVE_FOLLOW_FAILURE = "取消关注失败"
        const val STATE_GET_DYNAMIC_REPLY_SUCCESS = "获取回复数据成功"
        const val STATE_GET_DYNAMIC_REPLY_NONE = "无回复数据"
        const val STATE_GET_DYNAMIC_REPLY_FAILURE = "获取回复数据失败"
        const val STATE_DELETE_DYNAMIC_REQUEST = "请求删除动态"
        const val STATE_DELETE_DYNAMIC_SUCCESS = "删除动态成功"
        const val STATE_DELETE_DYNAMIC_FAILURE = "删除动态失败"
    }
    var dynamicModel = ObservableField<DynamicModel>() //可观察的用户模型数据
    var dynamicReplyModelList = ArrayList<DynamicReplyModel>() //回复模型的数组
    var response = MutableLiveData<String>() //请求的响应状态
    var replyListCount = 0 //回复数量
    var followFlag = "0" //关注标记
    var error = "" //错误信息
    //Int 类型转换为 String 类型。i 参数为 Int 类型数据，返回值类型为 String
    fun intToString(i:Int):String{
        return i.toString()
    }
    //获取动态。domainUrl 参数为域名，dynamicID 参数为动态 ID
    fun getDynamic(domainUrl:String,dynamicID:String){
    }
    //获取动态回复列表。domainUrl 参数为域名，startIndex 参数为开始序号，pageNum 参数为动态数量，dynamicReplyDynamicID 参数为被回复的动态 ID
    fun getDynamicReplyList(domainUrl:String,startIndex:Int, pageNum:Int, dynamicReplyDynamicID:Int){
    }
    //添加关注。followAccountID 参数为关注用户的账号 ID
    fun addFollow(followAccountID:String) {
    }
    //删除关注。followAccountID 参数为关注用户的账号 ID
    fun removeFollow(followAccountID:String) {
    }
    //删除动态
    fun deleteDynamic(){
    }
}
```

第 24 ~ 35 行声明响应状态的常量。第 37 行声明 dynamicModel 变量为 ObservableField
/namicModel>类型，该类型可以被视图的数据绑定观察数据变化并自动更新数据。第 39 行声明
onse 变量为 MutableLiveData<String>类型，可以被其他类观察其数据变化。第 47 ~ 49 行视图的数
阝定时将 Int 类型转换为 String 类型。

```
ava/com/pinkjiao/www/view/model/DynamicViewModel.kt
//获取动态。domainUrl 参数为域名，dynamicID 参数为动态 ID
fun getDynamic(domainUrl:String,dynamicID:String){
   DynamicRepository().getDynamic(domainUrl,baseUrl,   developerModel,   loginUserModel,
dynamicID, object : DynamicRepository.ResponseListener {
      override fun onCallBack(res: String, dynamicModel: DynamicModel?, followFlag: String) {
         if (res == Http.RESPONSE_SUCCESS) {//获取数据成功
            this@DynamicViewModel.dynamicModel.set(dynamicModel)
            this@DynamicViewModel.followFlag = followFlag
            response.value = STATE_GET_DYNAMIC_SUCCESS
         } else {//获取数据失败
            error = res
            response.value = STATE_GET_DYNAMIC_FAILURE
         }
      }
   })
}
```

第 56 行调用 DynamicRepository 类的 getDynamic()静态方法从服务器端获取指定 ID 的动态。第
~ 65 行重写回调方法，如果返回的结果为 Http.RESPONSE_SUCCESS 则表示获取动态数据成功，
则就是获取失败。由于 response 变量为可观察对象，不能对变量直接进行赋值，所以需要对该变量
value 属性进行赋值。response 变量将会在使用 DynamicViewModel 实例的类中观察其值的变化，从
寸 UI 进行相应的操作。

```
ava/com/pinkjiao/www/view/model/DynamicViewModel.kt
//获取动态回复列表。domainUrl 参数为域名，startIndex 参数为开始序号，pageNum 参数为动态数量，
dynamicReplyDynamicID 参数为被回复的动态 ID
fun getDynamicReplyList(domainUrl:String,startIndex:Int,pageNum:Int,dynamicReplyDynamicID:Int){
   DynamicReplyRepository().getDynamicReplyList(domainUrl, baseUrl, developerModel,
loginUserModel,startIndex,pageNum,dynamicReplyDynamicID, object: DynamicReplyRepository.
DynamicReplyListResponseListener{
      override fun onCallBack(res: String, count: Int, dynamicReplyModelList: ArrayList
<DynamicReplyModel>) {
         when (res) {
            FooterViewHolder.STATUS_FINISH -> {//获取数据成功
               this@DynamicViewModel.dynamicReplyModelList = dynamicReplyModelList
               replyListCount = count
               response.value = STATE_GET_DYNAMIC_REPLY_SUCCESS
            }
            FooterViewHolder.STATUS_NON_DATA -> {//没有获取到数据
               replyListCount = 0
               response.value = STATE_GET_DYNAMIC_REPLY_NONE
            }
            else -> {//获取数据失败
```

```
90                  error = res
91                  response.value = STATE_GET_DYNAMIC_REPLY_FAILURE
92              }
93          }
94      }
95    })
96 }
```

第 77 行调用 DynamicRepository 类的 getDynamicReplyList()静态方法从服务器端获取指定 ID 的回复列表。第 78 ~ 94 行重写回调方法，并根据返回结果对数据进行处理，修改 response 变量的值。response 变量将会在使用 DynamicViewModel 实例的类中观察其值的变化,从而对 UI 进行相应的操作。

/java/com/pinkjiao/www/view/model/DynamicViewModel.kt

```
    //添加关注。followAccountID 参数为关注用户的账号 ID
101 fun addFollow(followAccountID:String) {
102    FollowRepository().addFollow(baseUrl, developerModel, loginUserModel, followAccountID,
    object : FollowRepository.AddFollowResponseListener {
103        override fun onCallBack(res: String) {
104            if (res == Http.RESPONSE_SUCCESS) { //关注成功
105                response.value = STATE_ADD_FOLLOW_SUCCESS
106            } else { //关注失败
107                error = res
108                response.value = STATE_ADD_FOLLOW_FAILURE
109            }
110        }
111    })
112 }
    //取消关注。followAccountID 参数为关注用户的账号 ID。
117 fun removeFollow(followAccountID:String) {
118    //获取用户信息
119    FollowRepository().removeFollow(baseUrl, developerModel, loginUserModel, followAccountID,
    object : FollowRepository.RemoveFollowResponseListener {
120        override fun onCallBack(res: String) {
121            if (res == Http.RESPONSE_SUCCESS) {//取消关注成功
122                response.value = STATE_REMOVE_FOLLOW_SUCCESS
123            } else {//取消关注失败
124                error = res
125                response.value = STATE_REMOVE_FOLLOW_FAILURE
126            }
127        }
128    })
129 }
```

第 102 行调用 FollowRepository 类的 addFollow()静态方法向服务器端发送添加关注其他用户的求。第 103 ~ 110 行重写回调方法，并根据返回结果修改 response 变量的值。response 变量将会在使 DynamicViewModel 实例的类中观察其值的变化，从而对 UI 进行相应的操作。第 119 行调 FollowRepository 类的 removeFollow()静态方法向服务器端发送取消关注其他用户的请求。第 120 ~ 1 行重写回调方法，并根据返回结果修改 response 变量的值。

/java/com/pinkjiao/www/view/model/DynamicViewModel.kt

```
    //删除动态
```

```
133 fun deleteDynamic(){
134     DynamicRepository().deleteDynamic(baseUrl, developerModel, loginUserModel,dynamicModel.get()!!.id,object :DynamicRepository.DeleteResponseListener{
135         override fun onCallBack(res: String) {
136             when (res) {
137                 Http.CREATE -> {//正在发送删除动态请求
138                     response.value = STATE_DELETE_DYNAMIC_REQUEST
139                 }
140                 Http.RESPONSE_SUCCESS -> {//删除动态成功
141                     response.value = STATE_DELETE_DYNAMIC_SUCCESS
142                 }
143                 else -> {//删除动态失败
144                     error = res
145                     response.value = STATE_DELETE_DYNAMIC_FAILURE
146                 }
147             }
148         }
149     })
150 }
```

第 134 行调用 DynamicRepository 类的 deleteDynamic()静态方法向服务器端发送添加关注其他用户的请求。第 135 ~ 148 行重写回调方法，并根据返回结果修改 response 变量的值。res 变量值为 Http.CREATE 表示正在发送删除动态的请求。

下面介绍观察到 response 变量值等于 STATE_DELETE_DYNAMIC_REQUEST 常量时，显示等待对话框。

5. 显示动态的布局

fragment_dynamic 布局文件使用 DynamicViewModel 类绑定控件的属性，用于显示从服务器端获取的动态数据。

/res/layout/fragment_dynamic.xml

```xml
<?xml version = "1.0" encoding = "utf-8"?>
<layout xmlns:android = "http://schemas.android.com/apk/res/android"
    xmlns:app = "http://schemas.android.com/apk/res-auto"
    xmlns:tools = "http://schemas.android.com/tools">
    <!-- 绑定数据类 -->
    <data>
        <variable
            name = "dynamicViewModel"
            type = "com.pinkjiao.www.view.model.DynamicViewModel" />
    </data>
                    <TextView
                        android:id = "@+id/nameTextView"
                        android:text = "@{dynamicViewModel.dynamicModel.userName}"
                        android:textColor = "@color/black"
                        android:textSize = "16sp"
                        tools:text = "用户名" />
                    <TextView
                        android:id = "@+id/userAccountIDTextView"
                        android:layout_width = "wrap_content"
```

```
78                      android:layout_height = "wrap_content"
79                      android:layout_marginStart = "5dp"
80                      android:text = '@{"(ID:"+dynamicViewModel.dynamicModel.userAccountID + ")"}'
81                      tools:text = "(ID:6182739217)" />
177                 <TextView
178                     android:id = "@+id/replyPrivateNumTextView"
179                     android:layout_width = "wrap_content"
180                     android:layout_height = "match_parent"
181                     android:text = '@{"私回["+dynamicViewModel.intToString
    (dynamicViewModel.dynamicModel.replyPrivateNum)+"条]"}'
182                     android:textColor = "@color/purple_500"
183                     tools:text = "私回[1条]" />
203             <!-- 使用RelativeLayout 自动匹配 RecyclerView 的高度 防止 RecyclerView 出现滚动条 -->
204             <RelativeLayout
205                 android:layout_width = "match_parent"
206                 android:layout_height = "wrap_content">
208                 <androidx.recyclerview.widget.RecyclerView
209                     android:id = "@+id/replyRecyclerView"
210                     android:layout_width = "match_parent"
211                     android:layout_height = "wrap_content"
212                     android:layout_marginTop = "5dp" />
213             </RelativeLayout>
219 </layout>
```

第 06 ~ 10 行设置用于绑定的数据类。第 70 行将 nameTextView 控件的 text 属性与 dynamicViewModel.dynamicModel.userName 属性直接绑定。第 80 行 userAccountIDTextView 控件的 text 属性并没有与 dynamicViewModel.dynamicModel.userAccountID 属性直接绑定，而是与其他两个字符串直接通过+运算符连接起来合成新字符串作为属性值，**注意这里属性值被单引号引起来**。第 181 行 replyPrivateNumTextView 控件的 text 属性也不是直接绑定的，而是调用 dynamicViewModel 实例的 intToString(Int)方法将 dynamicViewModel.dynamicModel.replyPrivateNum 属性转为 String 类型后与其他字符串合成新字符串作为属性值。

提示：数据绑定时 Int 类型无法在布局文件中直接转换为 String 类型

当前版本在数据绑定时，无法调用 toString()方法将 Int 类型转换为 String 类型，所以需要在视图模型中自定义一个将 Int 类型转换为 String 类型的方法。

6. 回复列表的服务器端接口页面

回复列表的服务器端接口页面为 dynamicReplyList.php，使用 Get 方式接收数据，所需参数如表 14-16 所示。

表 14-16　dynamicReplyList.php 接收的参数

参　数	说　明
developerAccountNumber	开发者账号
developerKey	开发序列号
currentUserAccountID	登录用户的账号
currentUserPassword	登录用户的 MD5 算法加密后密码

续表

参　数	说　明
dynamicReplyDynamicID	被回复的动态 ID
start	开始加载的序号
num	加载数量

获取回复列表数据成功时，返回的JSON数据结构为{"state":"ok","count"=>"回复列表数据的JSON数组长度","dynamicReplyList":[回复列表数据的 JSON 数组]} ，回复列表数据的 JSON 数组的键名如表 14-17 所示。没有获取到回复列表数据时，返回的 JSON 数据结构为{"state":"ok","count" => "0"}。

表 14-17　回复列表数据的 JSON 数组的键名

键　名	说　明
userName	用户姓名
userAccountID	用户账号 ID
userFaceImage	用户头像图片地址
dynamicReplyID	回复 ID
dynamicReplyContent	回复内容
dynamicReplyCreateTime	回复发布时间
dynamicReplyDynamicID	回复的动态 ID
dynamicReplyRepliedDynamicReplyID	回复的回复 ID
dynamicReplyPrivateFlag	回复的私密标记(0 为回复，1 为私回)
repliedUserName	被回复的用户姓名
repliedUserAccountID	被回复的用户账号 ID

回复列表的 JSON 数据样例(服务器端实际返回单行形式的 JSON 数据)


```
{
    "state":"ok",
    "count":"2",
    "dynamicReplyList": [
        {
            "dynamicReplyID":"242",
            "dynamicReplyContent":"这是什么呀 这个造型好像..",
            "userID":"120",
            "userName":"九龙口",
            "userAccountID":"21805018",
            "userFaceImage":"common\/face\/202101\/599221805018.jpg","
            "dynamicReplyCreateTime":"2021-01-26 20:44:56",
            "dynamicReplyDynamicID":"60",
            "dynamicReplyRepliedDynamicReplyID":"242",
            "repliedUserName":null,
            "repliedUserAccountID":null,
            "dynamicReplyPrivateFlag":"0"
        },
        {
            "dynamicReplyID":"252",
            "dynamicReplyContent":"我觉得也是",
```

```
            "userID":"117",
            "userName":"NPC",
            "userAccountID":"42998725",
            "userFaceImage":"common\/face\/202101\/41722197019.jpg",
            "dynamicReplyCreateTime":"2021-01-30 18:29:08",
            "dynamicReplyDynamicID":"60",
            "dynamicReplyRepliedDynamicReplyID":"242",
            "repliedUserName":"九龙口",
            "repliedUserAccountID":"21805018",
            "dynamicReplyPrivateFlag":"0"
        }
    ]
}
```


7. 回复列表的子视图缓存类

回复列表的子视图缓存类包括 DynamicReplyListItemViewHolder 类（回复的子视图缓存类）和 DynamicReplyChildListItemViewHolder 类（回复其他回复的子视图缓存类），如图 14-28 所示。与 DynamicListAdapter 类中使用的内部子视图缓存类相比，降低了耦合度，并使用了视图绑定（视图绑定也可以在内部子视图缓存类中使用）。

图 14-28　回复列表的子视图的运行效果

```
/java/com/pinkjiao/www/view/hodler/DynamicReplyListItemViewHolder.kt
class DynamicReplyListItemViewHolder(
    private val viewBinding: ItemDynamicReplyListBinding, //视图绑定类
    val context: Context, //上下文
    val dynamicModel: DynamicModel //动态模型
) : RecyclerView.ViewHolder(viewBinding.root) {
fun bind(dynamicReplyModel:DynamicReplyModel) {
        viewBinding.itemConstraintLayout.setOnLongClickListener { //长按回复
            startReplyActivity(dynamicReplyModel, dynamicReplyModel.privateFlag)
            true
        }
        if (dynamicModel.userAccountID == dynamicReplyModel.userAccountID) { //是否是动态的作者
            viewBinding.authorTextView.text = "作者"
        } else {
            viewBinding.authorTextView.visibility = View.GONE
```

```
30          }
32          if (dynamicReplyModel.privateFlag == 1) { //私回
33              viewBinding.itemConstraintLayout.setBackgroundColor(context.getColor(R.color.bg_gray))
34              viewBinding.createTimeTextView.setTextColor(context.getColor(R.color.gray_600))
35              viewBinding.replyTextView.text = "私回"
36          }
            //其他控件的数据绑定可参见工程文件的源代码
56      }
        //回复。dynamicReplyModel 参数为回复模型数据，privateFlag 参数为私回标记
62      private fun startReplyActivity(dynamicReplyModel:DynamicReplyModel,privateFlag:Int){
63          val intent = Intent(context, DynamicReplyAddActivity::class.java)
64          intent.putExtra(DynamicReplyAddActivity.EXTRA_PRIVATE_FLAG, privateFlag)
65          intent.putExtra(DynamicReplyAddActivity.EXTRA_DYNAMIC_REPLAY_ID, dynamicReplyModel.id)
66          context.startActivity(intent)
67      }
68  }
```

第 15 ~ 17 行主构造方法中声明属性。第 21 ~ 24 行设置 itemConstraintLayout 的长按监听事件，当长按时启动回复动态的 Activity。第 26 ~ 30 行判断回复的发布者，如果是该动态的作者，则显示 authorTextView 控件作为提示，否则隐藏 authorTextView 控件。第 32 ~ 36 行判断该回复，如果是私回，则设置子视图布局的背景颜色为灰色，发布时间为深灰色，replyTextView 控件的文本修改为“私回”（因为对私回的回复仅被回复的用户和发布动态的用户可见）。第 62 ~ 67 行实现启动 DynamicReplyAddActivity，privateFlag 参数为 1 时表示私回。

```
/java/com/pinkjiao/www/view/hodler/DynamicReplyChildListItemViewHolder.kt
16  class DynamicReplyChildListItemViewHolder(
17      private val viewBinding: ItemDynamicReplyChildListBinding, //视图绑定类
18      val context: Context, //上下文
19      val dynamicModel: DynamicModel //动态模型
20  ) : RecyclerView.ViewHolder(viewBinding.root) {
21      fun bind(dynamicReplyChildModel: DynamicReplyModel) {
23          viewBinding.itemConstraintLayout.setOnLongClickListener { //长按回复
24              startReplyActivity(dynamicReplyChildModel)
25              true
26          }
28          if (dynamicReplyChildModel.repliedUserName != "") {  //是否是针对某个用户的回复
29              viewBinding.repliedUserTextView.text = dynamicReplyChildModel.repliedUserName
30              viewBinding.repliedUserTextView.setOnClickListener {
31                  val intent = Intent(context, UserHomeActivity::class.java)
32                  intent.putExtra(UserHomeActivity.EXTRA_USER_ACCOUNT_ID, dynamicReplyChildModel.
repliedUserAccountID)
33                  intent.putExtra(UserHomeActivity.EXTRA_USER_NAME, dynamicReplyChildModel.userName)
34                  context.startActivity(intent) //启动个人主页
35              }
36          } else {
37              viewBinding.repliedTextView.visibility = View.GONE
38              viewBinding.repliedUserTextView.visibility = View.GONE
39          }
            //其他控件的数据绑定可参见工程文件的源代码
71      }
```

```
    //回复。dynamicReplyChildModel 参数为回复模型数据
76      private fun startReplyActivity(dynamicReplyChildModel:DynamicReplyModel){
77          val intent = Intent(context, DynamicReplyAddActivity::class.java)
78          intent.putExtra(DynamicReplyAddActivity.EXTRA_PRIVATE_FLAG, dynamicReplyChildModel.privateFlag)
79          intent.putExtra(DynamicReplyAddActivity.EXTRA_DYNAMIC_REPLAY_ID, dynamicReplyChildModel.id)
80          context.startActivity(intent)
81      }
82  }
```

第 23～26 行设置 itemConstraintLayout 的长按监听事件，当长按时启动回复动态的 Activity。第 28～39 行判断该回复，如果是针对某个用户的回复，则使用 repliedUserTextView 控件显示被回复的用户，并添加单击事件(单击后启动个人主页)，否则隐藏 repliedTextView 控件和 repliedUserTextView 控件。第 76～81 行实现启动 DynamicReplyAddActivity，DynamicReplyAddActivity 类会根据 DynamicReplyAddActivity.EXTRA_PRIVATE_FLAG 参数传递的 dynamicReplyChildModel.privateFlag 变量值判断是回复还是私回。

8. 页脚子视图的缓存类

FooterViewHolder 类是页脚子视图的缓存类，通过构造方法的 messageOriginType 属性区分类型，以显示不同的提示内容。

```
/java/com/pinkjiao/www/view/hodler/FooterViewHolder.kt
class FooterViewHolder(
    private val viewBinding: FooterDynamicListBinding, //视图绑定类
    val context: Context, //上下文
    private val messageOriginType: Int //信息来源类型
) : RecyclerView.ViewHolder(viewBinding.root) {
    companion object {
        const val STATUS_REFRESHING = "上拉刷新状态"
        const val STATUS_LOADING_MORE = "正在加载状态"
        const val STATUS_FINISH = "加载完成状态"
        const val STATUS_NET_ERROR = "网络错误状态"
        const val STATUS_NON_DATA = "无数据状态"
        const val STATUS_ALL_DATA = "已经加载全部数据状态"
        const val STATUS_MAX_COUNT = "达到加载上限状态"
        const val TYPE_DEFAULT = 100 //通用
        const val TYPE_REPLY = 101 //回复
        const val TYPE_REMIND = 102 //提醒
        const val TYPE_FOLLOW = 103 //关注
        const val TYPE_FOLLOWER = 104 //粉丝
        const val TYPE_MESSAGE_RECEIVE = 0 //接收的私信
        const val TYPE_MESSAGE_SEND = 1 //发送的私信
    }
    //绑定。currentState 参数为当前状态
    fun bind(currentState: String) {
        when (currentState) { //根据当前状态显示页脚子视图的内容
            STATUS_FINISH -> viewBinding.linearLayout.visibility = View.GONE
            STATUS_REFRESHING -> viewBinding.linearLayout.visibility = View.GONE
            STATUS_LOADING_MORE -> {
```

```
39                  viewBinding.footerTextView.text. =
    context.resources.getString(R.string.recycler_view_footer_text_loading)
40                  viewBinding.progressBar.visibility = View.VISIBLE
41                  viewBinding.linearLayout.visibility = View.VISIBLE
42              }
                //其他状态可参见工程文件的源代码
73          }
74      }
75  }
```

第 10～12 行主构造方法中声明属性。第 15～21 行声明页脚子视图的状态常量。第 22～28 行声明页脚子视图的类型常量，用于不同的列表适配器。

9. 回复列表的适配器类

DynamicReplyListAdapter 类用于显示回复列表数据的 RecyclerView 控件的数据适配，需要与子视图缓存类搭配使用。

```
/java/com/pinkjiao/www/adapter/DynamicReplyListAdapter.kt
class DynamicReplyListAdapter(
    private val dynamicModel: DynamicModel, //动态数据模型
    private val dynamicReplyModelList: List<DynamicReplyModel> //回复数据模型的列表
) : BaseListAdapter() {
    private lateinit var context: Context //上下文
    var currentState = "" //当前状态
    override fun onCreateViewHolder(parent: ViewGroup, viewType: Int): RecyclerView.ViewHolder {
        context = parent.context
        val holder: RecyclerView.ViewHolder
        when (viewType) { //返回缓存子视图的 ViewHolder
            TYPE_REPLY_ITEM -> {
                val itemBinding = ItemDynamicReplyListBinding.inflate(LayoutInflater.from(parent.context),
parent, false)
                holder = DynamicReplyListItemViewHolder(itemBinding, context, dynamicModel)
            }
            TYPE_REPLY_CHILD_ITEM -> {
                val itemBinding = ItemDynamicReplyChildListBinding.inflate(LayoutInflater.from
(parent.context), parent, false)
                holder = DynamicReplyChildListItemViewHolder(itemBinding, context, dynamicModel)
            }
            else -> {
                val footerBinding = FooterDynamicListBinding.inflate(LayoutInflater.from
(parent.context), parent, false)
                (footerBinding.linearLayout.layoutParams as StaggeredGridLayoutManager.
LayoutParams).isFullSpan = true
                holder = FooterViewHolder(footerBinding, context, FooterViewHolder.TYPE_REPLY)
            }
        }
        return holder
    }
    override fun onBindViewHolder(holder: RecyclerView.ViewHolder, position: Int) {
        when (holder) { //为缓存子视图的 ViewHolder 匹配数据
```

```
47             is DynamicReplyListItemViewHolder -> { //回复项视图
48                 holder.bind(dynamicReplyModelList[position])
49             }
50             is DynamicReplyChildListItemViewHolder -> { //回复项子视图
51                 holder.bind(dynamicReplyModelList[position])
52             }
53             is FooterViewHolder -> { //页脚子视图
54                 holder.bind(currentState)
55             }
56         }
57     }
61     override fun getItemCount(): Int { //获取子视图的数量
62         return dynamicReplyModelList.size + 1//增加一个子视图用于显示加载提示子视图
63     }
68     override fun getItemViewType(position: Int): Int { //获取子视图的类型。position 参数为视图的序号
69         return when {
70             position + 1 == itemCount -> {
71                 FooterViewHolder.TYPE_DEFAULT
72             }
73             dynamicReplyModelList[position].id == dynamicReplyModelList[position].
   repliedDynamicReplyID -> {
74                 TYPE_REPLY_ITEM
75             }
76             else -> {
77                 TYPE_REPLY_CHILD_ITEM
78             }
79         }
80     }
85     fun changeStatus(state: String) { //改变状态。State 参数为状态
86         currentState = state
87         if (state != FooterViewHolder.STATUS_LOADING_MORE) {
88             notifyDataSetChanged() //通知数据集改变，刷新 RecyclerView
89         }
90     }
91 }
```

第 23 ~ 43 行根据子视图类型创建对应的缓存子视图实例，当子视图显示时对其调用。第 44 ~ 5 行根据不同类型的子视图缓存类进行数据绑定，三种类型的子视图缓存类都定义了 bind 方法用于数 绑定。第 60 ~ 63 行获取子视图数量。第 68 ~ 80 行获取子视图的类型，当 DynamicReplyModel 实例 id 属性和 repliedDynamicReplyID 属性相等时是普通回复，否则就是子回复(即回复其他用户回复的 复)。第 85 ~ 90 行改变当前状态，除 FooterViewHolder.STATUS_LOADING_MORE 类型外， RecyclerView 控件的数据集进行更新，刷新子视图显示的内容。

10. 显示动态的 Fragment 类

DynamicFragment 类用于显示动态及动态的回复，还可以关注或取消关注发布动态的用户。如 是用户自己发布的动态，还可以将动态删除。

```
/java/com/pinkjiao/www/fragment/DynamicFragment.kt
30  class DynamicFragment(private var backFlag: Boolean): RefreshFragment() {
```

```kotlin
    companion object {
        private const val ARG_PARAM_DYNAMIC_ID = "dynamicID"
        private const val ARG_PARAM_REFRESH_ENABLE = "refreshEnable"
        private lateinit var dynamicFragment:DynamicFragment
        //新建 DynamicFragment 实例。backFlag 参数为返回标记，dynamicID 参数为动态 ID，refreshEnable 参数为是否可以刷新
        fun newInstance(backFlag: Boolean, dynamicID: String = "", refreshEnable: Boolean = true): DynamicFragment {
            dynamicFragment = DynamicFragment(backFlag).apply {
                arguments = Bundle().apply {
                    putString(ARG_PARAM_DYNAMIC_ID, dynamicID)
                    putBoolean(ARG_PARAM_REFRESH_ENABLE, refreshEnable) }
            }
            return dynamicFragment
        }
    }
    private lateinit var viewBinding: FragmentDynamicBinding //视图绑定
    private var dynamicID: String = "" //动态 ID
    private var refreshEnable: Boolean = true //是否可以刷新
    private var maxNum: Int = 5000 //回复的最大加载数据量
    private var pageNum: Int = 50 //回复的每次加载的数据量
    private lateinit var adapter: DynamicReplyListAdapter //动态回复适配器
    private var dynamicReplyModelList = ArrayList<DynamicReplyModel>() //动态回复模型数组
    private val userModel = UserModel() //用户模型
    private var startIndex = 0 //开始加载的序号
    //重写 RefreshFragment 抽象类的方法
    override fun refresh(){
        viewBinding.swipeRefreshLayout.isRefreshing = true
        dynamicReplyModelList.clear()
        startIndex = 0
        viewBinding.dynamicViewModel!!.getDynamic(RequestUrl.getDomainUrl(requireContext()), dynamicID)
    }
    override fun onCreate(savedInstanceState: Bundle?) {
        super.onCreate(savedInstanceState)
        arguments?.let {
            dynamicID = it.getString(ARG_PARAM_DYNAMIC_ID).toString()
            refreshEnable = it.getBoolean(ARG_PARAM_REFRESH_ENABLE)
        }
    }
    override fun onCreateView(inflater: LayoutInflater, container: ViewGroup?, savedInstanceState: Bundle?): View {
        viewBinding = FragmentDynamicBinding.inflate(inflater, container, false) //实例化视图绑定
        initViewModel() //初始化视图模型
        initObserver() //初始化观察者
        initView() //初始化控件
        refresh() //加载数据
        return viewBinding.root
    }
    private fun initView(){ //初始化下拉刷新布局
```

```
97         }
101        private fun initViewModel() { //初始化视图模型
110        }
114        private fun setDynamicData(){ //设置动态数据
215        }
219        private fun initReplyView(){ //初始化回复视图
252        }
256        private fun initObserver(){ //初始化观察者
330        }
331    }
```

第32、33行声明两个常量作为Bundle对象方法参数的键名。第41～48行实例化DynamicFragment类，此处并没有使用构造方法传递数据，而是使用Bundle对象传递。第79～83行调用初始化视图模型、观察者和下拉刷新布局。在获取到动态数据后，会调用setDynamicData()方法对视图控件进行初始化，setDynamicData()方法则调用initReplyView()方法设置显示回复的控件并获取回复数据。

提示：构造方法和Bundle对象传递参数的区别

使用构造方法传递参数可能会导致异常。当Fragment重建时，会再次调用Fragment的构造方法。因此使用Fragment的构造方法传递参数，如果遇到横竖屏切换等需要重建Fragment的情况，那么将无法获取到构造方法传递的参数。因此，官方推荐使用Fragment.setArguments(Bundle)方法传递参数，而不推荐直接使用构造方法传递参数。

/java/com/pinkjiao/www/fragment/DynamicFragment.kt

```
89   private fun initView(){
91       viewBinding.swipeRefreshLayout.isEnabled = refreshEnable //设置是否启用下拉刷新
92       viewBinding.swipeRefreshLayout.isRefreshing = refreshEnable //显示刷新状态
93       viewBinding.swipeRefreshLayout.setOnRefreshListener { //刷新事件
94           refresh()
95       }
96       viewBinding.linearLayout.visibility = View.INVISIBLE
97   }
```

第91行设置swipeRefreshLayout控件是否可以下拉刷新。第92行设置swipeRefreshLayout控件是否显示刷新的旋转等待子控件。第93～95行设置swipeRefreshLayout控件的刷新监听器，刷新时调用refresh()方法重新加载动态数据。第96行设置包含显示动态内容控件的linearLayout布局不可见，因为此时还未获取到动态数据，待从服务器端获取到动态数据后再将其设置为可见。

/java/com/pinkjiao/www/fragment/DynamicFragment.kt

```
114  private fun setDynamicData(){
136      //判断是否是自己发布的动态。如果是，则能够进行删除操作
137      if (viewBinding.dynamicViewModel!!.dynamicModel.get()!!.userAccountID == userModel.accountID) {
139          viewBinding.deleteImageView.visibility = View.VISIBLE //显示删除按钮
140          viewBinding.followButton.visibility = View.GONE //隐藏关注按钮
142          viewBinding.deleteImageView.setOnClickListener { //删除单击事件
143              val builder = AlertDialog.Builder(context)
144              builder.setMessage(getString(R.string.dialog_delete))
145                  .setCancelable(false)
146                  .setPositiveButton(getString(R.string.dialog_delete_positive)) { _, _ ->
```

```
    viewBinding.dynamicViewModel!!.deleteDynamic() }
147                 .setNegativeButton(getString(R.string.dialog_delete_negative)) { dialog, _ -> dialog.cancel() }
148                 .show()
149         }
150     } else {
163         viewBinding.followButton.setOnClickListener {
164             val builder = AlertDialog.Builder(context)
165             if (viewBinding.dynamicViewModel!!.followFlag == "0") {
166                 builder.setMessage(getString(R.string.dialog_follow))
167                     .setCancelable(false)
168                     .setPositiveButton(getString(R.string.dialog_follow_positive)) { _, _ -> viewBinding.dynamicViewModel!!.addFollow(
169                         viewBinding.dynamicViewModel!!.dynamicModel.get()!!.userAccountID) }
170                     .setNegativeButton(getString(R.string.dialog_follow_negative)) { dialog, _ -> dialog.cancel() }
171                     .show()
172             } else {
173                 builder.setMessage(getString(R.string.dialog_follow_cancel))
174                     .setCancelable(false)
175                     .setPositiveButton(getString(R.string.dialog_follow_positive)) { _, _ -> viewBinding.dynamicViewModel!!.removeFollow(
176                         viewBinding.dynamicViewModel!!.dynamicModel.get()!!.userAccountID) }
177                     .setNegativeButton(getString(R.string.dialog_follow_negative)) { dialog, _ -> dialog.cancel() }
178                     .show()
179             }
180         }
181     }
210     //滚动监听：到达顶部允许下拉刷新
211     viewBinding.scrollView.viewTreeObserver.addOnScrollChangedListener {
212         viewBinding.swipeRefreshLayout.isEnabled = viewBinding.scrollView.scrollY == 0
213     }
214     initReplyView()
215 }
```

第 137～149 行如果是登录用户发布的动态，则显示删除按钮并隐藏关注按钮。然后添加删除单击事件，单击删除按钮时显示确认对话框。如果单击确认对话框的确定按钮，则执行viewBinding.dynamicViewModel!!.deleteDynamic()方法删除动态。第 163～180 行设置关注按钮的单击事件，未关注时单击后显示确认关注的对话框，关注时单击后显示取消关注的对话框。第 211～213 行设置scrollView 控件的滚动监听器，当滚动到该控件的顶部时允许进行刷新，否则不能刷新。这是因为 scrollView 控件包含在 swipeRefreshLayout 控件中，默认情况下，swipeRefreshLayout 控件下拉滑动时无论 scrollView 控件是否滑动到顶部都会刷新，因此需要判断 scrollView 控件滑动到顶部后才能开始刷新。

```
/java/com/pinkjiao/www/fragment/DynamicFragment.kt
private fun initReplyView(){
    //实例化适配器
    adapter = DynamicReplyListAdapter(viewBinding.dynamicViewModel!!.dynamicModel.get()!!, dynamicReplyModelList)
```

```
222         //设置瀑布流布局为纵向一列
223         val layoutManager = StaggeredGridLayoutManager(1, StaggeredGridLayoutManager.VERTICAL)
224         //设置回收视图
225         viewBinding.replyRecyclerView.layoutManager = layoutManager
226         viewBinding.replyRecyclerView.adapter = adapter
227         viewBinding.replyRecyclerView.addOnScrollListener(object :
228             OnScrollListener() { //添加滚动监听器
230             override fun onScrolled(recyclerView: RecyclerView, dx: Int, dy: Int) { //重写滚动完成事件
231                 super.onScrolled(recyclerView, dx, dy)
232                 //存储最后一行子视图位置的数组
233                 val positionArray = IntArray((recyclerView.layoutManager as StaggeredGridLayoutManager).spanCount)
234                 //获取最后一行显示的子视图位置的数组
235                 layoutManager.findLastVisibleItemPositions(positionArray)
236                 //判断显示的最后一个子视图是否是最后一个视图
237                 if (findMax(positionArray) + 1 == adapter.itemCount) {
238                     //避免重复加载
239                     if (adapter.currentState == FooterViewHolder.STATUS_LOADING_MORE) return
240                     //无数据不加载
241                     if (adapter.currentState == FooterViewHolder.STATUS_ALL_DATA) return
242                     //设置达到加载上限状态
243                     if (adapter.currentState == FooterViewHolder.STATUS_MAX_COUNT) return
244                     adapter.changeStatus(FooterViewHolder.STATUS_LOADING_MORE)//设置加载状态
245                     //获取回复
246                     val domain = RequestUrl.getDomainUrl(context!!)
247                     val dynamicReplyDynamicID = viewBinding.dynamicViewModel!!.dynamicModel.get()!!.id
248                     viewBinding.dynamicViewModel!!.getDynamicReplyList(domain, startIndex, pageNum, dynamicReplyDynamicID)
249                 }
250             }
251         })
252     }
```

第 221 行实例化回复列表的适配器。第 227 ~ 251 行设置显示回复列表的 replyRecyclerView 控件的滑动监听事件，与 DynamicListFragment 类中的 recyclerView 控件的滑动监听事件类似。由于添加了页脚子视图，在初始化后没有回复列表数据，所以页脚子视图就会在屏幕上，且调用第 248 行的 viewBinding.dynamicViewModel!!.getDynamicReplyList()方法加载回复列表数据。

```
/java/com/pinkjiao/www/fragment/DynamicFragment.kt
private fun initObserver(){
    //观察操作请求状态进行 UI 处理
    viewBinding.dynamicViewModel!!.response.observe(requireActivity(), {
        when (it) {
            DynamicViewModel.STATE_GET_DYNAMIC_SUCCESS -> { //获取动态数据成功
                viewBinding.swipeRefreshLayout.isRefreshing = false //取消刷新状态
                viewBinding.linearLayout.visibility = View.VISIBLE //显示动态内容布局
                setDynamicData()
            }
            DynamicViewModel.STATE_GET_DYNAMIC_REPLY_SUCCESS -> { //获取回复成功
```

```
293                 //遍历添加回复数据
294                 for (i in viewBinding.dynamicViewModel!!.dynamicReplyModelList) {
295                     this@DynamicFragment.dynamicReplyModelList.add(i)
296                 }
297                 when {
298                     viewBinding.dynamicViewModel!!.replyListCount < pageNum -> {//已经加载所有回复
299                         adapter.changeStatus(FooterViewHolder.STATUS_ALL_DATA)
300                     }
301                     startIndex + viewBinding.dynamicViewModel!!.replyListCount + pageNum > maxNum -> {
302                         adapter.changeStatus(FooterViewHolder.STATUS_MAX_COUNT)//达到加载回复数量的上限
303                     }
304                     else -> {
305                         adapter.changeStatus(FooterViewHolder.STATUS_FINISH)
306                     }
307                 }
308                 startIndex += viewBinding.dynamicViewModel!!.replyListCount //更新开始加载的序号
309             }
327         }
328     })
329 }
```

第 258 行对 MutableLiveData<String>类型的 response 变量进行观察，当请求状态改变时，会对 response 变量进行赋值，response 变量值发生变化时执行第 259 ~ 327 行的 lambda 表达式参数，该 lambda 表达式参数替代的是匿名函数重写的 Observer 接口的 onChanged()方法。第 259 ~ 263 行判断成功获取动态时，取消刷新状态、显示动态内容布局、设置动态数据。第 292 ~ 308 行获取回复列表数据成功时，使用遍历的方式将返回的回复列表数据添加到 dynamicReplyModelList 变量，显示在 replyRecyclerView 控件中。第 296 ~ 306 行根据服务器端返回的数据设置适配器的状态。

11. 偶遇的 Fragment 类

MeetFragment 类用于显示随机一条动态的 DynamicFragment 类实例，继承 RefreshFragment 类，实现 StatusBar 接口。

```
/java/com/pinkjiao/www/fragment/MeetFragment.kt
46 //添加显示动态的 Fragment
47 dynamicFragment = DynamicFragment.newInstance(false)
48 val transaction = childFragmentManager.beginTransaction()
49 transaction.add(R.id.dynamicFragment, dynamicFragment).commit()
```

第 47 行实例化 DynamicFragment 类赋给 dynamicFragment 变量，DynamicFragment.newInstance (backFlag: Boolean, dynamicID: String = "", refreshEnable: Boolean = true)方法的 dynamicID 参数和 refreshEnable 参数不设置，使其使用默认值。第 49 行将 dynamicFragment 变量存储的 DynamicFragment 实例显示在 dynamicFragment 控件中。

14.11.2 提醒模块

提醒模块显示提醒的消息列表，提醒的消息包括粉丝关注、粉丝取消关注、动态被回复和收到私信等，如图 14-29 所示。提醒模块涉及的类较多，为了帮助大家理解，核心类之间的调用关系如图 14-30 所示。由于提醒和私信公用消息列表适配器，所以在本模块中介绍私信消息列表的子视图缓存类。

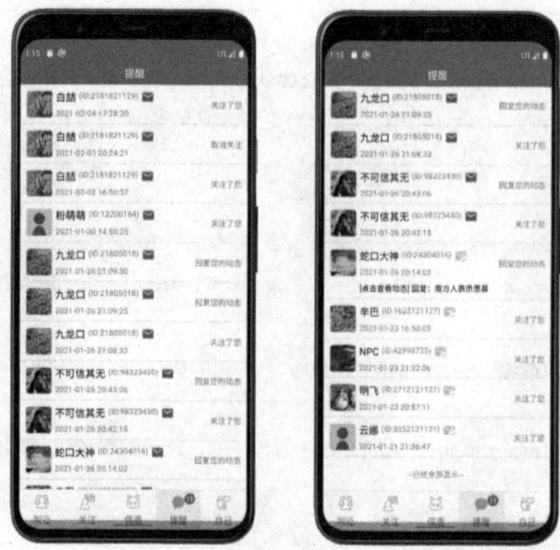

图 14-29　提醒模块的运行效果

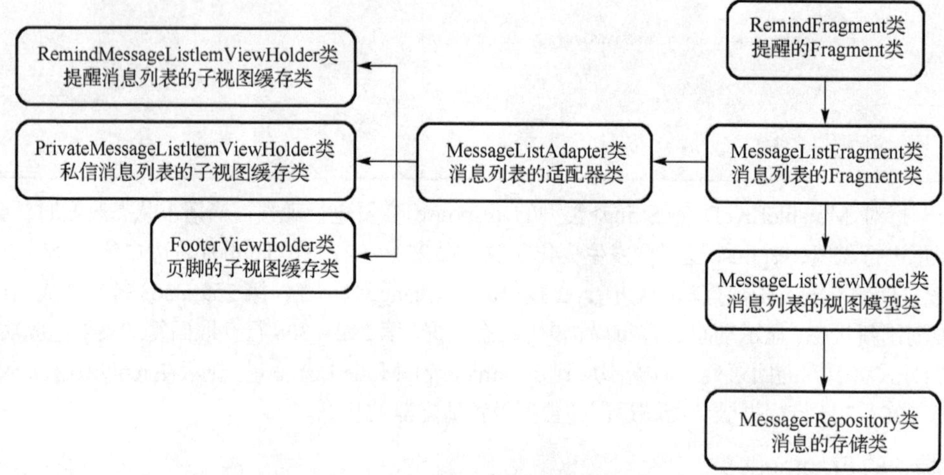

图 14-30　提醒模块核心类之间的调用关系

1. 消息列表的服务器端接口

消息列表的服务器端接口页面为 messageList.php，使用 Get 方式接收数据，所需参数如表 14-18 所示。

表 14-18　messageList.php 接收的参数

参　数	说　明
developerAccountNumber	开发者账号
developerKey	开发序列号
currentUserAccountID	登录用户的账号
currentUserPassword	登录用户的 MD5 算法加密后密码
messageOriginType	消息来源类型
messageType	消息类型
start	开始加载的序号
num	加载数量

获取消息列表数据成功时，返回的 JSON 数据结构为{"state":"ok","messageList":[消息列表数据

JSON 数组]}，如表 14-19 所示。获取消息列表数据失败时，返回的 JSON 数据结构为{"state":"错误码","info":"错误提示信息"}，如表 14-20 所示。

表 14-19 消息列表数据的 JSON 数组

键 名	说 明
messageType	消息类型
messageTypeParam	消息类型附带的参数
messageID	消息 ID
senderUserAccountID	发送者的用户账号 ID
senderUserName	发送者的用户姓名
senderUserFaceImage	发送者的用户头像地址
messageContent	消息的内容
messageCreateTime	消息的发送时间
receiverUserAccountID	接收者的用户账号 ID
receiverUserName	接收者的用户姓名
receiverUserFaceImage	接收者的用户头像地址
messageReadFlag	消息读取标记

表 14-20 返回的错误码

错 误 码	错误提示信息
error1205	暂无私信
error1301	暂无提醒

2. 设置消息已读取标记的服务器端接口页面

设置消息已读取标记的服务器端接口页面为 messageRead.php，使用 Get 方式接收数据，所需参数如表 14-21 所示。

表 14-21 messageRead.php 接收的参数

参 数	说 明
developerAccountNumber	开发者账号
developerKey	开发序列号
currentUserAccountID	登录用户的账号
currentUserPassword	登录用户的 MD5 算法加密后密码
messageID	消息 ID

设置消息已读取标记成功时，返回的 JSON 数据结构为{"state":"ok"}。设置消息已读取标记失败时，返回的 JSON 数据结构为{"state":"error1206","info":"消息不存在"}。

3. 消息的存储类

MessageRepository 类用于添加和查询动态的回复数据。下面先介绍该类主体结构，再介绍存储数据的方法。

```
/java/com/pinkjiao/www/repository/MessageRepository.kt
class MessageRepository {
    fun sendMessage(baseUrl: String, developerModel: DeveloperModel, loginUserModel: UserModel, messageModel: MessageModel, responseListener: ResponseListener){ //发送私信
```

```
53      }
65      fun getMessageList(domain:String,baseUrl:String,developerModel: DeveloperModel,
            loginUserModel: UserModel,messageOriginType:Int,messageType:Int,
            start:Int,num:Int,responseListener: ResponseListener){ //获取消息列表
106     }
115     fun readMessage(baseUrl: String, developerModel: DeveloperModel, loginUserModel: UserModel,
    messageID: String, responseListener: ResponseListener){ //读取消息
133     }
134     interface ResponseListener {
141         fun onCallBack(res: String, messageModelList:MutableList<MessageModel>? = null,
    unReadNum:Int = 0) //回调
142     }
143 }
```

第 134 ~ 142 行声明响应请求的监听接口，并且声明 onCallBack()方法接收响应数据。res 参数是 String 类型，接收响应的结果类型；messageModelList 参数是可空的 ArrayList<MessageModel>类型，接收消息列表数据；unReadNum 参数是 Int 类型，接收未读信息数量。

/java/com/pinkjiao/www/repository/MessageRepository.kt

```
//发送私信。baseUrl 参数为服务器端页面所在的文件夹 Url 路径，developerModel 参数为开发者，loginUserModel
参数为登录用户，messageModel 参数为私信，responseListener 参数为监听器接口
fun sendMessage(baseUrl: String, developerModel: DeveloperModel, loginUserModel: UserModel,
messageModel: MessageModel, responseListener: ResponseListener){
    val page = "messageSend.php"
    //post 参数
    val parameter: MutableMap<String, String>= HashMap()
    parameter["messageContent"] = messageModel.messageContent
    parameter["messageType"] = messageModel.messageType
    parameter["senderUserAccountID"] = messageModel.senderUserAccountID
    parameter["senderUserName"] = messageModel.senderUserName
    parameter["receiverUserAccountID"] = messageModel.receiverUserAccountID
    parameter["receiverUserName"] = messageModel.receiverUserName
    //post 请求对象
    val postRequest = PostRequest(baseUrl, page, developerModel, loginUserModel, parameter, null)
    //post 方式上传
    Http.post(postRequest.url, postRequest.parameter, postRequest.files, object : Http.
PostResponseListener {
        override fun onCreate() {
            responseListener.onCallBack(Http.CREATE)
        }
        override fun onResponse(res: String) {
            val jsonParser = JSONTokener(JSON.cleanBOM(res))
            val jsonObject = jsonParser.nextValue() as JSONObject
            if (jsonObject.getString("state") == "ok") {
                responseListener.onCallBack(Http.RESPONSE_SUCCESS)
            } else {
                responseListener.onCallBack(jsonObject.getString("info"))
            }
        }
        override fun onFailure(res: String) {
```

```
50              responseListener.onCallBack(res)
51          }
52      })
53  }
```

第 24 行设置服务器端接口页面。第 27 ~ 32 行将发送的信息添加到 parameter 对象。第 34 行实例化 Post 请求对象。第 36 行发送发送消息的请求。第 37 ~ 39 行重写 onCreate()方法，调用 onCallBack()方法回调请求创建的状态。第 40 ~ 48 行重写 onResponse()方法，解析请求返回的 JSON 数据。如果发送消息成功，通过 onCallBack()方法回调响应成功信息，否则通过 onCallBack()方法回调错误信息。第 49 ~ 52 行重写 onFailure()方法，调用 onCallBack()方法回调网络错误信息。

/java/com/pinkjiao/www/repository/MessageRepository.kt

```
//获取消息列表。baseUrl 参数为服务器端页面所在的文件夹 Url 路径，developerModel 参数为开发者，loginUserModel 参数为登录用户，messageOriginType 参数为信息来源类型，messageType 参数为信息类型，start 参数为开始加载的序号，num 参数为数量，responseListener 参数为监听器接口
fun getMessageList(domain:String,baseUrl:String,developerModel: DeveloperModel, loginUserModel: UserModel,messageOriginType:Int,messageType:Int,start:Int,num:Int,responseListener: ResponseListener){
    val page = "messageList.php"
    val parameter = "&messageOriginType=$messageOriginType&messageType=$messageType&start=$start&num=$num"
    val getRequest = GetRequest(baseUrl, page, developerModel, loginUserModel, parameter)
    Http.get(getRequest.url, object : Http.ResponseListener {
        override fun onResponse(res: String) {
            val jsonParser = JSONTokener(JSON.cleanBOM(res))
            val jsonObject = jsonParser.nextValue() as JSONObject
            if (jsonObject.getString("state") == "ok") {
                val messageJSON = JSONArray(jsonObject.getString("messageList"))
                val messageModelList = ArrayList<MessageModel>()
                var unReadNum = 0
                for (i in 0 until messageJSON.length()) { //遍历数据
                    val jo = messageJSON.opt(i) as JSONObject
                    val messageModel = MessageModel()
                    messageModel.messageType = jo.getString("messageType")
                    messageModel.messageTypeParam = jo.getString("messageTypeParam")
                    messageModel.messageID = jo.getString("messageID")
                    messageModel.senderUserAccountID = jo.getString("senderUserAccountID")
                    messageModel.senderUserName = jo.getString("senderUserName")
                    messageModel.senderUserFaceImage = Image.getUrl(domain, jo.getString("senderUserFaceImage"), Image.smallQuality)
                    messageModel.messageContent = jo.getString("messageContent")
                    messageModel.messageCreateTime = jo.getString("messageCreateTime")
                    messageModel.receiverUserAccountID = jo.getString("receiverUserAccountID")
                    messageModel.receiverUserName = jo.getString("receiverUserName")
                    messageModel.receiverUserFaceImage = Image.getUrl(domain, jo.getString("receiverUserFaceImage"), Image.smallQuality)
                    messageModel.messageReadFlag = jo.getString("messageReadFlag")
                    if (messageModel.messageReadFlag == "0") { unReadNum ++ }
                    messageModelList.add(messageModel)
                }
```

```
97                responseListener.onCallBack(Http.RESPONSE_SUCCESS,messageModelList,unReadNum)
98            } else {
99                responseListener.onCallBack(jsonObject.getString("info"),null)
100           }
101       }
102       override fun onFailure(res: String) {
103           responseListener.onCallBack(res,null)
104       }
105   })
106 }
```

第 66 行设置服务器端接口页面。第 67 行合成 Get 请求的参数。第 68 行实例化 Get 请求对象。第 69 行发送获取消息列表的请求。第 70 ~ 101 行重写 onResponse()方法，解析请求返回的 JSON 数据。如果获取消息列表数据成功，通过 onCallBack()方法回调响应成功信息、消息列表数据和未读信息数量，否则通过 onCallBack()方法回调错误信息。第 102 ~ 104 行重写 onFailure()方法，调用 onCallBack()方法回调网络错误信息。

```
/java/com/pinkjiao/www/repository/MessageRepository.kt
    //设置读取消息标记。baseUrl 参数为服务器端页面所在的文件夹 Url 路径，developerModel 参数为开发者，
    loginUserModel 参数为登录用户，messageID 参数为消息 ID，responseListener 参数为监听器接口
115 fun readMessage(baseUrl: String, developerModel: DeveloperModel, loginUserModel: UserModel,
    messageID: String, responseListener: ResponseListener){
116     val page = "messageRead.php"
117     val parameter = "&messageID=$messageID"
118     val getRequest = GetRequest(baseUrl, page, developerModel, loginUserModel, parameter)
119     Http.get(getRequest.url, object : Http.ResponseListener {
120         override fun onResponse(res: String) {
121             val jsonParser = JSONTokener(JSON.cleanBOM(res))
122             val jsonObject = jsonParser.nextValue() as JSONObject
123             if (jsonObject.getString("state") == "ok") {
124                 responseListener.onCallBack(Http.RESPONSE_SUCCESS)
125             } else {
126                 responseListener.onCallBack(jsonObject.getString("info"))
127             }
128         }
129         override fun onFailure(res: String) {
130             responseListener.onCallBack(res)
131         }
132     })
133 }
```

第 116 行设置服务器端接口页面。第 117 行合成 Get 请求的参数。第 118 行实例化 Get 请求对象。第 119 行发送设置读取消息标记的请求。第 120 ~ 128 行重写 onResponse()方法，解析请求返回的 JSON 数据。如果设置读取消息标记成功，通过 onCallBack()方法回调响应成功信息，否则通过 onCallBack()方法回调错误信息。第 129 ~ 131 行重写 onFailure()方法，调用 onCallBack 方法回调网络错误信息。

4. 提醒消息列表的子视图缓存类

RemindMessageListItemViewHolder 类用于缓存消息列表显示提醒消息时的子视图。

```
/java/com/pinkjiao/www/view/hodler/RemindMessageListItemViewHolder.kt
class RemindMessageListItemViewHolder (
    private val viewBinding: ItemRemindListBinding, //视图绑定类
    val context: Context, //上下文
    private val viewModel: MessageListViewModel //消息列表视图模型
) : RecyclerView.ViewHolder(viewBinding.root) {
    fun bind(messageModel: MessageModel) {
        //省略部分控件的绑定，详细代码可参见工程文件
        //还原读取状态图标
        If (messageModel.messageReadFlag == "1") {
            viewBinding.readImageView.setImageDrawable(
                context.getDrawable(R.drawable.ic_baseline_mark_email_read_24))
        } else if (messageModel.messageReadFlag == "0") {
            viewBinding.readImageView.setImageDrawable(
                context.getDrawable(R.drawable.ic_baseline_mail_24))
        }
        //还原内容为隐藏状态
        viewBinding.contentTextView.visibility = View.GONE
        viewBinding.constraintLayout.setOnClickListener {
            if (viewBinding.contentTextView.visibility == View.GONE) {
                viewBinding.contentTextView.visibility = View.VISIBLE
                //判断提醒信息是否已读
                if (messageModel.messageReadFlag == "0") { //向服务器端发送读取消息的请求
                    viewModel.readMessage(messageModel.messageID, object : MessageListViewModel.ResponseListener {
                        override fun onSuccess() {
                            viewBinding.readImageView.setImageDrawable(context.getDrawable(R.drawable.ic_baseline_mark_email_read_24))
                            messageModel.messageReadFlag = "1"
                        }
                    })
                }
            } else { viewBinding.contentTextView.visibility = View.GONE }
        }
        //省略部分控件的绑定，详细代码可参见工程文件
        //提醒类型
        when (messageModel.messageType) {
            "1" -> {
                viewBinding.typeTextView.text = "系统信息"
                viewBinding.typeTextView.setTextColor(
                    context.getColor(R.color.design_default_color_error))
                viewBinding.contentTextView.text = messageModel.messageContent
            }
            "2" -> {
                viewBinding.typeTextView.text = "私信"
                viewBinding.contentTextView.text = "[单击回信] : " + messageModel.messageContent
                viewBinding.contentTextView.setOnClickListener {
                    val intent = Intent(context, MessageSendActivity::class.java)
                    intent.putExtra(MessageSendActivity.EXTRA_RECEIVE_USER_ACCOUNT_ID, messageModel.receiverUserAccountID)
```

```
80                 intent.putExtra(MessageSendActivity.EXTRA_RECEIVE_USER_NAME, messageModel.
senderUserName)
81                 context.startActivity(intent)
82             }
83         }
           //省略提醒类型，详细代码可参见工程文件
104     }
105   }
106 }
```

第 18 ~ 20 行主构造方法中声明属性。第 40 ~ 44 行判断信息的阅读标记，显示对应的消息图标。第 47 ~ 60 行设置子视图的根布局单击事件，单击后展开显示消息内容并向服务器端发送已经读取信息的请求，再次单击后隐藏消息内容。第 68 ~ 104 行根据消息类型设置右侧 typeTextView 控件显示对应的文本内容，以及 contentTextView 控件的单击事件。

5. 私信消息列表的子视图缓存类

PrivateMessageListItemViewHolder 类用于缓存消息列表显示私信消息时的子视图。该缓存类会在消息列表的 Fragment 显示私信时使用，由于消息列表的适配器中会使用到该缓存类，所以先行介绍。

```
/java/com/pinkjiao/www/view/hodler/PrivateMessageListItemViewHolder.kt
class PrivateMessageListItemViewHolder (
    private val viewBinding: ItemMessageListBinding, //视图绑定类
    val context:Context, //上下文
    private var messageOriginType:Int, //消息来源类型
    private val viewModel: MessageListViewModel //消息列表视图模型
) : RecyclerView.ViewHolder(viewBinding.root) {
    fun bind(messageModel: MessageModel) {
        if(messageOriginType == FooterViewHolder.TYPE_MESSAGE_SEND) {
            viewBinding.accountIDTextView.text = messageModel.receiverUserAccountID
            viewBinding.nameTextView.text = messageModel.receiverUserName
            //省略部分控件的绑定，详细代码参见工程文件
            viewBinding.constraintLayout.setOnClickListener {
                if (viewBinding.contentTextView.visibility == View.GONE){
                    viewBinding.contentTextView.visibility = View.VISIBLE
                } else { viewBinding.contentTextView.visibility = View.GONE }
            }
        } else if (messageOriginType == FooterViewHolder.TYPE_MESSAGE_RECEIVE) {
            viewBinding.accountIDTextView.text = messageModel.senderUserAccountID
            viewBinding.nameTextView.text = messageModel.senderUserName
            //省略部分控件的绑定，详细代码参见工程文件
            viewBinding.constraintLayout.setOnClickListener {
                if (viewBinding.contentTextView.visibility == View.GONE){
                    viewBinding.contentTextView.visibility = View.VISIBLE
                    viewModel.readMessage(messageModel.messageID,object : MessageListViewModel.
ResponseListener{ //向服务器端发送读取消息请求的监听器
                        override fun onSuccess() {
                            viewBinding.readImageView.visibility = View.VISIBLE
                        }
                    })
                } else { viewBinding.contentTextView.visibility = View.GONE  }
```

```
81                 }
82             }
83             viewBinding.messageCreateTimeTextView.text = "发送时间: " + messageModel.messageCreateTime
84             viewBinding.contentTextView.text = messageModel.messageContent
85         }
86 }
```

第 16 ~ 19 行主构造方法中声明属性。第 23 ~ 47 行当消息来源类型为发送类型时，即私信的发件箱消息列表调用该缓存类时，设置子视图中控件显示的文本及单击事件。第 49 ~ 81 行当消息来源类型为接收类型时，即私信的收件箱消息列表调用该缓存类时，设置子视图中控件显示的文本及单击事件。

6. 消息列表的视图模型类

MessageListViewModel 类用于为显示消息列表的视图提供数据处理。因为没有使用 ObservableField<T>类型属性在布局文件中对控件属性进行绑定，所以对 ViewModelProvider 类进行实例化。

/java/com/pinkjiao/www/view/model/MessageListViewModel.kt

```
//消息列表视图模型。Domain参数为域名,baseUrl参数为基础Url,developerModel参数为开发者,loginUserModel
参数为当前登录用户
class MessageListViewModel (var domain:String, var baseUrl:String, var developerModel:
DeveloperModel, var loginUserModel: UserModel) : ViewModel() {
    class SharedViewModelFactory(var domain:String, var baseUrl:String, var developerModel:
DeveloperModel, var loginUserModel:UserModel) : ViewModelProvider.Factory {
        override fun <T : ViewModel?> create(modelClass: Class<T>): T {
            return MessageListViewModel(domain,baseUrl,developerModel,loginUserModel) as T
        }
    }
    companion object{
        const val STATE_GET_MESSAGE_LIST_SUCCESS = "获取消息列表数据成功"
        const val STATE_GET_MESSAGE_LIST_FAILURE = "获取消息列表数据失败"
        const val STATE_READ_MESSAGE_SUCCESS = "读取标记成功"
    }
    var messageModelList = mutableListOf<MessageModel>() //可观察的用户模型数据
    var response = MutableLiveData<String>() //请求的响应状态
    var unReadNum = 0 //未读消息数量
    var error = "" //错误信息
    //获取消息列表。messageOriginType参数为消息来源类型, messageType参数为消息类型, start参数为开始加
载的序号, num参数为数量
    fun getMessageList(messageOriginType:Int,messageType:Int,start:Int,num:Int){
        MessageRepository().getMessageList(domain, baseUrl, developerModel, loginUserModel,
messageOriginType, messageType, start, num, object : MessageRepository.ResponseListener {
            override fun onCallBack(res: String, messageModelList: MutableList<MessageModel>?,
unReadNum: Int) {
                if (res == Http.RESPONSE_SUCCESS) {
                    this@MessageListViewModel.messageModelList = messageModelList!!
                    this@MessageListViewModel.unReadNum = unReadNum
                    response.value = STATE_GET_MESSAGE_LIST_SUCCESS
                } else {
                    error = res
                    response.value = STATE_GET_MESSAGE_LIST_FAILURE
```

```
51                  }
52              }
53          })
54      }
        //设置消息为已读取状态。messageID 参数为消息 ID，responseListener 参数为监听器
60      fun readMessage(messageID: String, responseListener: ResponseListener){
61          MessageRepository().readMessage(baseUrl, developerModel, loginUserModel, messageID,
    object : MessageRepository.ResponseListener {
                override fun onCallBack(res: String, messageModelList: MutableList<MessageModel>?,
62  unReadNum: Int) {
                    if (res == Http.RESPONSE_SUCCESS) {
63                      responseListener.onSuccess()
64                      response.value = STATE_READ_MESSAGE_SUCCESS
65                  }
66              }
67          })
68      }
69      interface ResponseListener{
70          fun onSuccess()
71      }
72  }
```

第 20～24 行是使用工厂模式创建 MessageListViewModel 实例的 SharedViewModelFactory 类。第 26～28 行声明响应状态的常量。第 41～54 行获取消息列表数据。第 60～68 行将指定 ID 的消息设置为已读状态。第 69～71 行声明响应 Http 请求的监听接口。

7. 消息列表的适配器类

MessageListAdapter 类用于显示消息列表数据的 RecyclerView 控件的数据适配，需要与提醒消息子视图缓存类（RemindMessageListItemViewHolder 类）或私信消息子视图缓存类（PrivateMessageListItemViewHolder 类）搭配使用。

/java/com/pinkjiao/www/adapter/MessageListAdapter.kt

```
class MessageListAdapter(
    private val messageOriginType: Int, //消息来源类型
    private val viewModel: MessageListViewModel //消息数据模型的列表
) : BaseListAdapter() {
    private lateinit var context: Context //上下文
    var messageModelList: MutableList<MessageModel>= mutableListOf() //消息列表数据
    var currentState = "" //当前状态
    override fun onCreateViewHolder(parent: ViewGroup, viewType: Int): RecyclerView.ViewHolder {
        context = parent.context
        //返回缓存子视图的 ViewHolder
        return when (viewType) {
            TYPE_ITEM_MESSAGE -> { //私信子视图
                val itemBinding = ItemMessageListBinding.inflate(LayoutInflater.from
(parent.context), parent, false)
                MessageListItemViewHolder(itemBinding,context,messageOriginType,viewModel)
            }
            TYPE_ITEM_REMIND -> { //提醒子视图
                val itemBinding = ItemRemindListBinding.inflate(LayoutInflater.from
```

```
   (parent.context), parent, false)
34             MessageRemindItemViewHolder(itemBinding,context,viewModel)
35         }
36         else -> { //页脚子视图
37             val footerBinding = FooterDynamicListBinding.inflate(LayoutInflater.from
   (parent.context), parent, false)
38             (footerBinding.linearLayout.layoutParams as StaggeredGridLayoutManager.
   LayoutParams).isFullSpan = true
39             FooterViewHolder(footerBinding,context,messageOriginType)
40         }
41     }
42 }
43 override fun onBindViewHolder(holder: RecyclerView.ViewHolder, position: Int) {
44     //为缓存子视图的 ViewHolder 匹配数据
45     when (holder) {
46         is MessageListItemViewHolder -> { holder.bind(messageModelList[position]) } //私信子视图
47         is MessageRemindItemViewHolder -> { holder.bind(messageModelList[position]) } //提醒子视图
48         is FooterViewHolder -> { holder.bind(currentState) } //页脚子视图
49     }
50 }
   //获取子视图的数量
54 override fun getItemCount(): Int {
55     return messageModelList.size + 1
56 }
   //获取子视图的类型
61 override fun getItemViewType(position: Int): Int {
62     return when {
63         itemCount == position + 1 -> { FooterViewHolder.TYPE_DEFAULT } //页脚子视图类型
64         messageOriginType == FooterViewHolder.TYPE_REMIND -> { TYPE_ITEM_REMIND } //提醒子视图类型
65         else -> { TYPE_ITEM_MESSAGE } //私信子视图类型
66     }
67 }
   //改变状态
72 fun changeStatus(state: String) {
73     currentState = state
74     if (state != FooterViewHolder.STATUS_LOADING_MORE) {
75         notifyDataSetChanged() //通知数据集改变 刷新 RecyclerView
76     }
77 }
78 }
```

第 27 ~ 41 行根据子视图类型创建对应的缓存子视图实例作为返回值。第 43 ~ 50 行根据不同类型的子视图缓存类进行数据绑定。第 54 ~ 56 行获取子视图数量。第 61 ~ 67 行获取子视图的类型。第 72 ~ 77 行改变当前状态，除 FooterViewHolder.STATUS_LOADING_MORE 状态外，对 RecyclerView 控件的数据集更新进行刷新子视图的显示。

8. 消息列表的 Fragment 类

MessageListFragment 类用于显示提醒消息或私信消息的列表数据，根据传递的参数判断列表数据提醒消息还是私信消息。该类与 DynamicFragment 类（显示动态的 Fragment）的原理类似，由于只显示消息列表数据，结构更简单一些。初始化后从服务器端获取列表数据，通过适配器将列表数据显示在子视图中。

```
/java/com/pinkjiao/www/fragment/MessageListFragment.kt
private const val ARG_PARAM_MESSAGE_ORIGIN = "消息来源类型"
private const val ARG_PARAM_MESSAGE_TYPE = "消息类型"
class MessageListFragment : RefreshFragment() {
    companion object {
        private lateinit var messageListFragment:MessageListFragment
        //创建 MessageListFragment 实例。messageOriginType 参数为私信来源类型，返回值类型为MessageListFragment
        fun newInstance(messageOriginType:Int, messageType:Int): MessageListFragment {
            messageListFragment = MessageListFragment().apply {
                arguments = Bundle().apply {
                    putInt(ARG_PARAM_MESSAGE_ORIGIN, messageOriginType)
                    putInt(ARG_PARAM_MESSAGE_TYPE, messageType)
                }
            }
            return messageListFragment
        }
    }
    private var messageOriginType: Int = 1 //消息来源类型
    private var messageType: Int = 1 //消息类型
    private var maxNum: Int = 5000 //最大加载数据量
    private var pageNum: Int = 100 //每次加载的数据量
    private lateinit var viewBinding: FragmentMessageListBinding //视图绑定
    private lateinit var viewModel: MessageListViewModel //视图模型
    private lateinit var adapter: MessageListAdapter //适配器
    private val messageModelList = ArrayList<MessageModel>() //消息模型数组
    private var startIndex = 0 //开始加载的序号
    override fun onCreate(savedInstanceState: Bundle?) {
        super.onCreate(savedInstanceState)
        arguments?.let {
            messageOriginType = it.getInt(ARG_PARAM_MESSAGE_ORIGIN)
            messageType = it.getInt(ARG_PARAM_MESSAGE_TYPE)
        }
    }
    override fun onCreateView(inflater: LayoutInflater, container: ViewGroup?, savedInstanceState: Bundle?): View {
        //实例化ViewBinding
        viewBinding = FragmentMessageListBinding.inflate(inflater, container, false)
        //初始化
        initViewModel()
        initView()
        initObserver()
        //获取数据
        refresh()
        return viewBinding.root
    }

    override fun refresh(){ //重写RefreshFragment抽象类的方法，刷新读取数据
```

```
77            startIndex = 0
78            loadMessageListData(startIndex, pageNum)
79        }

83        private fun initViewModel() { //初始化视图模型
84            val domain = RequestUrl.getDomainUrl(requireContext())
85            val baseUrl = RequestUrl.getBaseUrl(requireContext())
86            val developerModel = DeveloperModel.getDeveloper(requireContext())
87            val loginUserModel = UserModel()
88            //从共享偏好设置获取用户数据
89            if (!UserModel.getSharedPreferences(requireContext(), loginUserModel)) {
90                startActivity(Intent(context, LoginActivity::class.java))
91                Toast.makeText(context, getString(
      R.string.toast_get_shared_preferences_error), Toast.LENGTH_LONG).show()
92            }
93            viewModel = ViewModelProvider(this, MessageListViewModel.SharedViewModelFactory(
      domain,baseUrl,developerModel,loginUserModel)).get(MessageListViewModel::class.java)
94        }
98        private fun initView(){ //初始化视图控件
128       }
132       private fun initObserver(){//初始化观察者
169       }
175       private fun loadMessageListData(start: Int, pageNum: Int) { //加载消息列表数据
176           if (start == 0) {
177               adapter.changeStatus(FooterViewHolder.STATUS_REFRESHING) //刷新状态
178           } else {
179               adapter.changeStatus(FooterViewHolder.STATUS_LOADING_MORE) //设置加载状态
180           }
181           viewModel.getMessageList(messageOriginType, messageType,start, pageNum) //获取消息列表数据
182       }
183   }
```

第 26、27 行声明两个常量作为 Bundle 对象方法参数的键名，设置消息来源类型和消息类型。第 36 ~ 44 行实例化 MessageListFragment 类，通过 Bundle 对象传递参数数据。第 62 ~ 72 行进行初始化，通过 refresh() 方法首次加载消息。第 93 行实例化视图模型，由于视图模型没有与视图布局进行绑定，需要通过 ViewModelProvider(ViewModelStoreOwner, Factory) 方法实例化。第 175 ~ 182 行根据状态设置页脚子视图，然后通过视图模型的 getMessageList() 方法获取消息列表数据。

9．提醒的 Fragment 类

RemindFragment 类用于显示提醒消息列表的 MessageListFragment 类实例，继承 RefreshFragment 类，实现 StatusBar 接口。

```
/java/com/pinkjiao/www/fragment/RemindFragment.kt
30   //添加显示消息列表的 Fragment
31   messageListFragment = MessageListFragment.newInstance(FooterViewHolder.TYPE_REMIND,0)
32   val transaction = childFragmentManager.beginTransaction()
33   transaction.add(R.id.messageListFragment, messageListFragment).commit()
```

第 31 行实例化 MessageListFragment 类赋给 messageListFragment 变量。第 33 行将 messageListFragment 变量存储的 MessageListFragment 实例显示在 messageListFragment 控件中。

14.11.3 自己模块

自己模块显示常用个人信息、发布的动态列表、关注数量、粉丝数量、动态数量和私信数量，以及退出登录按钮和设置按钮，如图 14-31 所示。自己模块核心类之间的调用关系如图 14-32 所示。

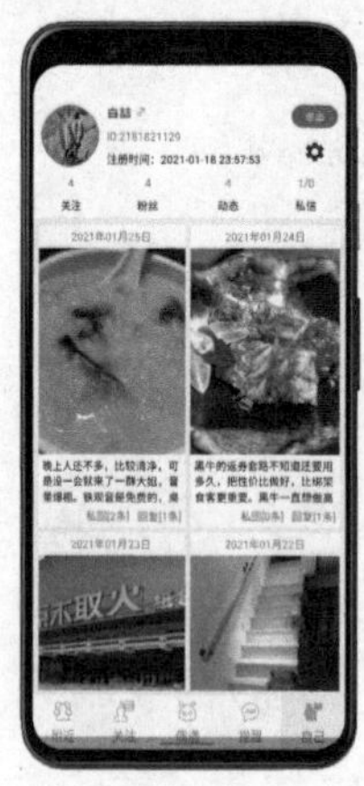

图 14-31 自己模块的运行效果

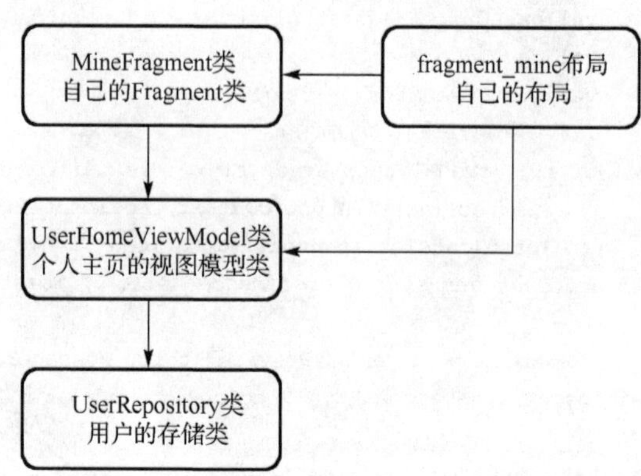

图 14-32 自己模块核心类之间的调用关系

1．获取用户信息的服务器端接口页面

获取用户信息的服务器端接口页面为 userInfo.php，使用 Get 方式接收数据，所需参数如表 14-22 所示。

表 14-22 userInfo.php 接收的参数

参数	说明
developerAccountNumber	开发者账号
developerKey	开发序列号
currentUserAccountID	登录用户的账号
currentUserPassword	登录用户的 MD5 算法加密后密码
userAccountID	用户账号 ID

获取用户数据成功时，返回的 JSON 数据结构为{"state":"ok", "followFlag":"是否关注当前用户（0 表示未关注，1 表示已关注）","user":[用户数据的 JSON 数组]} 。获取登录用户数据时，followFlag 键名和键值会省略；用户数据的 JSON 数组的键名如表 14-23 所示。获取用户数据失败时，返回的 JSON 数据结构为{"state":"error1001","info":"用户账号 ID 错误"}。

表 14-23 用户数据的 JSON 数组的键名

键名	说明
userAccountID	用户账号 ID
userName	用户姓名
userGender	用户性别(0 表示女性，1 表示男性)
userFaceImage	用户头像地址
userRegistrationTime	用户注册时间
userLock	用户锁定状态(0 表示未锁定，1 表示未激活，2 表示有期限锁定，3 表示无期限锁定)

续表

键　名	说　明
userNationalIDImageCheck	用户身份证验证状态(0 表示待提交，1 表示已提交，2 表示驳回验证，3 表示通过验证)
userFollowNum	用户的关注数量
userFollowerNum	用户的粉丝数量
userDynamicNum	用户发布的动态数量
sendMessageNum	用户发送的私信数量
receiveMessageNum	用户接收的私信数量

2. 用户的存储类

UserRepository 类用于查询用户数据、设置用户头像、设置用户密码和提交用户身份证信息，其中设置用户头像、设置用户密码和提交用户身份证信息在设置模块中使用。下面先介绍该类主体结构，再介绍存储数据的方法。

```
/java/com/pinkjiao/www/repository/UserRepository.kt
14  class UserRepository {
        //获取用户信息
23      fun getUserInfo(domain:String, baseUrl:String, developerModel:DeveloperModel, loginUserModel:UserModel, userAccountID:String?,responseListener:ResponseListener){
62      }
        //设置用户头像图片
72      fun setupFaceImage(baseUrl: String, developerModel: DeveloperModel, loginUserModel:UserModel, parameter: MutableMap<String, String>, files: Map<String, String>, responseListener:ResponseListener){
92      }
        //设置用户密码
101     fun setupPassword(baseUrl: String, developerModel: DeveloperModel, loginUserModel:UserModel, parameter: MutableMap<String, String>, responseListener: ResponseListener){
121     }
        //提交用户身份证
131     fun setupCheckNationalIDImage(baseUrl: String, developerModel: DeveloperModel, loginUserModel:UserModel, parameter: MutableMap<String, String>,files: Map<String, String>, responseListener:ResponseListener){
151     }
152     interface ResponseListener {
158         fun onCallBack(res: String, userModel:UserModel?) //回调
159     }
160 }
```

第 152 ~ 159 行声明响应请求的监听接口，并且声明 onCallBack() 方法接收响应数据。res 参数是 ring 类型，接收响应的结果类型；userModel 参数是可空的 UserModel 类型，接收用户数据。

```
/java/com/pinkjiao/www/repository/UserRepository.kt
//获取用户信息。baseUrl 参数为服务器端页面所在的文件夹 Url 路径，developerModel 参数为开发者，loginUserModel 参数为登录用户，userAccountID 参数为用户账号 ID，responseListener 参数为监听器接口
fun getUserInfo(domain:String, baseUrl:String, developerModel:DeveloperModel, loginUserModel:UserModel, userAccountID:String?,responseListener:ResponseListener){
    val page = "userInfo.php" //接口页面
    var parameter = "" //get 参数
```

```
26      if (userAccountID!= null) {
27          parameter = "&userAccountID = $userAccountID"
28      }
29      //get 请求对象
30      val getRequest = GetRequest(baseUrl, page, developerModel, loginUserModel, parameter)
31      //get 方法获取用户数据
32      Http.get(getRequest.url, object : Http.ResponseListener {
33          override fun onResponse(res: String) {
34              val jsonParser = JSONTokener(JSON.cleanBOM(res))
35              val jsonObject = jsonParser.nextValue() as JSONObject
36              if (jsonObject.getString("state") == "ok") {
37                  val userModel = UserModel()
38                  val userJsonObject = jsonObject.getJSONObject("user")
39                  userModel.accountID = userJsonObject.getString("userAccountID")
40                  userModel.name = userJsonObject.getString("userName")
41                  userModel.gender = userJsonObject.getString("userGender")
42                  userModel.faceImage = Image.getUrl(domain, userJsonObject.getString
    ("userFaceImage"), Image.smallQuality)
43                  userModel.registrationTime = userJsonObject.getString("userRegistrationTime")
44                  userModel.followNum = userJsonObject.getInt("userFollowNum")
45                  userModel.followerNum = userJsonObject.getInt("userFollowerNum")
46                  userModel.dynamicNum = userJsonObject.getInt("userDynamicNum")
47                  userModel.nationalIDImageCheck = userJsonObject.getInt("userNationalIDImageCheck")
48                  try {
49                      userModel.sendMessageNum = userJsonObject.getInt("sendMessageNum")
50                      userModel.receiveMessageNum = userJsonObject.getInt("receiveMessageNum")
51                  } catch (e:Exception){ }
52                  try {
53                      userModel.followFlag = jsonObject.getString("followFlag")
54                  } catch (e:Exception){ }
55                  responseListener.onCallBack(Http.RESPONSE_SUCCESS,userModel)
56              } else { responseListener.onCallBack(jsonObject.getString("info"), null) }
57          }
58          override fun onFailure(res: String) {
59              responseListener.onCallBack(res, null)
60          }
61      })
62  }
```

第 24 行设置服务器端接口页面。第 25 行初始化 Get 参数。第 26 ~ 28 行 userAccountID 参数(用户账号 ID)不为 null 时，将其加入 Get 请求的参数中。第 30 行实例化 Get 请求对象。第 32 行发送获取用户数据的请求。第 33 ~ 57 行重写 onResponse()方法，解析请求返回的 JSON 数据。如果获取用户数据成功，通过 onCallBack()方法回调响应成功信息及用户数据，否则通过 onCallBack()方法回调错误信息。第 48 ~ 51 行当获取非登录用户数据时，sendMessageNum 键名和 receiveMessageNum 键名不存在，所以需要捕获异常，但是无须捕获到的异常进行任何处理。第 52 ~ 54 行当获取登录用户数据时 followFlag 键名不存在，所以同样需要捕获异常且不进行任何处理。第 58 ~ 60 行重写 onFailure()方法调用 onCallBack()方法回调网络错误信息。

```
/java/com/pinkjiao/www/repository/UserRepository.kt
    //设置头像图片。baseUrl 参数为服务器端页面所在的文件夹 Url 路径，developerModel 参数为开发者，loginUserMode
```

```
参数为登录用户，parameter 参数为 Post 请求的参数，files 参数为上传文件，responseListener 参数为监听器接口
72 fun setupFaceImage(baseUrl: String, developerModel: DeveloperModel, loginUserModel: UserModel, parameter: MutableMap<String, String>, files: Map<String, String>, responseListener: ResponseListener){
73     val page = "setupFaceImage.php" //接口页面
74     //post 请求对象
75     val postRequest = PostRequest(baseUrl, page, developerModel, loginUserModel, parameter, files)
76     //post 方式上传
77     Http.post(postRequest.url, postRequest.parameter, postRequest.files, object : Http.PostResponseListener {
78         override fun onCreate() {
79             responseListener.onCallBack(Http.CREATE,null)
80         }
81         override fun onResponse(res: String) {
82             val jsonParser = JSONTokener(JSON.cleanBOM(res))
83             val jsonObject = jsonParser.nextValue() as JSONObject
84             if (jsonObject.getString("state") == "ok") {
85                 responseListener.onCallBack(Http.RESPONSE_SUCCESS,null)
86             } else { responseListener.onCallBack(jsonObject.getString("info"),null) }
87         }
88         override fun onFailure(res: String) {
89             responseListener.onCallBack(res,null)
90         }
91     })
92 }
```

第 73 行设置服务器端接口页面。第 75 行初始化 Post 参数。第 77 行发送修改用户头像图片的请求。第 78 ~ 80 行重写 onCreate() 方法，调用 onCallBack() 方法回调请求创建的状态。第 81 ~ 87 行重写 onResponse() 方法，解析请求返回的 JSON 数据。如果修改用户头像图片成功，通过 onCallBack() 方法回调响应成功信息，否则通过 onCallBack() 方法回调错误信息。第 88 ~ 90 行重写 onFailure() 方法，调用 onCallBack() 方法回调网络错误信息。

```
/java/com/pinkjiao/www/repository/UserRepository.kt
//设置密码。baseUrl 参数为服务器端页面所在的文件夹 Url 路径，developerModel 参数为开发者，loginUserModel 参数为登录用户，parameter 参数为 Get 请求的参数，responseListener 参数为监听器接口
fun setupPassword(baseUrl: String, developerModel: DeveloperModel, loginUserModel: UserModel, parameter: MutableMap<String, String>, responseListener: ResponseListener){
    val page = "setupPassword.php" //接口页面
    //post 请求对象
    val postRequest = PostRequest(baseUrl, page, developerModel, loginUserModel, parameter, null)
    //post 方式上传
    Http.post(postRequest.url, postRequest.parameter, postRequest.files, object : Http.PostResponseListener {
        override fun onCreate() {
            responseListener.onCallBack(Http.CREATE,null)
        }
        override fun onResponse(res: String) {
            val jsonParser = JSONTokener(JSON.cleanBOM(res))
            val jsonObject = jsonParser.nextValue() as JSONObject
            if (jsonObject.getString("state") == "ok") {
```

```
114                 responseListener.onCallBack(Http.RESPONSE_SUCCESS,null)
115             } else { responseListener.onCallBack(jsonObject.getString("info"),null) }
116         }
117         override fun onFailure(res: String) {
118             responseListener.onCallBack(res,null)
119         }
120     })
121 }
```

第 102 行设置服务器端接口页面。第 104 行初始化 post 参数。第 106 行发送修改用户密码的请求。第 107 ~ 109 行重写 onCreate()方法，调用 onCallBack()方法回调请求创建的状态。第 110 ~ 116 行重写 onResponse()方法，解析请求返回的 JSON 数据。如果修改用户密码成功，通过 onCallBack()方法回调响应成功信息，否则通过 onCallBack()方法回调错误信息。第 117 ~ 119 行重写 onFailure()方法，调用 onCallBack()方法回调网络错误信息。

/java/com/pinkjiao/www/repository/UserRepository.kt

```
    //提交用户身份证信息。baseUrl 参数为服务器端页面所在的文件夹 Url 路径，developerModel 参数为开发者，loginUserModel 参数为登录用户，parameter 参数为 Post 请求的参数，files 参数为 Post 请求的上传文件，responseListener 参数为监听器接口
131 fun setupCheckNationalIDImage(baseUrl: String, developerModel: DeveloperModel, loginUserModel: UserModel, parameter: MutableMap<String, String>,files: Map<String, String>, responseListener: ResponseListener){
132     val page = "setupCheckNationalIDImage.php" //接口页面
133     //post 请求对象
134     val postRequest = PostRequest(baseUrl, page, developerModel, loginUserModel, parameter, files)
135     //post 方式上传
136     Http.post(postRequest.url, postRequest.parameter, postRequest.files, object : Http.PostResponseListener {
137         override fun onCreate() {
138             responseListener.onCallBack(Http.CREATE,null)
139         }
140         override fun onResponse(res: String) {
141             val jsonParser = JSONTokener(JSON.cleanBOM(res))
142             val jsonObject = jsonParser.nextValue() as JSONObject
143             if (jsonObject.getString("state") == "ok") {
144                 responseListener.onCallBack(Http.RESPONSE_SUCCESS,null)
145             } else { responseListener.onCallBack(jsonObject.getString("info"),null) }
146         }
147         override fun onFailure(res: String) {
148             responseListener.onCallBack(res,null)
149         }
150     })
151 }
```

第 132 行设置服务器端接口页面。第 134 行初始化 post 参数。第 136 行发送提交用户身份证信的请求。第 137 ~ 139 行重写 onCreate()方法，调用 onCallBack()方法回调请求创建的状态。第 140 146 行重写 onResponse()方法，解析请求返回的 JSON 数据。如果提交用户身份证信息成功，通 onCallBack()方法回调响应成功信息，否则通过 onCallBack()方法回调错误信息。第 147 ~ 149 行重 onFailure()方法，调用 onCallBack()方法回调网络错误信息。

3. 个人主页的视图模型类

UserHomeViewModel 类用于为显示个人主页的视图提供绑定数据及数据处理。

/java/com/pinkjiao/www/view/model/UserHomeViewModel.kt

```
//用户主页视图模型。domain 参数为域名，baseUrl 参数为基础 Url，developerModel 参数为开发者，
loginUserModel 参数为当前登录用户
class UserHomeViewModel(private var domain:String, var baseUrl:String, var developerModel:
DeveloperModel, var loginUserModel: UserModel): ViewModel() {
    companion object{
        const val STATE_GET_USER_INFO_SUCCESS = "获取用户数据成功"
        const val STATE_GET_USER_INFO_FAILURE = "获取用户数据失败"
        const val STATE_ADD_FOLLOW_SUCCESS = "添加关注成功"
        const val STATE_ADD_FOLLOW_FAILURE = "添加关注失败"
        const val STATE_REMOVE_FOLLOW_SUCCESS = "取消关注成功"
        const val STATE_REMOVE_FOLLOW_FAILURE = "取消关注失败"
    }
    var userModel = ObservableField<UserModel>() //可观察的用户模型数据
    var response = MutableLiveData<String>() //请求的响应状态
    var error = "" //错误信息
    //获取用户信息的请求。userAccountID 参数为用户账号 ID。
    fun getUserInfo(userAccountID:String? = null) {
        UserRepository().getUserInfo(domain, baseUrl, developerModel, loginUserModel,
userAccountID, object : UserRepository.ResponseListener {
            override fun onCallBack(res: String, userModel: UserModel?) {
                if (res == Http.RESPONSE_SUCCESS) {
                    this@UserHomeViewModel.userModel.set(userModel)
                    response.value = STATE_GET_USER_INFO_SUCCESS
                } else {
                    error = res
                    response.value = STATE_GET_USER_INFO_FAILURE
                }
            }
        })
    }
    //添加关注的请求。followAccountID 参数为关注用户的账号 ID
    fun addFollow(followAccountID:String) {
        FollowRepository().addFollow(baseUrl, developerModel, loginUserModel, followAccountID,
object : FollowRepository.ResponseListener {
            override fun onCallBack(res: String, followModelList: ArrayList<FollowModel>?) {
                if (res == Http.RESPONSE_SUCCESS) {
                    response.value = STATE_ADD_FOLLOW_SUCCESS
                } else {
                    error = res
                    response.value = STATE_ADD_FOLLOW_FAILURE
                }
            }
        })
    }
    //取消关注的请求。followAccountID 参数为关注用户的账号 ID
fun removeFollow(followAccountID:String) {
```

```
70        FollowRepository().removeFollow(baseUrl, developerModel, loginUserModel, followAccountID,
object : FollowRepository.ResponseListener {
71            override fun onCallBack(res: String, followModelList: ArrayList<FollowModel>?) {
72                if (res == Http.RESPONSE_SUCCESS) {
73                    response.value = STATE_REMOVE_FOLLOW_SUCCESS
74                }else {
75                    error = res
76                    response.value = STATE_REMOVE_FOLLOW_FAILURE
77                }
78            }
79        })
80    }
81 }
```

第 22～27 行声明响应状态的常量。第 29 行声明 userModel 变量为 ObservableField<UserModel> 类型，该类型可以被视图的数据绑定观察数据变化并自动更新数据。第 30 行声明 response 变量为 MutableLiveData<String>类型，可以被其他类观察其数据变化。第 31 行 error 变量用于存储错误信息。第 36～48 行通过 Http 请求获取用户的个人信息数据。第 53～64 行通过 Http 请求添加关注指定 ID 的用户。第 69～80 行通过 Http 请求取消关注指定 ID 的用户。

4. 自己的布局

fragment_mine 布局文件使用 UserHomeViewModel 类绑定控件的属性，用于显示从服务器端获取的登录用户的部分个人数据。

```
/res/layout/fragment_mine.xml
06  <data>
07      <variable
08          name = "userHomeViewModel"
09          type = "com.pinkjiao.www.view.model.UserHomeViewModel" />
10  </data>
43  <com.google.android.material.imageview.ShapeableImageView
53      app:strokeColor = '@{userHomeViewModel.userModel.gender.equals("1") ? @color/blue : @color/pink}'
55      tools:srcCompat = "@tools:sample/avatars" />
57  <TextView
58      android:id = "@+id/nameTextView"
66      android:text = "@{userHomeViewModel.userModel.name}"
67      tools:text = "姓名" />
68  <TextView
69      android:id = "@+id/genderTextView"
73      android:text = '@{userHomeViewModel.userModel.gender.equals("1") ? "♂" : "♀"}'
74      android:textColor = '@{userHomeViewModel.userModel.gender.equals("1") ? @color/blue
@color/pink}'
77      tools:text = "♂/♀" />
267 <FrameLayout
269     android:id = "@+id/view"
277     app:layout_constraintTop_toBottomOf = "@+id/linearLayout1"/>
```

第 06～10 行设置用于绑定的数据类。第 53 行使用数据绑定类实例绑定根据用户性别显示不同色的头像边框颜色。第 66 行使用数据绑定类实例绑定显示用户姓名的控件。第 73、74 行使用数据定类实例绑定根据用户性别显示表示性别的字符及设置对应的文本颜色。第 267 ~ 277 行的 view 控件作为显示用户发布的动态列表。

5. 自己的 Fragment 类

MineFragment 类用于显示当前用户的部分数据及发布的动态列表，继承 RefreshFragment 类，实现 StatusBar 接口。下面先介绍该类主体结构，再分别介绍 initView(AppCompatActivity) 方法和 initObserver(AppCompatActivity) 方法。

```
/java/com/pinkjiao/www/fragment/MineFragment.kt
26  class MineFragment(private val baseActivity: BaseActivity) : RefreshFragment(),StatusBar{
27      companion object{
28          fun newFragment(parentContext: BaseActivity):MineFragment{
29              return MineFragment(parentContext)
30          }
31      }
32      private lateinit var viewBinding: FragmentMineBinding //视图绑定
33      private var loginUserModel:UserModel = UserModel() //登录用户
34      private lateinit var dynamicListFragment: DynamicListFragment //显示动态列表的 Fragment
35      private lateinit var fragmentTransaction:FragmentTransaction
36      private  var fragmentLoadedFlag = false //动态列表的 Fragment 的加载标记
37      override fun onCreateView(inflater: LayoutInflater, container: ViewGroup?, savedInstanceState:
    Bundle?): View {
39          viewBinding = FragmentMineBinding.inflate(inflater, container, false) //实例化视图绑定
41          initViewModel(baseActivity) //初始化视图模型
42          initObserver(baseActivity) //初始化观察者
43          initView(baseActivity) //初始化视图控件
45          refresh() //加载数据
46          return viewBinding.root
47      }
51      override fun refresh() { //刷新
52          viewBinding.swipeRefreshLayout.isRefreshing = true //显示刷新状态
53          viewBinding.userHomeViewModel!!.getUserInfo() //获取用户信息
54      }
58      override fun setStatusBar() { //设置状态栏颜色模式
59          baseActivity.setStatusBar(BaseActivity.STATUS_BAR_LIGHT)
60      }
65      private fun initViewModel(context: AppCompatActivity){ //初始化视图模型
73          val userHomeViewModel = UserHomeViewModel(url, domainUrl, developerModel, loginUserModel)
74          viewBinding.userHomeViewModel = userHomeViewModel //获取视图模型
75      }
80      private fun initView(context: AppCompatActivity){ //初始化视图控件
144     }
149     private fun initObserver(context: AppCompatActivity){ //初始化观察者
168     }
169 }
```

第 28 ~ 30 行实例化 MineFragment 类。第 36 行声明 fragmentLoadedFlag 变量存储动态列表的加载记。第 45 行使用 refresh() 方法首次加载个人信息。第 51 ~ 54 行重写 RefreshFragment 类的 refresh() 法，单击 Tab 按钮时刷新个人信息。第 58 ~ 60 行重写 StatusBar 接口的 setStatusBar() 方法，将状态设置为浅色模式。

```
/java/com/pinkjiao/www/fragment/MineFragment.kt
80  private fun initView(context: AppCompatActivity){
81      viewBinding.quitButton.setOnClickListener { //退出
82          val builder = AlertDialog.Builder(baseActivity)
83          builder.setMessage(getString(R.string.dialog_confirm_back))
84              .setCancelable(false)
85              .setPositiveButton(getString(R.string.dialog_confirm_back_positive)) { _, _ ->
86                  //清空缓存信息
87                  UserModel.clearSharedPreferences(context)
88                  //返回登录 Activity
89                  startActivity(Intent(baseActivity, LoginActivity::class.java))
90                  baseActivity.finish()
91                  //提示信息
92                  Toast.makeText(baseActivity.baseContext,   getString(R.string.toast_log_out), Toast.LENGTH_SHORT).show()
93              }
94              .setNegativeButton(getString(R.string.dialog_confirm_back_negative)) { dialog, _ ->
95                  dialog.cancel()
96              }
97              .show()
98      }
134     viewBinding.setUpImageView.setOnClickListener { //设置
135         val intent = Intent(context, SetupActivity::class.java)
136         startActivity(intent)
137     }
138     viewBinding.swipeRefreshLayout.setOnRefreshListener { //刷新事件
139         refresh()
140     }
141     //添加显示动态的 Fragment
142     dynamicListFragment = DynamicListFragment.newInstance("userDynamicList.php", loginUserModel.accountID,2,true, swipeRefreshLayoutEnable = false)
143     fragmentTransaction = context.supportFragmentManager.beginTransaction()
144 }
```

第 83 ~ 97 行显示确认退出的对话框。第 87 行清空用户存储在本地的数据。第 134 ~ 137 行设置 setUpImageView 控件的单击事件，单击后启动 SetupActivity。第 138 ~ 140 行设置 swipeRefreshLayou 布局的下拉刷新事件，下拉后调用 refresh()方法重新加载数据。第 142 行实例化 DynamicListFragme 类，显示用户发布的动态列表。

```
/java/com/pinkjiao/www/fragment/MineFragment.kt
149 private fun initObserver(context: AppCompatActivity){
150     viewBinding.userHomeViewModel!!.response.observe(context, {
151         when (it) {
152             UserHomeViewModel.STATE_GET_USER_INFO_SUCCESS -> {
153                 viewBinding.swipeRefreshLayout.isRefreshing = false //隐藏刷新状态
154                 //更新头像
155                 Image.loadFromUrl(viewBinding.faceImageView, viewBinding.userHomeViewModel!!.userModel.get()!!.faceImage, CacheDir.getFace(context))
156                 //个人动态
157                 if (fragmentLoadedFlag) { //刷新动态
```

```
158                     dynamicListFragment.refresh()
159                 } else { //首次运行时添加动态
160                     fragmentTransaction.add(R.id.view, dynamicListFragment)
161                     fragmentTransaction.show(dynamicListFragment)
162                     fragmentTransaction.commit() //提交事务
163                     fragmentLoadedFlag = true
164                 }
165             }
166         }
167     })
168 }
```

第 150 行使用 Lambda 参数的形式添加视图模型实例的 response 属性的观察者。第 152～165 行观察到 response 属性值为 UserHomeViewModel.STATE_GET_USER_INFO_SUCCESS（即获取个人主页数据成功）时，隐藏刷新状态，更新头像，加载个人发布的动态列表。第 157～164 行当非首次加载（即 fragmentLoadedFlag 变量值为 true）时，直接调用 refresh()方法刷新动态列表。当首次加载时，将 dynamicListFragment 实例显示到 View 控件中，并将 fragmentLoadedFlag 变量值设置为 true。

14.12 回复动态模块

回复动态模块将回复或私回数据发送到服务器端存储起来，如图 14-33 所示。回复所有人可见，私回只对动态作者和回复者可见。核心类之间的调用关系如图 14-34 所示。

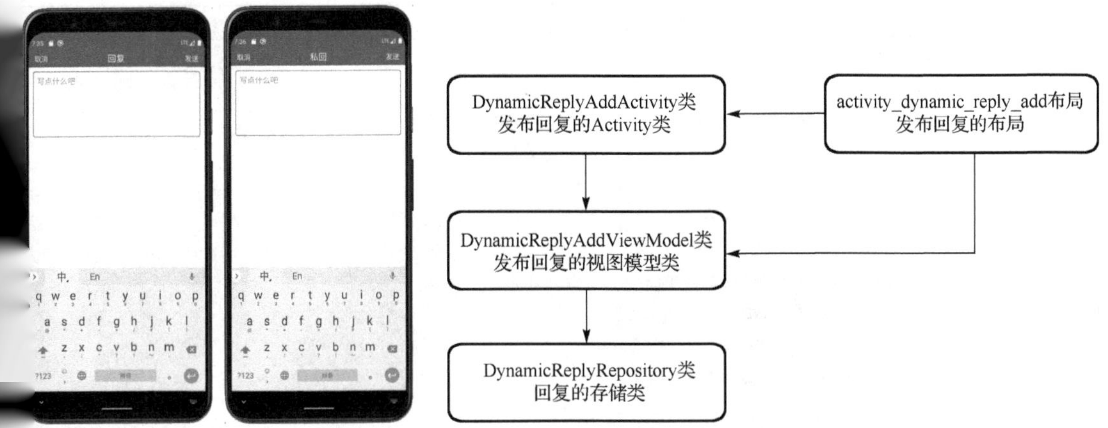

图 14-33 回复动态模块的运行效果

图 14-34 回复动态模块核心类之间的调用关系

4.12.1 发布回复的服务器端接口页面

发布回复的服务器端接口页面为 dynamicReplyAdd.php，使用 Post 方式接收数据，所需参数如 14-24 所示。

表 14-24 dynamicReplyAdd.php 接收的参数

参 数	说 明
developerAccountNumber	开发者账号
developerKey	开发序列号

续表

参　数	说　明
currentUserAccountID	登录用户的账号
currentUserPassword	登录用户的 MD5 算法加密后密码
dynamicReplyPrivateFlag	私回标记（0 表示回复，1 表示私回）
dynamicReplyContent	回复内容
dynamicReplyDynamicID	被回复的动态 ID
dynamicReplyRepliedDynamicReplyID	被回复的回复 ID

发布回复成功时，返回的 JSON 数据结构为{"state":"ok"}。发布回复失败时，返回的 JSON 数据结构为{"state":"错误码","info":"错误提示信息"}，如表 14-25 所示。

表 14-25　返回的错误码

错 误 码	错误提示信息
error901	回复数据有误
error902	发送回复失败
error903	被回复的回复 ID 不存在

14.12.2　发布回复的视图模型类

DynamicReplyAddViewModel 类用于存储回复的数据及通过 Http 请求发布回复。

```
/java/com/pinkjiao/www/fragment/view/model/DynamicReplyAddViewModel.kt
class DynamicReplyAddViewModel(
   var baseUrl: String, //基础 Url
   var dynamicReplyModel: DynamicReplyModel, //回复数据
   var developerModel: DeveloperModel, //开发者
   var loginUserModel: UserModel //登录用户
) : ViewModel() {
   companion object{
      const val STATE_CONTENT_EMPTY = "回复内容为空"
      const val STATE_SEND = "发布回复"
      const val STATE_SUCCESS = "回复成功"
      const val STATE_FAILURE = "回复失败"
   }
var content = ObservableField<String>() //回复内容
   var response = MutableLiveData<String>() //请求的响应状态
   var error = "" //错误信息
   fun publishDynamicReply() { //发布回复
      if (content.get() == "") {
         response.value = STATE_CONTENT_EMPTY
         return
      }
      //post 参数
      val parameter: MutableMap<String, String>= HashMap()
      parameter["dynamicReplyContent"] = content.get().toString()
      parameter["dynamicReplyDynamicID"] = dynamicReplyModel.dynamicID.toString()
      parameter["dynamicReplyRepliedDynamicReplyID"]==
```

```
   dynamicReplyModel.repliedDynamicReplyID.toString()
41         parameter["dynamicReplyPrivateFlag"] = dynamicReplyModel.privateFlag.toString()
42         //回复
43         DynamicReplyRepository().reply(baseUrl, developerModel, loginUserModel, parameter,
   object : DynamicReplyRepository.ResponseListener {
44             override fun onCallBack(res: String, count: Int, dynamicReplyModelList:
   ArrayList<DynamicReplyModel>?) {
45                 when (res) {
46                     Http.CREATE -> { response.value = STATE_SEND }
47                     Http.RESPONSE_SUCCESS -> { response.value = STATE_SUCCESS }
48                     else -> {
49                         error = res
50                         response.value = STATE_FAILURE
51                     }
52                 }
53             }
54         })
55     }
56 }
```

第 14 ~ 17 行在主构造方法中声明 4 个属性。第 20 ~ 23 行声明表示响应状态的 4 个常量。第 32 ~ 35 行当回复内容为空时，设置响应状态为 STATE_CONTENT_EMPTY，然后返回。第 38 ~ 41 行合成 Post 参数。第 44 ~ 53 行处理 Http 请求的回调数据。

14.12.3 发布回复的布局

activity_dynamic_reply_add 布局文件使用 DynamicReplyAddViewModel 类，虽然没有绑定任何控件的属性，但是 DynamicReplyAddViewModel 类可以不使用工厂模式进行实例化。与使用工厂模式实例化视图模型再通过 ViewModelProvider 类进行实例化相比，二者所实现的功能是一样的，但是后者增加了耦合度，因此不推荐使用。

/res/layout/activity_dynamic_reply_add.xml

```
<data>
    <variable
        name = "dynamicReplyAddViewModel"
        type = "com.pinkjiao.www.view.model.DynamicReplyAddViewModel" />
</data>
```

14.12.4 发布回复的 Activity 类

DynamicReplyAddActivity 类用于发布回复，根据 dynamicReplyPrivateFlag 参数判断是普通回复还是私回。

/java/com/pinkjiao/www/DynamicReplyAddActivity.kt

```
private fun initView(){
    viewBinding.publishTextView.setOnClickListener {
        viewBinding.dynamicReplyAddViewModel!!.publishDynamicReply() //发布
    }
    viewBinding.contentEditText.filters = arrayOf(Input.contentFilter())//设置过滤器
    viewBinding.contentEditText.addTextChangedListener(object : TextWatcher {
```

```
83         //输入文字产生变化时调用
84         override fun onTextChanged(s: CharSequence?, start: Int, before: Int, count: Int) {
85             //暂存输入的数据
86             viewBinding.dynamicReplyAddViewModel!!.dynamicReplyModel.content = s.toString()
87         }
88     })
90     Keyboard.show(this, viewBinding.contentEditText) //显示软键盘
91 }
```

第 75 行调用视图模型实例的 publishDynamicReply()方法发布回复。第 78 行设置 contentEditText 控件的过滤器，防止连续输入换行。第 84～87 行重写 contentEditText 控件监听器对象的 onTextChanged(CharSequence?,Int,Int,Int) 方法，当输入的内容发生变化时，将输入的文本赋给 dynamicReplyModel 对象的 content 属性。第 90 行显示软键盘并使 contentEditText 控件获取焦点。

```
/java/com/pinkjiao/www/DynamicReplyAddActivity.kt
95  private fun initObserver(){
96      //观察操作请求状态进行UI 处理
97      viewBinding.dynamicReplyAddViewModel!!.response.observe(this, {
98          var loadingProgressDialogFlag = false
99          var toastString = ""
100         when (it) {
101             DynamicReplyAddViewModel.STATE_CONTENT_EMPTY -> { //回复内容为空
102                 toastString = getString(R.string.dynamic_reply_content_hint)
103             }
104             DynamicReplyAddViewModel.STATE_SEND -> { //发布回复
105                 loadingProgressDialogFlag = true
106             }
107             DynamicReplyAddViewModel.STATE_SUCCESS -> { //回复成功
108                 toastString = getString(R.string.toast_dynamic_reply_publish_success)
109             }
110             DynamicReplyAddViewModel.STATE_FAILURE -> { //回复失败
111                 if(!NetStatus.isInternetAvailable(context)){
112                     Toast.makeText(context, "当前网络不可用", Toast.LENGTH_LONG).show()
113                 }else {
114                     toastString = viewBinding.dynamicReplyAddViewModel!!.error
115                 }
116             }
117         }
118         prompt(loadingProgressDialogFlag, toastString) //显示等待对话框或提示信息
119         if (it == DynamicReplyAddViewModel.STATE_SUCCESS) {
120             finish()
121         }
122     })
123 }
```

第 97～117 行观察 response 属性值的变化；显示等待对话框时将 loadingProgressDialogFlag 变量赋值为 true，否则赋值为 false；提示信息保存在 toastString 变量。第 118 行根据 loadingProgressDialogFl 变量值和 toastString 变量值，显示或隐藏等待对话框及显示提示信息。第 119～121 行当发布回复成时关闭当前 Activity。

14.13 关注和粉丝列表模块

关注和粉丝列表模块用于显示关注列表和粉丝列表，单击私信图标还可以发送私信，如图 14-35 所示。为了帮助大家理解，核心类之间的调用关系如图 14-36 所示。

图 14-35 关注和粉丝列表模块的运行效果

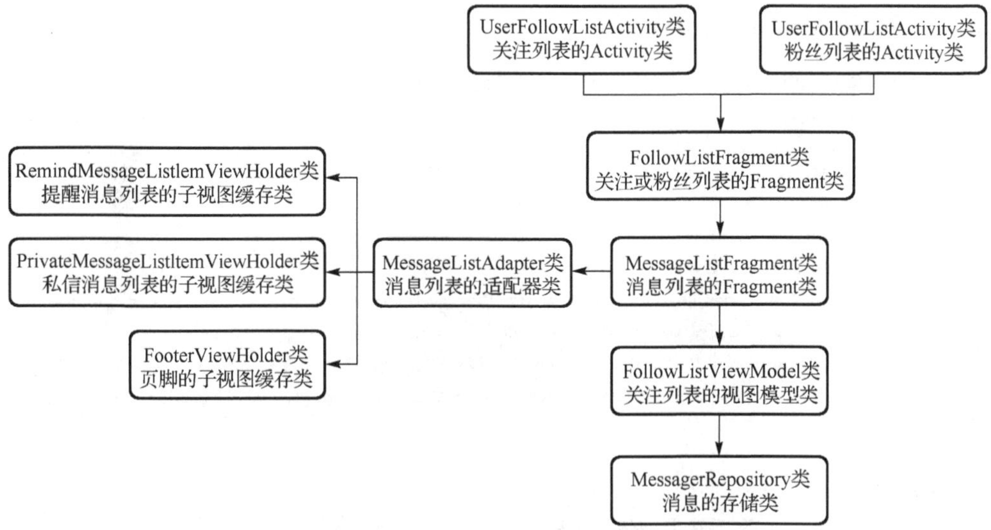

图 14-36 关注和粉丝列表模块核心类之间的调用关系

14.13.1 关注列表和粉丝列表的服务器端接口页面

关注列表的服务器端接口页面为 followList.php，粉丝列表的服务器端接口页面为 followerList.php，使用 Get 方式接收数据，所需参数如表 14-26 所示。

表 14-26 followList.php 和 followerList.php 接收的参数

参 数	说 明
developerAccountNumber	开发者账号

续表

参　数	说　明
developerKey	开发序列号
currentUserAccountID	登录用户的账号
currentUserPassword	登录用户的 MD5 算法加密后密码
start	开始加载的序号
Num	数量

获取关注或粉丝列表数据成功时，返回的 JSON 数据结构为{"state":"ok","followList":[关注或粉丝列表数据的 JSON 数组]}，关注或粉丝列表数据的 JSON 数组的键名如表 14-27 所示。没有关注或粉丝时，返回的 JSON 数据结构为{"state":"ok","followList" = >"[]"}。

表 14-27　关注或粉丝列表数据的 JSON 数组的键名

键　名	说　明
userName	用户姓名
userAccountID	用户账号 ID
userFaceImage	用户头像图片地址
followBecomeTime	关注时间

14.13.2　关注列表的视图模型类

FollowListViewModel 类用于请求关注或粉丝列表数据及请求的回调处理。

```
/java/com/pinkjiao/www/fragment/view/model/FollowListViewModel.kt
class FollowListViewModel(var domain:String,var baseUrl:String,var developerModel:DeveloperModel,
var loginUserModel:UserModel) : ViewModel() {
    //创建 FollowListViewModel 实例。domain 参数为域名，baseUrl 参数为基础 Url，developerModel 参数为
开发者，loginUserModel 参数为当前登录用户
    class SharedViewModelFactory(var domain:String, var baseUrl:String, var developerModel:
DeveloperModel, var loginUserModel:UserModel) : ViewModelProvider.Factory {
        override fun <T : ViewModel?> create(modelClass: Class<T>): T {
            return FollowListViewModel(domain,baseUrl,developerModel,loginUserModel) as T
        }
    }
    companion object{
        const val STATE_GET_FOLLOW_LIST_SUCCESS = "获取关注或粉丝用户列表数据成功"
        const val STATE_GET_FOLLOW_LIST_NONE = "暂无关注或粉丝用户列表数据"
        const val STATE_GET_FOLLOW_LIST_FAILURE = "获取关注或粉丝用户列表数据失败"
    }
    var followModelList = mutableListOf<FollowModel>() //用户模型数据
    var response = MutableLiveData<String>() //请求的响应状态
    var error = "" //错误信息
    //获取关注列表。start 参数为开始加载的序号，num 参数为数量
    fun getFollowList(start:Int,num:Int){
        FollowRepository().getFollowList(domain,baseUrl,  developerModel,  loginUserModel
start, num, object : FollowRepository.ResponseListener {
            override fun onCallBack(res: String, followModelList: ArrayList<FollowModel>?) {
                callBack(res, followModelList)
```

```
50              }
51          })
52      }
        //获取粉丝列表。start 参数为开始加载的序号，num 参数为数量
58      fun getFollowerList(start:Int,num:Int){
59          FollowRepository().getFollowerList(domain,baseUrl, developerModel, loginUserModel,
    start, num, object : FollowRepository.ResponseListener {
60              override fun onCallBack(res: String, followModelList: ArrayList<FollowModel>?) {
61                  callBack(res, followModelList)
62              }
63          })
64      }
        //回调。res 参数为请求的结果，followModelList 参数为关注或粉丝的数据
70      fun callBack(res: String, followModelList: ArrayList<FollowModel>?){
71          when (res) {
72              Http.RESPONSE_SUCCESS -> {
73                  this@FollowListViewModel.followModelList = followModelList!!
74                  response.value = STATE_GET_FOLLOW_LIST_SUCCESS
75              }
76              FooterViewHolder.STATUS_NON_DATA -> {
77                  response.value = STATE_GET_FOLLOW_LIST_NONE
78              }
79              else -> {
80                  error = res
81                  response.value = STATE_GET_FOLLOW_LIST_FAILURE
82              }
83          }
84      }
85  }
```

第 28 ~ 32 行是使用工厂模式创建 FollowListViewModel 实例的 SharedViewModelFactory 类。第 34 ~ 36 行声明请求响应状态的常量。第 46 ~ 52 行获取关注列表数据后，通过 callBack(String,ArrayList<FollowModel>?)方法处理回调数据。第 58 ~ 64 行获取粉丝列表数据后，也通过 callBack(String,ArrayList<FollowModel>?)方法处理回调数据。第 70 ~ 84 行处理 Http 请求回调的数据。

14.13.3 关注或粉丝列表的子视图缓存类

FollowListItemViewHolder 类用于缓存关注或粉丝列表的子视图。

```
/java/com/pinkjiao/www/view/hodler/FollowListItemViewHolder.kt
class FollowListItemViewHolder(private val viewBinding: ItemFollowListBinding, val context:
Context) : RecyclerView.ViewHolder(viewBinding.root) {
    //绑定数据。followModel 参数为关注或粉丝的数据
    fun bind(followModel: FollowModel) {
        if (followModel.userFaceImage != "") {
            Image.loadFromUrl(viewBinding.faceImageView, followModel.userFaceImage, CacheDir.getFace
(context))
        }
        viewBinding.faceImageView.setOnClickListener {
```

```
23            val intent = Intent(context, UserHomeActivity::class.java)
24            intent.putExtra(UserHomeActivity.EXTRA_USER_ACCOUNT_ID, followModel.userAccountID)
25            intent.putExtra(UserHomeActivity.EXTRA_USER_NAME, followModel.userName)
26            context.startActivity(intent)
27        }
28        viewBinding.nameTextView.text = followModel.userName
29        viewBinding.accountIDTextView.text = "(ID:" + followModel.userAccountID + ")"
30        viewBinding.messageCreateTimeTextView.text = "关注时间: "+followModel.followBecomeTime
31        viewBinding.messageImageView.setOnClickListener {
32            val intent = Intent(context, MessageSendActivity::class.java)
33            intent.putExtra(MessageSendActivity.EXTRA_RECEIVE_USER_ACCOUNT_ID, followModel.
userAccountID)
34            intent.putExtra(MessageSendActivity.EXTRA_RECEIVE_USER_NAME, followModel.userName)
35            context.startActivity(intent)
36        }
37    }
38 }
```

第 13 行通过主构造方法声明 viewBinding 属性和 context 属性，分别获取视图控件和上下文。第 31 ~ 36 行设置 messageImageView 控件的单击事件，单击后通过 Intent 对象保存接收者的用户账号 ID 和接收者的用户姓名，传递给启动后的 MessageSendActivity 作为发送私信的参数。

14.13.4 关注或粉丝列表的 Fragment 类

FollowListFragment 类用于显示关注或粉丝的列表数据，通过主构造方法的 followFlag 属性区分关注列表和粉丝列表。initView()方法、initViewModel()方法和 initObserver()方法与消息列表的 Fragmen中同名的方法实现的功能类似，都是初始化视图控件、视图模型和观察者。

```
/java/com/pinkjiao/www/fragment/FollowListFragment.kt
class FollowListFragment(private var followFlag:Boolean) : Fragment() {
    companion object {
        fun newInstance(followFlag:Boolean = true) = FollowListFragment(followFlag)
    }
    override fun onCreateView(inflater: LayoutInflater, container: ViewGroup?,
savedInstanceState: Bundle?): View {
        //实例化 ViewBinding
        viewBinding = FragmentFollowListBinding.inflate(inflater, container, false)
        //初始化
        initView()
        initViewModel()
        initObserver()
        return viewBinding.root
    }
    private fun initView(){ //初始化视图控件
        if (followFlag) {
            adapter = FollowListAdapter(FooterViewHolder.TYPE_FOLLOW) //实例化关注列表适配器
        } else {
            adapter = FollowListAdapter(FooterViewHolder.TYPE_FOLLOWER) //实例化粉丝列表适配器
        }
    }
```

```
94      private fun initViewModel() { //初始化视图模型
104     }
108     private fun initObserver(){ //初始化观察者
141     }
147     private fun loadFollowListData(start: Int, pageNum: Int) {
154         if (followFlag == TYPE_FOLLOW) {
155             viewModel.getFollowList(start, pageNum) //加载关注列表数据
156         } else {
157             viewModel.getFollowerList(start, pageNum) //加载粉丝列表数据
158         }
159     }
160 }
```

第 36 行通过静态方法实例化 FollowListFragment 对象。第 63 ~ 67 行根据 followFlag 属性值实例化关注列表或粉丝列表的适配器。第 154 ~ 158 行根据 followFlag 属性值通过 Http 请求获取关注列表或粉丝列表的数据。

14.13.5 关注列表的 Activity 类

UserFollowListActivity 类是显示关注列表的 FollowListFragment 实例的容器。

```
/java/com/pinkjiao/www/UserFollowListActivity.kt
21 //添加显示关注的 Fragment
22 val followListFragment: Fragment = FollowListFragment.newInstance(FollowListFragment.TYPE_FOLLOW)
23 val transaction = supportFragmentManager.beginTransaction()
24 transaction.add(viewBinding.fragmentFrameLayout.id, followListFragment).commit()
```

第 22 行实例化显示关注列表的 FollowListFragment 实例。第 24 行 add(int, Fragment)方法的第一个参数是显示 Fragment 的容器的 id，由于使用视图绑定，所以通过 viewBinding. fragmentFrameLayout.id 属性获取 fragmentFrameLayout 布局的 id。

14.13.6 粉丝列表的 Activity 类

UserFollowerListActivity 类是显示粉丝列表的 FollowListFragment 实例的容器，与 UserFollowListActivity 类的主要区别是实例化 FollowListFragment 对象时传递的参数不同。

```
/java/com/pinkjiao/www/UserFollowerListActivity.kt
22 val followerListFragment: Fragment = FollowListFragment.newInstance
   (FollowListFragment.TYPE_FOLLOWER)
```

14.14 搜索动态模块

搜索动态模块根据关键字搜索相关动态，输入部分关键字后会在服务器端搜索关键字提示列表，帮助用户进行快速选择，如图 14-37 所示。核心类之间的调用关系如图 14-38 所示。

14.14.1 搜索关键字提示的服务器端接口

搜索关键字提示的服务器端接口页面为 searchDynamicKeys.php，根据关键字查找其他相关的关键字提示，使用 Get 方式接收数据，所需参数如表 14-28 所示。

图 14-37　搜索动态模块的运行效果

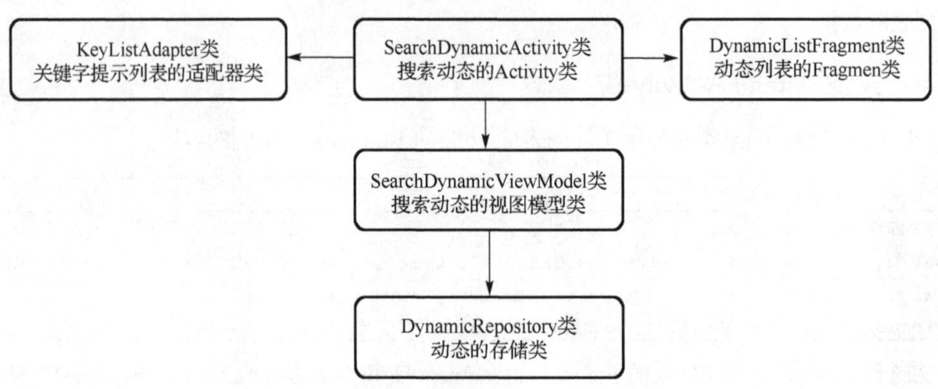

图 14-38　搜索动态模块核心类之间的调用关系

表 14-28　searchDynamicKeys.php 接收的参数

参　数	说　明
developerAccountNumber	开发者账号
developerKey	开发序列号
currentUserAccountID	登录用户的账号
currentUserPassword	登录用户的 MD5 算法加密后密码
searchKey	搜索关键字

搜索到关键字提示时，返回的 JSON 数据结构为{"state":"ok","searchKeyList":[关键字提示列表数据的 JSON 数组]}，关键字提示列表数据的 JSON 数组的键名如表 14-29 所示。没有搜索到关键字提示时，返回的 JSON 数据结构为{"state":"error802","info" = >"动态关键字不存在"}。

表 14-29　关键字提示列表数据的 JSON 数组的键名

键　名	说　明
searchKey	关键字
count	与关键字相关的动态数量

14.14.2 搜索动态的服务器端接口页面

搜索动态的服务器端接口页面为 searchDynamicList.php，根据关键字搜索动态，使用 Get 方式接收数据，所需参数如表 14-30 所示。

表 14-30 searchDynamicList.php 接收的参数

参 数	说 明
developerAccountNumber	开发者账号
developerKey	开发序列号
currentUserAccountID	登录用户的账号
currentUserPassword	登录用户的 MD5 算法加密后密码
searchKey	搜索关键字
start	开始加载的序号
num	数量

搜索到关键字提示时，返回的 JSON 数据结构为{"state":"ok","dynamicList":[动态列表数据的 JSON 数组]}。没有搜索到动态时，返回的 JSON 数据结构为{"state":"error801","info" = >"动态不存在"}。

14.14.3 关键字提示列表的适配器类

KeyListAdapter 类用于显示关键字提示列表数据的 RecyclerView 控件的数据适配。由于关键字列表的子视图较为简单，所以使用了内嵌的子视图缓存类，没有使用视图绑定。

```
/java/com/pinkjiao/www/adapter/KeyListAdapter.kt
20 override fun onCreateViewHolder(parent: ViewGroup, viewType: Int): ItemViewHolder { //创建缓存子视图
21     return ItemViewHolder(LayoutInflater.from(parent.context).inflate(R.layout.item_key_list, parent, false))
22 }
26 override fun onBindViewHolder(holder: ItemViewHolder, position: Int) { //绑定缓存子视图
27     holder.constraintLayout.setOnClickListener { onItemClickListener.onClick(position) }
28     holder.keyTextView.text = keyList.keys.elementAt(position)
29     holder.countTextView.text = keyList.values.elementAt(position)
30 }
34 override fun getItemCount(): Int { //获取子视图数量
35     return keyList.size
36 }
41 open class ItemViewHolder(view: View) : RecyclerView.ViewHolder(view) { //动态子视图的缓存
42     var constraintLayout: ConstraintLayout = view.findViewById(R.id.constraintLayout)
43     var keyTextView: TextView = view.findViewById(R.id.keyTextView)
44     var countTextView: TextView = view.findViewById(R.id.countTextView)
45 }
49 interface OnItemClickListener{ //子视图单击监听器接口
54     fun onClick(position: Int) //单击。position 参数为子视图序号
55 }
```

第 27 行设置子视图的单击事件，单击后调用 OnItemClickListener 接口的 onClick(Int)方法。第 35 由于显示关键字提示列表没有页脚子视图，所以子视图的数量设置为关键字提示列表数据的长度。49 ~ 55 行声明监听子视图单击的接口及单击事件调用的方法。

14.14.4 搜索动态的视图模型类

SearchDynamicViewModel 类用于获取关键字列表数据及数据处理。显示搜索结果使用 DynamicListFragment 类实例，所以搜索动态的请求使用 DynamicListFragment 类的 searchDynamic (String, Boolean)方法，没有在 SearchDynamicViewModel 类中获取搜索到的动态列表数据。

/java/com/pinkjiao/www/fragment/view/model/SearchDynamicViewModel.kt

```
   //搜索动态的视图模型。baseUrl 参数为基础 Url，developerModel 参数为开发者，loginUserModel 参数为登录用户。
16 class SearchDynamicViewModel(var baseUrl:String, var developerModel: DeveloperModel, private
   var loginUserModel: UserModel): ViewModel() {
17     companion object{
18         const val STATE_GET_KEY_SUCCESS = "获取搜索关键字提示列表成功"
19         const val STATE_GET_KEY_FAILURE = "获取搜索关键字提示列表失败"
20     }
21     var keyList = mutableMapOf<String,String>() //关键字提示列表
22     var response = MutableLiveData<String>() //请求的响应状态
23     var error = "" //错误信息
       //获取关键字。searchKey 参数为关键字
28     fun getKeys(searchKey:String){
29         //获取用户信息
30         DynamicRepository().getKeys(baseUrl, developerModel, loginUserModel, searchKey,
   object : DynamicRepository.KeysResponseListener {
31             override fun onCallBack(res: String, keyList: MutableMap<String,String>?) {
32                 if (res == Http.RESPONSE_SUCCESS) {
33                     this@SearchDynamicViewModel.keyList = keyList!!
34                     response.value = STATE_GET_KEY_SUCCESS
35                 } else {
36                     error = res
37                     response.value = STATE_GET_KEY_FAILURE
38                 }
39             }
40         })
41     }
42 }
```

第 18、19 行声明响应状态的常量。第 21 行声明 keyList 变量为 mutableMapOf<String,String>类型 键名存储关键字，键值存储关键字对应的动态数量。第 22 行声明 response 变量 MutableLiveData<String>类型，可以被其他类观察其数据变化。第 23 行 error 变量存储错误信息。第 28～41 行通过 Http 请求获取关键字提示列表数据，使用回调方法处理请求结果数据。

14.14.5 搜索动态的 Activity 类

SearchDynamicActivity 类用于搜索动态，并且根据输入的关键字自动从服务器端搜索相关的关 字提示。单击某个关键字提示后，会搜索该关键字的动态，并通过 DynamicListFragment 类实例显 出来。下面先介绍该类主体结构，再介绍包含的部分方法。

/java/com/pinkjiao/www/SearchDynamicActivity.kt

```
class SearchDynamicActivity : HideKeyboardActivity() {
    private val context: Context = this //上下文
```

```
19      private lateinit var viewBinding: ActivitySearchDynamicBinding //视图绑定
20      private lateinit var dynamicListFragment:DynamicListFragment //显示动态列表的 Fragment
21      private lateinit var adapter: KeyListAdapter //关键字提示列表适配器
22      private var keyStr = "" //关键字
23      private var keyItemClick = false //关键字提示子视图单击状态
24      override fun onCreate(savedInstanceState: Bundle?) {
25          super.onCreate(savedInstanceState)
26          viewBinding = ActivitySearchDynamicBinding.inflate(layoutInflater)
27          setContentView(viewBinding.root)
28          //初始化
29          initViewModel()
30          initView()
31          initDynamicListFragment()
32          initObserver()
33      }
36      private fun initViewModel(){ //初始化视图模型
45      }
50      private fun initView(){ //初始化视图控件
94      }
98      private fun initDynamicListFragment(){ //初始化搜索结果的动态列表
99          dynamicListFragment = DynamicListFragment.newInstance("searchDynamicList.php","",2,false)
100         val transaction = supportFragmentManager.beginTransaction()
101         transaction.add(R.id.dynamicListFragment, dynamicListFragment)
102         transaction.show(dynamicListFragment) //显示 Fragment
103         transaction.commit() //提交事务
104     }
108     private fun initObserver(){ //初始化请求响应状态的观察者
121     }
126     private fun searchKey(key:String){ //搜索关键字提示
127         viewBinding.searchDynamicViewModel!!.getKeys(key)
128     }
132     private fun clearKey(){ //清除关键字提示
133         viewBinding.searchDynamicViewModel!!.keyList.clear()
134         adapter.notifyDataSetChanged() //更新关键字提示视图
135     }
136 }
```

第 22 行声明 keyStr 变量保存输入的关键字字符串。第 23 行声明 keyItemClick 变量记录搜索到的关键字提示列表子视图是否被单击，用于判断是否隐藏关键字提示列表。第 101 行初始化显示搜索到的动态列表的 DynamicListFragment 实例。第 127 行和第 133 ~ 136 行将搜索关键字提示和清除关键字提示抽象为两个独立的方法。

```
/java/com/pinkjiao/www/SearchDynamicActivity.kt
private fun initView(){
    viewBinding.backImageView.setOnClickListener { finish() }
    viewBinding.searchView.isIconified = false
    //获取查询文本焦点改变监听器
    viewBinding.searchView.setOnQueryTextFocusChangeListener { _, hasFocus ->
        if(hasFocus) { searchKey(keyStr) } else { clearKey() }
    }
```

```
57    //查询文本监听器
58    viewBinding.searchView.setOnQueryTextListener(object : SearchView.OnQueryTextListener {
60        override fun onQueryTextSubmit(query: String?): Boolean { //提交事件
61            viewBinding.searchView.clearFocus() //清除焦点
62            if (keyItemClick) { keyItemClick = false } //重置关键字提示子视图单击状态
63            dynamicListFragment.searchDynamic(query!!) //根据关键字搜索动态
64            return false
65        }
67        override fun onQueryTextChange(newText: String?): Boolean { //文本内容改变事件
68            if (keyItemClick) {
69                keyItemClick = false //重置关键字提示子视图单击状态
70            } else {
71                keyStr = newText!!
72                if (keyStr == "") {
73                    clearKey() //清除关键字提示
74                } else {
75                    searchKey(keyStr) //不为空时，搜索关键字提示
76                }
77            }
78            return false
79        }
80    })
81    //设置关键字提示列表的适配器
82    adapter = KeyListAdapter(mutableMapOf(),object :KeyListAdapter.OnItemClickListener{
83        override fun onClick(position: Int) {
84            keyItemClick = true
85            val searchKey = adapter.keyList.keys.elementAt(position) //获取单击的关键字提示
86            viewBinding.searchView.setQuery(searchKey,true) //搜索关键字
87        }
88    })
89    //设置瀑布流布局为纵向单列
90    val layoutManager = StaggeredGridLayoutManager(1, StaggeredGridLayoutManager.VERTICAL)
92    viewBinding.recyclerView.layoutManager = layoutManager //设置回收视图的布局管理
93    viewBinding.recyclerView.adapter = adapter //设置数据适配器
94 }
```

第 54 ~ 56 行 searchView 控件设置查询文本焦点改变监听器，当获取焦点时搜索关键字提示，当失去焦点时清空关键字提示列表。第 60 ~ 65 行 searchView 控件提交查询文本时清空关键字提示列表并根据关键字搜索动态。第 67 ~ 79 行 searchView 控件的文本内容变为空时，清空关键字提示列表，否则搜索关键字提示。第 82 ~ 88 行设置关键字提示列表的适配器，并重写子视图单击事件监听器的 onClick(Int)方法。第 90 ~ 94 行设置显示关键字提示列表的 recyclerView 控件。

```
/java/com/pinkjiao/www/SearchDynamicActivity.kt
private fun initObserver(){
    viewBinding.searchDynamicViewModel!!.response.observe(this, { value -> //观察视图模型的响应
        when (value) {
            SearchDynamicViewModel.STATE_GET_KEY_SUCCESS -> {
                adapter.keyList = viewBinding.searchDynamicViewModel!!.keyList
                adapter.notifyDataSetChanged() //更新关键字提示视图
            }
```

```
116             SearchDynamicViewModel.STATE_GET_KEY_FAILURE -> {
117                 clearKey() //清除关键字提示
118             }
119         }
120     })
121 }
```

第 112 ~ 115 行获取到关键字提示列表数据时，更新适配器的数据并更新子视图。第 116 ~ 118 行未获取到关键字提示列表数据时，虽然没有搜索到关键字提示，但是之前输入的关键字可能搜索到了关键字提示，所以要清除关键字提示。

14.15 私 信 模 块

私信模块包括发送私信和私信箱的功能，私信箱包含收件箱和发件箱，如图 14-39 所示。

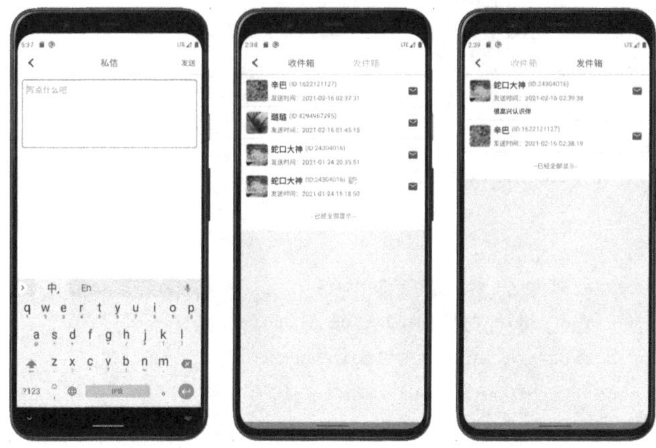

图 14-39 私信模块的运行效果

14.15.1 发送私信的服务器端接口页面

发送私信的服务器端接口页面为 messageSend.php，用于发送私信。使用 Get 方式接收数据，所需参数如表 14-31 所示。

表 14-31 messageSend.php 接收的参数

参 数	说 明
developerAccountNumber	开发者账号
developerKey	开发序列号
currentUserAccountID	登录用户的账号
currentUserPassword	登录用户的 MD5 算法加密后密码
receiverUserAccountID	接收者的用户账号 ID
receiverUserName	接收者的用户姓名
messageContent	私信内容

发送私信成功时，返回的 JSON 数据结构为{"state":"ok"} 。发送私信失败时，返回的 JSON 数据构为{"state":"错误码","info" =>"错误提示信息"}，如表 14-32 所示。

表 14-32 返回的错误码

错 误 码	错误提示信息
error1201	私信发送失败
error1202	私信数据错误
error1203	不能向自己发送私信
error1204	私信接收者不存在

14.15.2 发送私信的视图模型类

MessageSendViewModel 类用于发送私信及回调数据处理。由于是开放平台，所以发送私信的同时使用接收者的用户账号 ID 和用户姓名确定接收者，增加恶意使用接口发送私信的难度。

```
/java/com/pinkjiao/www/fragment/view/model/MessageSendViewModel.kt
26 var content = ObservableField<String>() //私信内容
27 var response = MutableLiveData<String>() //请求的响应状态
28 var error = "" //错误信息
32 fun sendMessage() { //发送私信
33     if (content.get() == null) {
34         response.value = STATE_CONTENT_EMPTY
35         return
36     }
37     messageModel.messageType = "1" //1表示私信
38     messageModel.messageContent = content.get().toString()
39     messageModel.senderUserAccountID = loginUserModel.accountID
40     messageModel.senderUserName = loginUserModel.name
41     //回复
42     MessageRepository().sendMessage(baseUrl, developerModel, loginUserModel, messageModel,
   object : MessageRepository.ResponseListener {
43         override fun onCallBack(res: String, messageModelList: MutableList<MessageModel>?,
   unReadNum: Int) {
44             when (res) {
45                 Http.CREATE -> { response.value = STATE_SEND }
46                 Http.RESPONSE_SUCCESS -> { response.value = STATE_SUCCESS }
47                 else -> {
48                     error = res
49                     response.value = STATE_FAILURE
50                 }
51             }
52         }
53     })
54 }
```

第 26 行声明 content 属性，绑定输入私信内容的 contentEditText 控件的 text 属性值。第 37 行设消息的类型为 1。由于私信使用消息的数据结构，所以需要通过类型进行区分。第 38 行获取绑定私内容数据。第 42 行调用消息存储类的 sendMessage(String,DeveloperModel,UserModel,MessageMod ResponseListener) 方法发送私信。

14.15.3 发送私信的 Activity 类

MessageSendActivity 类用于发送私信，通过 Intent 类对象传递私信接收者的用户账户 ID 和用户姓名，发送私信时作为私信的参数。

```
/java/com/pinkjiao/www/MessageSendActivity.kt
26 override fun onCreate(savedInstanceState: Bundle?) {
32     //获取 intent 传递的回复数据
33     val messageModel = MessageModel()
34     messageModel.receiverUserAccountID = intent.getStringExtra(EXTRA_RECEIVE_USER_ACCOUNT_ID)!!
35     messageModel.receiverUserName = intent.getStringExtra(EXTRA_RECEIVE_USER_NAME)!!
40 }
62 private fun initView(){ //初始化视图控件
66     viewBinding.publishTextView.setOnClickListener {
67         viewBinding.messageSendViewModel!!.sendMessage() //发送私信
68     }
73 }
```

第 34、35 行接收 Intent 类对象发送过来的用户账号 ID 和用户姓名。第 67 行调用视图模型的 sendMessage()方法发送私信。

14.15.4 私信箱的 Activity 类

MessageBoxActivity 类用于显示收件箱和发件箱，可以通过左右滑动或单击顶部菜单进行切换。收件箱和发件箱的私信列表都是通过 MessageListFragment 实例显示的。

```
/java/com/pinkjiao/www/MessageBoxActivity.kt
29 fun initView(){ //初始化视图控件
33     val titleViewList = mutableListOf(viewBinding.receiveMessageTextView, viewBinding.
   sendMessageTextView) //顶部菜单的 TextView 控件集合
35     val fragmentList = mutableListOf<Fragment>() //与顶部菜单对应的 Fragment 集合
36     fragmentList.add(MessageListFragment.newInstance(FooterViewHolder.TYPE_MESSAGE_RECEIVE,2))
37     fragmentList.add(MessageListFragment.newInstance(FooterViewHolder.TYPE_MESSAGE_SEND,2))
39     val fragmentPagerAdapter = MessageFragmentPagerAdapter(supportFragmentManager,
   FragmentPagerAdapter.BEHAVIOR_RESUME_ONLY_CURRENT_FRAGMENT,fragmentList) //实例化分页适配器
40     viewBinding.viewPager.adapter = fragmentPagerAdapter
42     val onPageChangeListener = OnPageChangeListener(this,titleViewList) //实例化分页改变监听器
43     viewBinding.viewPager.addOnPageChangeListener(onPageChangeListener)
45     for (i in titleViewList.indices) { //通过遍历设置顶部菜单单击事件
46         titleViewList[i].setOnClickListener { viewBinding.viewPager.setCurrentItem(i,false) }
47     }
48 }
```

第 33 行初始化顶部菜单按钮控件集合。第 36、37 行添加收件和发件的私信列表的实例到 Fragment 合。第 39～40 行实例化并设置分页适配器。第 42、43 行实例化并设置分页改变监听器。第 45～47 遍历顶部菜单按钮，并设置单击监听器，单击后切换显示分页。

/java/com/pinkjiao/www/MessageBoxActivity.kt

```
   //自定义的分页适配器。fragmentManager 参数为 Fragment 管理器，behavior 参数为确定是否只有当前 Fragment
   处于恢复状态，fragmentList 参数为分页视图的数组列表
56 class MessageFragmentPagerAdapter (fragmentManager:FragmentManager, behavior:Int, private val
   fragmentList: MutableList<Fragment>) : FragmentPagerAdapter(fragmentManager,behavior) {
57     override fun getCount(): Int {
58         return fragmentList.size
59     }
60     override fun getItem(position: Int): Fragment {
61         return fragmentList[position]
62     }
63 }
```

第 56 行声明自定义的分页适配器，继承 FragmentPagerAdapter 抽象类，其中 fragmentList 属性是用于保存分页 Fragment 的集合。第 57 ~ 59 行重写 getCount()方法，获取分页的数量。第 60 ~ 62 行重写 getItem(Int)方法，获取对应位置的分页。

/java/com/pinkjiao/www/MessageBoxActivity.kt

```
//自定义的分页改变监听器。context 参数为上下文，titleViewList 参数为标题控件列表。
class OnPageChangeListener(var context:Context, var titleViewList:MutableList<TextView>):
ViewPager.OnPageChangeListener{
    var prePositionOffset:Float = 0f //缓存位置偏移
    override fun onPageScrolled(position: Int, positionOffset: Float, positionOffsetPixels: Int) {
        //根据偏移量修改顶部菜单文字的颜色
        if (prePositionOffset < positionOffset) { //左滑动
            if (positionOffset > 0.6f) {
                titleViewList[position].setTextColor(context.getColor(R.color.gray_400))
                titleViewList[position + 1].setTextColor(context.getColor(R.color.purple_500))
            } else {
                titleViewList[position].setTextColor(context.getColor(R.color.purple_500))
                titleViewList[position + 1].setTextColor(context.getColor(R.color.gray_400))
            }
        } else if (prePositionOffset > positionOffset && position<titleViewList.size-1){//右滑动
            if (positionOffset < 0.4f) {
                titleViewList[position].setTextColor(context.getColor(R.color.purple_500))
                titleViewList[position + 1].setTextColor(context.getColor(R.color.gray_400))
            } else {
                titleViewList[position].setTextColor(context.getColor(R.color.gray_400))
                titleViewList[position + 1].setTextColor(context.getColor(R.color.purple_500))
            }
        }
        prePositionOffset = positionOffset //保存偏移值
    }
    override fun onPageSelected(position: Int) {
        for (i in titleViewList.indices) {
            if (i == position) {
                titleViewList[i].setTextColor(context.getColor(R.color.purple_500))
            } else {
                titleViewList[i].setTextColor(context.getColor(R.color.gray_400))
            }
```

```
99              }
100         }
101         override fun onPageScrollStateChanged(state: Int) {}
102     }
```

第 69 行声明自定义的分页改变监听器，实现 ViewPager.OnPageChangeListener 接口。第 71 ~ 91 行重写 onPageScrolled(Int,Float,Int) 方法，滑动超过 60%偏移量时修改顶部按钮菜单项的颜色。第 92 ~ 100 行重写 onPageSelected(Int) 方法，切换分页时同步修改顶部按钮菜单项的颜色。

14.16 设置模块组

设置模块用于设置头像、重置密码和验证身份证信息，如图 14-40 所示。设置头像和重置密码是实时生效的，而验证身份证信息需要后台人工审核。

图 14-40 设置模块的运行效果

4.16.1 设置的 Activity 类

SetupActivity 类用于汇总设置功能。为了便于操作，并没有为显示每项设置功能的 TextView 件添加单击事件监听器，而是为每项设置功能的布局添加单击事件监听器，单击后启动对应 Activity。

```
/java/com/pinkjiao/www/SetupActivity.kt
23  viewBinding.faceImageLayout.setOnClickListener {
24      startActivity(Intent(context, SetupFaceImageActivity::class.java))
25  }
26  viewBinding.passwordLayout.setOnClickListener {
27      startActivity(Intent(context, SetupPasswordActivity::class.java))
28  }
29  viewBinding.checkNationalIDImageLayout.setOnClickListener {
30      startActivity(Intent(context, SetupCheckNationalIDImageActivity::class.java))
31  }
```

第 23 ~ 25 行添加设置头像的单击事件监听器，单击后启动 SetupFaceImageActivity。第 26 ~ 28 行添加重置密码的单击事件监听器，单击后启动 SetupPasswordActivity。第 29 ~ 31 行添加验证身份证信息的单击事件监听器，单击后启动 SetupCheckNationalIDImageActivity。

14.16.2 设置头像的服务器端接口页面

设置头像的服务器端接口页面为 setupFaceImage.php，用于修改用户头像图片。使用 Post 方式接收数据，所需参数如表 14-33 所示。

表 14-33 setupFaceImage.php 接收的参数

参　数	说　明
developerAccountNumber	开发者账号
developerKey	开发序列号
currentUserAccountID	登录用户的账号
currentUserPassword	登录用户的 MD5 算法加密后密码
uploadInput	用户头像图片

设置头像成功时，返回的 JSON 数据结构为{"state":"ok"} 。设置头像失败时，返回的 JSON 数据结构为{"state":"error1401","info" = >"设置头像失败"}。

14.16.3 设置头像的视图模型类

SetupFaceImageViewModel 类用于设置头像及回调数据处理，使用工厂模式进行实例化。

```
/java/com/pinkjiao/www/fragment/view/model/SetupFaceImageViewModel.kt
companion object{
    const val STATE_SEND = "修改头像"
    const val STATE_SUCCESS = "修改成功"
    const val STATE_FAILURE = "修改失败"
}
var response = MutableLiveData<String>() //请求的响应状态
var error = "" //错误信息
fun submit(postFiles: MutableMap<String, String>) { //提交修改头像。postFiles 参数为头像图片。
    val parameter: MutableMap<String, String>= HashMap() //post 参数
    //设置头像
    UserRepository().setupFaceImage(baseUrl, developerModel, loginUserModel, parameter
postFiles, object : UserRepository.ResponseListener {
        override fun onCallBack(res: String, userModel: UserModel?) {
```

```
42              when (res) {
43                  Http.CREATE -> { response.value = STATE_SEND }
44                  Http.RESPONSE_SUCCESS -> { response.value = STATE_SUCCESS }
45                  else -> {
46                      error = res
47                      response.value = STATE_FAILURE
48                  }
49              }
50          }
51      })
52  }
```

第 26～28 行声明请求响应状态的常量。第 36～52 行通过 Post 请求上传头像图片到服务器端，并处理回调数据。

14.16.4 设置头像的 Activity 类

SetupFaceImageActivity 类用于设置头像图片，单击“选择头像”按钮调用系统图库选择头像图片，单击“上传头像”按钮上传头像图片到服务器端。

/java/com/pinkjiao/www/SetUpFaceImageActivity.kt

```
57  private fun initView(){
59      Image.loadFromUrl(viewBinding.faceImageView, loginUserModel.faceImage, CacheDir.getFace(this))
61      viewBinding.backImageView.setOnClickListener { finish() } //返回按钮的单击事件
63      viewBinding.selectImageButton.setOnClickListener { //选取按钮的单击事件
64          val intent = Intent(Intent.ACTION_PICK, MediaStore.Images.Media.EXTERNAL_CONTENT_URI)
65          intent.type = "image/*"
66          startActivityForResult(intent, REQUEST_CODE_SELECT)
67      }
69      viewBinding.submitButton.setOnClickListener { viewModel.submit(postFiles) } //提交按钮的单击事件
70  }
```

第 64～66 行调用系统图库选择文件，设置可以选择的文件类型为图片，并设置 REQUEST_CODE_SELECT 常量为请求码。第 69 行设置 submitButton 控件的单击监听事件，单击后调用视图模型实例的 submit(MutableMap<String, String>)方法将头像图片上传到服务器端。

/java/com/pinkjiao/www/SetUpFaceImageActivity.kt

```
override fun onActivityResult(requestCode: Int, resultCode: Int, intent: Intent?) {
    super.onActivityResult(requestCode, resultCode, intent)
    if (requestCode == REQUEST_CODE_SELECT && resultCode == RESULT_OK) {
        //新建保存裁剪照片的文件，需要使用外部缓存(内部缓存无法写入剪裁图片)
        photoFile = CacheFile().createCacheFile(context.externalCacheDir, "face", "cacheFace.jpg")
        val cropIntent = Intent("com.android.camera.action.CROP") //裁剪照片
        cropIntent.setDataAndType(intent!!.data, "image/*")
        cropIntent.putExtra("crop", "true")
        cropIntent.putExtra("aspectX", 1) //设置裁剪的比例
        cropIntent.putExtra("aspectY", 1)
        cropIntent.putExtra("outputX", 1000) //设置裁剪后保存图片的尺寸
        cropIntent.putExtra("outputY", 1000)
        cropIntent.putExtra("outputFormat", Bitmap.CompressFormat.JPEG.toString())//设置文件格式
        cropIntent.putExtra(MediaStore.EXTRA_OUTPUT, photoFile.absoluteFile.toUri())//设置文件路径
        cropIntent.putExtra("return-data", false)
```

```
109             startActivityForResult(cropIntent, REQUEST_CODE_CROP)
110         } else if (requestCode == REQUEST_CODE_CROP && resultCode == RESULT_OK) {
111             val imagesUri = arrayOf(photoFile.absoluteFile.toUri())
112             val imagesDegree = IntArray(1)
113             imagesDegree[0] = 0
114             postFiles = Image.cacheFaceImage(contentResolver, "uploadInput", CacheDir.getFace(context),
    "cacheFace", imagesUri)
116             viewBinding.faceImageView.setImageURI(null) //清除显示照片
117             viewBinding.faceImageView.setImageURI(photoFile.absoluteFile.toUri()) //显示照片
118         }
119     }
```

第 91 行根据请求码和结果码判断是否选取完图片。第 92 ~ 109 行创建剪裁照片的缓存文件，启动剪裁图片窗口。第 110 行根据请求码和结果码判断是否剪裁完图片。第 111 ~ 117 行获取剪裁后的头像图片，并获取缓存文件显示在 faceImageView 控件中。

14.16.5 重置密码的服务器端接口页面

重置密码的服务器端接口页面为 setupPassword.php，用于修改用户密码。使用 Get 方式接收数据，所需参数如表 14-34 所示。

表 14-34 setupPassword.php 接收的参数

参 数	说 明
developerAccountNumber	开发者账号
developerKey	开发序列号
currentUserAccountID	登录用户的账号
currentUserPassword	登录用户的 MD5 算法加密后密码
nationalIDNumber	身份证号
currentPassword	MD5 算法加密后的当前密码
newPassword	MD5 算法加密后的新密码

重置密码成功时，返回的 JSON 数据结构为{"state":"ok"} 。重置密码失败时，返回的 JSON 数据结构为{"state":"error1403","info" = >"修改密码失败"}。

14.16.6 重置密码的视图模型类

SetupPasswordViewModel 类用于设置头像及回调数据处理，使用工厂模式进行实例化。

```
/java/com/pinkjiao/www/fragment/view/model/SetupPasswordViewModel.kt
fun submit(nationalIDNumber:String,currentPassword:String, newPassword: String) {
    val parameter: MutableMap<String, String>= HashMap() //post 参数
    parameter["nationalIDNumber"] = nationalIDNumber
    parameter["currentPassword"] = MD5.encode(currentPassword)
    parameter["newPassword"] = MD5.encode(newPassword)
    //回复
    UserRepository().setupPassword(baseUrl,  developerModel,  loginUserModel,  parameter
object : UserRepository.ResponseListener {
        override fun onCallBack(res: String, userModel: UserModel?) {
            when (res) {
                Http.CREATE -> { response.value = STATE_SEND }
                Http.RESPONSE_SUCCESS -> { response.value = STATE_SUCCESS }
```

```
51                 else -> {
52                     error = res
53                     response.value = STATE_FAILURE
54                 }
55             }
56         }
57     })
58 }
```

第 42 ~ 44 行设置 Post 传递的参数值。第 46 行调用 UserRepository 类的 setupPassword() 方法上传重置密码所需的数据。第 47 ~ 56 行处理回调数据。

14.16.7 重置密码的 Activity 类

SetupPasswordActivity 类用于修改密码，需要输入身份证号、当前密码、新密码和确认密码，发送到服务器端之前验证输入的内容。

```
/java/com/pinkjiao/www/SetupPasswordActivity.kt
53  private fun initView(){
56      with(viewBinding) {
93          submitButton.setOnClickListener{ //提交按钮的单击事件
94              val nationalIDNumber = nationalIDNumberEditText.editableText.toString()
95              val currentPassword = currentPasswordEditText.editableText.toString()
96              val newPassword = newPasswordEditText.editableText.toString()
97              val confirmPassword = confirmPasswordEditText.editableText.toString()
98              when {
99                  nationalIDNumber == "" -> {
100                     Toast.makeText(context, getText(R.string.setup_password_national_id_number_
    none), Toast.LENGTH_LONG).show()
101                 }
114                 else -> {
115                     viewModel.submit(nationalIDNumber, currentPassword, newPassword) //提交
116                 }
117             }
118         }
121     }
122 }
```

第 93 行 submitButton 控件设置单击事件监听器。第 94 ~ 97 行获取输入的身份证号、当前密码、新密码和确认密码等内容。第 98 ~ 117 行验证获取输入的内容是否符合规范。如果符合规范，则调用视图模型的 submit(String,String,String) 方法发送修改密码请求。

14.16.8 提交验证身份证的服务器端接口页面

提交验证身份证的服务器端接口页面为 setupCheckNationalIDImage.php，用于提交身份证图片。使用 Post 方式接收数据，所需参数如 14-35。所示。

表 14-35 setupCheckNationalIDImage.php 接收的参数

参 数	说 明
developerAccountNumber	开发者账号

续表

参　数	说　明
developerKey	开发序列号
currentUserAccountID	登录用户的账号
currentUserPassword	登录用户的 MD5 算法加密后密码
uploadInput	身份证图片

提交验证身份证成功时，返回的 JSON 数据结构为{"state":"ok"} 。提交验证身份证失败时，返回的 JSON 数据结构为{"state":"错误码","info" = >"错误提示信息"}，如表 14-36 所示。

表 14-36　返回的错误码

错 误 码	错误提示信息
error1404	未上传身份证图片
error1205	身份证图片正在审核中
error1206	身份证图片已经通过验证
error1207	上传身份证图片失败

14.16.9　验证身份证的视图模型类

SetupNationalIDImageViewModel 类用于提交身份证图片及回调数据处理，使用工厂模式进行实例化。

/java/com/pinkjiao/www/fragment/view/model/SetupNationalIDImageViewModel.kt

```
var userModel = UserModel() //用户数据
var response = MutableLiveData<String>() //请求的响应状态
var error = "" //错误信息
fun getUserInfo(domain:String) { //获取用户数据的请求。Domai 参数为域名
   UserRepository().getUserInfo(domain,     baseUrl,     developerModel,     loginUserModel,
loginUserModel.accountID, object : UserRepository.ResponseListener {
      override fun onCallBack(res: String, userModel: UserModel?) {
         if (res == Http.RESPONSE_SUCCESS) {
            this@SetupNationalIDImageViewModel.userModel = userModel!! //保存用户数据
            response.value = STATE_GET_USER_INFO_SUCCESS
         } else {
            error = res
            response.value = STATE_GET_USER_INFO_FAILURE
         }
      }
   })
}
fun submit(postFiles: MutableMap<String, String>) { //提交身份证图片。postFiles 参数为身份证图片。
   val parameter: MutableMap<String, String>= HashMap() //post 参数
   UserRepository().setupCheckNationalIDImage(baseUrl,    developerModel,    loginUserModel,
parameter,postFiles, object : UserRepository.ResponseListener { //发送提交身份证图片的请求
      override fun onCallBack(res: String, userModel: UserModel?) {
         when (res) {
            Http.CREATE -> { response.value = STATE_SEND }
            Http.RESPONSE_SUCCESS -> { response.value = STATE_SUCCESS }
            else -> {
```

```
66                  error = res
67                  response.value = STATE_FAILURE
68              }
69          }
70      }
71  })
72 }
```

第 32 行 userModel 属性主要用于保存从服务器端获取的用户身份证状态。第 39 ~ 51 行通过 Get 请求获取用户信息，并处理回调数据。提交验证身份证时，只需要调用用户的 userNationalIDImageCheck 属性判断提交状态。由于提交验证身份证属于低频操作，所以并没有单独开发一个服务器端接口，而是公用获取用户信息的服务器端接口。第 56 ~ 72 行通过 Post 请求提交身份证图片，并处理回调数据。

14.16.10 验证身份证的 Activity 类

SetupCheckNationalIDImageActivity 类用于提交身份证图片，服务器端进行审核。暂时已验证身份证的用户与未验证的用户权限相同，后续会对未未验证身份证的用户进行权限限制。

```
/java/com/pinkjiao/www/SetupCheckNationalIDImageActivity.kt
private fun submit(){ //提交身份证图片
    val postFileUri: Array<Uri?>= Common.removeEmptyArrayElement(imagesUri)//清空 Uri 数组中的空元素
    val postFiles: MutableMap<String, String>
    postFiles = if (postFileUri.isNotEmpty()) {
        LoadingProgressDialog.showProgressDialog(context, getString(R.string.setup_
national_id_image_dialog)) //显示等待对话框
        Image.cacheImage(contentResolver, "uploadInput",
            CacheDir.getImage(context), cacheImagesPrefixName,
            cacheImageMaxHeight, cacheImageMaxWidth, postFileUri, imagesDegree) //缓存图片
    } else {
        mutableMapOf()
    }
    viewModel.submit(postFiles)
}
//回调。requestCode 参数为请求码，resultCode 参数为结果码，intent 参数为意图
override fun onActivityResult(requestCode: Int, resultCode: Int, intent: Intent?) {
    super.onActivityResult(requestCode, resultCode, intent)
    if (requestCode == REQUEST_CODE_FRONT && resultCode == RESULT_OK) {
        imagesUri[0] = intent!!.data //图片的 Uri
        imagesDegree[0] = Image.getImageOrientation(contentResolver, imagesUri[0]!!) //图片的旋转角度
        imagesPath[0] = Common.uriToPath(contentResolver, imagesUri[0]!!) //图片的物理路径
        //获取并设置缩略图
        val thumbnailBitmap = Image.loadThumbnail (contentResolver, imagesUri[0]!!, viewBinding.
frontNationalIDImageView.height, viewBinding.frontNationalIDImageView.width)
        viewBinding.frontNationalIDImageView.setImageBitmap(thumbnailBitmap)
    } else if (requestCode == REQUEST_CODE_BACK && resultCode == RESULT_OK){
        imagesUri[1] = intent!!.data //图片的 Uri
        imagesDegree[1] = Image.getImageOrientation(contentResolver, imagesUri[1]!!) //图片的旋转角度
        imagesPath[1] = Common.uriToPath(contentResolver, imagesUri[1]!!) //图片的物理路径
        //获取并设置缩略图
        val thumbnailBitmap = Image.loadThumbnail (contentResolver, imagesUri[1]!!, viewBinding.
```

```
    frontNationalIDImageView.height, viewBinding.frontNationalIDImageView.width)
170         viewBinding.backNationalIDImageView.setImageBitmap(thumbnailBitmap)
171     }
172 }
```

第 138 ~ 146 行根据 postFileUri 数组是否为空的两种情况，分别对 postFiles 变量进行初始化。第 157 行判断是否选择了身份证正面图片。第 158 ~ 160 行通过数组保存选择的身份证正面图片数据。第 162、163 行获取选择身份证正面图片的缩略图，并显示在 frontNationalIDImageView 控件中。第 164 行判断是否选择了身份证背面图片。第 165 ~ 167 行通过数组保存选择的身份证背面图片数据。第 169、170 行获取选择身份证背面图片的缩略图，并显示在 backNationalIDImageView 控件中。

14.17 应用程序发布

Android 应用程序可以发布成两种格式文件，分别是 APK（Android Package）和 AAB（Android App Bundle）。APK 格式文件用户下载后可以直接安装；AAB 格式需要上传到 Google Play 中，针对不同用户设备由 Google Play 进行签名，生成优化后文件量更小的 APK 格式。Google 宣布从 2021 年下半年开始 Google Play 上发布的应用都需要使用 AAB 格式进行发布。对国内用户而言 APK 格式是首选，所以下面演示发布 APK 格式文件的方法。

提示：Android App Bundle

Android App Bundle 是一种发布格式，无法直接安装，目前压缩下载大小限制为 150MB。该格式文件包含 App 所有经过编译的代码和资源，由 Google Play 完成 APK 的生成及签名。

14.17.1 生成 APK 文件

选择【Build】→【Generate Signed Bundle/APK】命令，打开“Generate Signed Bundle or APK”对话框，选择“APK”单选按钮，如图 14-41 所示。

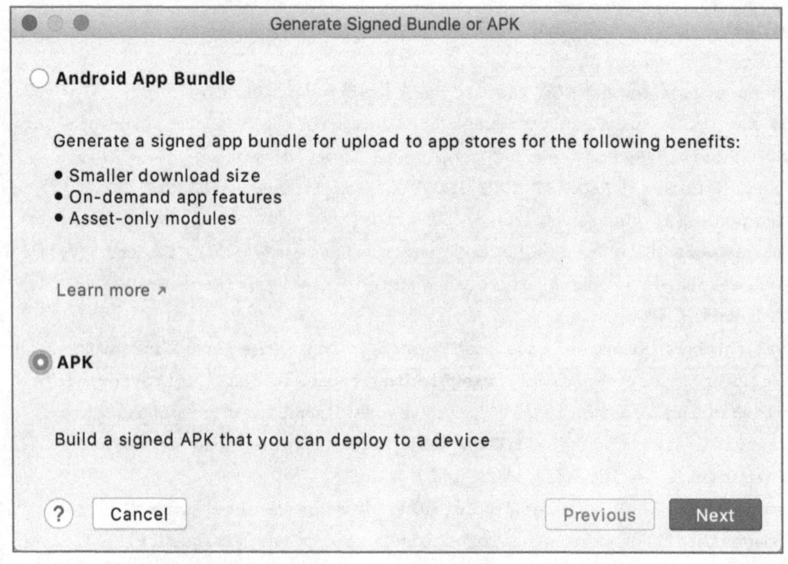

图 14-41 “Generate Signed Bundle or APK”对话框（1）

单击“Next”按钮，默认会显示之前选择的密钥库文件，如图 14-42 所示。首次使用时需要创建密钥库文件，单击“Create new”按钮，打开“New Key Store”对话框。

Generate Signed Bundle or APK

Module　粉色辣椒.app

Key store path　ers/baizhe/Desktop/Kotlin开发Android移动应用程序/工程/ApkKey

Create new...　Choose existing...

Key store password　•••••••••••••

Key alias　pinkjiaokey

Key password　•••••••••••••

Remember passwords

?　Cancel　Previous　Next

图 14-42　“Generate Signed Bundle or APK”对话框(2)

在“New Key Store”对话框中，输入创建密钥库文件所需的信息，如图 14-43 所示。为了增加暴力破解密码的难度，密钥库密码和密钥密码不可相同。单击“Key alias”文本框内的文件夹图标，打开“Choose Key”对话框。

New Key Store

Key store path:　he/Desktop/Kotlin开发Android移动应用程序/工程/ApkKey　←密钥库路径

Password:　••••••　Confirm:　•••••　←密钥库密码

Key

Alias:　pinkjiaokey　←密钥别名

Password:　••••••　Confirm:　••••••　←密钥密码

Validity (years):　25　←密钥有效年限

Certificate

First and Last Name:　白喆

Organizational Unit:

Organization:　←证书内容

City or Locality:　shenyang

State or Province:　china

Country Code (XX):　86

Cancel　OK

图 14-43　“New Key Store”对话框

在“Choose Key”对话框中，可以在当前密钥库中选择密钥或创建密钥，如图 14-44 所示。新建钥和创建密钥库时，需要输入的信息一致。

选择完密钥后，单击“Next”按钮进入设置输出 APK 文件的步骤，设置输出路径(Destination der)、构建版本(Build Variants)和签名版本(Signature Versions)，如图 14-45 所示

图 14-44 “Choose Key”对话框

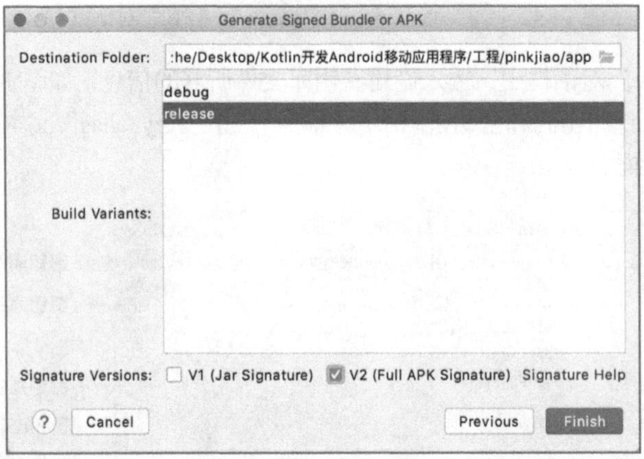

图 14-45 “Generate Signed Bundle or APK”对话框

单击“Finish”按钮，等待生成 APK 文件。完成后，会在 Event Log 窗口中显示日志，如图 14-4 所示。单击“locate”链接会打开输出文件夹，如图 14-47 所示。无法在“Generate Signed Bundle APK”对话框中设置输出的 APK 文件名称，APK 文件会被默认命名为“app-release.apk”，需要手 修改为所需的文件名。

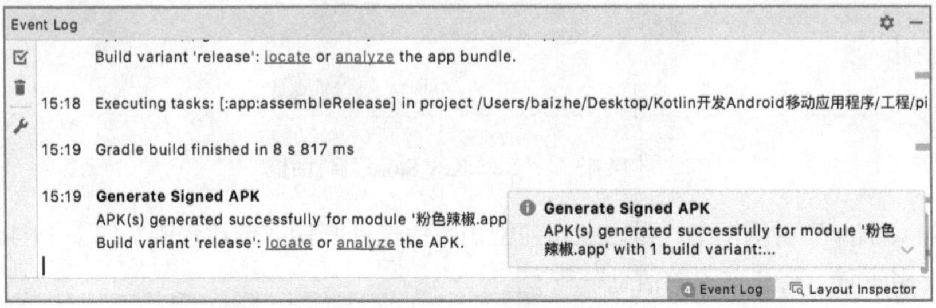

图 14-46 Event Log 窗口

名称	修改日期	大小	种类
app-release.aab	今天 14:55	13.7 MB	文稿
app-release.apk	今天 15:19	14 MB	Android Application Package
output-metadata.json	今天 15:19	337 字节	纯文本文稿

图 14-47　输出文件夹

提示：lintVitalRelease

打包时出现“Android--Execution failed for task ':app:lintVitalRelease'”错误时，可以在 build.gradle(Module) 中的 android 标签下添加以下配置代码:

```
lintOptions {
    checkReleaseBuilds false
    abortOnError false
}
```

14.17.2　发布到网站或应用市场

在国内 APK 文件一般会放在网站上或发布到应用市场供用户下载使用。各大 Android 手机品牌都有自己的应用市场，此外还有腾讯、阿里、百度等应用市场。每个应用市场都需要单独注册、发布、等待审核等步骤，需要消耗大量的时间，因此可以通过 App 分发平台进行代发，如酷传(https://www.kuchuan.com)和蒲公英(https://www.pgyer.com)。

参考文献

[1] 吕娜. 生而 Geek——Android 之父 Andy Rubin[J]. 程序员, 2010, 09: 7-8.

[2] 雷擎. 基于 Android 平台的移动互联网开发[M]. 北京: 清华大学出版社, 2014.

[3] 马宁, 高丽娜. Oracle 诉 Google 版权案解读——兼评对 Android 开源政策的影响[J]. 中国版权, 2013(03): 50-53.

[4] 于泽源, 张丽滨. Java 中两种有益的类—StringBuffer、Vector 及它们在编辑器设计中的应用[J]. 小型微型计算机系统, 1997(01): 66-69.

[5] 闵军, 罗泓. 用 lambda 表达式和 std: : function 类模板改进泛型抽象工厂设计[J]. 软件工程, 2017, 20(09): 9-14.

[6] 周建儒. 基于单继承的“动态多态性”的分析与应用[J]. 信息技术, 2014(03): 162-164.

[7] 赵智. 方法重载与重写机制运用经验与技巧[J]. 软件导刊, 2007(15): 148-150.

[8] 刘冬梅, 蒋立源. 面向对象程序的抽象类测试之研究[J]. 微处理机, 2008(04): 126-128+132.

[9] 张志祥, 李庆华, 贲可荣. 行为抽象方法及其在设计模式中的应用[J]. 计算机科学, 2004(09): 135-136+151.

[10] 赵合计, 石冰. 面向对象软件系统开发中的软件重用[J]. 计算机应用研究, 1996(05): 35-37.

[11] 丁俊华, 董桓, 吴定豪, 等. 软件互操作研究与进展[J]. 计算机研究与发展, 1998(07): 2-8.

[12] 廖旭, 张力. 工作流管理系统中一种基于任务的委托模式[J]. 计算机工程与应用, 2005(07): 44-46+50.

[13] 兰娟. 树枚举算法的设计与研究[D]. 华东师范大学, 2013.

[14] 肖睿, 龙浩, 孙琳, 等. Java 高级特性编程及实战[M]. 北京: 人民邮电出版社, 2018.

[15] 刘书健. 基于协程的高并发的分析与研究[D]. 昆明理工大学, 2016.

[16] 张富为, 杨秋翔, 宋超峰. 基于适配器的构件组装技术[J]. 计算机工程与设计, 2018, 39(04): 1039-1046.

[17] 吕晓刚. 基于数据绑定的界面设计模式[D]. 浙江大学, 2008.

[18] 耿乙超. Android 平台下通讯录的研究与开发[D]. 西安电子科技大学, 2012.

[19] 陈施卫. 基于 XML 的异构数据库数据交换的研究与实现[D]. 电子科技大学, 2012.

[20] 张春祥, 赵春蕾, 陈超, 等. 基于手机传感器的人体活动识别综述[J]. 计算机科学, 2020, 47(10): 1-8.

[21] 叶坤. Android Jetpack 应用指南[M]. 北京: 电子工业出版社, 2020.